V. Uma Devi

Solid State Physics

The Manchester Physics Series

General Editors

F. MANDL: R. J. ELLISON: D. J. SANDIFORD

Physics Department, Faculty of Science,
University of Manchester

Properties of Matter:	B. H. Flowers and E. Mendoza
Optics: *Second Edition*	F. G. Smith and J. H. Thomson
Statistical Physics: *Second Edition*	F. Mandl
Solid State Physics: *Second Edition*	J. R. Hook and H. E. Hall
Electromagnetism: *Second Edition*	I. S. Grant and W. R. Phillips
Electromagnetic Radiation:	F. H. Read
Statistics:	R. J. Barlow

In preparation

Particle Physics:	B. R. Martin and G. Shaw
Quantum Mechanics:	F. Mandl

SOLID STATE PHYSICS
Second Edition

J. R. Hook
H. E. Hall

Department of Physics,
University of Manchester

John Wiley & Sons

CHICHESTER   NEW YORK   BRISBANE   TORONTO   SINGAPORE

Copyright © 1974, 1991 by John Wiley & Sons Ltd.
Baffins Lane, Chichester
West Sussex PO19 1UD, England

First published 1974
Second edition 1991

Other Wiley Editorial Offices

John Wiley & Sons, Inc., 605 Third Avenue,
New York, NY 10158-0012, USA

Jacaranda Wiley Ltd, G.P.O. Box 859, Brisbane,
Queensland 4001, Australia

John Wiley & Sons (Canada) Ltd, 22 Worcester Road,
Rexdale, Ontario M9W 1L1, Canada

John Wiley & Sons (SEA) Pte Ltd, 37 Jalan Pemimpin 05-04,
Block B, Union Industrial Building, Singapore 2057

Library of Congress Cataloging-in-Publication Data:

Hook, J. R. (John R.)
    Solid state physics / J. R. Hook, H. E. Hall.—2nd ed.
        p.   cm.—(The Manchester physics series)
    Rev. ed. of: Solid state physics / H. E. Hall. 1st ed. 1974.
    Includes bibliographical references and index.
    ISBN 0 471 92804 6 (cloth)—ISBN 0 471 92805 4 (paper)
    1. Solid state physics. I. Hall, H. E. (Henry Edgar), 1928-
II. Hall, H. E. (Henry Edgar), 1928-   Solid state physics.
III. Title. IV. Series.
QC176.H66   1991
530.4'1—dc20   90-20571
                                                   CIP

British Library Cataloguing in Publication Data:
Hook, J. R.
    Solid state physics.
    1. Solids. Structure & physics properties
    I. Title II. Hall, H. E. (Henry Edgar) III. Series
    530.41

ISBN 0 471 92804 6 (cloth)
ISBN 0 471 92805 4 (paper)

Typeset by APS, Salisbury Wilts.
Printed and bound in Great Britain by
Biddles Ltd, Guildford, Surrey

To my family in partial compensation for taking so much of
my time away from them

Contents

★ Starred sections may be omitted as they are not required later in the book.

xiv   Contents

Editors' preface to the Manchester Physics Series

The first book in the Manchester Physics Series was published in 1970, and other titles have been added since, with total sales world-wide of more than a quarter of a million copies in English language editions and in translation. We have been extremely encouraged by the response of readers, both colleagues and students. The books have been reprinted many times, and some of our titles have been rewritten as new editions in order to take into account feedback received from readers and to reflect the changing style and needs of undergraduate courses.

The Manchester Physics Series is a series of textbooks at undergraduate level. It grew out of our experience at Manchester University Physics Department, widely shared elsewhere, that many textbooks contain much more material than can be accommodated in a typical undergraduate course and that this material is only rarely so arranged as to allow the definition of a shorter self-contained course. In planning these books, we have had two objects. One was to produce short books: so that lecturers should find them attractive for undergraduate courses; so that students should not be frightened off by their encyclopaedic size or their price. To achieve this, we have been very selective in the choice of topics, with the emphasis on the basic physics together with some instructive, stimulating and useful applications. Our second aim was to produce books which allow courses of different length and difficulty to be selected, with emphasis on different applications. To achieve such flexibility we have encouraged authors to use flow diagams showing the logical connections between different chapters and to put some topics in starred sections. These cover more advanced and alternative material which is not required for the understanding of later parts of each volume. Although these books were conceived as a series, each of them is self-contained and can be used independently of the others. Several of them are suitable for wider use in other sciences. Each author's preface gives details about the level, prerequisites, etc., of his volume.

We are extremely grateful to the many students and colleagues, at Manchester and elsewhere, whose helpful criticisms and stimulating comments have led to many improvements. Our particular thanks go to the authors for all the work they have done, for the many new ideas they have contributed, and for discussing patiently, and often accepting, our many suggestions and requests. We would also like to thank the publishers, John Wiley & Sons, who have been most helpful.

F. MANDL
R. J. ELLISON
D. J. SANDIFORD

January, 1987

The story of the creation was told in 200 words. Look it up if you don't believe me.—*Edgar Wallace*

Foreword

When the time came to consider a second edition of *Solid State Physics* I felt that I had already said what I had to say on the subject in the first edition. I also felt that the book was rather too idiosyncratic for many students. For these reasons I thought it would be better if the revision and updating were undertaken by another hand, and the editors shared this view.

We therefore approached Dr John Hook, a friend and colleague for many years, and I think the result justifies the decision. The new edition is, in my opinion, a substantial improvement on the old one, but it would not have occurred to me to write it like that.

September 1990   HENRY HALL

Author's preface to second edition

I accepted the invitation of the editors of the Manchester Physics Series to write a second edition of *Solid State Physics* for two main reasons. Firstly I felt that, although the approach adopted in the first edition had much to commend it, some re-ordering and simplification of the material was required to make the book more accessible to undergraduate students. Secondly there was a need to take account of some of the important developments that have occurred in solid state physics since 1973.

To achieve re-ordering and simplification it has been necessary to rewrite most of the first edition. A major change has been to introduce the idea of mobile electron states in solids through the free electron theory of metals rather than through the formation of energy bands by overlap of atomic states on neighbouring atoms. The latter approach was used in the first edition because it could be applied first to the dilute electron gas in semiconductors where an independent particle model might be expected to work. Although this was appealing to the experienced physicist, it proved difficult to the undergraduate student, who was forced to assimilate too many new ideas at the beginning. One feature of the first edition that I have retained is to delay for as long as possible a *formal* discussion of the reciprocal lattice and Brillouin zones in a three-dimensional crystal. Although these concepts provide an elegant general frame-work for describing many of the properties of crystalline solids, they are, like Maxwell's equations in electromagnetism, more likely to be appreciated by students if they have met some of the ideas earlier in a simpler context. The use of the formal framework is avoided in the early chapters by using one- and two-dimensional geometries whenever necessary.

To take account of recent developments the amount of material on semi-conductor physics and devices has been substantially increased, a chapter has been added on the two-dimensional electron gas and quantum Hall effect, and sections on quasi-crystals, high-T_C superconductors and the use of electrons to probe surfaces have been included. A chapter on the electrical properties of insulators has also been added.

I have tried to conform to the aim of the Manchester Physics Series by producing a book of reasonable length (and thus cost), from which it is possible to define self-contained undergraduate courses of different length and difficulty. The problem with solid state physics in this context is that it contains many diverse topics so that many quite different courses are possible. I have had to be very selective therefore in my choice of subjects, which has been strongly influenced by the third year undergraduate solid state physics courses at Manchester. We currently have a basic course of 20 lectures, which is given at two levels; the courses cover material from Chapters 1–5 of this book and the higher level course also incorporates appropriate sections of Chapters 11–13. A further course of 20 lectures on selected topics in solid state physics currently covers magnetism, superconductivity and ferroelectricity (Chapters 7–10). The flow diagram inside the front cover can be used as an aid to the design of courses based on this book.

Important subjects that are not covered in this book are crystal defects and disordered solids. I would have liked to include a chapter on each of these topics but would have exceeded the length limit set by the publishers and editors had I done so.

Like the first edition, this book presupposes a background knowledge of properties of matter (interatomc potentials and their relation to binding energies and elastic moduli, kinetic theory), quantum mechanics (Schrödinger's equation and its solution to find energy eigenvalues and eigenfunctions), electricity and magnetism (Maxwell's equations and some familiarity with electric and magnetic fields in matter) and thermal physics (the Boltzmann factor and the Fermi and Bose distributions). Books in which this background information can be found are listed in the bibliography along with selected general reference books on solid state physics and some books and articles that provide further information on specific topics.

This book includes some more advanced and detailed material, which can be omitted without loss of continuity. Complete sections in this category are identified by starring and parts of sections are printed on a grey background.

The use of **bold** type for a technical term in the text, normally when the term is first encountered, indicates that a definition or explanation of the term can be found there. *Italic* type is used for emphasis.

I am very grateful to David Sandiford and Henry Hall for their helpful advice and constructive criticism. I would also like to thank Manchester undergraduate Colin Lally, who read the manuscript from the point of view of a prospective consumer; his reaction reassured me that the level was appropriate. Ian Callaghan's draughtmanship and photography was invaluable in producing many of the figures, and my son James helped willingly with some of the more mundane manuscript-preparation tasks.

September 1990   JOHN HOOK

Beauty when uncloth'd is clothèd best.—*Phineas Fletcher*
(1582–1650)

CHAPTER

Crystal structure

1.1 INTRODUCTION

The aim of solid state physics is to explain the properties of solid materials as found on Earth. For almost all purposes the properties are expected to follow from Schrödinger's equation for a collection of atomic nuclei and electrons interacting with electrostatic forces. The fundamental laws governing the behaviour of solids are therefore known and well tested. It is nowadays only in cosmology, astrophysics and high-energy physics that the fundamental laws are still in doubt.

In this book we shall be concerned almost entirely with crystalline solids, that is solids with an atomic structure based on a regular repeated pattern, a sort of three-dimensional wallpaper. Many important solids are crystalline in this sense, although this is not always manifest in the external form of the material. Because calculations are easier, more progress has been made in understanding the behaviour of crystalline than of non-crystalline materials. Many common solids—for example, glass, plastics, wood, bone—are not so highly ordered on an atomic scale and are therefore non-crystalline. Only recently has progress been made in understanding the behaviour of non-crystalline solids at a fundamental level.†

† The December 1988 issue of *Physics Today* contains articles describing some of the progress towards an understanding of disordered materials.

Even in the restricted field of crystalline solids the most remarkable thing is the great variety of *qualitatively different* behaviour that occurs. We have insulators, semiconductors, metals and superconductors—all obeying different macroscopic laws: an electric field causes an electric dipole moment in an insulator (Chapter 9), a steady current in a metal or semiconductor (Chapters 3 to 6) and a steadily accelerated current in a superconductor (Chapter 10). Solids may be transparent or opaque, hard or soft, brittle or ductile, magnetic or non-magnetic.

In this chapter we first introduce in section 1.2 the basic ideas of crystallography. In section 1.3 we describe some important crystal structures and in section 1.4 we explain how x-ray diffraction is used to determine crystal structure. In section 1.5 we discuss quasi-crystals, ordered solids that challenge much of the conventional wisdom concerning crystalline materials. Section 1.6 contains a qualitative description of the interatomic forces responsible for binding atoms into solids.

1.2   ELEMENTARY CRYSTALLOGRAPHY

A basic knowledge of crystallography is essential for solid state physicists. They must know how to specify completely, concisely and unambiguously any crystal structure and they must be aware of the way that structures can be classified into different types according to the symmetries they possess; we shall see that the symmetry of a crystal can have a profound influence on its properties. Fortunately we will be concerned in this book only with solids with simple structures and we can therefore avoid the sophisticated group theoretical methods required to discuss crystal structures in general.

1.2.1   The crystal lattice

We will use a simple example to illustrate the methods and nomenclature used by crystallographers to describe the structure of crystals. Graphite is a crystalline form of carbon in which hexagonal arrays of atoms are situated on a series of equally spaced parallel planes. The arrangement of the atoms on one such plane is shown in Fig. 1.1(*a*). We choose graphite for our example because a single two-dimensional plane of atoms in this structure illustrates most of the concepts that we need to explain. Solid state physicists often resort to the device of considering a system of one or two dimensions when confronted with a new problem; the physics is often (but not *always*) the same as in three dimensions but the mathematics and understanding can be much easier.

To describe the structure of the two-dimensional graphite crystal it is necessary to establish a set of coordinate axes within the crystal. The origin can in principle be anywhere but it is usual to site it upon one of the atoms. Suppose we choose the atom labelled O in Fig 1.1(*a*) for the origin. The next step is a very

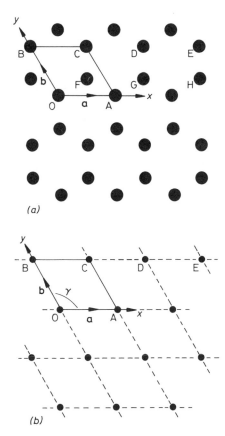

Fig. 1.1   Two-dimensional crystal of carbon atoms in graphite: (*a*) shows how the atoms are situated at the corners of regular hexagons; (*b*) shows the crystal lattice obtained by identifying all the atoms in (*a*) that are in identical positions to that at O. The crystal axes, lattice vectors and conventional unit cell are shown in both figures

important one; we must proceed to identify all the positions within the crystal that are identical in all respects to the origin. To be identical it is necessary that an observer situated at the position should have exactly the same view in any direction as an observer situated at the origin. Clearly for this to be possible we must imagine that the two-dimensional crystal is infinite in extent. Readers should convince themselves that the atoms at A, B, C, D and E (and eight others in the diagram) are identical to the atom at the origin but that the atoms at F, G and H are not; compare for example the directions of the three nearest neighbours of the atom at O with the directions of the three nearest neighbours of the atom at F. The set of identical points identified in this way is shown in Fig. 1.1(*b*) and is called the **crystal lattice**; comparison of Figs. 1.1(*a*) and (*b*) illustrates clearly that the lattice is not in general the same as the structure. Readers should convince themselves that, apart from an unimportant shift in position, the lattice is independent of the choice of origin. Having identified the crystal lattice in this way the coordinate axes are simply obtained by joining the lattice point at the origin to two of its neighbours.

There are many ways of doing this but the **conventional** choice for graphite is to take OA and OB for the x and y axes as shown in Fig. 1.1(b). Note that the coordinate axes for graphite are not orthogonal. An example of an unconventional choice of coordinate axes for graphite would be to take OA for the x axis as before but to take the OD direction for the y axis. The distances and directions of the nearest lattice points along the x and y axes are specified by the **lattice vectors a** and **b** respectively (Fig. 1.1.(b)). The crystal lattice is completely defined by giving the lengths of **a** and **b** and the angle γ between them. For graphite we have $a = b = 2.46$ Å, $\gamma = 120°$ (1 Å = 1 ångstrom = 10^{-10} m). The conventional choice of axes for graphite therefore clearly reflects the hexagonal symmetry of the structure; this is not the case for the unconventional choice discussed above.

The positions of all the lattice points of the two-dimensional graphite crystal are reached by drawing all possible vectors of the form

$$\mathbf{r} = u\mathbf{a} + v\mathbf{b} \tag{1.1}$$

from the origin, where u and v take on all possible integer values, positive, negative and zero. That the crystal appears identical when viewed from all the positions given by this equation is an indication that it possesses the important property of **translational invariance**.

The generalization of the above ideas to a three-dimensional crystal is straightforward. An origin is chosen and all the points within the crystal that are identical to it are identified; this set of points constitutes the three-dimensional crystal lattice. The directions of the crystal coordinate axes are then defined by joining the lattice point at the origin to *three* of its near neighbours (Fig. 1.2). The choice of neighbours is sometimes obvious but, where this is not the case, convention usually dictates the choice that most clearly reflects the symmetry of the lattice. The distances and directions of the nearest lattice points along the crystallographic x, y and z axes are specified by the *three* lattice vectors **a**, **b** and

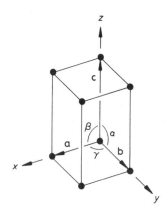

Fig. 1.2   Crystallographic axes and unit cell for a three-dimensional crystal lattice

c. The lattice is completely specified by giving the lengths of **a**, **b** and **c**, and the angles α, β and γ between them (Fig. 1.2). The positions of all the lattice points are reached by drawing all possible vectors of the form

$$\mathbf{r} = u\mathbf{a} + v\mathbf{b} + w\mathbf{c} \qquad (u, v \text{ and } w \text{ are integers}) \qquad (1.2)$$

from the origin. The ability to express the positions of the points in this way, with a suitable choice of **a**, **b** and **c**, may be taken as a definition of a lattice in crystallography. A crystalline material may be defined as a material that possesses a lattice of this kind; the translational invariance property of the crystal is that it appears identical from all positions of the form of Eq. (1.2). Note that the only effect of a shift in choice of origin on a crystal lattice is a shift in the lattice as a whole by the same amount.

The lattice vectors also define the **unit cell** of a crystal. This concept is most easily explained by returning to the two-dimensional graphite crystal of Fig. 1.1, for which the unit cell is the parallelogram OACB defined by the vectors **a** and **b**. It is so called because stacking such cells together generates the entire crystal lattice, as is indicated by the broken lines in Fig. 1.1(b). The analogous three-dimensional object in Fig. 1.2, defined by lattice vectors **a**, **b** and **c**, is called a parallelopiped and is the unit cell for the three-dimensional lattice. The unit cell obtained from the conventional choice of lattice vectors is known as the **conventional unit cell**.

The concept of the unit cell as a building block allows us to understand the remarkable similarities between different crystals of the same material. In particular we can explain the law of constancy of angle (first stated by Nicolaus Steno in 1761) that: *In all crystals of the same substance the angles between corresponding faces have a constant value*. Fig. 1.3 is an illustration from an early book on mineralogy showing how macroscopically plane faces in various orientations can be built up by using cubic unit cells as building blocks. We shall see in Chapter 12 that the surfaces of crystals are *not* in fact constructed in the manner suggested by this illustration.

The reader will have noticed that the two-dimensional lattice of graphite (Fig. 1.1(b)) possesses symmetry properties other than the translational invariance indicated by Eq. (1.1). The lattice is transformed into itself, for example, by a rotation of $60°$ about an axis perpendicular to the xy plane through a lattice point; this axis is the crystallographic z axis of graphite, which is therefore a six-fold rotation axis of the *lattice*. In 1845 Bravais deduced that any three-dimensional lattice of the form of Eq. (1.2) could be classified into one of 14 possible types according to the symmetry that it possessed. The 14 **Bravais lattices** contain only one-, two-, three-, four- and six-fold rotation axes.

We will not describe all 14 Bravais lattices since only a few will feature in this book, but to illustrate the principle of the classification of lattices by symmetry we consider the corresponding two-dimensional problem. A two-dimensional lattice is specified by a, b and the angle γ between **a** and **b**. A lattice with

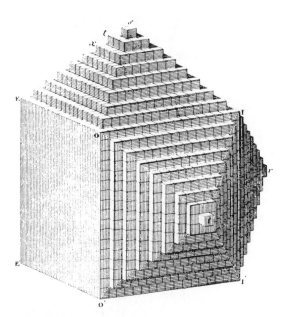

Fig. 1.3   The way in which the stacking of cubic unit cells can produce crystal faces of different orientations (Hauy, *Traite de crystallographie*)

translational symmetry only is shown in Fig. 1.4(*a*) with three possible choices of primitive unit cell. Lattices of higher symmetry are shown in Figs. 1.4(*b*)–(*e*). The **rectangular lattice** in Fig. 1.4(*b*) has $\gamma = 90°$. Alternatively with a general value of γ we may have $a = b$, giving the **rhombic lattice** shown in Fig. 1.4(*c*). This latter example is interesting in that it shares some symmetries with the rectangular lattice and it can also be described by the rectangular unit cell defined by **a′** and **b′**. This rectangular unit cell has a lattice point at the centre as well as at the corners and the rhombic lattice may therefore also be referred to as a **centred rectangular lattice**. The unit cell defined by **a′** and **b′** has an area twice that defined by **a** and **b**. The latter is the smallest possible unit cell of the lattice and is said therefore to be a **primitive unit cell**; the former unit cell is consequently a **non-primitive unit cell**. We will encounter examples of both primitive and non-primitive three-dimensional unit cells in section 1.3. To complete our survey of two-dimensional lattices we must consider the possibility $a = b$ combined with a special value of γ. Two cases arise: $\gamma = 60°$ (or 120°) gives the **triangular lattice** of Fig. 1.4(*d*) with each lattice point surrounded by six neighbours at the corners of a regular hexagon; and $\gamma = 90°$ gives the **square lattice** of Fig. 1.4(*e*). The two-dimensional graphite lattice of Fig. 1.1(*b*) is a triangular lattice.

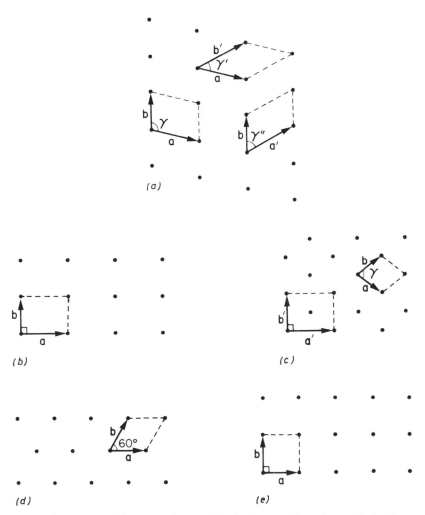

Fig. 1.4   The five possible types of crystal lattice in two dimensions. (a) Lattice with translational symmetry only, showing three possible primitive unit cells. (b) Rectangular lattice, $\gamma = 90°$. (c) Rhombic lattice, $a = b$, equivalent to a centred rectangular lattice with the non-primitive unit cell defined by $\mathbf{a}'$ and $\mathbf{b}'$. (d) Triangular lattice, $a = b$, $\gamma = 60°$. (e) Square lattice, $a = b$, $\gamma = 90°$

1.2.2 The basis

Once the crystal lattice has been determined in the way described in the previous section and used to identify suitable coordinate axes and a unit cell, the description of the crystal structure is completed by specifying the contents of the unit cell. This is accomplished by identifying the group of atoms which, when associated with each lattice point, completely generates the structure. This group of atoms is known as the **basis** of the structure. The basis is specified by giving the position and chemical type of all the atoms within it. We again use the two-dimensional graphite structure of Fig. 1.1(*a*) to illustrate this procedure. The unit cell OACB of this structure contains the atom F, and readers should convince themselves that a suitable (but not unique) choice of basis for the two-dimensional graphite crystal can be obtained by associating the carbon atoms at O and F with the lattice point at O. This is so because the association of the corresponding pair of atoms with each lattice point (atoms G and A with the lattice point at A, for example) does indeed generate the entire structure. The position of an atom within the cell is most easily described by using the **basis vector r**, which connects the atom to the origin. Thus the position of the atom at F may be written

$$\mathbf{r} = \tfrac{2}{3}\mathbf{a} + \tfrac{1}{3}\mathbf{b}.$$

This atom is said to be at position $(\tfrac{2}{3}, \tfrac{1}{3})$. Our choice of basis for the two-dimensional graphite crystal can therefore be written concisely as

$$C(0, 0), \ C(\tfrac{2}{3}, \tfrac{1}{3})$$

where the chemical type of the atom (carbon in this case) is specified by giving its chemical symbol. That a basis of two atoms is required to specify completely the two-dimensional graphite structure is an indication that each primitive unit cell contains just two atoms. In a three-dimensional crystal the basis vector of an atom can always be written as

$$\mathbf{r} = x\mathbf{a} + y\mathbf{b} + z\mathbf{c}$$

and this atom is therefore said to be at (x, y, z).

Taking the symmetry of the basis as well as that of the lattice into consideration allows any crystal to be sorted into one of 32 possible **point symmetry groups** (sometimes referred to as the 32 **crystal classes**) and one of 230 possible **space symmetry groups**. A knowledge of these classifications is not assumed in this book, and readers requiring an understanding of them are recommended to consult one of the standard texts on crystallography.

1.2.3   Crystal planes and directions

Within a crystal lattice it is possible to identify sets of equally spaced parallel planes. Two examples of sets of **lattice planes** for a two-dimensional lattice are illustrated in Fig. 1.5. The density of lattice points on each plane of a set is the same and all the lattice points are contained on each set of planes. Planes of lattice points play an important role in the physics of the diffraction of waves by crystals, and it is necessary therefore to have a method of identifying the different

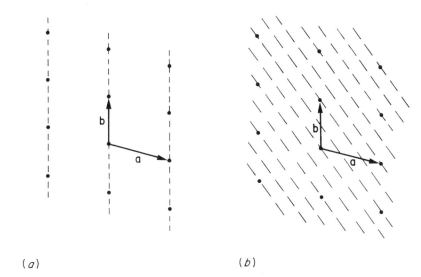

Fig. 1.5   (a) The (1 0) set of planes in a two-dimensional lattice. (b) The (3 2) set of planes in a two-dimensional lattice

sets. **Miller indices** are used for this purpose. These are derived from the intercepts made on the crystal axes by the plane that is nearest to the origin (but not the one that actually passes through the origin). Thus in Fig. 1.5(b) the nearest plane to the origin has intercepts $a/3$ and $b/2$, and this set of planes is therefore referred to by the Miller indices (3 2); note that it is the *reciprocal* of the intercept that determines the Miller index so that a large index indicates a small intercept. Fig. 1.5(a) illustrates a special case in which one intercept is infinite so that the corresponding Miller index is zero; thus the planes (1 0) are parallel to the y axis.

For a set of lattice planes in a three-dimensional lattice the plane nearest the origin will have intercepts a/h, b/k and c/l and the set is referred to by Miller indices $(h\ k\ l)$. Some three-dimensional examples are illustrated in Fig. 1.6. The planes (1 0 0) are parallel to both the y and z axes, and hence to the yz plane. Negative intercepts are indicated by a bar over the corresponding index, as in $(1\ \bar{1}\ 1)$ and $(2\ \bar{1}\ 0)$. The $(\bar{h}\ \bar{k}\ \bar{l})$ set of planes is however identical to the $(h\ k\ l)$ set. Note that if **a**, **b** and **c** define a primitive unit cell, then the Miller indices do not have a common factor. To see why this is, consider the special case of a set of planes with Miller indices (6 4) on Fig. 1.5(b); such a set would be parallel to the (3 2) planes shown but would have half the spacing. The (6 4) planes would therefore have lattice points only on alternate planes.

For crystals of high symmetry certain sets of planes may be related by symmetry and thus be equivalent from an atomic point of view. Thus for crystals of cubic symmetry, in which the unit cell sides **a**, **b** and **c** are equal in magnitude

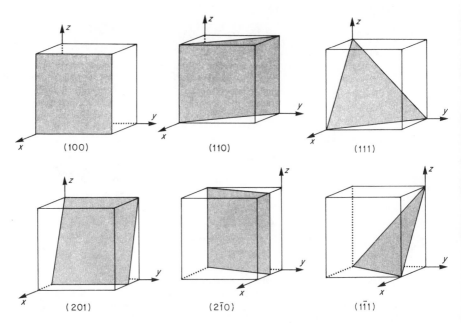

Fig. 1.6   Some crystal planes inscribed in a unit cell, with their Miller indices

and mutually perpendicular, the three sets of planes (1 0 0), (0 1 0) and (0 0 1) are related by symmetry; they are said to belong to the **form** $\{1\ 0\ 0\}$ where the curly brackets mean *all planes equivalent by symmetry to the given plane*.

It is often necessary to specify the direction of a vector **r** in a crystal. The vector can always be written as $\mathbf{r} = u\mathbf{a} + v\mathbf{b} + w\mathbf{c}$ and hence the direction is referred to as 'the $[u\ v\ w]$ direction' using square brackets. If the direction corresponds to that between two lattice points, as is usually the case, then it follows from Eq. (1.2) that u, v and w take integer values. It is important to remember that indices in square brackets are not Miller indices, although for cubic crystals it is a consequence of the symmetry that the direction $[u\ v\ w]$ is normal to the planes of Miller indices $(u\ v\ w)$ (problem 1.3). In this case therefore, which we shall mainly use for practical examples, the direction $[u\ v\ w]$ and the planes $(u\ v\ w)$ are simply related.

1.3   TYPICAL CRYSTAL STRUCTURES

1.3.1   Cubic and hexagonal close-packed structures

The crystal structure adopted by a particular material depends on the nature of the forces between the atoms within it. In some solids, particularly the inert-gas solids and many metals, the forces are such that, to a good approximation, the atoms look like attracting hard spheres. For minimum energy in such cases it is necessary that the spheres should be packed as closely as possible. In two dimensions this principle leads to the close-packed layer structure shown in Fig. 1.7; this is a two-dimensional crystal in which the centres of the spheres lie on a triangular lattice like that of Fig. 1.4(d). The close packing can be extended to three dimensions if a second close-packed layer is placed over the first such that the spheres in the second layer are centred over interstices in the first. Let us suppose, for example, that the second layer occupies the positions marked B in Fig. 1.7. In this way each sphere in the second layer will touch three spheres in the first and the packing will be as close as possible. Inspection of Fig. 1.7 shows that such packing may be continued in various ways, for a third layer can occupy either positions C or A; both of these sets of positions mark interstices in the second layer.

A very common stacking sequence is ABCABC..., which gives a structure known as **cubic close-packed** (ccp) or **face-centred cubic** (fcc). The cubic unit cell of this structure is shown in Fig. 1.8; there are atoms at the corners of the cell and at the centres of each face. To make clear the relation to Fig. 1.7 a close-packed layer of atoms is shaded in Fig. 1.8; it is a (1 1 1) plane, normal to a body diagonal of the cube. It follows from symmetry that all planes of the form {1 1 1} are close-packed planes. In the fcc structure the environment of every atom is identical so that the crystal lattice corresponds to the atomic structure in this case. The rhombohedral *primitive* unit cell of the lattice is shown in Fig. 1.8(b).

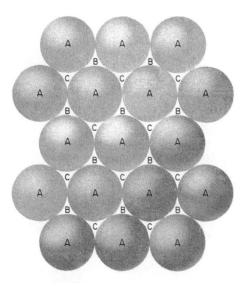

Fig. 1.7   A close-packed layer of spheres occupying positions A. The adjacent layers can occupy positions B or C

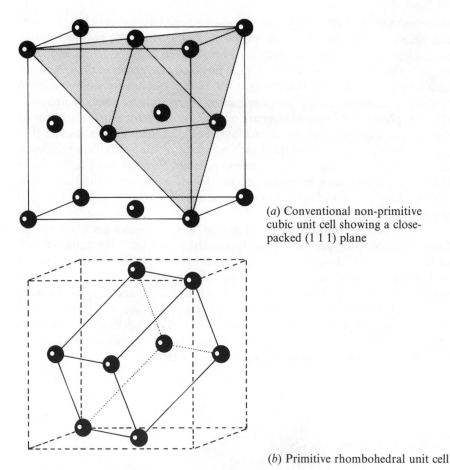

(*a*) Conventional non-primitive cubic unit cell showing a close-packed (1 1 1) plane

(*b*) Primitive rhombohedral unit cell

Fig. 1.8   The cubic close-packed (ccp) or face-centred cubic (fcc) structure

The conventional *non-primitive* choice of unit cell is however the cubic unit cell of Fig. 1.8(*a*) because it more obviously shows the full cubic symmetry. The conventional cell has a volume four times that of the primitive cell and thus contains four lattice points. Not surprisingly the lattice of the fcc structure is denoted an fcc lattice in the Bravais classification. Examples of elements that crystallize into the fcc structure are aluminium, calcium, nickel, copper, silver, gold, lead, neon, argon, krypton and xenon.

The environment of an atom in the fcc structure is best visualized by looking at the **atomic coordination polyhedron**. This is the figure formed from planes which are the perpendicular bisectors of lines joining an atom to its neighbours. Suppose you built a model of the structure out of Plasticine spheres and then

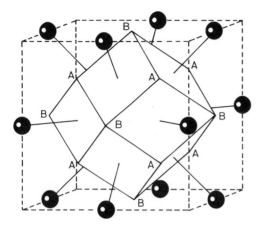

Fig. 1.9   Coordination polyhedron of the ccp structure: the rhombic dodecahedron. The origin of the crystallographic axes has been shifted so that there is an atom at the centre of the unit cell. The bonds from this atom to its nearest neighbours are shown

compressed it; if you then picked it apart you would find that the spheres had deformed into coordination polyhedra.†

The coordination polyhedron thus represents the 'sphere of influence' of an atom. The coordination polyhedron of the fcc structure is shown in Fig. 1.9, together with the positions of the nearest neighbour atoms. Fig. 1.9 shows a cubic unit cell with the origin shifted from that of Fig. 1.8 by half a cell side along a cube axis, the [1 0 0] direction. This polyhedron is called a rhombic dodecahedron; it has 12 faces corresponding to contact with the 12 nearest neighbours; hence each atom is said to have a **coordination number** of 12. On this polyhedron the symmetries characteristic of the cubic structure (and of the cube itself) may be identified. The rhombic dodecahedron has four three-fold axes of symmetry through opposite pairs of corners A ([1 1 1] directions); it looks the same after rotation by 120° about any of these axes. There are also three four-fold axes through opposite pairs of corners B ([1 0 0] directions) and six two-fold axes through the centres of opposite faces ([1 1 0] directions).

The coordination polyhedra of the fcc structure all stack together in the same orientation in such a way as to fill the whole of space. The polyhedron therefore constitutes an alternative choice of primitive unit cell for the crystal. This type of unit cell is known as a **Wigner–Seitz cell** after those who first used it for a quantum mechanical problem. A Wigner–Seitz cell is defined for a general lattice as the smallest polyhedron bounded by planes that are the perpendicular bisectors of vectors joining one lattice point to the others; Fig. 1.10 illustrates this method of construction of the Wigner–Seitz cell for a two-dimensional lattice. It follows from the definition that the interior of the cell is the locus of points which are nearer to the given lattice point than any other. Only for the special case that each atom represents a lattice point is the Wigner–Seitz cell also the atomic coordination polyhedron.

---

† This method has actually been used by Bernal to study disordered (liquid) structures.

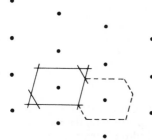

Fig. 1.10 The Wigner–Seitz cell (broken lines) of a two-dimensional lattice obtained by drawing the perpendicular bisectors (full lines) of the lines joining a lattice point to its neighbours

Another common stacking sequence of the close-packed layers in Fig. 1.7 is ABABAB. . . ; this gives the **hexagonal close-packed** (hcp) structure illustrated in Fig. 1.11(a). All the A-plane atoms in this structure have an identical environ- ment and can therefore be taken as lattice points. The environment of the B- plane atoms is different to that of the A-plane atoms, so that the B-plane atoms do not then lie on lattice points. The resulting lattice is denoted a hexagonal lattice in the Bravais classification. The conventional choice of crystallographic axes for the lattice is shown in Fig. 1.11(a) and the resulting primitive unit cell is identified by the thicker lines. Fig. 1.11(b) is a simpler two-dimensional way of depicting this three-dimensional unit cell. It shows a plan view as seen along the z axis; the z coordinate of the atom in the cell is indicated as a fraction of the side c of the unit cell. The unit cell contains a basis of an A atom at (0, 0, 0) and a B atom at $(\frac{2}{3}, \frac{1}{3}, \frac{1}{2})$. The close-packed A planes in the hcp structure are the (0 0 1) planes, with the B planes sandwiched half-way between.

Since the environments of the A and B atoms are different, their coordination polyhedra have the same shape but differ in orientation. Because there are two atoms in each primitive unit cell, the Wigner–Seitz unit cell of the hcp lattice is double the volume of each coordination polyhedron and does not bear a simple relation to it.

The ratio c/a of the lattice constants for the 'ideal' hexagonal close packing of hard spheres can be shown by elementary geometry to equal $(8/3)^{1/2} = 1.633$ (problem 1.1). This ideal value is not however imposed by the hexagonal symmetry of the lattice. Since atoms are not really hard spheres, therefore, the unit cell axial ratio c/a differs a little from the ideal value. Examples of elements that crystallize into the hcp structure with the corresponding c/a values in brackets are magnesium (1.623), titanium (1.586), zinc (1.861), cadmium (1.886), cobalt (1.622) and helium (1.633).

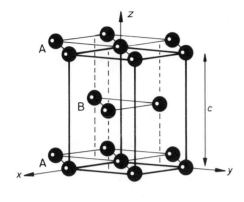

(a) The hexagonal close-packed structure with the primitive unit cell indicated by thicker lines

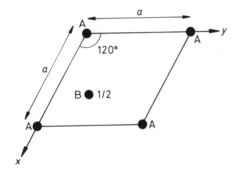

(b) Plan view of the primitive unit cell of the hexagonal close-packed (hcp) structure. The 1/2 indicates the height of the atom along the z axis as a fraction of the lattice spacing c. The basis has atoms at $(0,0,0)$ and $(\frac{2}{3}, \frac{1}{3}, \frac{1}{2})$

Fig. 1.11

Other more complicated stacking sequences of close-packed layers are sometimes found, ABACABAC... for example, particularly in the rare-earth metals. These more exotic possibilities will not concern us in this book.

1.3.2   The body-centred cubic structure

A cubic structure only slightly less close-packed than fcc is the **body-centred cubic** (bcc) structure, three cubic unit cells of which are shown in Fig. 1.12(a). The environments of all the atoms are identical in the bcc structure so that the lattice is the same as the structure. The non-primitive cubic unit cell is the conventional choice for this lattice and contains two lattice points; the lattice vectors of the primitive cell are shown in Fig. 1.12(a). The eight *hexagonal* faces of the coordination polyhedron shown in Fig. 1.12(b) represent the 'contact' of

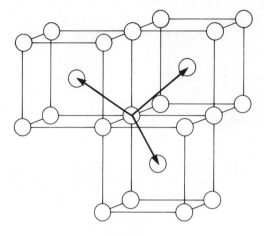

(a) Three conventional cubic unit cells of the body-centred cubic (bcc) structure. The lattice vectors of the primitive cell are indicated

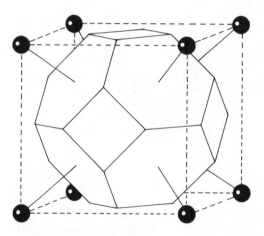

(b) Coordination polyhedron of the bcc structure. The hexagonal faces represent 'contact' with eight nearest neighbours; the square faces represent 'contact' with six second nearest neighbours

Fig. 1.12

an atom with its eight nearest neighbours; the coordination number of the bcc structure is thus 8. This is smaller than the coordination number (12) of the fcc and hcp structures, but the existence of the six *square* faces of the coordination polyhedron indicates that an atom in the bcc crystal has six second nearest neighbours not much further away than the first nearest neighbours. The metallic elements lithium, sodium, potassium, chromium, barium and tungsten crystallize into the bcc structure.

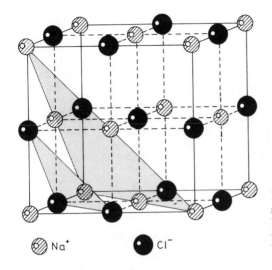

Fig. 1.13   Unit cell of the NaCl structure, with alternate planes of Na^+ and Cl^- ions perpendicular to the [1 1 1] direction indicated by shading

1.3.3   Structures of ionic solids

We now progress to consider examples of the simplest type of crystals containing atoms of more then one element, namely ionic solids. Since ions of opposite charge approximate to attracting hard spheres, the crystal structure of these is also often dominated by close-packing considerations. The electrons in the negatively charged anion are in general less tightly bound than those in the positively charged cation and the anion is therefore normally larger. In crystals containing equal numbers of positive and negative ions such as sodium chloride, NaCl, the structure is then likely to be determined by the number of larger anions that will pack tightly around the cation. For NaCl this is only six, and this leads to the three-dimensional chessboard structure shown in Fig. 1.13. All the Na^+ ions have identical environments within the crystal so that these can be taken to represent the crystal lattice, which is therefore an example of an fcc Bravais lattice. The Cl^- ions also lie on an fcc lattice displaced by half a unit cell in the [1 0 0] direction.

It can also be seen from Fig. 1.13 that planes of the form $\{1\ 1\ 1\}$ of the Na^+ lattice contain all the Na^+ ions and no Cl^- ions; the Cl^- ions are contained on parallel planes mid-way between the (1 1 1) planes; this fact was crucial in the elucidation of the structure of NaCl which was the first to be determined by x-ray diffraction (see section 1.4). When this structure was first discovered some chemists were horrified to find that it contained no identifiable NaCl molecule. We now know that the absence of an identifiable molecule is very general in inorganic crystals and we have become used to the idea of a crystal as a single giant molecule.

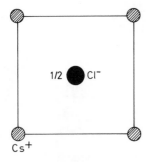

Fig. 1.14   Primitive cubic unit cell of CsCl shown in plan view. The basis is Cs^+ $(0,0,0)$ and Cl^- $(\frac{1}{2}, \frac{1}{2}, \frac{1}{2})$

The caesium chloride, CsCl, structure of Fig. 1.14 is an alternative structure for an ionic solid containing equal numbers of anions and cations; the cubic unit cell has Cs^+ ions at the corners and a Cl^- ion at the centre. Each ion in the structure has a coordination number of 8 so that this is the structure likely to be adopted when just eight anions will pack tightly around each cation. All cations (or alternatively all anions) in this structure have an identical environment so that their positions form a crystal lattice. In this case the conventional cubic unit cell is primitive and is designated as simple cubic in the Bravais classification.

1.3.4   The diamond and zincblende structures

A very important structure in solid state physics is that adopted by carbon atoms in diamond. In this structure each carbon atom is covalently bonded to four nearest neighbours arranged at the corners of a regular tetrahedron as in Fig. 1.15. Fig. 1.15 shows that only half of the atoms have identical environments and these lie on an fcc Bravais lattice. The other atoms form an fcc lattice displaced by one-quarter of a unit cell in the [1 1 1] direction. The two types of atom differ only in the orientation of the bonds to the nearest neighbours. The small coordination number (4) of the diamond structure indicates that it is very far from being a close-packed structure and that the interatomic forces are very different in nature to those in most metallic, ionic and 'inert-gas' solids. Two other elements from group IV of the periodic table, the semiconducting elements silicon and germanium, crystallize into the diamond structure and this explains its importance in solid state physics.

Group III–Group V semiconducting compounds (see Chapter 5), such as gallium arsenide, GaAs, and indium antimonide, InSb, crystallize into the closely related zincblende, ZnS, structure. This structure differs from that of diamond only in that one type of carbon atom is replaced by zinc atoms and the other by sulphur atoms.

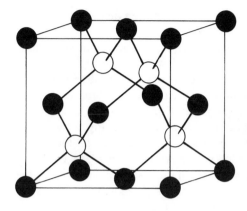

(a) The conventional cubic unit cell of the diamond and zincblende structures. In diamond both types of atomic site are occupied by carbon atoms. In zincblende one type of site is occupied by zinc atoms and the other by sulphur atoms

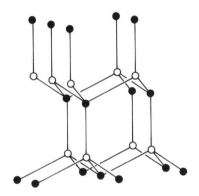

(b) The tetrahedral arrangement of covalent bonds in the diamond and zincblende structures. The vertical direction is [1 1 1]

Fig. 1.15

1.4 X-RAY CRYSTALLOGRAPHY

1.4.1 The Bragg law

The wavelength of x-rays is typically 1 Å, comparable to the interatomic spacings in solids. This means that a crystal behaves as a three-dimensional diffraction grating for x-rays. In an optical diffraction experiment it is possible to deduce the spacing of the lines on the grating from the separation of the diffraction maxima; by measuring the relative intensities of different orders information about the structure of the lines on the grating can be obtained. In an exactly similar way, measurement of the separation of the x-ray diffraction maxima from a crystal allows us to determine the size of the unit cell, and from the intensities of the diffracted beams we obtain information on the arrangement of atoms within the cell.

The general laws of diffraction as formulated by von Laue will be considered in Chapter 11. For the present, the simpler and more physical formulation discovered by Bragg and used by him in his earliest structure determinations will suffice. Bragg derived the condition for constructive interference of the x-rays scattered from a set of parallel lattice planes. Consider x-rays incident at a glancing angle θ on one of the planes of the set as shown in Fig. 1.16(a). The figure illustrates that there will be constructive interference of the waves scattered from the two successive lattice points A and B in the plane if the distances AC and DB are equal. This is the case if the scattered wave makes the same angle θ to the plane as the incident wave; the diffracted wave thus looks as though it has been reflected from the plane. The use of the glancing angle θ rather than the angle of incidence is conventional in x-ray crystallography; the reflection condition implies that the x-ray beam is deflected through an angle 2θ. Note that we consider the scattering associated with lattice points rather than atoms because it is the basis of atoms associated with each lattice point that is the true repeat unit of the crystal; the lattice point is the analogue of the line on an optical diffraction grating and the basis represents the structure of the line.

Coherent scattering from a single plane is not sufficient to obtain a diffraction maxium. It is also necessary that successive planes should scatter in phase. This will be the case if the path difference for scattering off two adjacent planes is an

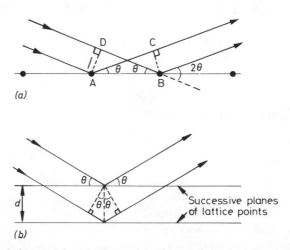

Fig. 1.16   Proof of Bragg's law. (a) Scattering of x-rays from the adjacent lattice points A and B in the plane will be in phase if AC = DB and thus if the scattered beam makes the same angle θ to the plane as the incident beam. (b) Scattering of x-rays off successive planes is in phase if the path difference $2d \sin \theta$ is an integral number of wavelengths $n\lambda$

integral number of wavelengths. From Fig. 1.16(*b*) we see that this is so if

$$2d \sin \theta = n\lambda \tag{1.3}$$

where *d* is the spacing of the planes and *n* is an integer. This is **Bragg's law**.

The diffracted beams (often referred to as reflections) from any set of lattice planes can only occur at the particular angles predicted by the Bragg law. X-ray crystallographers use the Miller indices of the planes to label the reflections. A beam corresponding to a value of *n* greater than 1 could be identified by a statement such as 'the *n*th-order reflection from the (*h k l*) planes'. This however is rather cumbersome and such a beam is described instead more concisely as the (*nh nk nl*) reflection. Thus the third-order reflection from the (1 1 1) planes is described as the (3 3 3) reflection. This notation is justified by rewriting the Bragg law as

$$2(d/n) \sin \theta = \lambda$$

which makes *n*th-order diffraction off (*h k l*) planes of spacing *d* look like first-order diffraction off planes of spacing *d/n*. Planes of this reduced spacing would have Miller indices (*nh nk nl*).

To illustrate the general principles of x-ray structure analysis we will explain the way that Bragg deduced the structure of NaCl and KCl (*Proc. R. Soc.* A **89**, 248 (1914)) in the same series of experiments as he established the existence of x-ray spectral lines. A more general treatment of structure determination by diffraction methods can be found in Chapter 11. Bragg used an arrangement like an ordinary spectrometer to measure the intensity of specular reflection from a cleaved face of a crystal and found six values of θ for which a sharp peak in intensity occurred, corresponding to three characteristic wavelengths (K, L and M x-rays) in first and second order (*n* = 1 and *n* = 2 in Eq. (1.3)). By repeating the experiment with a different crystal face he could use Eq. (1.3) to find for example the ratio of the (1 0 0) and (1 1 1) plane spacings, information that confirmed the cubic symmetry of the atomic arrangement.

The details of the structure were then deduced from the differences between the diffraction patterns for NaCl and KCl. The major differences was the absence of the (1 1 1) reflection in KCl compared to a weak but clearly detectable (1 1 1) reflection in NaCl. This arises because the K^+ and Cl^- ions both have the argon electron shell structure and hence scatter x-rays almost equally whereas the Na^+ and Cl^- ions have different scattering strengths. The (1 1 1) reflection in NaCl corresponds to one wavelength of path difference between neighbouring (1 1 1) planes, and thus to half a wavelength difference between the alternate planes of Na^+ and Cl^- ions that make up the crystal structure of Fig. 1.13. The difference in scattering of x-rays by the Na^+ and Cl^- ions is necessary therefore to prevent elimination of the (1 1 1) reflection by destructive interference. Bragg was able to deduce that the structures of NaCl and KCl corresponded to alternate planes of positive and negative ions perpendicular to the [1 1 1] direction and the structure of Fig. 1.13 followed from this.

1.4.2   Experimental arrangements for x-ray diffraction

Since the pioneering work of Bragg, x-ray diffraction has developed into a routine technique for the determination of crystal structure. In most experiments the x-rays are produced by accelerating electrons through a potential difference of order 30 keV and allowing them to collide with a metal target; the x-ray emission is a mixture of the characteristic lines (K, L, M, etc.) of the metal atoms and a continuous background which varies smoothly with wavelength. By changing the accelerating voltage it is possible to vary the relative amounts in the mixture to obtain either almost monochromatic x-rays or a broadened *white* spectrum.

If a higher-intensity source of x-rays is required, the intense radiation emitted by the charged particles (usually electrons) in a synchrotron can be used. The particles radiate predominantly in a direction tangential to their path as a result of the acceleration associated with their orbits. The intensity of the synchrotron radiation normally varies smoothly with increasing wavelength above a minimum cut-off value, which depends on the radius of curvature of the path and the energy of the particles. The intensity can be made to peak at a particular wavelength by placing bending magnets at regular intervals along a straight section of the synchrotron. This configuration is known as an undulator and wavelength selection occurs because of constructive interference between the radiation emitted in the vicinity of successive magnets. Examples of situations in which synchrotron radiation is used are for the determination of the structure of very small crystals and of crystals containing biological molecules where the unit cell may contain thousands of atoms. In the latter case it is necessary to measure the intensities of a large number of closely spaced diffracted beams in order to determine the structure.

Many types of x-ray camera have been invented to sort out the reflections from different crystal planes. We shall describe only three very common types of x-ray photograph that are widely used for the simple structures that we study in this book.

For a **Laue photograph**, historically the first type, a single crystal is illuminated with a collimated beam of 'white' (i.e. continuous spectrum) x-rays as in Fig. 1.17(*a*). Each set of crystal planes will satisfy the Bragg condition, Eq. (1.3), for some wavelength (perhaps several wavelengths if the spread in wavelength is large enough for different orders of diffraction to occur) and the resulting diffracted beams generate a pattern of spots on the photographic film as in Fig. 1.17(*b*). The symmetry of the spot pattern reflects the symmetry of the crystal when viewed along the direction of the incident beam. The deduction of the symmetry of the crystal is one of the main uses of the Laue method; it is often

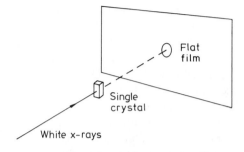

(*a*) Experimental geometry for a Laue photograph

(*b*) A Laue photograph of Si with a (1 1 1) face normal to the x-rays; note the three-fold symmetry. (Reproduced with permission from *Elements of X-ray Crystallography* by Leonid V. Azaroff. © 1968 McGraw-Hill Book Company Inc.)

Fig. 1.17

used to determine the orientation of single crystals that do not have well developed external faces.

When a *single* crystal is exposed to a collimated *monochromatic* beam of x-rays no diffraction takes place in general because no set of lattice planes is at the correct angle to satisfy Bragg's law. If the crystal is rotated about a fixed axis perpendicular to the x-ray beam, then the glancing angle θ varies for sets of planes that are not perpendicular to the rotation axis. A set of such planes is likely to satisfy the Bragg condition for some orientation of the crystal. This is the basis of the **rotating crystal method**; the crystal is typically surrounded by a photographic film in the form of a cylinder parallel to the rotation axis and the resulting pattern of diffraction spots is analysed to obtain the structure.

An alternative method of ensuring that there are sets of lattice planes in the specimen at the correct angles to satisfy Bragg's law for a monochromatic incident beam of x-rays is to use a sample in the form of many small crystalline grains glued together. If the orientation of the grains is random then, for any set of lattice planes, some of the grains will be oriented at the Bragg angle θ to the incident x-rays. The locus of the beams reflected from the same set of planes in different grains will be a cone of half-angle 2θ with the incident beam as axis as shown in Fig. 1.18(a); intersection of the x-ray cone with the film produces a line on the photograph. A typical example of a **powder photograph** is shown in Fig. 1.18(b); each line represents diffraction from a different set of lattice planes. The structure is determined from the measured θ values and the relative intensities of the reflections. Another application of the powder method arises because of the very high resolution that can be obtained for the radiation that is almost back-scattered, as is evidenced by the resolution of the Co Kα doublet in Fig. 1.18(b). Eq. (1.3) shows that when θ is close to 90° it is very sensitive to the precise value of d. Very accurate unit cell dimensions can therefore be obtained from almost back-scattered radiation and this provides a valuable method of measuring thermal expansion.

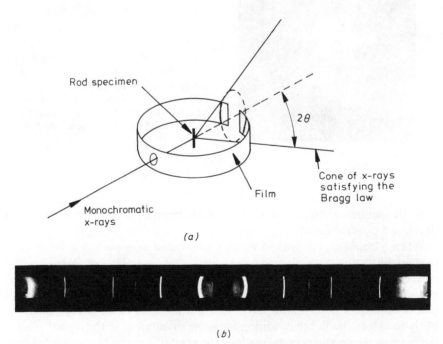

Fig. 1.18   (a) Experimental geometry for a power photograph. (b) A powder photograph of molybdenum taken with Co Kα radiation. The x-rays enter the camera through the hole in the centre of the film and leave between the ends of the film. Note that the Kα_1–Kα_2 x-ray doublet (wavelengths 1.789 and 1.793 Å) is resolved for the back-scattered radiation near the entrance hole. (Courtesy of H. Lipson)

★1.5   QUASI-CRYSTALS

To the crystallographer devoted to the view of ordered structures described earlier in this chapter the x-ray Laue photograph of Fig. 1.19 presents a seemingly insuperable problem. A diffraction pattern of this type was first observed in 1984 from a sample of Al/Mn alloy, cooled so rapidly from the molten state that the first solid structure to form was 'frozen in'. Similar patterns have subsequently been observed for other materials. The existence of such sharp spots in the pattern indicates a highly ordered atomic arrangement, containing presumably parallel planar structures capable of scattering x-rays coherently. The spot pattern in Fig. 1.19 however has a tenfold symmetry axis, indicating that the atomic structure must possess similar symmetry,† thus contradicting one of the fundamental theorems of crystallography that any lattice of the form of Eq. (1.2) can contain only two-, three-, four- and sixfold symmetry axes. A tenfold axis fails to appear in the list for essentially the same reason as it is impossible to tile a two-dimensional area with tiles shaped like regular decagons, as is apparent from Fig. 1.20. As the materials giving rise to diffraction patterns like that of Fig. 1.19 cannot therefore possess the property of translational invariance that is expected in crystals, they have been designated as **quasi-crystals**.

When diffraction patterns for different angles of incidence are viewed, quasi-crystals often appear to have the same symmetry as the icosahedron shown in Fig. 1.21(*a*). This figure has 20 faces each of which is an equilateral triangle. The line AA is one of the six fivefold axes of symmetry of the icosahedron which lead to the tenfold symmetry of the diffraction pattern. Icosahedral arrangements of atoms arise quite naturally in attempts at close packing if the tetrahedral

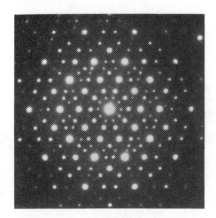

Fig. 1.19   Spot diffraction pattern with tenfold symmetry obtained from a rapidly cooled Al(86 at%)–Mn(14 at%) alloy. (Reproduced with permission from D. Schechtman *et al.*, *Phys. Rev. Lett.* **53**, 1951 (1984))

---

† Subsequently magnified images of quasi-crystallites have been observed which confirm the existence of a tenfold symmetry axis.

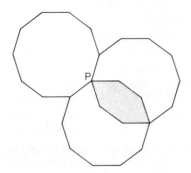

Fig. 1.20   Three regular decagons
sharing a vertex, P, demonstrating the
impossibility of tiling a two-dimensional
area with unit cells of this shape

arrangement of atoms of Fig. 1.21(*b*) rather than the close-packed plane of
Fig. 1.7 is taken as the basic building block. The icosahedron is formed by
allowing 20 tetrahedra to share a common vertex. In order to achieve this each
tetrahedron has to distort slightly; an atom is about 5% further from its
neighbours on the surface of the icosahedron than it is from the atom at the
centre. It is this feature that prevents the attainment of long-range close packing
by continued stacking of tetrahedra outwards from the original shared vertex.

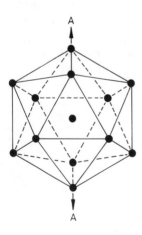

(*a*) The icosahedron: the line AA is one
of six five fold symmetry axes that pass
through the 12 vertices

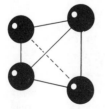

(*b*) Tetrahedral arrangement of atoms;
if 20 slightly distorted tetrahedra share
a common vertex the icosahedral
arrangement of atoms of (*a*) is obtained

Fig. 1.21

All solids discovered prior to 1984 with a local icosahedral arrangement of atoms had relieved the distortion by incorporating additional atoms in such a way as to regain a structure with translational invariance.

It is generally believed that an understanding of the structure of quasi-crystals will be obtained by generalizing to three dimensions the two-dimensional tiling pattern invented by Roger Penrose in 1974, which is shown in Fig. 1.22. In contrast to the stacking of identical parallelogram unit cells, which generates a two-dimensional crystal lattice like that of Fig. 1.1(b), Penrose tiling uses the two building blocks of Fig. 1.22(a). Both the basic tiles are rhombuses, like the unit cell of the rhombic lattice in Fig. 1.4(c), but they have values of the angle γ of 144° and 108°. In the pattern (Fig. 1.22(b)) tiles of the former angle occur exactly $(1 + \sqrt{5})/2$ times as often as those of the latter. Despite the lack of translational invariance, the pattern contains regular decagons all with the same orientation and also sets of almost straight lines intersecting at angles of 72°

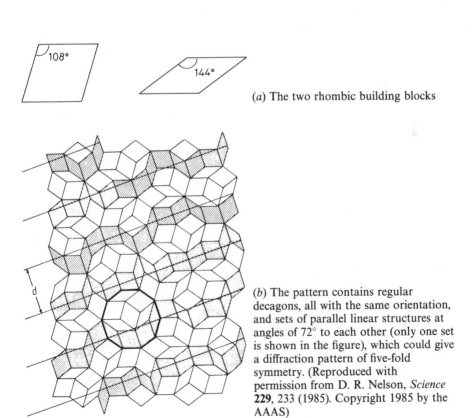

(a) The two rhombic building blocks

(b) The pattern contains regular decagons, all with the same orientation, and sets of parallel linear structures at angles of 72° to each other (only one set is shown in the figure), which could give a diffraction pattern of five-fold symmetry. (Reproduced with permission from D. R. Nelson, *Science* **229**, 233 (1985). Copyright 1985 by the AAAS)

Fig. 1.22   Penrose tiling

(Fig. 1.22(*b*) shows one of the sets) which could give a diffraction pattern of fivefold symmetry. Three-dimensional Penrose tiling can be achieved by using two different rhombohedrons (squashed cubes) as building blocks, but although it is likely that the structure of quasi-crystals can be explained in this way there is as yet no unambiguous structure determination.

The difficulties encountered in elucidating the structure of quasi-crystals arise because of the lack of translational invariance, and it is worth stressing at this point just how useful this property is to the crystallographer and solid state physicist. For a translationally invariant crystal it is necessary to specify only the position and orientation of one unit cell in order to specify the position of all the atoms in the crystal. So valuable is this property that one theoretical approach to quasi-crystals has been to investigate the possibility that they may be represented by a translationally invariant lattice in a six-dimensional space; the actual structure is then seen on a three-dimensional 'surface' in this space!†

1.6 INTERATOMIC FORCES

The binding energy of the atom in all solids results from the reduction in energy of the atomic electrons due to the proximity of the neighbouring atoms. To give a quantitative description it is necessary to calculate the electron states; some indication of the nature of these and the factors that affect the binding energy are given in Chapters 3 and 4. Here we give a qualitative discussion of interatomic forces, most of which is probably already familiar to the reader. It is conventional to classify the bonds between atoms into different types: van der Waals, ionic, covalent, metallic and hydrogen bonds are the types discussed further below. We must always bear in mind that these terms are inventions of the human imagination, introduced as an aid to thought. *All bonding is a consequence of the electrostatic interaction between nuclei and electrons obeying Schrödinger's equation.* In most cases also the bonds must be regarded as being intermediate between the extreme types described below.

1.6.1 Van der Waals bonding

The simplest examples of van der Waals bonding are the inert-gas solids. The spherically symmetric filled-shell electronic configurations of isolated inert-gas atoms are very stable and little affected when the atoms come together to form the solid. The interaction energy of two inert-gas atoms depends only on the distance between them and is represented by the well known interatomic potential curve of Fig. 1.23; the force between the atoms is the negative of the slope of this curve. The attractive force at large separations arises because even a

---

† For more information on quasi-crystals, see the article by D. R. Nelson in the August 1986 issue of *Scientific American*.

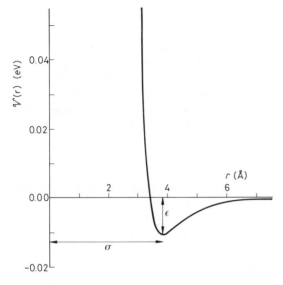

Fig. 1.23   Lennard-Jones potential

$$\mathscr{V}(r) = \varepsilon \left[\left(\frac{\sigma}{r} \right)^{12} - 2 \left(\frac{\sigma}{r} \right)^{6} \right]$$

for the interaction of two argon atoms. The parameters ε and σ which specify the potential were obtained from measurements on gaseous argon

spherically symmetric atom has a *fluctuating* electric dipole moment; this *induces* a dipole moment in the other atom and the two dipoles then attract each other. The attractive force is known as the **van der Waals** or **London force**. The repulsive force at small separations is associated with the overlap of the outer electron shells on the two atoms. An important contribution to the repulsion arises because of the Pauli principle, which prevents two electrons from occupying the same quantum state; overlap of the filled shells means that electrons must be promoted to higher atomic energy levels in order to comply with this requirement. The repulsive force increases very rapidly with increasing overlap, and this explains why inert-gas atoms behave like attracting hard spheres and form close-packed structures.

The binding energies, interatomic forces and related properties of the inert-gas crystals can be calculated with reasonable accuracy by assuming that the interaction of any two atoms within the crystal is given by an interatomic potential of the form of Fig. 1.23. A reader who is unfamiliar with this approach is recommended to read Chapter 3 of Flowers and Mendoza.[1]

1.6.2   Ionic bonding

A similar situation to that in the inert-gas solids arises in ionic solids such as sodium chloride. Transfer of an electron from a sodium atom to a chlorine atom produces Na^+ and Cl^- ions with the stable electronic configurations of the inert gases neon and argon respectively. The electron states are little affected by the coming together of the ions to form the solid and the interaction of any two ions within the solid can be represented by the interionic potential curve of two isolated ions. At large separations the interionic potential is dominated by the long-range electrostatic interaction $\pm e^2/4\pi\varepsilon_0 r$; the $+$ sign is for two ions of the same sign, the $-$ sign for ions of opposite sign. As for the inert-gas atoms, the potential at small separations is dominated by the repulsive interaction associated with overlap of the electron shells. The use of the interionic potential in calculating the binding energy and related properties of ionic solids is also discussed in Chapter 3 of Flowers and Mendoza.[1]

1.6.3 Covalent bonding

In covalently bonded crystals such as diamond, silicon and germanium the bonding energy is associated with the sharing of valence electrons between atoms. The states of the valence electrons are profoundly changed by the approach of the atoms to form the solid, and where an atom forms more than one bond the energy depends strongly on their relative orientation. Covalent bonds are thus said to be **directed**; for the carbon atoms in diamond the minimum energy occurs when the four covalent bonds are directed towards the corners of a regular tetrahedron as in Fig. 1.15. In section 4.3 we present a simple approach to the construction of trial electron wavefunctions to describe the covalent bonding in diamond and other solids. The directed nature of covalent bonds means that the energy of a covalently bonded solid cannot be written simply as a sum of the interatomic potentials for isolated pairs of atoms.

Since a pair of electrons are an essential feature of a covalent bond, an atom cannot in general form more covalent bonds than it has valence electrons. Because of the limited number of bonds per atom, covalent binding is said to be **saturable**; the structures of covalently bonded materials are determined by this and the directed nature of the bonds rather than by close-packing considerations.

In an ideal covalent bond between two atoms the two electrons are equally shared. Symmetry considerations suggest that the atoms must be identical for this to be so. This is the case for the elemental semiconductors Si and Ge but not for the semiconductor compounds such as GaAs, which have the zincblende structure (Fig. 1.15); the shared electrons reside nearer to the group V element, As, than to the group III element, Ga. The covalent bonds in such materials are thus intermediate in nature between a pure covalent bond with equally shared electrons and a pure ionic bond with one electron completely transferred from one atom to the other. A measure of the ionicity of a bond is obtained by expressing the electric dipole moment associated with the bond as a fraction of the dipole moment for a pure ionic bond; the latter is equal to the electronic charge times the interatomic separation. The 'covalent' bond in GaAs turns out to be 32% ionic; for comparison the 'ionic' bond in NaCl is actually 94% ionic according to the same criterion.

1.6.4   Metallic bonding

Metal atoms have fewer than the four valence electrons required to form a three-dimensional covalently bonded structure. One way of looking at metallic bonding is as a type of covalent bonding in which some of the bonds are missing. There are many possible arrangements of the omitted bonds and we can imagine a ground state wavefunction for the crystal that is a linear combination of all possible ways of leaving out the prescribed fraction of bonds. This leads naturally to the idea that the electrons are not localized as is necessary for electrical conductivity. Alternatively metallic bonding may be regarded as a limiting case of ionic bonding in which the negative ions are just electrons. Thus, sodium chloride contains equal numbers of Na^+ and Cl^- ions, and metallic sodium contains equal numbers of Na^+ and e^-. The crucial difference is that the very small mass of an electron means that its zero point energy is so large that it is not localized to vibrating with a small amplitude about a fixed position in the crystal. The reduction in the kinetic energy of the electrons resulting from their delocalization makes a significant contribution to the binding energy (section 3.2.5). A metallic structure is determined largely by the packing of the positive ions alone; the electron fluid is just a sort of negatively charged glue. In Chapter 4 we introduce a formalism for discussing the electron states in solids which encompasses both metallic and covalent bonding.

1.6.5   Hydrogen bonding

Hydrogen bonding arises because a hydrogen atom is usually a somewhat positively charged region of a molecule. This can, by electrostatic attraction, form a weak bond to a negatively charged region of another (or the same) molecule. Hydrogen bonding is important in ice and in many organic solids; the helical form of the DNA molecule is due to hydrogen bonding between different parts of the same long molecule.

1.6.6   Mixed bonding

More than one type of bonding can exist simultaneously in the same solid. In graphite for example the carbon atoms within the hexagonal planes (Fig. 1.1) are covalently bonded whereas the weaker forces between the planes are similar in origin to the forces between the inert-gas atoms; the weakness of the interplanar forces explains the ease with which the planes slide relative to each other and is thus responsible for the lubricating properties of graphite. That graphite is an electrical conductor for current flow parallel to the planes and a non-conductor for current flow perpendicular to the planes is another conse- quence of the mixed bonding. The forces between the covalently bonded molecules in many organic solids are also similar in origin to those between the inert-gas atoms. In some cases the molecules possess a permanent electric dipole moment, which gives rise to an attractive potential varying as $1/R^3$ rather than the $1/R^6$ obtained for fluctuating dipole moments.

PROBLEMS 1

1.1 Show that $c/a = (8/3)^{1/2}$ for hexagonal close packing of hard spheres.

1.2 Sketch a few cubic unit cells and draw the following lattice planes within them (as in Fig. 1.6): (0 0 1), (1 0 1), (0 1 1), (0 2 1), (2 1 0), (2 1 1) and (1 2 2).

1.3 Prove that in a lattice of cubic symmetry the direction $[h\ k\ l]$ is perpendicular to the plane $(h\ k\ l)$ with the same indices.

1.4 Show that the spacing d of the $(h\ k\ l)$ set of lattice planes in a cubic lattice of side a is

$$d = a/(h^2 + k^2 + l^2)^{1/2}.$$

(Remember that the sum of the squares of the direction cosines of the normal to a plane is 1.)

1.5 Consider the following pattern:

q p	d b	q p	d b	q p	d b	
d b	q p	d b	q p	d b	q p	⋯
q p	d b	q p	d b	q p	d b	

Indicate:
(a) a rectangular unit cell,
(b) a primitive unit cell, and
(c) the basis of letters associated with each lattice point.

1.6 Consider the fcc, bcc, hcp and diamond structures.
(a) Draw *plans* of the conventional unit cells of these structures, indicating the height of the atoms as a fraction of the unit cell height.
(b) Give coordinates of the atoms in the basis of each structure.
(c) If the structures are formed by spheres in contact, calculate the fraction of space occupied by spheres.

1.7 A crystal has a basis of one atom per lattice point and a set of primitive translation vectors (in Å):

$$\mathbf{a} = 3\mathbf{i}, \qquad \mathbf{b} = 3\mathbf{j}, \qquad \mathbf{c} = 1.5(\mathbf{i} + \mathbf{j} + \mathbf{k}),$$

where $\mathbf{i}, \mathbf{j}$ and $\mathbf{k}$ are unit vectors in the x, y and z directions of a Cartesian coordinate system. What is the Bravais lattice type of this crystal, and what are the Miller indices of the set of planes most densely populated with atoms? Calculate the volumes of the primitive unit cell and the conventional unit cell.

1.8 For the fcc and bcc structures it is possible to choose a primitive unit cell that is a rhomb, i.e. the primitive translation vectors $\mathbf{a}$, $\mathbf{b}$ and $\mathbf{c}$ are equal in magnitude as are the angles α, β and γ. Draw diagrams showing $\mathbf{a}$, $\mathbf{b}$ and $\mathbf{c}$ for each structure and calculate α for each case.

1.9 The Bragg angle for a certain reflection from a powder specimen of copper is 47.75° at a temperature of 293 K and 46.60° at 1273 K. Calculate the coefficient of linear thermal expansion of copper.

I'm picking up good vibrations. She's giving me excitations.
—Beach Boys, 1967 pop song

CHAPTER

Crystal dynamics

2.1 INTRODUCTION

The picture of a crystal given in the previous chapter as a regular arrangement of *stationary* atoms cannot be entirely correct. It conflicts with the Heisenberg uncertainty principle that it is not possible to know simultaneously and exactly both the position and momentum of a particle. Thus, at the absolute zero of temperature, the atoms in a crystal must vibrate about their equilibrium positions. The energy they possess as a result of this **zero point motion** is known as **zero point energy**. At higher temperatures the amplitude of the motion increases as the atoms gain thermal energy. In this chapter we discuss the nature of the atomic motions, sometimes referred to as **lattice vibrations**.

Initially our calculations will be restricted to lattice vibrations of small amplitude. Since the solid is then close to a position of stable equilibrium its motion can be calculated by a generalization of the method used to analyse a simple harmonic oscillator. The small amplitude limit is known as the **harmonic limit**. The **anharmonic effects** that occur at larger amplitudes are the subject of sections 2.7 and 2.8.

The atomic motions are governed by the forces exerted on atoms when they are displaced from their equilibrium positions. To calculate the forces in detail it is necessary to determine the wavefunctions and energies of the electrons within the crystal. Fortunately we can deduce many important properties of the atomic motions without doing this calculation.

2.2 SOUND WAVES

We begin our investigation of the dynamics of crystals by recalling our everyday experience that sound waves propagate through solids. This tells us that wavelike lattice vibrations of wavelength long compared to the interatomic spacing are possible. The detailed atomic structure is unimportant for these waves and their propagation is governed by the macroscopic elastic properties of the crystal. We discuss sound waves since they must correspond to the low-frequency, long-wavelength limit of the more general lattice vibrations considered later in this chapter.

At a given frequency and in a given direction in a crystal it is possible to transmit *three* sound waves, differing in their direction of polarization and in general also in their velocity. For sound travelling in a direction of high symmetry, e.g. [1 0 0] in a crystal of cubic symmetry, one of the waves is *longitudinally* polarized and the other two are *transversely* polarized in mutually perpendicular directions. For waves travelling in the [1 0 0] direction the two transverse waves have the same velocity.

To illustrate the method for calculating the velocity of sound waves we suppose that a longitudinal wave is travelling along the x axis. Consider the element of the crystal that is between the planes x and $x + \delta x$ before the wave arrives. Suppose that as the wave propagates the plane x is displaced by $\xi(x, t)$ as shown in Fig. 2.1(*a*). The plane $x + \delta x$ is displaced by $\xi(x + \delta x, t)$, which if δx is small compared to a wavelength may be written $\xi(x, t) + (\partial \xi / \partial x)\delta x$. The extension of the element δx is therefore $(\partial \xi / \partial x)\partial x$ and the strain ($=$ extension/

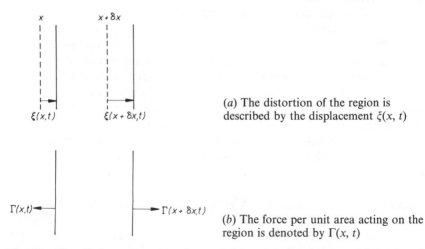

(*a*) The distortion of the region is described by the displacement $\xi(x, t)$

(*b*) The force per unit area acting on the region is denoted by $\Gamma(x, t)$

Fig. 2.1   The effect on the region of a crystal between the planes x and $x + \delta x$ of the passage of a longitudinally polarized sound wave travelling in the x direction

original length) suffered by this element is just $\partial\xi/\partial x$. According to Hooke's law (strain proportional to stress) the associated stress Γ (force per unit area) is of the form

$$\Gamma = C\frac{\partial\xi}{\partial x}, \tag{2.1}$$

where C is an elastic modulus. Note that C is not quite the same as Young's modulus because it refers to extension of the element δx in which lateral contraction is prevented, whereas Young's modulus is for a situation of no lateral constraint. For longitudinal waves of wavelength short compared to the dimensions of the solid the rest of the material prevents lateral contraction.

To calculate the motion of the element δx of the crystal we must allow for the small difference in Γ between the two ends of the element as indicated in Fig, 2.1(b). The net force per unit area $d\Gamma$ can be written as $(\partial\Gamma/\partial x)\delta x$ if δx is small compared to a wavelength, and the equation of motion of the element is therefore

$$\text{force} = \frac{\partial\Gamma}{\partial x}\delta x = \text{mass} \times \text{acceleration} = \rho\,\delta x\,\frac{\partial^2\xi}{\partial t^2},$$

where ρ is the mass density of the crystal. Using Eq. (2.1) for the stress this can be written

$$\frac{C}{\rho}\frac{\partial^2\xi}{\partial x^2} = \frac{\partial^2\xi}{\partial t^2}, \tag{2.2}$$

which is the wave equation for longitudinal waves of velocity

$$v_{\text{L}} = (C/\rho)^{1/2}. \tag{2.3}$$

The velocity of a sound wave can always be written in this form, with an elastic modulus C that depends in general on the propagation and polarization directions. Velocities of sound waves in solids are of order $1000\ \text{m s}^{-1}$.

Our derivation of sound wave propagation is valid only if δx is both much less than the wavelength λ and much larger than an atomic spacing a; the latter requirement is necessary for the use of macroscopic elastic properties. The calculation thus fails for the important lattice vibrations, which have a wavelength comparable to the interatomic spacing. We can nevertheless use a typical sound wave velocity of $1000\ \text{m s}^{-1}$ and a wavelength of $3\ \text{Å}$ to estimate a frequency for these vibrations of $3 \times 10^{12}\ \text{Hz}$. This suggests that lattice vibrations are responsible for the strong absorption by some solids of electromagnetic radiation at frequencies of this order, which are in the infrared part of the spectrum (see section 9.1.4). In the following section we consider a very simple crystal for which it is possible to generalize the above calculation to vibrations of wavelength comparable to an interatomic spacing.

2.3 LATTICE VIBRATIONS OF ONE-DIMENSIONAL CRYSTALS

2.3.1 Chain of identical atoms

The simplest crystal is the one-dimensional chain of identical atoms of equilibrium lattice spacing a shown in Fig. 2.2. We assume that the atoms can move only in the direction parallel to the chain, that they interact via an interatomic potential of the form shown in Fig. 1.23 and that the forces between them are of such short range that only nearest neighbour forces are important. In this limit the small amplitude motions of the chain of atoms are identical to those of the chain of identical masses M connected by identical springs of spring constant K shown on Fig. 2.3, as we shall now demonstrate. The interaction $\mathscr{V}(r)$ between nearest neighbours of separation r may, for a small deviation of r from its equilibrium value a, be expanded as a Taylor series about $r = a$. Thus

$$\mathscr{V}(r) = \mathscr{V}(a) + \frac{(r-a)^2}{2}\left(\frac{\mathrm{d}^2\mathscr{V}}{\mathrm{d}r^2}\right)_{r=a} + \cdots. \tag{2.4}$$

No term linear in $r - a$ appears in the Taylor series because the first derivative of $\mathscr{V}$ must vanish at the equilibrium spacing where $\mathscr{V}$ is a minimum. If the

Fig. 2.2   One-dimensional crystal consisting of a chain of identical atoms

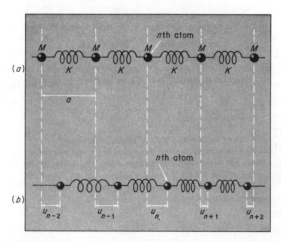

Fig. 2.3   A chain of identical masses M connected by springs: (a) at equilibrium spacings $x_n^0 = na$; (b) at displaced positions $x_n = na + u_n$

higher-order terms are ignored (this is the harmonic limit referred to in the introduction to this chapter), Eq. (2.4) looks like the potential energy associated with a spring of spring constant

$$K = \left(\frac{d^2 \mathscr{V}}{dr^2}\right)_{r=a}. \tag{2.5}$$

The spring constant can be simply related to the elastic modulus C of the one-dimensional crystal, defined by writing the force required to increase the interatomic distance from a to r as $C(r-a)/a$ (i.e. force = elastic modulus × strain). The model of Fig. 2.3 predicts a force $K(r-a)$ and this identifies the following relation between C and K:

$$C = Ka. \tag{2.6}$$

We now proceed to calculate the lattice vibrations of the one-dimensional chain of Fig. 2.3, which exhibit many of the important qualitative features of lattice vibrations in general. We will use the laws of classical mechanics and postpone until section 2.5 a discussion of the difference in the results of a quantum mechanical calculation. We shall suppose that the chain consists of a very large number of atoms and that the last is joined to the first so as to make a ring. This last assumption is simply a device to make the chain endless so that all the atoms have an identical environment, and has no important effect on the problem for a long chain, where end effects are unimportant anyway.

If the displacements of the atoms from their equilibrium positions are u_n as shown in Fig. 2.3, the force on the nth atom consists of:

(i) $K(u_n - u_{n-1})$ to the left, from the spring on its left; and
(ii) $K(u_{n+1} - u_n)$ to the right, from the spring on its right.

Equating the total force to the right, (ii) − (i), to the product of mass and acceleration we have

$$M\ddot{u}_n = K(u_{n+1} - 2u_n + u_{n-1}). \tag{2.7}$$

The equations of motion of all atoms are of this form, only the value of n varies. To solve Eqs. (2.7) we try a wavelike solution in which all the atoms oscillate at the same amplitude A. Thus we substitute†

$$u_n = A \exp [i(kx_n^0 - \omega t)] \tag{2.8}$$

where $x_n^0 = na$ is the undisplaced position of the nth atom, to obtain

$$-\omega^2 M A \, e^{i(kna - \omega t)} = KA(e^{i[k(n+1)a - \omega t]} - 2e^{i(kna - \omega t)} + e^{i[k(n-1)a - \omega t]}),$$

---

† As usual when solving vibration problems by means of complex exponentials, it is the real part of u_n that we interpret physically as the atomic displacement.

or, on cancelling $Ae^{i(kna-\omega t)}$ from each term,

$$-\omega^2 M = K(e^{ika} - 2 + e^{-ika})$$
$$= 2K[\cos(ka) - 1].$$

Hence

$$\omega^2 M = 4K \sin^2 (\tfrac{1}{2}ka). \tag{2.9}$$

Fig. 2.4 shows the dispersion relation (relation between frequency ω and wavenumber k) given by Eq. (2.9) for our wavelike lattice vibrations. The maximum value of $\sin(\tfrac{1}{2}ka)$ is 1 so that the maximum possible frequency of the waves is $2(K/M)^{1/2}$. This is known as the cut-off frequency of the lattice.

We notice that n has cancelled out in Eq. (2.9), so that the equations of motion of *all* atoms lead to the same algebraic relation between ω and k. This shows that our trial function for u_n is indeed a solution of Eqs. (2.7). It is also important to notice that we started from the equations of motion of N *coupled* harmonic oscillators (Eqs. (2.7)). If one atom starts vibrating it does not continue with constant amplitude, but transfers energy to the others in a complicated way; the vibrations of individual atoms are not simple harmonic because of this energy exchange among them. Our wavelike solutions on the other hand are *uncoupled* oscillations called **normal modes**; each k has a definite ω given by Eq. (2.9) and oscillates independently of the other modes. We should expect the number of modes to be the same as the number of equations N that we started with; let us now see whether this is the case.

To do so we must establish which wavenumbers are possible for our one-dimensional chain. Not all values are allowed because the nth atom is the same as the $(N + n)$th as the chain is joined on itself. This means that the wave (2.8) must satisfy the **periodic boundary condition**

$$u_n = u_{N+n} \tag{2.10}$$

which requires that there should be an integral number of wavelengths in the

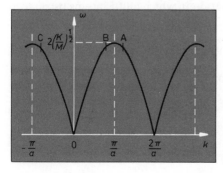

Fig. 2.4   Normal mode frequencies for a chain of identical atoms. Note that the modes with wavenumbers at A, B and C all have the same frequency and correspond to the same instantaneous atomic displacements (see Fig. 2.5). Point B represents a wave moving to the right, points A and C a wave moving to the left

length of our ring of atoms

$$Na = p\lambda.$$

Hence

$$k = \frac{2\pi}{\lambda} = \frac{2\pi p}{Na} \tag{2.11}$$

where p is an integer. There are thus N possible k values in a range $2\pi/a$ of k, say the range

$$-\pi/a < k \leqslant \pi/a.$$

Fig. 2.4 shows that this restricted range of k does indeed include all possible values of the frequency ω and the group velocity $(d\omega/dk)$; it also gives the N normal modes we expect for N atoms. What, if anything, is the physical significance of wavenumbers outside this range?

To understand this, consider the instantaneous atomic displacements shown in Fig. 2.5; we are really considering longitudinal waves, but the displacements are shown as transverse in Fig. 2.5 because this makes their wavelike nature easier to visualize. Fig. 2.5(a) shows the displacements for $k = \pi/a$, which gives

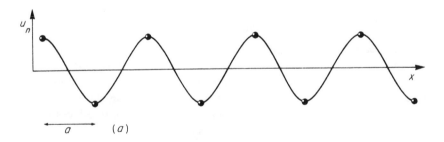

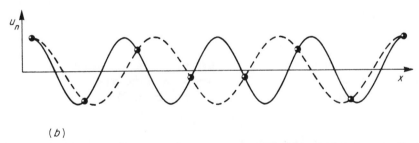

Fig. 2.5 (a) Atomic displacements (shown as transverse for clarity) for wavelength $\lambda = 2a$, wavenumber $k = \pi/a$. (b) The atomic displacements for a wave with $\lambda = 7a/4$, $k = 8\pi/7a$, as given by the full curve, are identical to those for a wave with $\lambda = 7a/3$, $k = 6\pi/7a$, as given by the broken curve

the maximum frequency; alternate atoms oscillate in antiphase and the waves at this value of k are essentially standing waves. The midpoint of each spring is at rest and each mass therefore behaves as if held by two springs each of spring constant $2K$, giving the frequency $2(K/M)^{1/2}$ that we have calculated.

Now consider the displacements, shown by the full curve in Fig. 2.5(b), for the slightly larger value $8\pi/7a$ of k corresponding to the point A on Fig. 2.4. The displacements can also be represented by the longer wave, shown broken in Fig. 2.5(b), for which $|k| = 6\pi/7a$; this corresponds to points B or C in Fig. 2.4. Thus, points A, B and C correspond to the same instantaneous atomic displacements as well as the same frequency. At B the group velocity $d\omega/dk > 0$, so we have a wave travelling to the right; A and C both represent a wave travelling to the left and are thus completely equivalent. The k values of points A and C differ by $2\pi/a$ and we therefore conclude that adding any multiple of $2\pi/a$ to k does not alter the atomic displacements or the group velocity and is without physical significance. We need only consider the range $-\pi/a < k \leqslant \pi/a$, which contains just the N modes we expected.

Further insight into what is special about the k values $\pm \pi/a$ is gained by writing down Bragg's law (Eq. (1.3)) for the one-dimensional crystal:

$$n\lambda = n2\pi/k = 2d \sin \theta = 2a$$

or

$$k = \pi n/a, \qquad (2.12)$$

where we have taken $\theta = 90°$ and $d = a$ as appropriate to waves travelling along a one-dimensional chain. Waves with $k = \pm \pi/a$ will thus undergo Bragg reflection. The standing wave pattern that occurs at these two k values can be pictured as occurring because of Bragg reflection of running waves.

We note that in the long-wavelength limit, $ka \ll 1$, Eq. (2.9) reduces to

$$M\omega^2 = Kk^2a^2$$

so that in this limit the waves are dispersionless with group velocity and phase velocity (ω/k) both being equal to

$$v = a(K/M)^{1/2} \qquad (2.13)$$

These waves are long-wavelength sound waves and a calculation of their velocity from the macroscopic elastic properties of the crystal by the method used in the previous section yields a velocity (cf Eq. (2.3))

$$v_S = (C/\rho)^{1/2} \qquad (2.14)$$

where ρ $(= M/a)$ and C are respectively the mass per unit length and the elastic modulus of the crystal. Using Eq. (2.6), we confirm that Eqs. (2.13) and (2.14) are identical and thus that our more general calculation of lattice vibrations gives the correct answer in the long-wavelength limit. Note that, since there is only one possible propagation direction and one polarization direction, the one-dimensional crystal has only one sound velocity. Given the velocity of sound and the lattice spacing it is possible to draw the dispersion relation for our simple crystal. This illustrates our statement in the introduction to this chapter that many of the properties of lattice vibrations can be deduced without detailed knowledge of the interatomic forces.

The inclusion of only nearest neighbour forces in our calculation appears very restrictive. Although it is a good approximation for the inert-gas solids, it is not a good assumption for many solids. The effects of removing this restriction can be investigated by using a model in which each atom is attached by springs of different spring constant to neighbours at different distances (see problem 2.1). When this is done, many of the features of the above calculation are preserved. The wave solution of Eq. (2.8) still satisfies the equations of motion. The detailed form of the dispersion relation is changed but ω is still a periodic function of k with period $2\pi/a$ and the group velocity vanishes at $k = \pm\pi/a$. There are still N distinct normal modes, which can be represented by the N possible k values in the range $-\pi/a < k \leqslant +\pi/a$. Furthermore the motion at long wavelengths corresponds to sound waves with a velocity given by Eq. (2.14).

2.3.2   Chain of two types of atom

We now consider the lattice vibrations of a chain containing two types of atom, of masses M and m, connected by identical springs of spring constant K as shown in Fig. 2.6. This is the simplest possible model of an ionic crystal,

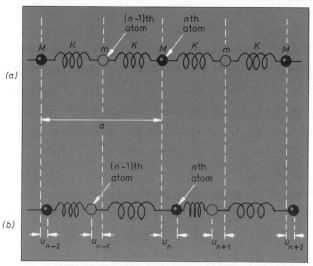

Fig. 2.6   A chain containing two unequal masses connected by springs: (a) at equilibrium positions $x_n^0 = na/2$; (b) at displaced positions $x_n = na/2 + u_n$. Here a is the length of the unit cell as indicated

although the assumption of only nearest neighbour forces, implicit in the model, is a poor approximation for ionic crystals because of the long range of the Coulomb interaction between ions. Fortunately the simple model again produces the important qualitative features of the lattice vibrations of ionic solids. In section 9.1.4 we discuss the changes to the vibrations when the long-range effects of the Coulomb force are included.

To emphasize the more complicated motions that are possible when there is more than one type of atom we show in Fig. 2.6(b) a configuration in which the two types of atom are displaced in opposite directions. Note that we use a to denote the unit cell length (lattice spacing) of the crystal; the nearest neighbour separation of the undisplaced atoms is $a/2$.

The equations of motion can be written down in the same way as Eq. (2.7) but there are now two distinct types of equation: those for masses M,

$$M\ddot{u}_n = K(u_{n+1} - 2u_n + u_{n-1})\qquad(2.15)$$

and those for masses m,

$$m\ddot{u}_{n-1} = K(u_n - 2u_{n-1} + u_{n-2}).\qquad(2.16)$$

For the masses M we may assume as before a solution of the form (2.8), i.e.

$$u_n = A \exp\left[i(kx_n^0 - \omega t)\right]$$

where $x_n^0 = na/2$ is the undisplaced atomic position. There is now an extra unknown quantity, the relative amplitude and phase of the vibrations of the two types of atom; this we allow for by taking for the masses m

$$u_n = \alpha A \exp\left[i(kx_n^0 - \omega t)\right]\qquad(2.17)$$

where α is a complex number giving the relative amplitude and phase.

Substitution in Eqs. (2.15) and (2.16) then gives

$$-\omega^2 M e^{i(kna/2 - \omega t)} = K(\alpha e^{i[k(n+1)a/2 - \omega t]} - 2e^{i(kna/2 - \omega t)} + \alpha e^{i[k(n-1)a/2 - \omega t]})$$

and

$$-\alpha\omega^2 m e^{i[k(n-1)a/2 - \omega t]} = K(e^{i(kna/2 - \omega t)} - 2\alpha e^{i[k(n-1)a/2 - \omega t]} + e^{i[k(n-2)a/2 - \omega t]})$$

or, by cancelling common factors as before,

$$\begin{aligned}-\omega^2 M &= 2K[\alpha \cos\left(\tfrac{1}{2}\, ka\right) - 1]\\ -\alpha\omega^2 m &= 2K[\cos\left(\tfrac{1}{2}\, ka\right) - \alpha]\end{aligned}\qquad(2.18)$$

Thus instead of a single algebraic equation for ω as a function of k, we now have a pair of algebraic equations for α and ω as functions of k. As before the fact that n does not appear in Eqs. (2.18) indicates that our assumed solution is of the correct form. Eqs. (2.18) may be rewritten in the form

$$\alpha = \frac{2K\cos(\tfrac{1}{2}ka)}{2K - \omega^2 m} = \frac{2K - \omega^2 M}{2K\cos(\tfrac{1}{2}ka)}, \tag{2.19}$$

from which by cross-multiplication we obtain a quadratic equation for ω^2,

$$mM\omega^4 - 2K(M + m)\omega^2 + 4K^2\sin^2(\tfrac{1}{2}ka) = 0, \tag{2.20}$$

with solutions

$$\omega^2 = \frac{K(M + m)}{Mm} \pm K\left[\left(\frac{M + m}{Mm}\right)^2 - \frac{4}{Mm}\sin^2(\tfrac{1}{2}ka)\right]^{1/2}. \tag{2.21}$$

The two roots are sketched in Fig. 2.7. As there are two values of ω for each value of k the dispersion relation is said to have two **branches** and the upper and lower branches in Fig. 2.7 corresponds to the $+$ and $-$ signs in Eq. (2.21) respectively. We see that chains containing two types of atom share with those containing one the property that the dispersion relations are periodic in k with period $2\pi/a = 2\pi/($unit cell length); this result remains valid for a chain containing an arbitrary number of atoms per unit cell.

If the crystal continues N unit cells we would expect to find $2N$ normal modes of vibration as this is the total number of atoms and hence the total number of equations of motion (Eqs. (2.15) and (2.16)). Joining the ends of the crystal to form a ring requires the atomic displacements to satisfy the periodic boundary condition $u_n = u_{2N+n}$, leading to the same expression for the possible k values,

$$k = 2\pi p/Na,$$

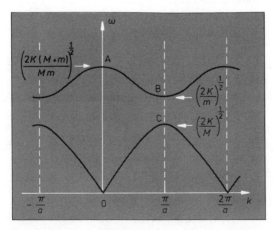

Fig. 2.7   Normal mode frequencies of a chain of two types of atom. At A, the two types are oscillating in antiphase with their centre of mass at rest; at B, the lighter mass m is oscillating and M is at rest; at C, M is oscillating and m is at rest

as for the crystal containing a single type of atom. Thus there are again exactly N allowed values of k in the range $-\pi/a < k \leqslant \pi/a$; also as in the previous section, adding any multiple of $2\pi/a$ to k does not alter the atomic displacements, and we deduce that all the allowed motions can be described by k values in this range. Hence the two branches of the dispersion relation contain $2N$ normal modes as required.

It is instructive to examine the limiting solutions of Eq. (2.21) near the points O, A, B and C in Fig. 2.7. For $ka \ll 1$, $\sin(\tfrac{1}{2}ka) \approx \tfrac{1}{2}ka$ and

$$\omega^2 \approx \frac{K(M+m)}{Mm}\left[1 \pm \left(1 - \frac{mM}{(M+m)^2}k^2a^2\right)^{1/2}\right]$$

$$\approx \frac{K(M+m)}{Mm}\left[1 \pm \left(1 - \frac{mM}{2(M+m)^2}k^2a^2\right)\right]$$

$$\approx \frac{2K(M+m)}{Mm} \quad \text{or} \quad \frac{Kk^2a^2}{2(M+m)}.$$

By substituting these values of ω in Eq. (2.19) and using $\cos(\tfrac{1}{2}ka) \approx 1$ for $ka \ll 1$ we find the corresponding values of α as

$$\alpha \approx -M/m \quad \text{or} \quad 1.$$

The first solution corresponds to point A in Fig. 2.7; this value of α corresponds to M and m oscillating in antiphase with their centre of mass at rest, and the frequency is therefore given by the spring constant $2K$ and the reduced mass $M^* = Mm/(M+m)$. The second solution represents long-wavelength sound waves in the neighbourhood of point O in Fig. 2.7; the two types of atom oscillate with the same amplitude and phase, and the velocity of sound is

$$v_S = \frac{\omega}{k} = a\left(\frac{K}{2(M+m)}\right)^{1/2}.$$

This sound velocity must agree with that, namely $(C/\rho)^{1/2}$ (Eq. (2.14)), predicted from the macroscopic elastic properties of the crystal; substituting values of $(M+m)/a$ and $Ka/2$ (cf. Eq. (2.6), recalling that the definition of a has changed) respectively for the mass per unit length ρ and the elastic modulus C into Eq. (2.14) confirms that this is so.

The other limiting solutions of Eq. (2.21) are for $ka = \pi$, i.e. $\sin(\tfrac{1}{2}ka) = 1$. In this case

$$\omega^2 = \frac{K(M+m)}{Mm} \pm K\left[\left(\frac{M+m}{Mm}\right)^2 - \frac{4}{Mm}\right]^{1/2}$$

$$= \frac{K(M+m) \pm K(M-m)}{mM}$$

$$= 2K/m \quad \text{or} \quad 2K/M$$

with, from Eq. (2.19), the corresponding amplitude ratios $\alpha = \infty$ or $\alpha = 0$ respectively. In this limit the half-wavelength is a, the spacing between atoms of the same kind. In the first solution m oscillates and M is at rest (point B on Fig. 2.7 if $M > m$), and the frequency therefore depends only on m; in the second solution M oscillates and m is at rest (point C on Fig. 2.7).

It is instructive to compare our present results with those of section 2.3.1 for a chain of one type of atom. In Fig. 2.8 we plot the lower-frequency branch of Fig. 2.7 in the region $k < \pi/a$ and the higher-frequency branch in the region $\pi/a < k < 2\pi/a$. If we now let $m \to M$ the points B and C in Fig. 2.8 come together and we recover Fig. 2.4, as we must. (Do not forget that the value of a in Fig. 2.4 is half that in Fig. 2.8.)

There is thus a certain arbitrariness about how we assign k values to the modes of a diatomic lattice. The most direct comparison with a monatomic lattice is obtained with the assignment shown in Fig. 2.8, where there is only one ω for each k and there are $2N$ modes in the range $-2\pi/a < k \leqslant 2\pi/a$. It is more usual, however, to assign the lowest possible k to each mode as in Fig. 2.7; there are now two branches with N modes on each branch in the range $-\pi/a < k \leqslant \pi/a$. This latter approach is the usual one and has the useful feature that the range of k values is $2\pi/(\text{unit cell side})$, independently of the number of atoms in a unit cell.

Although the dispersion relation is no longer given by Eq. (2.21) when the restriction to nearest neighbour forces is removed, most of the qualitative conclusions concerning the nature of the dispersion relation that we have deduced above remain valid. In particular the dispersion relation has two

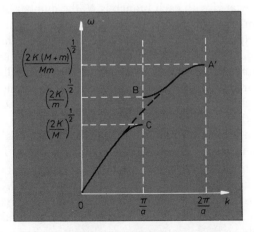

Fig. 2.8   This shows Fig. 2.7 replotted for comparison with Fig. 2.4; in the limit $m \to M$ points B and C come together and the dispersion relation is given by the broken line. The atomic displacements at A' are the same as those at point A on Fig. 2.7

branches, both periodic in k with period $2\pi/a$. Only one of the branches has the limiting form of sound waves at long wavelength ($\omega/k \to$ constant as $k \to 0$). This branch, the lower branch on Fig. 2.7, is consequently known as the **acoustic branch**. The other branch is called the **optical branch** because as $k \to 0$ on this branch the vibrations of the two types of atom are in antiphase and the resulting charge oscillations in an ionic crystal give a strong coupling to electromagnetic waves at the frequency of point A on Fig. 2.7. An estimate of this frequency is obtained from Fig. 2.8 by extrapolating the linear ω/k relation near $k = 0$ up to $k = 2\pi/a$; this is essentially the method we used to estimate the frequency of lattice vibrations at the end of section 2.2, and therefore gives an answer in the infrared region of the electromagnetic spectrum.

2.4 LATTICE VIBRATIONS OF THREE-DIMENSIONAL CRYSTALS

We now comment briefly on the way in which the properties of lattice vibrations in one-dimensional crystals manifest themselves in real three-dimensional crystals. In a particular direction in a three-dimensional crystal there are similar dispersion relations for lattice vibration waves to those in Figs. 2.4 and 2.7. Some measured dispersion relations can be seen later in Fig. 12.7. On any branch of the dispersion relation ω is a periodic function of the wavevector $\mathbf{k}$; the manner in which the periodicity is determined by the crystal lattice is explained in Chapter 11.

For a unit cell containing only one atom the major difference to Fig. 2.4 in the three-dimensional case is the existence of three branches of the dispersion relation; as the wavenumber $|\mathbf{k}| \to 0$ each branch tends to one of the three possible sound waves so that the three branches are all acoustic. The number of lattice modes associated with each branch is N, the number of unit cells in the crystal, just as for one-dimensional crystals, so that altogether there are $3N$ modes. Thus, as expected, the number of modes equals the number of equations of motion (one equation for each Cartesian coordinate of N atoms).

For a three-dimensional crystal with a primitive unit cell containing two atoms there are three acoustic branches and three optical branches. The general result for a unit cell containing s atoms is three acoustic branches and $3(s - 1)$ optical branches; the number of lattice modes associated with any branch of the dispersion relation is always equal to the number of unit cells so that the total number of modes is always three times the number of atoms in the crystal.

We will now indicate briefly the way in which, at least in principle, the motion of the atoms in a three-dimensional crystal may be determined. Since most of the mass of the atom is in the nucleus it is the motion of the nucleus that concerns us. The large mass of the nucleus enables us to make a very useful approximation, which simplifies the problem both conceptually and mathematically. This is the **adiabatic approximation** and it asserts that the motion of electrons and nuclei can be decoupled to a good approximation. Because the nuclei are more massive than the electrons they move more slowly and the electrons thus behave at any instant almost as though the nuclei were stationary in their instantaneous positions. In other words we can, to a good approximation, think of an electronic wavefunction that is an eigenstate for nuclei fixed in their instantaneous positions; as the nuclei move, this wavefunction adjusts itself smoothly to the changing boundary conditions, but remains an eigenstate. Such a slow perturbation of the boundary conditions (ideally infinitely slow) is called an **adiabatic perturbation**, and it is a principle of quantum mechanics that such a perturbation does not cause transitions between quantum states (see, for example, R. Becker, *Theory of Heat*, Springer, Berlin, 1967, pp. 170–3). The wavefunction and energy alter during an adiabatic perturbation but the quantum state does not.

This enables us to split the calculation of the energy of a solid into two stages. First we calculate the electronic energy $E_e(\mathbf{R}_1, \mathbf{R}_2, \ldots, \mathbf{R}_N)$ for the nuclei fixed in their instantaneous positions $\mathbf{R}_1, \mathbf{R}_2, \ldots, \mathbf{R}_N$. We then use the adiabatic approximation described above to assert that $E_e(\mathbf{R}_1, \mathbf{R}_2, \ldots, \mathbf{R}_N)$ so calculated is the electronic contribution to the total energy of the system even when the nuclei are allowed to move. The total energy of the solid is then given by

$$E_{\text{tot}} = \sum_{i=1}^{N} \frac{P_i^2}{2M_i} + \frac{1}{2} \sum_{i=1}^{N} \sum_{\substack{j=1 \\ i \neq j}}^{N} \frac{Q_i Q_j}{4\pi\varepsilon_0 |\mathbf{R}_i - \mathbf{R}_j|} + E_e(\mathbf{R}_1, \mathbf{R}_2, \ldots, \mathbf{R}_N) \quad (2.22)$$

where P_i, M_i and Q_i are the momentum, mass and charge of the ith nucleus. The first term of Eq. (2.22) is the nuclear kinetic energy, the second term the electrostatic potential energy of the nuclei and the final term the electronic energy. We see that $E_e(\mathbf{R}_1, \mathbf{R}_2, \ldots, \mathbf{R}_N)$ appears like an extra potential energy of interaction between the nuclei.

A full quantum calculation of lattice vibrations proceeds by solving the Schrödinger equation for the nuclear motions with a Hamiltonian obtained by replacing the momenta P_i in Eq. (2.22) by their equivalent quantum mechanical operators $-i\hbar\nabla_i$. The classical treatment proceeds by writing down Newton's law for each of the N ions moving in the potential $\mathscr{V}(R_1, R_2, \ldots, R_N)$ given by the final two terms in Eq. (2.22). In both treatments progress can only be made by making the harmonic approximation in which the potential $\mathscr{V}(R_1, R_2, \ldots, R_N)$ is expanded as a Taylor series to second order in the displacements of the nuclei from their equilibrium positions. The first-order terms vanish because the potential is a minimum in the equilibrium position. The terms quadratic in the displacement resemble the potential energy of simple harmonic oscillators and lead to lattice waves with the properties we have indicated earlier. The existence of the potential $\mathscr{V}(R_1, R_2, \ldots, R_N)$ for any crystal is an important result of the adiabatic approximation; only in simple inert-gas and ionic crystals can $\mathscr{V}(R_1, R_2, \ldots, R_N)$ be written as the sum of the interatomic potentials for pairs of atoms. A comprehensive account of the calculation of lattice vibrations in a three-dimensional crystal can be found in chapters 22 and 23 of Ashcroft and Mermin.[11]

In performing the calculations E_e is normally taken to be the ground state energy of the electrons although Eq. (2.22) holds also for an excited state. The nuclear motions are not in general significantly affected by the existence of excited electronic states.

2.5 PHONONS

So far we have considered the mechanics of lattice vibrations in a completely classical way. To the extent that the normal modes we have found are truly harmonic and independent, the transition to quantum mechanics is easily made by supposing that a lattice vibration mode of frequency ω will behave like a simple harmonic oscillator and will thus be restricted to energy values

$$\varepsilon_n = (n + \tfrac{1}{2})\hbar\omega. \tag{2.23}$$

Since Eq. (2.23) represents a set of uniformly spaced energy levels, it is possible to regard the state ε_n as constructed by adding n 'excitation quanta' each of energy $\hbar\omega$ to the ground state. You may already have met this viewpoint in the context of electromagnetic radiation of angular frequency ω; there we say that the state ε_n corresponds to the presence of n **photons** each of energy $\hbar\omega$. The reality of such energy-carrying particles is shown, for example, in the photoelectric effect.

It is often convenient to treat lattice vibrations in an analogous way, and to introduce the concept of **phonons** of energy $\hbar\omega$ as quanta of excitation of the lattice vibration mode of angular frequency ω. Our normal modes are plane waves extending throughout the crystal, and correspondingly the phonons are not localized particles; the uncertainty principle demands that the position cannot be determined because the momentum $\hbar k$ is exact. However, just as with photons or electrons, one can construct a fairly localized wavepacket by combining modes of slightly different frequency and wavelength. Thus, if we take waves with a spread of $\mathbf{k}$ of order $\pi/10a$ we can make a wavepacket localized to within about 10 unit cells. Such a wavepacket represents a fairly localized phonon moving with group velocity $d\omega/d\mathbf{k}$. We can therefore treat phonons as localized particles within the limits of the uncertainty principle. The $\omega(\mathbf{k})$ curve for lattice vibrations can be interpreted, if both axes are multiplied by $\hbar$, as a relation between energy and momentum for phonons ($E = \hbar\omega$, $\mathbf{p} = \hbar\mathbf{k}$).

Although it is convenient to interpret $\hbar k$ as the momentum of a phonon (and we will continue to do so), we should be aware that it is not the true kinematic momentum. To see this it is only necessary to recall from section 2.3 that, in a one-dimensional crystal, a lattice mode of wavenumber k can be equally well represented by a wavenumber $k + 2\pi n/a$. It is thus not possible to ascribe a unique value of $\mathbf{k}$ to a phonon. We shall find that the quantity $\hbar k$ does possess many of the properties of momentum; to make it clear that it is not the true momentum it is often referred to as the **crystal momentum**.

Like photons, phonons are bosons and are not conserved; they can be created or destroyed in collisions. Thus in Eq. (2.23) n can take any value and can change with time. We shall meet examples of the usefulness of the idea of phonons later. In section 12.4 we shall see that phonons can be created and absorbed when neutrons are scattered from a solid, leading to a direct experimental measurement of $\omega(\mathbf{k})$, and in sections 2.8 and 3.3 we shall see that the thermal conductivity of insulators and the electrical resistivity of metals can be understood by using a model of a crystal containing a gas of phonons.

2.6   HEAT CAPACITY FROM LATTICE VIBRATIONS

In most solids the energy given to lattice vibrations is the dominant contribution to the heat capacity; in non-magnetic insulators it is the only contribution. Other contributions arise in metals from the conduction electrons, and in magnetic materials from magnetic ordering.

We have seen in our examples of one-dimensional crystals that the coupling together of atomic vibrations leads to a band of normal mode frequencies from zero up to some maximum value. Calculation of the lattice energy and heat capacity of a solid therefore falls into two parts: the evaluation of the contribution of a single mode, and the summation over the frequency distribution of the modes.

2.6.1   Energy and heat capacity of a harmonic oscillator

The average energy $\bar{\varepsilon}$ of a harmonic oscillator and hence of a lattice mode of angular frequency ω at temperature T is given by

$$\bar{\varepsilon} = \sum_n p_n \varepsilon_n$$

where ε_n is an energy level of the oscillator, as given by Eq. (2.23), and p_n is the probability of the oscillator being in this level as given by the Boltzmann factor $\exp\left(-\varepsilon_n/k_B T\right)$. Thus

$$\bar{\varepsilon} = \frac{\sum_{n=0}^{\infty} (n + \tfrac{1}{2})\hbar\omega \exp\left[-(n + \tfrac{1}{2})\hbar\omega/k_B T\right]}{\sum_{n=0}^{\infty} \exp\left[-(n + \tfrac{1}{2})\hbar\omega/k_B T\right]}.$$

A neat way of evaluating this expression is to note that it can be written

$$\bar{\varepsilon} = k_B T^2 \frac{1}{Z} \frac{\partial Z}{\partial T} = k_B T^2 \frac{\partial(\ln Z)}{\partial T}, \tag{2.24}$$

where

$$Z = \sum_{n=0}^{\infty} \exp\left[-(n + \tfrac{1}{2})\hbar\omega/k_B T\right]$$

$$= e^{-\hbar\omega/2k_B T}\left(1 + e^{-\hbar\omega/k_B T} + e^{-2\hbar\omega/k_B T} + \ldots\right)$$

$$= e^{-\hbar\omega/2k_B T}\left(1 - e^{-\hbar\omega/k_B T}\right)^{-1}. \tag{2.25}$$

Z is in fact the partition function of the oscillator (Mandl[2], Chapter 6) and has been calculated by noting that the sum is a geometric series with the ratio of successive terms $\exp(-\hbar\omega/k_B T)$. Hence, using Eq. (2.24),

$$\bar{\varepsilon} = \tfrac{1}{2}\hbar\omega + \frac{\hbar\omega}{e^{\hbar\omega/k_B T} - 1}. \tag{2.26}$$

This mean energy is readily interpreted in terms of phonons. For bosons of energy $\hbar\omega$, which are not conserved, the average number present in thermal equilibrium at temperature T is given by the Bose–Einstein distribution function

$$n(\omega) = \frac{1}{e^{\hbar\omega/k_B T} - 1}, \tag{2.27}$$

as in the case of photons in black-body radiation (Mandl[2], Chapter 10). Multiplication of Eq. (2.27) by $\hbar\omega$ gives the second term in Eq. (2.26) as the contribution of phonons to the energy. The first term in Eq. (2.26), $\tfrac{1}{2}\hbar\omega$, is the zero point energy, which cannot be frozen out because of the uncertainty principle. Fig. 2.9(a) shows that the mean energy tends to this value in the low-temperature limit, $k_B T \ll \hbar\omega$. At high temperatures, $k_B T \gg \hbar\omega$, we can expand the exponential to obtain

$$\bar{\varepsilon} = \tfrac{1}{2}\hbar\omega + \hbar\omega\left[\frac{\hbar\omega}{k_B T} + \frac{1}{2}\left(\frac{\hbar\omega}{k_B T}\right)^2 + \ldots\right]^{-1}$$

$$= \tfrac{1}{2}\hbar\omega + k_B T\left[1 - \frac{1}{2}\frac{\hbar\omega}{k_B T} + \ldots\right]$$

$$= k_B T\left[1 + O\left(\frac{\hbar\omega}{k_B T}\right)^2\right] \approx k_B T, \tag{2.28}$$

so that the classical equipartition value is obtained in this limit; note that a classical one-dimensional harmonic oscillator has thermal energy $k_B T$ not $\tfrac{1}{2}k_B T$, because the potential energy, as well as the kinetic energy, is a quadratic contribution to the total energy. The heat capacity C is found by differentiating Eq. (2.26) with respect to temperature, i.e.

$$C = \frac{d\bar{\varepsilon}}{dT} = k_B\left(\frac{\Theta}{T}\right)^2 \frac{e^{\Theta/T}}{(e^{\Theta/T} - 1)^2}, \tag{2.29}$$

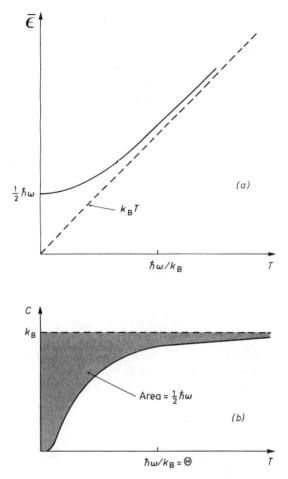

Fig. 2.9   (a) Mean energy and (b) heat capacity of a simple harmonic oscillator as functions of temperature

where $\Theta = \hbar\omega/k_B$. C is plotted as a function of T in Fig. 2.9(b); it vanishes exponentially at low temperatures and tends to the classical value k_B at high temperatures.

The general features of Fig. 2.9 are common to all quantum systems: the energy tends to the zero point energy at low temperatures and to the classical equipartition value at high temperatures; the heat capacity tends to zero at absolute zero and rises to the classical value in such a way that the 'missing area' under the classical heat capacity curve (shaded in Fig. 2.9(b)) is equal to the zero point energy.

The first quantum theory of the heat capacity of solids was due to Einstein. He made the simplifying assumption that all $3N$ vibrational modes of a three-dimensional solid of N atoms had the same frequency, so that the whole solid had a heat capacity $3N$ times Eq. (2.29). Even this very crude model gave the correct limit at high temperatures, a heat capacity of $3Nk_B$ ($=3R$ for a mole of atoms as first found empirically by Dulong and Petit); this result depends only on the classical theorem of the equipartition of energy. Einstein's model also gave correctly a specific heat tending to zero at absolute zero, but the temperature dependence near $T = 0$ did not agree with experiment. We shall now account for this discrepancy by taking into account the actual distribution of vibration frequencies in a solid, using the one-dimensional model of section 2.3 as out starting point.

2.6.2   The density of states

We saw in section 2.3 that, for a one-dimensional crystal containing N unit cells of side a, the application of periodic boundary conditions gives the allowed wavenumbers

$$k = \frac{2\pi p}{Na} = \frac{2\pi p}{L}$$

where p is an integer and $L = Na$ is the length of the crystal. The allowed k values are therefore uniformly distributed in k at a density $\rho_R(k)$ (Fig. 2.10(a)) such that the number of values in the range $k \rightarrow k + dk$ is given by

$$\rho_R(k)\,dk = \frac{L}{2\pi}\,dk. \tag{2.30}$$

These allowed k values correspond to running waves, so that both positive and negative k values are significant.

It is sometimes more convenient to consider a chain with *fixed* ends, in which case the normal modes are standing waves and we have an integral number of half-wavelengths in the chain, so that $L = n\lambda/2$, $k = 2\pi/\lambda = n\pi/L$. The standing

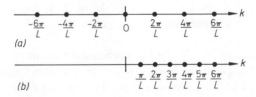

Fig. 2.10   Allowed k values for (a) running waves and (b) standing waves on a one-dimensional chain of length L Note that a running wave with $k = 0$ corresponds to a rigid displacement of the crystal as a whole; a standing wave with $k = 0$ does not exist

wave states are thus uniformly distributed in k at a density $\rho_S(k)$ (Fig. 2.10(b)) and the number of states in the range $k \to k + dk$ is

$$\rho_S(k) \, dk = \frac{L}{\pi} dk. \tag{2.31}$$

The density of standing wave states is therefore twice that of running wave states. However since only *positive* k values are used for standing waves the total number of states within a range dk of the *magnitude* of k is the same for both running and standing waves. The standing waves have the same dispersion relation as running waves, and for a chain containing N atoms there are exactly N distinct states with k values in the range 0 to π/a.

To calculate an energy or heat capacity by summing over normal modes we need the **density of states per unit frequency range** $g(\omega)$, which is defined such that the number of modes with frequencies $\omega \to \omega + d\omega$ is $g(\omega)d\omega$. We can write $g(\omega)$ in terms of $\rho_S(k)$ by noting that if there are dn modes with frequency ω to $\omega + d\omega$ corresponding to the wavenumber range k to $k + dk$ then

$$dn = \rho_S(k) \, dk = g(\omega) \, d\omega,$$

so that

$$g(\omega) = \rho_S(k) \frac{dk}{d\omega}. \tag{2.32}$$

Thus, using Eq. (2.9) for $\omega(k)$ for a chain of identical atoms, we obtain

$$\frac{d\omega}{dk} = a \left(\frac{K}{M} \right)^{1/2} \cos\left(\tfrac{1}{2}ka\right),$$

so that

$$g(\omega) = \frac{L}{\pi a} \left(\frac{M}{K} \right)^{1/2} \sec\left(\tfrac{1}{2}ka\right) = \frac{N}{\pi} \left(\frac{M}{K} \right)^{1/2} \frac{2(K/M)^{1/2}}{(4K/M - \omega^2)^{1/2}}$$

$$= \frac{2N}{\pi} \left(\frac{4K}{M} - \omega^2 \right)^{-1/2}. \tag{2.33}$$

This density of states is plotted in Fig. 2.11; we notice that it tends to infinity as the cut-off frequency $2(K/M)^{1/2}$ is approached from below, because the group velocity $d\omega/dk$ tends to zero there. If we had ignored dispersion of sound at wavelengths comparable to an atomic spacing we would have obtained the constant density of states shown by the broken line in Fig. 2.11, for which the total number of states does not reach the value N until a frequency $\omega = \pi(K/M)^{1/2}$. The reader should check that the same value of $g(\omega)$ as that of Eq. (2.33) is obtained for running waves.

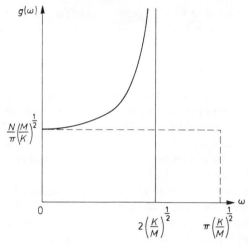

Fig. 2.11   Density of states for a chain of identical atoms. The full curve gives the true density of states; the broken curve gives the density of states obtained if the dispersion of sound is ignored

The energy of the lattice vibrations is obtained by integrating the energy of a single oscillator, Eq. (2.26), over the distribution of vibration frequencies. Thus

$$E = \int_0^\infty \left(\frac{1}{2}\hbar\omega + \frac{\hbar\omega}{e^{\hbar\omega/k_B T} - 1} \right) g(\omega)\, d\omega. \tag{2.34}$$

We shall not proceed further with our one-dimensional example since it cannot be compared with experiment. Instead we will attempt to generalize our calculation of $g(\omega)$ to three dimensions.

Our first step will be to calculate the number of lattice modes in a given range of the wavevector $\mathbf{k}$. In this book we will find that the need to count $\mathbf{k}$ states is a problem common to many types of wave motion in crystals. We will therefore discuss it in some detail here and refer to the results of this section when we encounter other types of waves. Although the choice of boundary conditions, as we discovered in the one-dimensional case, determines whether we are working with running waves or standing waves, the number of states in a given range of the magnitude of $\mathbf{k}$ is always independent of the choice of boundary conditions for macroscopic crystals.

We begin by considering a two-dimensional crystal of side L as shown in Fig. 2.12(a). If we suppose that the boundary condition at the edges of the crystal is that the vibration amplitude should vanish, then the lattice vibration waves will be standing waves of the form

$$u = u_o \sin(k_x x) \sin(k_y y). \tag{2.35}$$

This already satisfies the boundary condition at the edges of the crystal at $x = 0$ and $y = 0$ and will do so at the edges at $x = L$ and $y = L$ if we choose k_x and k_y

such that

$$k_x = p\pi/L \qquad \text{and} \qquad k_y = q\pi/L, \qquad (2.36)$$

where p and q are *positive* integers (changing the sign of p or q does not give a different solution). The allowed values of the wavevector components can be represented as a lattice in a space with axes k_x and k_y as shown on Fig. 2.12(b); a space of this type is known as **k-space** and extensive use is made of this concept in solid state physics. We see that the allowed **k** values lie on a square lattice of side π/L in the *positive* quadrant of k-space and are therefore uniformly distributed throughout this quadrant with a density $(L/\pi)^2$ per unit area.

This result can be extended easily to three dimensions. For a crystal in the form of a cube of side L, generalization of Eq. (2.36) shows that the allowed *standing wave* states are represented by a simple cubic lattice of side π/L in the

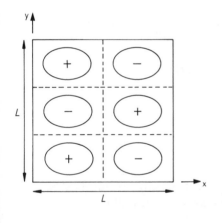

(*a*) Standing wave pattern in a two-dimensional box of side L

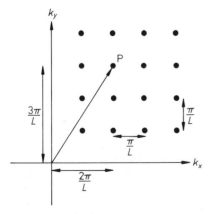

(*b*) Diagram in **k**-space of the allowed modes; the mode illustrated in (*a*) is indicated as point P

Fig. 2.12

octant of a three-dimensional **k**-space in which k_x, k_y and k_z are all positive. The number of *standing wave* states inside a volume element $d^3\mathbf{k}$ in this octant of **k**-space is therefore

$$\rho_S(\mathbf{k})d^3\mathbf{k} = \left(\frac{L}{\pi}\right)^3 d^3\mathbf{k} = \frac{V}{\pi^3} d^3\mathbf{k} \qquad (2.37)$$

where $V = L^3$ is the volume of the crystal and $\rho_S(\mathbf{k}) = V/\pi^3$ is the density of states for *standing waves* in a three-dimensional crystal.

When summing a property of the waves that depends only one the magnitude of **k** and not its direction (this is not strictly true for the frequencies of the lattice vibration modes but is often used as an approximation in this case), it is appropriate to take the spherical shell between k and $k + dk$ as the volume element in **k**-space as shown in Fig. 2.13. Thus

$$d^3\mathbf{k} = \tfrac{1}{8} \times 4\pi k^2 dk$$

where the factor $\tfrac{1}{8}$ arises because the standing wave **k** values are restricted to the octant of **k**-space in which k_x, k_y and k_z are all positive. Eq. (2.37) becomes

$$\rho_S(\mathbf{k})\, d^3\mathbf{k} = \frac{Vk^2}{2\pi^2}\, dk = g(k)\, dk,$$

where

$$g(k) = \frac{Vk^2}{2\pi^2} \qquad (2.38)$$

is a new density-of-states function defined as the number of states per unit *magnitude* of k (in the one-dimensional case $g(k)$ and $\rho_S(k)$ are identical because the volume element in the one-dimensional k-space is just dk).

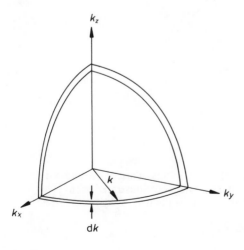

Fig. 2.13   Spherical shell in **k**-space that defines an increment dk in the magnitude of k. Only the portion of the shell in the octant of **k**-space where k_x, k_y and k_z are all positive is shown. This is appropriate for standing waves but for running waves the complete spherical shell is required

Eq. (2.38) can also be derived using running waves. Running waves in a cubic box of side L are obtained by using the periodic boundary conditions†

$$u(x + L, y, z) = u(x, y + L, z) = u(x, y, z + L) = u(x, y, z). \qquad (2.39)$$

An extension of the argument leading to Eq. (2.11) gives the allowed wavevector components for the running waves as

$$k_x = 2\pi p/L, \qquad k_y = 2\pi q/L, \qquad k_z = 2\pi r/L, \qquad (2.40)$$

where p, q and r are *positive* or *negative* integers. The allowed **k** values lie on a simple cubic lattice in **k**-space of side $2\pi/L$. The density of points is therefore only one-eighth of that for standing wave boundary conditions so that the number of allowed *running wave* **k** values in a volume d^3k of **k**-space is

$$\rho_R(\mathbf{k})\, d^3\mathbf{k} = \frac{V}{8\pi^3}\, d^3\mathbf{k}, \qquad (2.41)$$

where $\rho_R(\mathbf{k})$ is now the density of states for running waves. As the running wave points occupy the *whole* of **k**-space the total number of points inside a spherical shell between k and $k + dk$ is now $4\pi k^2 dk\, \rho_R(\mathbf{k})$, and Eq. (2.38) is again obtained. Eqs. (2.37), (2.38) and (2.41) are the general results that are applicable to any wave motions in the crystal. Although we have derived the densities of states only for a crystal in the shape of a cube we should not worry that properties derived from them might depend on the shape; experiment tells us that the properties of samples much bigger than atomic dimensions, when expressed in a suitable way (e.g. the heat capacity *per unit mass*), are independent of shape and size provided that the crystal is much larger than atomic dimensions.

Returning to the lattice vibration problem, if ω is a function only of the magnitude of **k** and $d\omega$ is the range of ω corresponding to the range dk in k, then from Eq. (2.38) we obtain the result

$$g(\omega) = g(k)\frac{dk}{d\omega} = \frac{Vk^2}{2\pi^2}\frac{dk}{d\omega}, \qquad (2.42)$$

analogous to Eq. (2.32). The density of states $g(\omega)$ must be evaluated by using Eq. (2.42) for each branch of the dispersion relation and then the total density of states obtained by summing over all branches. The energy of the lattice vibrations can then be evaluated using Eq. (2.34). This is a task that is best left to a computer. We will content ourselves with evaluating the energy in the limits of high and low temperatures, and presenting an approximate method for interpolating between these limits that is adequate for many purposes.

---

† Note that periodic boundary conditions are merely a convenient fiction in three dimensions since a three-dimensional object cannot be deformed so as to join up on itself in all three directions at once. The fiction is nevertheless useful since it enables running waves to be considered as normal modes.

2.6.3 The high- and low-temperature limits

As we have already discussed in section 2.6.1 each of the $3N$ lattice modes of a crystal containing N atoms contributes $k_B T$ to the energy at high temperatures, leading to a heat capacity

$$C = 3Nk_B. \tag{2.43}$$

This will only be the case if T is much higher than the characteristic temperature $\Theta = \hbar\omega/k_B$ (as defined after Eq. (2.29)) of *all* the lattice modes.

At low temperatures only lattice modes of low frequency will be excited from their ground state. These are the long-wavelength acoustic modes (sound waves) for which the dispersion relation is of the form $\omega = v_S k$, where v_S is the sound velocity. From Eq. (2.42) the density of states associated with this dispersion relation is

$$g(\omega) = \frac{Vk^2}{2\pi^2} \frac{dk}{d\omega} = \frac{V\omega^2}{2\pi^2 v_S^3} \tag{2.44}$$

where, in general, v_S depends on the direction of propagation so that we must regard the factor $1/v_S^3$ as an average over all directions. In fact the dispersion relation has three acoustic branches, one longitudinal and two transverse, so that altogether

$$g(\omega) = \frac{V\omega^2}{2\pi^2}\left(\frac{1}{v_L^3} + \frac{2}{v_T^3}\right) \tag{2.45}$$

where v_L and v_T are the sound velocities of the longitudinal and transverse modes and the quantity in brackets must again be regarded as an average over all directions.†

Inserting the density of states of Eq. (2.45) into Eq. (2.34) for the energy of the lattice vibrations gives

$$E = E_Z + \frac{V}{2\pi^2}\left(\frac{1}{v_L^3} + \frac{2}{v_T^3}\right)\int_0^\infty \left(\frac{\hbar\omega}{e^{\hbar\omega/k_B T} - 1}\right)\omega^2 \, d\omega \tag{2.46}$$

where E_Z is the zero point energy. This expression for E will be valid at sufficiently low temperatures that the Bose–Einstein distribution function ensures that only the long-wavelength acoustic modes make significant contributions to the energy. Changing the variable in the integral in Eq. (2.46) to $x = \hbar\omega/k_B T$ gives

$$E = E_Z + \frac{V}{2\pi^2}\left(\frac{1}{v_L^3} + \frac{2}{v_T^3}\right)\frac{(k_B T)^4}{\hbar^3}\int_0^\infty \frac{x^3 dx}{e^x - 1}. \tag{2.47}$$

---

† Although the vibrations are not exactly transverse or longitudinal in general, we can identify the polarization associated with a mode by noting the polarization it acquires in a propagation direction of high symmetry.

The integral is just a number, $\pi^4/15$ in fact, so that differentiating E with respect to T to obtain the heat capacity C gives

$$C = \frac{2V\pi^2 k_B}{15}\left(\frac{1}{v_L^3} + \frac{2}{v_T^3}\right)\left(\frac{k_B T}{\hbar}\right)^3. \tag{2.48}$$

The lattice heat capacity of solids thus varies as T^3 at low temperatures; this is often referred to as the Debye T^3 law. Fig. 2.14 illustrates the excellent agreement of this prediction with experiment for a non-magnetic insulator; we shall see later that there are extra contributions to the low-temperature heat capacity of other substances. The heat capacity vanishes more slowly than the exponential behaviour of a single harmonic oscillator because the vibration spectrum extends down to zero frequency.

It is interesting to note that the energy of black-body radiation varies as T^4 at all temperatures (Mandl,[2] Chapter 10). This is because the vacuum has no atomic structure and the dispersion relation $\omega = ck$ applies up to infinite frequency. Another difference between photons and phonons is that photons have no state of longitudinal polarization, only two transverse states.

2.6.4   The Debye interpolation scheme

The calculation of $g(\omega)$ from the interatomic forces is a very heavy calculation for a three-dimensional crystal. Debye obtained a good approximation to the resulting heat capacity by neglecting the dispersion of the acoustic waves, i.e. assuming $\omega = v_s k$ for arbitrary wavenumber. In a one-dimensional crystal this is equivalent to taking $g(\omega)$ as given by the broken line on Fig. 2.11 rather than the full curve. In a three-dimensional crystal the Debye interpolation method consists of assuming that the density of states of Eq. (2.45) is valid at all frequencies up to a cut-off value ω_D, the **Debye frequency**, above which there are

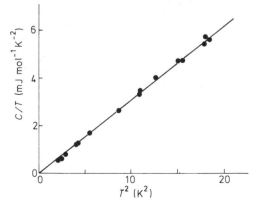

Fig. 2.14   Low-temperature heat capacity of KCl plotted so as to demonstrate the T^3 law at low temperature. The fact that the graph of C/T versus T^2 goes through the origin indicates the absence of a term linear in T. (Reproduced with permission from P. H. Keesom and N. Pearlman, *Phys. Rev.* **91**, 1354 (1953))

no modes. The cut-off frequency is chosen to make the total number of lattice modes correct. Since there are $3N$ lattice vibration modes in a crystal containing N atoms, we choose ω_D so that

$$\int_0^{\omega_D} g(\omega)\, d\omega = 3N. \qquad (2.49)$$

Hence, using Eq. (2.45),

$$\frac{V}{2\pi^2}\left(\frac{1}{v_L^3} + \frac{2}{v_T^3}\right)\int_0^{\omega_D} \omega^2 d\omega = 3N$$

or

$$\frac{V}{6\pi^2}\left(\frac{1}{v_L^3} + \frac{2}{v_T^3}\right)\omega_D^3 = 3N. \qquad (2.50)$$

By substituting the density of states of Eq. (2.45) and using Eq. (2.50), the lattice vibration energy (Eq. (2.34)) becomes

$$E = \frac{9N}{\omega_D^3}\int_0^{\omega_D}\left(\frac{1}{2}\hbar\omega + \frac{\hbar\omega}{e^{\hbar\omega/k_BT} - 1}\right)\omega^2\, d\omega$$

$$= \frac{9}{8}N\hbar\omega_D + \frac{9N}{\omega_D^3}\int_0^{\omega_D}\frac{\hbar\omega^3 d\omega}{e^{\hbar\omega/k_BT} - 1}. \qquad (2.51)$$

The first term is the estimate of the zero point energy provided by the Debye interpolation scheme. The heat capacity is obtained by differentiating Eq. (2.51) with respect to temperature. Thus

$$C = 9Nk_B\left(\frac{T}{\Theta_D}\right)^3\int_0^{\Theta_D/T}\frac{x^4 e^x dx}{(e^x - 1)^2}, \qquad (2.52)$$

where we have introduced the variable $x = \hbar\omega/k_BT$, as in the previous section, and we have defined the **Debye temperature** Θ_D by

$$\Theta_D = \hbar\omega_D/k_B. \qquad (2.53)$$

At high temperatures, $T \gg \Theta_D$, x is always small and by expanding the exponential the integrand reduces to x^2 so that the heat capacity becomes $3Nk_B$ as already calculated in the previous section (Eq. (2.43)). At low temperatures, $T \ll \Theta_D$, the upper limit of the integral is essentially infinite; the integral is then just a number $(4\pi^4/15)$ and we obtain the Debye T^3 law in the form

$$C = \frac{12Nk_B\pi^4}{5}\left(\frac{T}{\Theta_D}\right)^3 \qquad (2.54)$$

which reduces to Eq. (2.48) if Eqs. (2.53) and (2.50) are used.

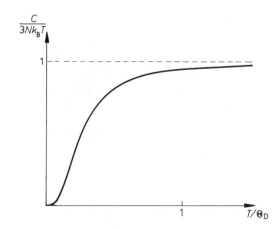

Fig. 2.15   Lattice heat capacity
of a solid as predicted by the
Debye interpolation scheme

Fig. 2.15 shows the heat capacity between these two limits as predicted by the Debye interpolation formula (2.52). Because it is exact in both high- and low-temperature limits the Debye formula gives quite a good representation of the heat capacity of most solids, even though the actual phonon density-of-states curve may differ appreciably from the Debye assumption; this is illustrated for copper in Fig. 2.16. The true density of states, deduced from $\omega(k)$ curves determined by neutron scattering (section 12.4), is compared with the Debye assumption in Fig. 2.16(a). The difference arises from two main effects: (i) dispersion of sound lowers the cut-off frequency and causes a rise in the density of states just below it, as in our one-dimensional example (Fig. 2.11); (ii) the variation of cut-off wavenumber with crystallographic orientation blurs the sharp cut-off in the density of states so that the actual maximum frequency is raised somewhat. The general result, as in Fig. 2.16(a), is that the maximum frequency can be quite close to the Debye value, but the centre of gravity of the true frequency distribution is lower. This means that the main rise in the heat capacity comes at a lower temperature than one would expect from Eq. (2.52). Fig. 2.16(a) illustrates that the true density of states contains far more structure than that obtained using the Debye interpolation scheme. This is not too surprising since it is clearly impossible to contain all the information on the atomic structure and forces in the one parameter Θ_D that the Debye theory uses to distinguish one material from another.

Departures from the Debye theory are best investigated by using Eq. (2.52) backwards to calculate $\Theta_D(T)$ at each temperature from the measured heat capacity;† a non-constant Θ_D indicates departures from Debye's interpolation scheme. In Fig. 2.16(b) the $\Theta_D(T)$ deduced from the measured heat capacity is compared with the $\Theta_D(T)$ calculated from the true density-of-states curve of

---

† Contributions to the heat capacity other than from the lattice vibrations must be deducted.

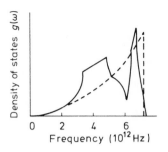

(a) Density of states for copper deduced from neutron scattering experiments (full curve) compared to the Debye density of states (broken curve) calculated using sound velocities obtained from the measured elastic moduli

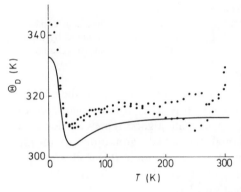

(b) Temperature dependence of Θ_D for copper deduced from the density-of-states curve in (a) (full curve) compared with experimental points deduced from heat capacity measurements. (Reproduced with permission from E. C. Svensson, B. N. Brockhouse and J. M. Rowe, *Phys. Rev.* **155**, 619 (1967))

Fig. 2.16

Fig. 2.16(a). It can be seen that the general trend is very similar; the small systematic difference is attributable to the fact that the neutron experiments measured $\omega(\mathbf{k})$ at room temperature, not at the temperature of the heat capacity measurements. The temperature dependence of $\omega(\mathbf{k})$ arises because of the anharmonic effects that we have ignored (see section 2.7).

From Eq. (2.50) we see that the Debye frequency and hence the Debye temperature scale with the velocity of sound in the solid. Thus solids with low densities and large elastic moduli have high Θ_D. Values of Θ_D for various solids can be found in Table 2.1. We will use the **Debye energy** $\hbar\omega_D$ when we need an estimate for the maximum phonon energy in a solid.

TABLE 2.1　Values of Debye temperature Θ_D for various solids obtained from the low-temperature heat capacity using Eq. (2.54)

Solid	Ar	Na	Cs	Fe	Cu	Pb	C (diamond)	KCl
$\Theta_D(K)$	93	158	38	457	343	105	2230	235

(Data from the *American Institute of Physics Handbook*, 3rd edn, McGraw-Hill, New York (1972))

2.7   ANHARMONIC EFFECTS

Any real crystal resists compression to a volume smaller than the equilibrium value more strongly than expansion to a larger volume. This is a consequence of the shape of interatomic potential curves such as that of Fig. 1.23, and it represents a departure from Hooke's law since positive and negative stresses do not produce strains of equal magnitude. The harmonic approximation used in our previous discussion of lattice vibrations in this chapter does *not* produce this property since it amounts to replacing the interatomic potentional curve by a parabola fitted at the minimum (Eq. (2.4)). Apart from departures from Hooke's law, there are other properties of solids that are not predicted if the harmonic approximation is made. Such properties are classified as **anharmonic effects** and the higher-order terms in the potential that are ignored in making the harmonic approximation are known as **anharmonic terms**. An important anharmonic effect is thermal expansion; for the symmetric harmonic approximation to the interatomic potential the mean separation of the atoms does not change as their amplitude of vibration increases with increasing temperature.

The harmonic approximation is necessary for the separation of the lattice motions into independent normal modes. Inclusion of higher-order terms in the Taylor series expansion of Eq. (2.4) leads to coupling of the modes. This coupling can be pictured as collisions between the phonons associated with the modes. The collisions limit the thermal conductivity associated with the flow of phonons. In the harmonic approximation the phonons do not interact with each other and, in the absence of boundaries, lattice defects and impurities (which also scatter phonons), the thermal conductivity is infinite.

2.7.1   Thermal expansion

The volume coefficient of thermal expansion is defined as†

$$\beta = \frac{1}{V}\left(\frac{\partial V}{\partial T}\right)_p. \tag{2.55}$$

In order to explain why β vanishes in the harmonic limit and to discuss its value more generally, it will be helpful to rewrite this as

$$\beta = -\frac{1}{V}\left(\frac{\partial V}{\partial p}\right)_T\left(\frac{\partial p}{\partial T}\right)_V = \frac{1}{B}\left(\frac{\partial p}{\partial T}\right)_V, \tag{2.56}$$

where

$$B = -V\left(\frac{\partial p}{\partial V}\right)_T \tag{2.57}$$

---

† The linear coefficient of expansion is $\beta/3$.

is the **bulk modulus**, the elastic modulus that determines the volume change produced by the application of pressure. To evaluate the expansion coefficient it is therefore necessary to determine the volume and temperature dependence of the pressure. The pressure can be calculated from the Helmholtz free energy $F = U - TS$, using

$$p = -\left(\frac{\partial F}{\partial V}\right)_T. \qquad (2.58)$$

In the harmonic approximation

$$F = E_{\text{pot}} + F_{\text{modes}}, \qquad (2.59)$$

where E_{pot} is the temperature-independent potential energy associated with the interatomic interactions and F_{modes} is the free energy associated with the lattice vibrations. The contribution f of one lattice mode to F_{modes} can be calculated from the partition function (Mandl,[2] chapter 6) of the simple harmonic oscillator (Eq. (2.25))

$$f = -k_B T \ln Z = \tfrac{1}{2}\hbar\omega + k_B T \ln\left[1 - \exp\left(-\hbar\omega/k_B T\right)\right]. \qquad (2.60)$$

In the harmonic approximation the frequency of a lattice mode is independent of volume;† thus f and hence F_{modes} do not depend on the volume and the lattice vibrations do not contribute to the pressure (Eq. (2.58)) and thus not to thermal expansion (Eq. (2.56)). Although the term E_{pot} in F does depend on the volume and hence contributes to the pressure, its temperature independence ensures that it does not contribute to the thermal expansion. Thermal expansion thus vanishes in the harmonic limit.

One effect of anharmonic terms is to cause the frequencies of the lattice vibration modes to depend on the volume, and this is the important effect as far as thermal expansion is concerned. We thus ignore the coupling of the modes and assume that F is still of the form of Eq. (2.59) with the contribution of each mode given by Eq. (2.60). From Eq. (2.58) the pressure is therefore

$$\begin{aligned} p &= -\frac{dE_{\text{pot}}}{dV} - \sum_{\text{modes}} \frac{\partial f}{\partial V} \\ &= -\frac{dE_{\text{pot}}}{dV} - \sum_{\text{modes}} \hbar \frac{\partial \omega}{\partial V}\left(\frac{1}{2} + \frac{1}{\exp\left(\hbar\omega/k_B T\right) - 1}\right) \end{aligned} \qquad (2.61)$$

so that the volume dependence of the lattice mode frequencies appears explicitly through the derivative $\partial\omega/\partial V$.

---

† This can be checked explicitly for the one-dimensional crystal of section 2.3.1 by recalculating the vibrations using the harmonic approximation for the situation where the crystal is subject to a tension so that the average spacing is increased from a. A simpler calculation which expresses the same physical principle is to show that the frequency of vibration of a mass suspended from a spring is the same on the Earth as on the Moon.

The simplest assumption is to assume that the volume dependence of all lattice mode frequencies is the same and can be represented by the simple power law $\omega \propto V^{-\gamma}$ which is more normally written

$$\frac{\mathrm{d}\,(\ln \omega)}{\mathrm{d}\,(\ln V)} = -\gamma. \qquad (2.62)$$

The dimensionless exponent γ is known as the **Gruneisen parameter** and can be regarded as a measure of the strength of the anharmonic effects; we indicate below how this parameter may be calculated. From Eq. (2.62) we obtain $\partial \omega / \partial V = -\gamma \omega / V$ and inserting this in Eq. (2.61) gives

$$p = -\frac{\mathrm{d}E_{\mathrm{pot}}}{\mathrm{d}V} + \frac{\gamma}{V} \sum_{\mathrm{modes}} \left(\frac{1}{2}\hbar\omega + \frac{\hbar\omega}{\exp\,(\hbar\omega/k_{\mathrm{B}}T) - 1} \right)$$

$$= -\frac{\mathrm{d}E_{\mathrm{pot}}}{\mathrm{d}V} + \frac{\gamma E_{\mathrm{modes}}}{V} \qquad (2.63)$$

where the energy E_{modes} of the lattice vibrations has been identified using Eq. (2.26). Inserting this form for the pressure into Eq. (2.56) and recalling that E_{pot} does not depend on temperature, we obtain the expansion coefficient

$$\beta = \frac{\gamma}{BV}\left(\frac{\partial E_{\mathrm{modes}}}{\partial T}\right)_V = \frac{\gamma C_V}{BV}, \qquad (2.64)$$

where C_v is the lattice heat capacity at *constant volume*; another anharmonic effect associated with the volume dependence of the lattice vibration frequencies is to cause the heat capacity measured at constant volume to differ from that measured at constant pressure.

Since elastic moduli such as the bulk modulus depend only weakly on temperature, Eq. (2.64) predicts that the temperature dependence of the expansion coefficient is approximately the same as that of the heat capacity, a feature first noted empirically by Gruneisen. Eq. (2.64) is therefore known as **Gruneisen's law** and is reasonably well obeyed by most solids. The value of $\beta BV/C_V$, which should be constant according to Eq. (2.64), is typically between 1 and 3 and is slightly temperature-dependent.

To estimate a theoretical value for γ we note that in our simple one-dimensional models of lattice vibrations the frequencies of the modes scale as $K^{1/2}$ (see for example Eqs. (2.9) and (2.21)), where K is the spring constant of the springs joining the atoms. By curtailing the Taylor series expansion of the interatomic potential (Eq. (2.4)) at the second-order term, one obtains a spring constant that does not depend on the mean separation between the atoms. If however the Taylor series of Eq. (2.4) is continued to include the first anharmonic term

$$\mathscr{V}(r) = \mathscr{V}(a) + \frac{(r-a)^2}{2}\left(\frac{d^2\mathscr{V}}{dr^2}\right)_{r=a} + \frac{(r-a)^3}{6}\left(\frac{d^3\mathscr{V}}{dr^3}\right)_{r=a}, \qquad (2.65)$$

then the spring constant for vibrations around a mean separation a' is

$$K = \left(\frac{d^2\mathscr{V}}{dr^2}\right)_{r=a'} = \left(\frac{d^2\mathscr{V}}{dr^2}\right)_{r=a} + (a'-a)\left(\frac{d^3\mathscr{V}}{dr^3}\right)_{r=a} \qquad (2.66)$$

The Gruneisen parameter, which describes the effect of the anharmonic term on the volume dependence of the lattice vibration frequencies, can therefore be obtained from Eq. (2.62) as follows:

$$\gamma = -\frac{V}{\omega}\frac{\delta\omega}{\delta V} = -\frac{a}{3\omega}\frac{\delta\omega}{\delta a} = -\frac{a}{6K}\frac{\delta K}{\delta a}$$

$$= -\frac{a}{6}\left(\frac{d^3\mathscr{V}}{dr^3}\right)_{r=a}\bigg/\left(\frac{d^2\mathscr{V}}{dr^2}\right)_{r=a}, \qquad (2.67)$$

where we have used $V \propto a^3$, $\omega \propto K^{1/2}$ and Eq. (2.66) to determine $\delta K/\delta a$. Problem 2.7 is a calculation of γ for an inert-gas crystal. The curvature of the interatomic potential essentially determines the elastic moduli of the crystal and we therefore see that another anharmonic effect is to cause the elastic moduli to depend on volume and hence pressure.

That anharmonicity does not in fact have the same effect on all modes is clear from the existence of **soft modes** in some crystalline solids; a soft mode is one for which the anharmonicity causes the frequency to vanish at a particular finite temperature. When this happens the atomic displacements associated with the mode become time-independent and a permanent displacement of the atoms therefore occurs. This typically happens for a transverse optic mode of zero wavenumber (infinite wavelength); in this case each unit cell is subject to the same change (although the displacement of different atoms inside each cell is different). The soft mode thus provides the mechanism for a phase transition from one crystal structure to another; such a transition is referred to as a **displacive phase transition**. The measured variation of the frequency of a soft mode near a displacive phase transition can be found in Fig. 9.12.

In all cases the high-temperature phase has a higher symmetry; as the temperature decreases through the critical value the displacement associated with the soft mode begins to grow. In ionic solids the opposite displacement of positive and negative ions associated with a zero-wavenumber transverse optic soft mode causes the low-temperature phase to possess a permanent electric polarization; the low-temperature phase is then **ferroelectric** (see section 9.2).

2.7.2 Phonon–phonon collisions

The coupling of normal modes by the anharmonic terms in the interatomic forces can be pictured as collisions between the phonons associated with the modes. A typical collision process is shown in Fig. 2.17(a); a phonon of wavenumber k_1 and frequency ω_1 coalesces with a phonon of wavenumber k_2 and frequency ω_2 to produce a phonon of wavenumber k_3 and frequency ω_3. We show below that in a one-dimensional crystal this process is described by the equations

$$\omega_3 = \omega_1 + \omega_2 \tag{2.68}$$

$$k_3 = k_1 + k_2. \tag{2.69}$$

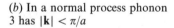

(a) A collision in which phonons 1 and 2 coalesce to give phonon 3

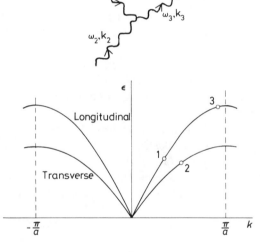

(b) In a normal process phonon 3 has $|\mathbf{k}| < \pi/a$

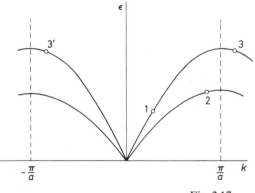

(c) In an Umklapp process phonon 3 has $|\mathbf{k}| > \pi/a$ and is equivalent to phonon 3' which has $|\mathbf{k}| < \pi/a$

Fig. 2.17

These equations have a simple physical interpretation; when multiplied by $\hbar$ they look like laws of conservation of energy and momentum for the collision of phonons.† It is necessary to modify Eq. (2.69) to take account of the convention that phonons are represented by wavenumbers in the range $-\pi/a < k < \pi/a$. If k_3 lies outside this range then we must add a suitable multiple of $2\pi/a$ to bring it back within the range. Eq. (2.69) then becomes

$$k_3 \pm n2\pi/a = k_1 + k_2 \tag{2.70}$$

where k_1, k_2 and k_3 are all in the above range. The three-dimensional generalization of the $\pm n2\pi/a$ term is given in section 11.4 (see also Eq. (12.8)). We will find that Eq. (2.70) holds also for the collisions of other types of particle within a crystal, electrons for example.

It is usual to make a distinction between processes for which $n = 0$, called **normal processes**, and those for which $n \neq 0$, called **Umklapp processes**. The two types of process are illustrated in Figs. 2.17(b) and (c). We have had to invent one-dimensional transverse phonons in order to satisfy Eqs. (2.68) and (2.70) simultaneously; for a single $\omega(k)$ curve that is everywhere concave downwards, two phonons cannot combine to give enough momentum for a phonon of their combined energy. The inclusion of the additional term in Eq. (2.69) to give Eq. (2.70) indicates again that the crystal momentum $\hbar k$ is not the true momentum of a phonon. Readers prepared to accept Eqs. (2.68) and (2.69) on trust can omit the proof given below.

The mechanism for coupling lattice modes can be understood if we recall from section 2.7.1 that anharmonic effects cause the elastic constants of a crystal to depend on the density or, more generally, the state of strain of the crystal; the velocity of sound is thus also affected by the state of strain. Hence, if a sound wave passes through a crystal in which another sound wave is already present, its wavefront will become phase modulated by this effect. To see this, we consider a wave of frequency ω_1 and wavenumber k_1 which is phase modulated at frequency ω_2 and wavenumber k_2:

$$A = \exp\{i[k_1 x + C\cos(k_2 x - \omega_2 t) - \omega_1 t]\} \tag{2.71}$$

where C specifies the amount of the modulation. The phase velocity of this wave is obtained by writing the equation for a wavefront, or surface of constant phase, as

$$k_1 x + C\cos(k_2 x - \omega_2 t) - \omega_1 t = \text{const.},$$

and differentiating with respect to time to obtain

$$k_1 \frac{dx}{dt} - k_2 \frac{dx}{dt} C\sin(k_2 x - \omega_2 t) + \omega_2 C\sin(k_2 x - \omega_2 t) - \omega_1 = 0,$$

---

† In a three-dimensional crystal the wavenumbers in Eq. (2.69) must be replaced by the wavevectors associated with the phonons.

so that the phase velocity is

$$\frac{dx}{dt} = \frac{\omega_1 - \omega_2 C \sin(k_2 x - \omega_2 t)}{k_1 - k_2 C \sin(k_2 x - \omega_2 t)}. \tag{2.72}$$

Hence, in the absence of the phase modulation, the phase velocity is ω_1/k_1; the additional terms in the numerator and denominator of Eq. (2.72) represent modulation of the phase velocity at a frequency ω_2 and wavenumber k_2, appropriate to a crystal containing a sound wave of this frequency and wavenumber. The strength C of the modulation depends on the amplitude of the ω_2, k_2 wave and on the strength of the anharmonicity.

To investigate the implications of the phase modulation we assume that C is sufficiently small that Eq. (2.71) can be expanded in powers of C, i.e.

$$\begin{aligned} A &= \exp[i(k_1 x - \omega_1 t)][1 + iC\cos(k_2 x_2 t) + \cdots] \\ &= \exp[i(k_1 x - \omega_1 t)] + \tfrac{1}{2}iC \exp\{i[(k_1 + k_2)x - (\omega_1 + \omega_2)t]\} \\ &\quad + \tfrac{1}{2}iC \exp\{i[(k_1 - k_2)x - (\omega_1 - \omega_2)t]\} + \text{terms of order } C^2. \end{aligned} \tag{2.73}$$

The first term is our original ω_1, k_1 sound wave; the second is a new wave of frequency ω_3 given by Eq. (2.68) and wavenumber k_3 given by Eq. (2.69). If $\omega_3(k_3)$ is a point on the phonon dispersion curve this new wave can propagate and the second term in Eq. (2.73) thus represents the process shown in Fig. 2.17(a). The third term in Eq. (2.73) corresponds to a process in which a phonon of wavenumber k_1 *emits* a phonon of wavenumber k_2 to become a phonon of wavenumber k_3. These two processes are therefore the inverse of each other: either two phonons coalesce to form one, or one phonon splits into two; they are both **three-phonon processes**. The terms of order C^2 and higher in Eq. (2.73) correspond to processes in which four or more phonons are involved; we shall not discuss these higher-order processes further. We see that Eq. (2.69) is more correctly regarded as a geometrical interference condition on wavenumbers than as a conservation law for momentum.

2.8 THERMAL CONDUCTION BY PHONONS

When there is a temperature gradient in a solid a flow of heat takes place from a hotter region to a cooler region. In an electrically insulating solid the most important contribution to thermal conduction comes from the flow of phonons. Thermal conduction is an example of a **transport property**, a term used to describe any process in which the flow of some quantity occurs. The coefficient such as the thermal conductivity which describes the flow is known as a **transport coefficient**. We will use the elementary kinetic theory of the transport coefficients of gases to calculate the thermal conductivity of the phonon gas in a solid.

2.8.1 Kinetic theory

In the elementary kinetic theory of gases it is shown, by assuming a constant average speed $\bar{v}$ for the molecules, that the steady state flux of a property P in the z direction is given by

$$\text{flux} = \tfrac{1}{3}l\bar{v}\frac{dP}{dz} \qquad (2.74)$$

where l is the mean free path, and the factor $\tfrac{1}{3}$ arises from an angular average (see, for example, Flowers and Mendoza[1]). In the simplest case where P is the number density of particles the transport coefficient obtained from Eq. (2.74) is the diffusion coefficient $D = \tfrac{1}{3}\bar{v}l$. If P is the energy density E then the flux W is the heat flow per unit area so that

$$W = \tfrac{1}{3}\bar{v}l\frac{dE}{dz} = \tfrac{1}{3}\bar{v}l\frac{dE}{dT}\frac{dT}{dz}.$$

Now dE/dT is the specific heat C *per unit volume*, so that the thermal conductivity is given by

$$K = \tfrac{1}{3}\bar{v}lC. \qquad (2.75)$$

Particle conservation is not used anywhere in the derivation of Eq. (2.75) so it may be applied to a phonon gas just as to a real gas; in fact it works well for a phonon gas, because $\bar{v}$ *is* almost a constant (the velocity of sound) for phonons of not too large a wavenumber. For a real gas of atoms the application of Eq. (2.75) is not so straightforward, for several reasons. First, $\bar{v}$ depends on temperature and so should really be included in the derivative in Eq. (2.74); secondly the conservation of atoms imposes the constraint that there is no net particle flux; finally, hydrostatic equilibrium requires that the pressure is uniform. A satisfactory theory of heat conduction in a real gas is therefore quite hard, and the correct numerical factor in Eq. (2.75) turns out to be rather different from $\tfrac{1}{3}$ in this case. For phonons the simple theory is much better.

The essential differences between the processes of heat conduction in a phonon and a real gas are illustrated in Fig. 2.18. For phonons (Fig. 2.18(a)) the speed is approximately constant, but both the number density and the energy density are greater at the hot end; heat flow is primarily due to phonon flow with phonons being created at the hot end and destroyed at the cold end. For a real gas (Fig. 2.18(b)) there is, in contrast, no flow of particles. The average velocity and the kinetic energy per particle are greater at the hot end, but the number density is greater at the cold end, and the energy density is in fact *uniform* (because the pressure is uniform). Heat flow is solely by transfer of kinetic energy from one particle to another in collisions; this is a rather minor effect in the phonon case.

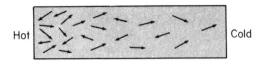

(a) In a phonon gas there is a net phonon flow and there are more phonons at the hot end

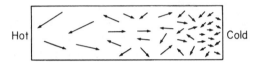

(b) In a real gas there is no net flow of atoms; the atoms are fewer but faster at the hot end

Fig. 2.18   Heat conduction in a phonon gas and a real gas

We will now proceed to use Eq. (2.75) to discuss the temperature dependence of the thermal conductivity of a phonon gas. We assume that $\bar{v}$ is approximately the velocity of sound and therefore temperature-independent. The temperature dependence of the lattice heat capacity C_V is discussed in section 2.6. We must therefore determine the temperature dependence of the phonon mean free path l. Except at very low temperatures this is determined by phonon–phonon collisions. Since the heat flow is associated with a flow of phonons, the most effective collisions for limiting the flow are those in which the phonon group velocity is reversed. It is the Umklapp processes that conspicuously have this property (Fig. 2.17), and these are therefore important in limiting the thermal conductivity. However, the rigid distinction between normal and Umklapp processes is a somewhat artificial one, since phonons with k just less than π/a and k just greater than π/a are really very similar; both have a small group velocity and contribute little to the energy flow. The energy flow is proportional to the sum over all phonons of

phonon energy × group velocity = $\hbar\omega \, d\omega/dk$,

and this is reduced by *both* the three-phonon processes shown in Fig. 2.17.

It is however true that if there were no Umklapp processes the energy flow would be *statistically* steady even in the absence of a temperature gradient; consequently the thermal conductivity would be infinite! To see this we note that in the absence of Umklapp processes the collisions of phonons are described by equations like (2.69) (or their three-dimensional equivalents), which express the conservation of crystal momentum in any collision. Thus in any state where there is a flow of phonons the total phonon crystal momentum

$$\mathbf{P} = \sum n(\mathbf{k})\hbar\mathbf{k}$$

is conserved when collisions occur. The flow of heat associated with the phonon flow is likewise unchanged. The existence of the Umklapp processes described by Eq. (2.70) with $n \neq 0$ (and its three-dimensional equivalent) is therefore necessary for finite conductivity.

2.8.2   Conduction at high temperatures

At temperatures much greater than the Debye temperature Θ_D the heat capacity is given by the temperature-independent classical result, Eq. (2.43). We would naively expect that the rate of collisions of two phonons would be proportional to the phonon density.† If collisions involving larger numbers of phonons are important, however, then the scattering rate will increase more rapidly than this with phonon density. At high temperatures the average phonon energy is constant and the total lattice energy is proportional to T; this means that the phonon number is proportional to T. consequently we expect a scattering rate proportional to T and a mean free path l which varies as T^{-1}. The thermal conductivity from Eq. (2.75) should therefore vary as T^{-1} (or T^{-x} with $x \geqslant 1$ if collisions involving larger numbers of phonons are important). Fig. 2.19(a) shows that the experimental results do tend towards this behaviour at high temperatures.

2.8.3   Conduction at intermediate temperatures

Fig. 2.19(a) also shows that at temperatures below about Θ_D the conductivity rises more steeply with falling temperature, despite the fact that the heat capacity is falling in this region. This can be understood if we recall that Umklapp processes are essential for phonon–phonon collisions to be effective in limiting the thermal conductivity. We see from Fig. 2.17 that Umklapp processes will only occur if there are phonons of sufficient energy to create a phonon with $k_3 > \pi/a$. This requires phonons with an energy comparable to the Debye energy $k\Theta_D$. The energy of the relevant phonons is thus not sharply defined but we might expect their number to vary roughly as $\exp(-\Theta_D/bT)$ when $T \ll \Theta_D$, where b is a number of order unity. The mean free path would be expected therefore to vary as $\exp(+\Theta_D/bT)$; this exponential factor dominates any low power of T in the thermal conductivity, such as a factor T^3 from the heat capacity. Experimentally the variation is of this form over a fair range of temperature; the empirical values of b are of the order of 2 or 3.

---

† This is supported by Eq. (2.73) in which the 'three-phonon' scattered waves are proportional to C and hence to the amplitude of the wave causing the scattering; the scattered intensity is thus proportional to the intensity of the scattering wave and hence to the number of phonons associated with it. The higher-order processes in Eq. (2.73) depend on higher powers of C and the resulting scattered intensity therefore depends on a higher power of the phonon density.

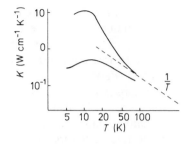

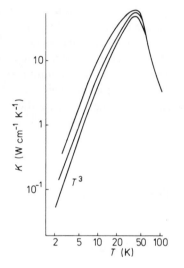

(a) Thermal conductivity of a quartz crystal; the lower curve shows the effect of lattice defects in a neutron irradiated sample. (Reproduced with permission from R. Berman, *Proc. R. Soc.* A **208**, 90 (1951))

(b) Thermal conductivity of artificial sapphire rods of different diameters. (Reproduced with permission from R. Berman, E. L. Foster and J. M. Ziman, *Proc. R. Soc.* A **231**, 130 (1955))

Fig. 2.19

2.8.4   Conduction at low temperatures

Because of the exponentially decaying population of the high-energy phonons necessary for Umklapp processes to occur, the mean free path for phonon–phonon collisions becomes very long at low temperatures and eventually exceeds the size of the solid. The phonon mean free paths in a good quality single crystal are then limited by collision with the specimen surface, and the flow of phonons becomes analogous to the flow of a real gas in the Knudsen regime. The effective phonon mean free path to be used in Eq. (2.75) is then of the order of the specimen diameter; it may even be larger if the specimen surface is smooth enough for appreciable specular reflection of phonons to occur, since specular reflection does not contribute to thermal resistance. In this limit there is no true thermal conductivity since Eq. (2.75) predicts a value that depends on the size of the specimen; the thermal *conductance* becomes proportional to the cube of the specimen diameter instead of the square. The variation of effective conductivity with diameter is illustrated in Fig. 2.19(b).The only temperature dependence of the conductivity now comes from the heat capacity, which obeys the Debye T^3 law (Eq. (2.54)) in this region. The dependence of a transport coefficient on the shape and size of a crystal, which occurs when the mean free path becomes comparable to the sample dimensions, is known as a **size effect**.

If the specimen is not a perfect single crystal and contains imperfections such as dislocations, grain boundaries and impurities, then these will also scatter phonons. At the very lowest temperatures the dominant phonon wavelength becomes so long that these imperfections are not effective scatterers, so the thermal conductivity always has a T^3 dependence at these temperatures. The maximum conductivity between the T^3 region and the exp (Θ_D/bT) region is however largely controlled by imperfections. For an impure or polycrystalline specimen the maximum can be broad and low (Fig. 2.19(a)), whereas for a carefully prepared single crystal, as illustrated in Fig. 2.19(b), the maximum is quite sharp and the conductivity reaches a very high value, of the order of that of metallic copper in which the conductivity is predominantly due to the conduction electrons (section 3.3).

PROBLEMS 2

2.1 Show that the dispersion relation for the lattice vibrations of a chain of identical masses M, in which each is connected to its first and second nearest neighbours by springs of spring constant K and K_2 respectively, is

$$M\omega^2 = 2K[1 - \cos(ka)] + 2K_2[1 - \cos(2ka)]$$

where a is the equilibrium spacing.
   Show that:
(a) this dispersion relation reduces to that for sound waves in the long-wavelength limit (ensure that the velocity corresponds to that predicted by the elastic modulus of the crystal);
(b) the group velocity vanishes at $k = \pm\pi/a$; and
(c) ω is periodic in k with period $2\pi/a$.
   Explain why you would expect (a), (b) and (c) to remain valid if forces between neighbours of even higher order are included.
2.2 We may make a model of the stretching vibrations of a polyethylene chain $-CH=CH-CH=CH-\cdots$ by considering a linear chain of identical masses M connected by springs of alternating force constants K_1 and K_2. Show that the characteristic frequencies of such a chain are given by

$$\omega^2 = \frac{K_1 + K_2}{M}\left[1 \pm \left(1 - \frac{4K_1 K_2 \sin^2(\tfrac{1}{2}ka)}{(K_1 + K_2)^2}\right)^{1/2}\right]$$

where a is the repeat distance of the chain. (Note that the relative lengths of the single and double bonds are irrelevant; why? By obtaining values for ω as $k \to 0$ and $k \to \pm\pi/a$, sketch the dispersion curves for the optical and acoustic branches of the vibration spectrum.

2.3 Obtain expressions for the heat capacity due to longitudinal vibrations of a chain of identical atoms;
(a) in the Debye approximation;
(b) using the exact density of states (Eq. (2.33)).
With the same constants K and M, which expression gives the greater heat capacity and why?
Show that at low temperatures both expressions give the same heat capacity, proportional to T.

2.4 The relation between frequency v and wavelength λ for surface tension waves on a liquid of density ρ and surface tension σ is

$$v^2 = \frac{2\pi\sigma}{\rho\lambda^3}.$$

Use this result to construct a 'Debye theory' of the surface contribution to the internal energy of a liquid. Obtain the analogue of the Debye T^3 law for the surface contribution to the heat capacity of liquid helium very near to absolute zero.
Given that σ is the surface free energy $(F = U - TS)$, how does σ vary with temperature near absolute zero?

2.5 Use Eq. (2.26) to show that in thermal equilibrium at temperature T the average energy of a sufficiently long-wavelength mode is $k_B T$.
At temperatures much less than the Debye temperature Θ_D, approximately how many modes will be excited?
Use your answer to show that for $T \ll \Theta_D$ the heat capacity due to atomic vibrations is of order $Nk_B(T/\Theta_D)^3$ for a solid containing N atoms.

2.6 Estimate the zero point energy per atom of the lattice vibrations of solid argon $(\Theta_D = 92 \text{ K})$ and compare this with the measured binding energy of solid argon of 0.090 eV per atom.

2.7 Estimate the value of Gruneisen's constant γ for an inert-gas crystal. Use the Lennard-Jones form of the interatomic potential (see caption to Fig. 1.23).

2.8 From the data of Fig. 2.19(b) estimate:
(a) the diameters of the sapphire rods; and
(b) the value of b that enters the temperature dependence $\exp(\Theta_D/bT)$ of the phonon mean free path at intermediate temperatures.
For sapphire $\Theta_D = 1000 \text{ K}$, speed of sound $= 10^4 \text{ m s}^{-1}$, and for $T \ll \Theta_D$, $C = 10^{-1} T^3 \text{ J m}^{-3} \text{ K}^{-1}$.)

Science may be described as the art of systematic oversimplification.—*Sir Karl Popper (1982)*

CHAPTER

Free electrons in metals

3.1 INTRODUCTION

Many solids conduct electricity; this is usually an indication that there are electrons that are not bound to atoms but are able to move through the whole crystal. Conducting solids fall into two main classes; metals and semiconductors. The room-temperature resistivity of metals is typically in the range 10^{-6} to 10^{-8} Ω m and is usually *increased* by the addition of small amounts of impurity; the resistivity normally decreases monotonically with decreasing temperature. For pure semiconductors the room-temperature resitivity is very much larger than that of metals but can be *reduced* by many orders of magnitude towards that of metals by the addition of small amounts of impurity; the resitivity does not always vary monotonically with temperature but semiconductors tend to become insulators at the lowest temperatures.

We will set aside until the following chapter the question of why mobile electrons appear in some solids and not others; this is a very difficult question, particular if the interactions between the electrons are considered. In this chapter we will calculate the properties of metals using the assumption that conduction electrons exist and consist of all the valence electrons from all the atoms; thus metallic sodium, magnesium and aluminium will be assumed to have one, two and three mobile electrons per atom respectively. We will describe a simple theory, the free electron model, which works remarkably well in explaining the properties of many metals. Semiconductors are the subject of Chapter 5.

3.2   THE FREE ELECTRON MODEL

The simplest possible approach is to assume that the electrons in a metal behave like a gas of free particles; this is the **free electron model**. The removal of the valence electrons from an atom leaves a positively charged **ion core**. The free electron model assumes that the charge density associated with the ion cores is spread uniformly throughout the metal so that the electrons move in a constant electrostatic potential. Note that all the details of the crystal structure are lost when this assumption is made; in the next chapter we investigate the effect of using a more realistic potential for the positive ion cores. The free electron model also ignores the repulsive interaction between the conduction electrons. The model therefore considers the electrons as moving independently in a square potential well of finite depth, the edges of the well corresponding to the boundaries of the metal.

Since the *bulk* properties of a macroscopic piece of metal, such as the specific heat capacity or the resistivity, are independent of the shape, we will for convenience consider a cube of metal of side L with faces perpendicular to the x, y and z axes. We must solve the time-independent Schrödinger equation

$$-\frac{\hbar^2}{2m}\nabla^2\psi = \varepsilon\psi \qquad (3.1)$$

for the wavefunctions ψ and energies ε of the electrons inside the cube; we have taken the uniform potential inside the cube to be zero. The wavefunctions depend on the boundary condition at the surfaces. One possibility is to use $\psi = 0$ there, which leads to standing wave solutions of Schrödinger's equation within the metal. Although this boundary condition is not strictly appropriate at a finite discontinuity in the potential, the wavefunctions decay to zero within about one atomic spacing of the surface (see problem 3.1) and, as this is much less than L, the error involved in using it is negligible. Experimentally the bulk properties of metals do not depend on the condition of the surfaces and, correspondingly, the calculated properties ought not to depend on the boundary condition assumed. This feature makes it attractive to use a periodic boundary condition, as the electron wavefunctions are then running waves. This approach was adopted in the discussion of lattice vibration waves in Chapter 2 (Eq. (2.39)) and has essentially the same consequences here as there.

Thus by imposing a periodic boundary condition in the form

$$\psi(x + L, y + L, z + L) = \psi(x, y, z), \qquad (3.2)$$

we find that the solutions of Schrödinger's equation are plane waves

$$\psi(x, y, z) = \frac{1}{V^{1/2}}\, e^{i\mathbf{k}\cdot\mathbf{r}} = \frac{1}{V^{1/2}}\, e^{i(k_x x + k_y y + k_z z)}, \qquad (3.3)$$

where $V = L^3$ is the volume of the cube and the $1/V^{1/2}$ factor ensures that the wavefunction is normalized. The wavevector components must satisfy

$$k_x = \frac{2\pi p}{L}, \qquad k_y = \frac{2\pi q}{L}, \qquad k_z = \frac{2\pi r}{L} \tag{3.4}$$

(cf. Eq. (2.40)), with p, q and r taking any integer values, positive, negative or zero. The wavefunction (3.3) corresponds to an energy

$$\varepsilon = \frac{\hbar^2 k^2}{2m} = \frac{\hbar^2}{2m}(k_x^2 + k_y^2 + k_z^2) \tag{3.5}$$

and a momentum

$$\mathbf{p} = \hbar \mathbf{k} = \hbar(k_x, k_y, k_z). \tag{3.6}$$

From section 2.6.2 we know that the number of allowed $\mathbf{k}$ values inside a spherical shell of $\mathbf{k}$-space of radius k and thickness dk, centred on the origin (see Fig. 2.13), is given by Eq. (2.38) as

$$g(k)\,dk = \frac{Vk^2}{2\pi^2}\,dk,$$

where $g(k)$ is the density of states per unit magnitude of k. We can use this result to calculate the number of allowed electron states per unit energy range $g(\varepsilon)$. To do this we must take account of the fact that electrons have a spin of $\frac{1}{2}$. Each $\mathbf{k}$ state therefore represents two possible electron states, one for each of the two possible spin polarizations. In the absence of an applied magnetic field these states have the same energy. Thus the number of electron states in the spherical shell between k and $k + dk$, which corresponds to energies between ε and $\varepsilon + d\varepsilon$ say, can be written as

$$g(\varepsilon)\,d\varepsilon = 2g(k)\,dk.$$

That is

$$g(\varepsilon) = 2g(k)\frac{dk}{d\varepsilon}.$$

Hence, using Eqs. (3.5) and (2.38),

$$g(\varepsilon) = \left(2\frac{V}{2\pi^2}\right)k\left(k\frac{dk}{d\varepsilon}\right) = \left(\frac{V}{\pi^2}\right)\left(\frac{2m\varepsilon}{\hbar^2}\right)^{1/2}\left(\frac{m}{\hbar^2}\right)$$

$$= \frac{V}{2\pi^2\hbar^3}(2m)^{3/2}\varepsilon^{1/2}. \tag{3.7}$$

3.2.1   Ground state of the free electron gas

Since electrons have half-integral spin they are fermions and must obey the Pauli exclusion principle; accordingly each state can accommodate only one electron. The lowest-energy state of N free electrons is therefore obtained by filling the N states of lowest energy. Thus all states are filled up to an energy ε_F, known as the **Fermi energy**, determined by the requirement that the number of states with $\varepsilon < \varepsilon_F$, obtained by integrating the density of states (Eq. (3.7)) between 0 and ε_F, should equal N. Hence

$$N = \int_0^{\varepsilon_F} g(\varepsilon)\,d\varepsilon = \frac{V}{3\pi^2\hbar^3}(2m\varepsilon_F)^{3/2} \qquad [= \tfrac{2}{3}\varepsilon_F g(\varepsilon_F)], \qquad (3.8)$$

so that

$$\varepsilon_F = \frac{\hbar^2}{2m}\left(\frac{3\pi^2 N}{V}\right)^{2/3}. \qquad (3.9)$$

The occupied states are those inside the **Fermi sphere** in **k**-space as shown in Fig. 3.1; the surface of the sphere is the **Fermi surface** and the radius is the **Fermi wavenumber** k_F. From Eq. (3.5), k_F is given by $\varepsilon_F = \hbar^2 k_F^2/2m$; hence

$$k_F = (3\pi^2 N/V)^{1/3}. \qquad (3.10)$$

k_F may be evaluated more simply from the density of running wave states in **k**-space of Eq. (2.41) by calculating the radius of the sphere that contains $(N/2)$ **k** states; thus

$$\tfrac{4}{3}\pi k_F^3 \rho_R(k) = \tfrac{4}{3}\pi k_F^3 \frac{V}{8\pi^3} = \frac{N}{2}$$

which reduces to Eq. (3.10).

Typical values may be obtained by using monovalent potassium metal as an example; for potassium the atomic density and hence the valence electron

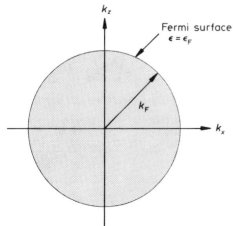

Fig. 3.1 Section through the Fermi sphere in **k**-space. The Fermi surface is the surface of the sphere and it marks the boundary between the occupied (shaded) and unoccupied **k** states at absolute zero for the free electron gas

density N/V is 1.402×10^{28} m^{-3}, so that

$$\varepsilon_F = 3.40 \times 10^{-19} \text{ J} = 2.12 \text{ eV} \tag{3.11}$$

$$k_F = 0.746 \text{ Å}^{-1}. \tag{3.12}$$

Thus ε_F is of the order of atomic ionization energies and k_F is of the order of the reciprocal of an atomic spacing. It is instructive to define the **Fermi temperature** or **degeneracy temperature** T_F by $\varepsilon_F = k_B T_F$; for potassium

$$T_F = 2.46 \times 10^4 \text{ K}. \tag{3.13}$$

The significance of T_F is that it is only at a temperature of this order that the particles in a classical gas attain kinetic energies as high as ε_F; only at temperatures above T_F will the free electron gas behave like a classical gas. In practice metals vaporize before the temperature T_F is reached. At ordinary temperatures where $T \ll T_F$ the behaviour of the free electron gas is dominated by the Pauli exclusion principle and is said to be **highly degenerate**. The large kinetic energy of the electrons makes a significant contribution to the bulk modulus of most metals (see problem 3.3). Two other parameters that we will use are the **Fermi momentum** p_F $(= \hbar k_F)$ and the **Fermi velocity** v_F $(= p_F/m)$; these are respectively the momentum and velocity of electrons in states on the Fermi surface. For potassium $v_F = 0.86 \times 10^6$ m s^{-1}. As the reader will probably have guessed from the use of so many parameters labelled Fermi, the Fermi surface plays an important role in the behaviour of metals.

3.2.2 The free electron gas at finite temperature

At a temperature T the probability of occupation of an electron state of energy ε is given by the Fermi distribution function

$$f(\varepsilon, T) = \frac{1}{e^{(\varepsilon - \mu)/k_B T} + 1}. \tag{3.14}$$

where μ is the chemical potential (Mandl,[2] chapter 11). This is plotted in Fig. 3.2(a) at absolute zero and at a finite temperature. At absolute zero $f(\varepsilon)$ is a step function: $f(\varepsilon) = 1$ for $\varepsilon < \mu$ and $f(\varepsilon) = 0$ for $\varepsilon > \mu$. Therefore $\varepsilon = \mu$ represents the boundary between occupied and unoccupied states and, from our discussion of the previous section, we see that μ is equal to ε_F at $T = 0$. The large degeneracy temperature of the free electron gas has the consequence that very few electrons are thermally excited; the number of electrons per unit energy range in thermal equilibrium is given by multiplying the density-of-states function $g(\varepsilon)$ by the probability $f(\varepsilon, T)$ that a state is occupied

$$n(\varepsilon, T) = g(\varepsilon)f(\varepsilon, T). \tag{3.15}$$

This is plotted in Fig. 3.2(b) for $T = 0$ and for a finite temperature $T \ll T_F$. The

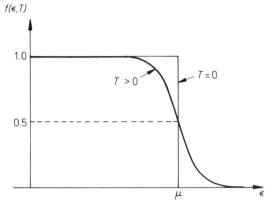

$f(\epsilon,T)$

1.0

0.5

$T > 0$

$T = 0$

μ

ϵ

(a) The Fermi function $f(\varepsilon, T)$ at $T = 0$ and at a finite temperature

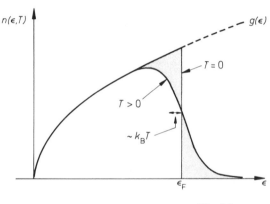

$n(\epsilon,T)$

$g(\epsilon)$

$T = 0$

$T > 0$

$\sim k_B T$

ε_F

ϵ

(b) Number of electrons per unit energy range according to the free electron model. The shaded area shows the change in distribution between absolute zero and a finite temperature

Fig. 3.2

number of electrons in any range of energy is just the area under the $n(\varepsilon, T)$ graph in that energy range. At finite temperatures the Fermi distribution function (Fig. 3.2(a)) decreases from 1 to 0 in a range of temperature of order $k_B T$ centred on $\varepsilon = \mu$. The effect of the finite temperatures is therefore to shift the few electrons in the shaded area with $\varepsilon < \varepsilon_F$ in Fig. 3.2(b) to the shaded area with $\varepsilon > \varepsilon_F$.

3.2.3 Heat capacity of the free electron gas

We can use Fig. 3.2(b) to obtain an estimate of the thermal energy and hence of the heat capacity of the electrons. If the shaded areas are approximated as triangles of height $\frac{1}{2}g(\varepsilon_F)$ and base $2k_B T$ then the implications of Fig. 3.2(b) are that approximately $\frac{1}{2}g(\varepsilon_F)k_B T$ electrons have their energies increased by about

$k_B T$, so that their thermal energy (difference in internal energy from the value at $T = 0$) is

$$E(T) - E(0) \sim \tfrac{1}{2}g(\varepsilon_F)(k_B T)^2.$$

Differentiating with respect to T gives the heat capacity at constant volume,

$$C_V \sim g(\varepsilon_F)k_B^2 T = \tfrac{3}{2}(N/\varepsilon_F)k_B^2 T = \tfrac{3}{2}Nk_B(T/T_F) \qquad (3.16)$$

where we have used the expression in brackets in Eq. (3.8).

To obtain the exact result it is necessary *first* to calculate the chemical potential at temperature T using the $T \neq 0$ generalization of Eq. (3.8), namely

$$N = \int_0^\infty n(\varepsilon, T)\, d\varepsilon \qquad (3.17)$$

(μ is the only unknown in this equation), and *secondly* to evaluate the energy

$$E(T) = \int_0^\infty \varepsilon n(\varepsilon, T)\, d\varepsilon \qquad (3.18)$$

This is a rather tedious mathematical process; for $T \ll T_F$, μ decreases very slightly from ε_F with increasing T (for most purposes the T dependence can be ignored) and C_V is given by

$$C_V = \frac{\pi^2}{3} g(\varepsilon_F)k_B^2 T = \frac{\pi^2}{2} Nk_B\left(\frac{T}{T_F}\right). \qquad (3.19)$$

This is the same as our estimate (3.16) except for the numerical coefficient.

The free electron theory was first introduced before quantum theory, and one of the problems encountered was that the heat capacity of the electrons according to classical equipartition theory was $\tfrac{3}{2}Nk_B$, which was not observed experimentally; from Eq. (3.19) we see that quantum theory solved this problem by reducing the expected heat capacity by a factor of order T/T_F. At room temperature the *lattice* heat capacity of most metals is close to its classical equipartition value ($3Nk_B$ for monovalent metals according to Eq. (2.43)) and therefore completely dominates the electronic contribution. At temperatures low compared to the Debye temperature Θ_D, however, the lattice contribution falls off rapidly with the T^3 dependence of Eq. (2.54) and the electronic contribution becomes important. The total heat capacity at low temperatures is therefore of the form

$$C_V = \gamma T + \beta T^3. \qquad (3.20)$$

The constants γ and β can be determined by plotting C_V/T as a function of T^2; this is done for potassium in Fig. 3.3. We see that the two contributions are comparable at temperatures of order 1 K. Note that the heat capacity of potassium chloride in Fig. 2.14 is also plotted in this way and, as expected, the

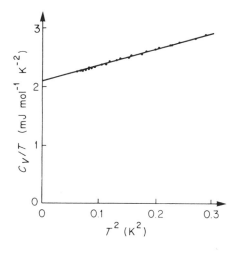

Fig. 3.3   Separation of the electronic and lattice heat capacities of potassium at low temperatures (Reproduced with permission from W. H. Lien and N. E. Phillips, *Phys. Rev.* **133**, A1370 (1964))

electronic contribution is absent in this material. The value of the **electronic specific heat constant** γ for potassium is given by the intercept on Fig. 3.3 as 2.08 mJ mol^{-1} K^{-2}; the predicted value obtained using Eq. (3.19) and the value of T_F for potassium of Eq. (3.13) is 1.67 mJ mol^{-1} K^{-2}. This 25 % discrepancy is not surprising in view of the very simple-minded approach adopted in the free electron model.

The discrepancy is often interpreted as arising because the electrons have an **effective mass** m^* which differs from their *bare* mass m. Replacing m by m^* in Eqs. (3.1) and (3.5) and carrying through the calculation as before, we find that ε_F, T_F and v_F are changed by a factor m/m^* from their free electron values whereas $g(\varepsilon_F)$ and hence C_V are changed by m^*/m; p_F and k_F are unchanged (the allowed **k** values are all unchanged). For potassium therefore $m^*/m = 1.25$; the corresponding values for magnesium and aluminium are 1.3 and 1.48, indicating an effective-mass correction of similar magnitude for all metals (note that for some metals, for example zinc and cadmium, $m^*/m < 1$).

A theoretical calculation of m^* must take into account the true distribution of positive charge density within the crystal and must include electron–electron interactions. The positive charge distribution affects m^* in two distinct ways: the periodic potential associated with the stationary ion cores at their crystalline sites causes a change which is discussed further in the following chapter; in addition the motion of an electron causes nearby ion cores to move and this **electron–phonon interaction** also contributes to the effective mass (this is probably the dominant effect in the alkali metals). That electron–electron interactions change the effective mass is evident if the conduction electrons are pictured as a fluid. Repulsion of the fluid by an electron moving through it causes the fluid in its path to move out of its way. The resulting **backflow** of fluid around the moving electron is similar to that of a real fluid around a solid object moving through it, for which hydrodynamic theory predicts an effective inertial mass for the object that is larger than the bare mass (the **hydrodynamic virtual mass**).

One problem with the effective-mass concept is that different effective masses are required to explain the departures of different properties of the conduction electrons of a particular metal from their free electron values. We will give a conclusive demonstration of this fact in section 4.4. Nevertheless the effective mass associated with any property provides a useful way of quantifying departures of that property from the free electron prediction and is a concept that we will use frequently.

3.2.4 Soft x-ray emission spectrum

The electronic heat capacity of metals depends only on the properties of electrons with energies close to the Fermi energy. The x-ray emission spectrum provides a method for studying conduction electrons of all energies. The spectrum is obtained when electrons are removed from the K and L shells of the ion cores by bombardment with high-energy electrons from an external source; conduction electrons fall into the vacant states and soft x-rays are emitted. The energy range of the x-rays should reflect that of the conduction electrons; we would expect the spectrum to be roughly similar in shape to the $n(\varepsilon, T)$ curve of Fig. 3.2(b); the shape will not be identical since the spectrum depends on an energy-dependent transition probability as well as on $n(\varepsilon, T)$. Inspection of the x-ray emisssion spectrum of sodium in Fig. 3.4 shows that this is the case.

Particularly noticeable on Fig. 3.4 is the sharp cut-off of the spectrum due to the sharp decrease of $n(\varepsilon, T)$ at the Fermi surface. The sharpness of the cut-off is consistent with the width $k_B T$ of the rounding of the Fermi distribution at room temperature, and we can draw a very important conclusion from this. It implies that the energy levels of the conduction electrons are well defined on an energy

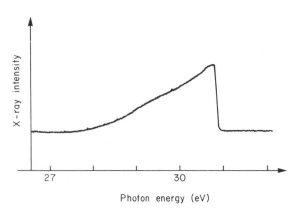

Fig. 3.4   Experimental L x-ray emission spectrum for sodium. (Reproduced with permission from H. W. B. Skinner, *Phil. Trans. R. Soc.* **239**, 95 (1940))

scale of order $k_B T$, and by using the energy–time uncertainty relation this enables us to put a *lower* limit on the lifetime τ of a conduction electron state near the Fermi surface. Denoting the uncertainty in energy of the state by $\Delta\varepsilon$ we have $\Delta\varepsilon < k_B T$ so that

$$\tau \sim \frac{\hbar}{\Delta\varepsilon} > \frac{\hbar}{k_B T} = 2 \times 10^{-14} \, s.$$

This is a long time, sufficient for an electron in a state on the Fermi surface to travel about 100 atomic spacings. One factor limiting the lifetime of the electron states is the electron–electron interaction, but the evidence from Fig. 3.4 is that the effects of this interaction are not sufficiently strong to prevent an independent particle model (such as the free electron model) from being a good approximation. Although it is difficult to locate with precision the onset of x-ray emission at the low-energy end of the spectrum, Fig. 3.4 indicates conduction electrons in an energy range greater than or equal to 2.5 eV. This is comparable with the free electron value of ε_F for sodium of 3.23 eV.

3.2.5   Metallic binding

The reduction in kinetic energy associated with the delocalization of the conduction electrons contributes to the binding energy of metals. To illustrate this we consider the one-dimensional free electron metal of Fig. 3.5(b) con-

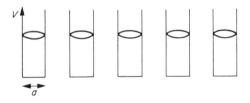

(a) A row of square-well 'atoms' and their ground-state wavefunctions

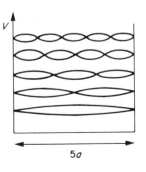

(b) The same row assembled into a 'crystal' showing the five lowest-energy wavefunctions

Fig. 3.5

structed by assembling the very special (and unrealistic) 'atoms' shown in Fig. 3.5(a). Each atom consists of one electron in an infinite square potential well of width a; the ground state energy of an electron in such an atom is $h^2/8ma^2$ (measured from the bottom of the well). The five lowest-energy wavefunctions of the one-dimensional free electron metal of length $5a$ formed from five such atoms are shown in Fig. 3.5(b) and we can see that the *highest* of these has the same wavelength (and hence the same energy) as the ground state of a single 'atom'. The mean energy of the five electrons is therefore lowered by forming the crystal even without allowing for electron spin, which permits two electrons in each energy level. This quantum mechanical effect, the reduction of kinetic energy by delocalizing electrons, is an important contribution to metallic binding; though in a real metal the changes in the electrostatic potential energy of the electrons and ions are of comparable importance.

3.3   TRANSPORT PROPERTIES OF THE CONDUCTION ELECTRONS

In the presence of an electric field or a temperature gradient the occupation probability of the electron states is no longer given by the Fermi distribution function, Eq. (3.14); instead the distribution is such as to give rise to the *transport* of electric charge and of heat respectively. The transport coefficients that describe these flows are the electrical and thermal conductivities, which we now proceed to calculate.

3.3.1   The equation of motion of the electrons

In the absence of collisions we take the electrons to obey the acceleration equation

$$m_e \, dv/dt = -eE - ev \times B \qquad (3.21)$$

where v is the electron velocity, E the electric field and B the magnetic field. This is just Newton's law for particles of mass m_e and charge $-e$. We use an effective mass m_e in the belief that by so doing we will take account of some of the factors ignored in our free electron model; we show in section 4.4 that this takes account for example of the interaction of the electrons with the periodic potential of the stationary ion cores. In section 4.4 we explain why the effective mass m_e that appears in Eq. (3.21) is different from the heat capacity effective mass m^* discussed in section 3.2.3.

The use of the classical equation of motion of a particle to describe the behaviour of electrons in plane wave states (Eq. (3.3)), which extend throughout the crystal, requires justification. A particle-like entity can be obtained by superposing the plane wave states to form a wavepacket; it is possible to show that Eq. (3.21) is the equation of motion of the wavepacket. The velocity of the wavepacket is the group velocity of the waves. Thus

$$\mathbf{v} = \frac{d\omega}{dk} = \frac{1}{\hbar}\frac{d\varepsilon}{dk} = \frac{\hbar\mathbf{k}}{m_e} = \frac{\mathbf{p}}{m_e} \qquad (3.22)$$

where we have used Eq. (3.5) with m replaced by m_e; Eq. (3.22) gives the usual relation between velocity and momentum $\mathbf{p}$ for a particle of mass m_e.

To behave like a classical particle our wavepacket ought to have reasonably well defined position and momentum. To obtain a wavepacket localized in position to about 10 atomic spacings requires the use of a wavenumber range of order $k_F/10$ in the plane wave superposition; this follows because k_F is of the order of an inverse atomic spacing. The uncertainty in momentum for such a wavepacket is therefore of order $p_F/10$. For Eq. (3.21) to be valid we might expect that the wavepacket would have to be smaller than both the length scale associated with the variation of $\mathbf{E}$ and $\mathbf{B}$ (the wavelength if we are considering electromagnetic waves) and the mean free path between collisions for the electrons. A lower limit on the size of the wavepacket is provided by the requirement that it should be much larger than an atomic spacing in order that the interaction of the electron with the ion cores can be described by an effective mass.

In the absence of an applied magnetic field, Eq. (3.21) predicts that a dc electric field will cause a constant acceleration of the electrons, giving a steadily increasing electric current. This does not happen in practice because the electrons suffer collisions with thermal vibrations of the ion cores† and with imperfections in the crystal such as impurity atoms. We allow for collisions by modifying Eq. (3.21) to

$$m_e\left(\frac{d\mathbf{v}}{dt} + \frac{\mathbf{v}}{\tau}\right) = -e\mathbf{E} - e\mathbf{v} \times \mathbf{B}. \qquad (3.23)$$

The effect of the additional term is to cause $\mathbf{v}$ to decay exponentially to zero with a time constant τ on removal of the applied fields; $\mathbf{v}$ in Eq. (3.23) must therefore be interpreted as the **drift velocity** of the electrons, that is the additional velocity associated with the departure from the thermal equilibrium state given by the Fermi distribution function. Eq. (3.23) then says that the electron distribution relaxes to the Fermi distribution when the fields are removed. If we suppose that an electron loses all its drift velocity in each collision then τ is the mean time between collisions; in any case it is a quantity of the same order of magnitude.

3.3.2 The electrical conductivity

In the presence of a dc electric field only, Eq. (3.23) has the *steady state* solution

$$\mathbf{v} = -\frac{e\tau}{m_e}\mathbf{E}. \qquad (3.24)$$

---

† The effect of stationary ion cores can be completely included in the effective mass (section 4.4).

The constant of proportionality between $|\mathbf{v}|$ and $|\mathbf{E}|$ is known as the **mobility** μ_e of the electrons. Thus

$$\mu_e = +\frac{e\tau}{m_e}. \tag{3.25}$$

The electric current density $\mathbf{j}$ is $n(-e)\mathbf{v}$ where $n = N/V$ is the electron density. Hence

$$\mathbf{j} = \frac{ne^2\tau}{m_e}\mathbf{E} = \sigma\mathbf{E} \tag{3.26}$$

which is Ohm's law with an electrical conductivity

$$\sigma = ne^2\tau/m_e = ne\mu_e. \tag{3.27}$$

In a perfect crystal of a pure metal the dominant collisions of the conduction electrons are with thermally excited lattice vibrations;† the collision can be pictured as the scattering of an electron by a phonon, a picture that is pursued further in section 3.3.4. This electron–phonon scattering gives a temperature-dependent collision time $\tau_{ph}(T)$ which tends to infinity as $T \to 0$. In a real metal the electrons also collide with impurity atoms, vacancies (missing atoms) and other structural defects; this results in a finite scattering time τ_0 even at $T = 0$. At finite temperature the collision rate for electrons in a slightly imperfect crystal is obtained to a good approximation by addition the *zero-temperature* rate $1/\tau_0$ for scattering by imperfections to the phonon scattering rate $1/\tau_{ph}(T)$ for a *perfect* crystal. Thus

$$\frac{1}{\tau} = \frac{1}{\tau_{ph}(T)} + \frac{1}{\tau_0}. \tag{3.28}$$

This assumption is valid if the two scattering mechanisms operate independently, that is if the scattering by imperfections is temperature-independent and there are insufficient imperfections to affect significantly the phonon scattering. Eq. (3.28) implies that the electrical resistivity ρ $(= 1/\sigma)$ can be written as the sum of two terms,

$$\rho = \frac{m_e}{ne^2\tau} = \frac{m_e}{ne^2\tau_{ph}(T)} + \frac{m_e}{ne^2\tau_0} = \rho_i(T) + \rho_0. \tag{3.29}$$

Eq. (3.29) explains **Mattheisen's rule**, illustrated for sodium in Fig. 3.6, that the resistivity versus temperature graphs for *different* specimens of the same material differ only by a displacement; the displacement is associated with the variation

---

† Since a free electron metal does not possess a lattice it may seem absurd to refer to the lattice vibrations within it. In the following chapter we explain why conduction electrons can continue to behave like free electrons in the presence of the periodic potential of the positive ions. The electrons are scattered by any departure from the perfect periodicity such as that caused by lattice vibrations.

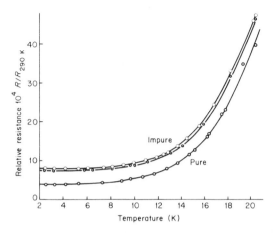

Fig. 3.6   Resistivity–temperature curves for sodium specimens of differing purity. (Reproduced with permission from D. K. C. Macdonald and K. Mendelssohn, *Proc. R. Soc.* A **202**, 103 (1950))

in ρ_0 due to different imperfection densities. The temperature-dependent part of the resistivity $\rho_i(T)$ is known as the **ideal resistivity** and ρ_0 is called the **residual resistivity**; all but the purest specimens effectively attain their residual resistivity on cooling to the boiling point of liquid helium, 4.2 K. The ratio of room-temperature resistivity to the residual resistivity is known as the **residual resistance ratio** and it can be as high as 10^6 for highly purified, annealed single crystals.

The room-temperature conductivity of sodium is $2.0 \times 10^7 \, \Omega^{-1} \, \mathrm{m}^{-1}$ so that the residual conductivity of the purest specimen of Fig. 3.6 is $5.3 \times 10^{10} \, \Omega^{-1} \, \mathrm{m}^{-1}$; we can deduce estimates of the collision time τ from these data. Taking $n = 2.7 \times 10^{28} \, \mathrm{m}^{-3}$, $m_\mathrm{e} = m$ and using Eq. (3.27) gives

$$\tau = m\sigma/ne^2 \sim 2.6 \times 10^{-14} \, \mathrm{s} \qquad \text{at room temperature}$$
$$\sim 7.0 \times 10^{-11} \, \mathrm{s} \qquad \text{at } T = 0. \tag{3.30}$$

The free electron Fermi velocity of sodium is $1.1 \times 10^6 \, \mathrm{m \, s^{-1}}$ so that the corresponding electron mean free paths are 29 nm at room temperature and 77 μm at $T = 0$; these are much longer than the interatomic spacing, confirming that the electrons do not collide with the atoms themselves. We postpone until section 3.3.4 a discussion of the temperature dependence of the ideal resistivity.

3.3.3   The thermal conductivity

Because of heat transport by the conduction electrons, the thermal conductivity of metals is normally much greater than that of non-metals (but see section

2.8.4).† Electrons coming from a hotter region of the metal carry more thermal energy than those from a cooler region, resulting in a net flow of heat. This is also the mechanism of heat conduction in a gas and, as in discussing the conduction by phonons in section 2.8, we will use the elementary kinetic theory result (Eq. (2.75)) for the thermal conductivity. Thus

$$K = \tfrac{1}{3}C_V v_F l. \tag{3.31}$$

where C_V is the specific heat *per unit volume* and l is the mean free path. We have taken v_F as the mean speed of the electrons responsible for thermal conductivity since only electron states within about $k_B T$ of ε_F change their occupation number as the temperature varies. The mean free path of these electrons is $v_F \tau$ and therefore using $\varepsilon_F = \tfrac{1}{2}m_e v_F^2$ and Eq. (3.19) for C_V we find

$$K = \tfrac{1}{3}C_V v_F^2 \tau = \frac{1}{3}\frac{\pi^2}{2}\frac{N}{V}k_B\left(\frac{T}{T_F}\right)\frac{2}{m_e}\varepsilon_F\tau = \frac{\pi^2 n k_B^2 T\tau}{3m_e}. \tag{3.32}$$

It is interesting to note that the final result in Eq. (3.32) is also true in order of magnitude for a *classical* electron gas: the specific heat is larger by a factor of order (T_F/T) but the square of the thermal velocity is smaller by the same factor.

3.3.4   The Wiedemann–Franz law and the temperature dependence of the electrical and thermal conductivities

The combination $n\tau/m_e$ appears in both Eqs. (3.27) and (3.32) so that by dividing them we find that the ratio of electrical and thermal conductivities is independent of the electron gas parameters. We obtain

$$\frac{K}{\sigma T} = \frac{\pi^2}{3}\left(\frac{k_B}{e}\right)^2 = 2.45 \times 10^{-8}\ \text{W}\,\Omega\,\text{K}^{-2}, \tag{3.33}$$

which is known as the **Wiedemann–Franz law**. The ratio $K/\sigma T$ is called the **Lorenz number** and is denoted by L. For copper at 0°C, $L = 2.23 \times 10^{-8}\ \text{W}\,\Omega\,\text{K}^{-2}$, so that the Wiedemann–Franz law works reasonably well, as indeed it does for most metals at temperatures of this order. Fig. 3.7 shows a plot of the Lorenz number for sodium as a function of temperature from 0 to 100 K; we see that L is significantly below the predicted value over most of this temperature range although it does tend towards the predicted value in the $T \to 0$ limit where collisions with impurities are the dominant scattering mechanism.

The breakdown of the Wiedemann–Franz law results from the failure of our assumption that the collision times limiting the flow of electric and heat currents are the same. To explain this we must investigate the collision processes in more

---

† The thermal conductivity due to the phonons in metals is in general less than that in non-metals because the phonons are scattered by the conduction electrons.

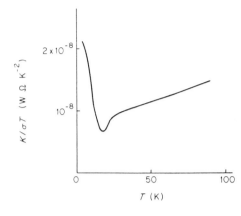

Fig. 3.7   Lorenz number $L = K/\sigma T$ for sodium at low temperatures. (Reproduced with permission from R. Berman and D. K. C. Macdonald, *Proc. R. Soc.* A **209**, 368 (1951))

detail, and we begin by considering the change in occupation of electron states in **k**-space associated with the current-carrying states.

From Eqs. (3.22) and (3.24) the drift velocity **v** associated with an *electric* current corresponds to a change δ**k** in the wavevector of each electron, where

$$\delta \mathbf{k} = \frac{m_e}{\hbar} \mathbf{v} = -\frac{e\tau}{\hbar} \mathbf{E}. \tag{3.34}$$

The electric-current-carrying state therefore corresponds to a shift by this amount of the whole electron distribution in **k**-space, that is of the whole Fermi sphere, as shown in Fig. 3.8; the blurring of the Fermi surface due to a finite temperature makes no essential difference to the rearrangement of the electrons in **k**-space associated with a finite electric current. The displacement shown on Fig. 3.8(*b*) corresponds to an electric current flowing in the $-x$ direction since there are more electrons with momentum in the $+x$ direction than the $-x$ direction. For a current density of 10^7 A m^{-2} (about the largest normally used), the electron drift velocity is

$$v = \frac{j}{ne} \sim \frac{10^7}{10^{28} \times 10^{-19}} \text{ m s}^{-1} \sim 10 \text{ mm s}^{-1}$$

for a typical electron density in a metal; this is about $10^{-8} v_F$, so the displacement of the Fermi sphere in Fig. 3.8(*b*) is actually minute. For the current to decay to zero the Fermi sphere must relax to the unshifted state of Fig. 3.8(*a*), and the type of collision that is effective in achieving this is one in which an electron is removed from the right side of the Fermi sphere and added to the left side; typical transitions are shown in Fig. 3.8(*b*) and it can be seen that they involved a change in the **k** vector of the electron of magnitude comparable to the diameter of the Fermi sphere, that is a change in momentum of order $2p_F$.

Fig. 3.8(*c*) shows the electron distribution in the presence of a temperature gradient. Because of the finite temperature there will be some vacant states

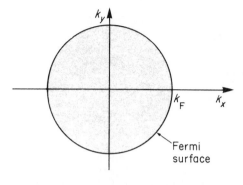

(a) Equilibrium Fermi sphere at $T = 0$.

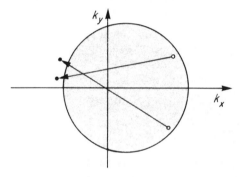

(b) Electric-current carrying state with typical relaxation processes

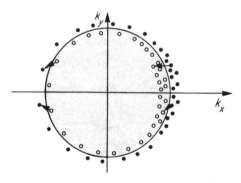

(c) Heat-current-carrying state in a temperature gradient with small-momentum-change relaxation processes

Fig. 3.8

(open circles) below the Fermi surface and some occupied states (full circles) above. If the specimen is hotter at the left-hand end, electrons moving from the left (i.e. those with $k_x > 0$) will have a distribution corresponding to a higher temperature than those coming from the right (i.e. those with $k_x < 0$). This is the situation shown in Fig. 3.8(c), where the Fermi surface is more blurred for $k_x > 0$ than for $k_x < 0$. Note that the blurring shown is much exaggerated; the total blurring at room temperature is of order 1 % of k_F, and the difference in blurring

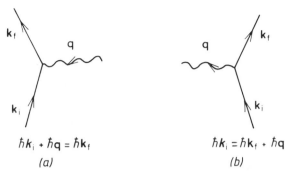

$\hbar k_i + \hbar q = \hbar k_f$ $\hbar k_i = \hbar k_f + \hbar q$

(a)   (b)

Fig. 3.9   Scattering of an electron by (a) phonon absorption and (b) phonon emission. Straight lines indicate electrons, wavy lines phonons. The initial and final wavenumbers of the electron are k_i and k_f respectively; q is the phonon wavenumber. In each case the equation for conservation of momentum is given

on the two sides due to a temperature gradient corresponds to the temperature difference in a mean free path (typically 100 nm at room temperature). Effective scattering processes in limiting thermal conductivity are ones in which the blurring of the Fermi surface is evened out; although this can be achieved in scattering events involving a large momentum change such as those shown in Fig. 3.8(b), processes involving a small momentum change as indicated in Fig. 3.8(c) are also effective. Since more scattering processes are effective in limiting thermal conduction the relaxation time for thermal conductivity is shorter and the Lorenz number falls below the Wiedemann–Franz value.

To investigate this further and to explain the temperature dependence of the electrical and thermal conductivities we must investigate the scattering of electrons by lattice vibrations; this can be pictured as collisions between electrons and phonons with conservation of energy and momentum.† The two most important processes are the absorption and emission of a phonon by an electron as shown in Fig. 3.9; the changes in momentum and energy of the electron are equal to the momentum and energy of the absorbed or emitted phonon. The maximum energy change of an electron in such a process is therefore the maximum phonon energy, which is of order $k_B \Theta_D$ where Θ_D is the Debye temperature and is of the order of room temperature for a typical metal (Table 2.1). The maximum energy change of an electron is therefore much less than the energy $k_B T_F$ of an electron on the Fermi surface; since a colliding electron must scatter into a vacant state and vacant states occur only close to the Fermi surface, this has the important consequence that only electrons close to the surface can be scattered by phonons.

---

† This statement can be proved by following the same procedure as that used in section 12.4 to establish conservation of energy and momentum in the scattering of neutrons by phonons. The subtleties associated with the distinction between true momentum and crystal momentum need not worry us here.

At high temperatures ($T \gg \Theta_D$) a typical phonon has energy $k_B\Theta_D$ and thus a wavelength of the order of an interatomic spacing; since electrons on the Fermi surface also have wavelengths of this order, we see that typical phonons have sufficient momentum to cause the large-momentum-transfer collisions of Fig. 3.8(b) that are required to produce electrical resistance. Consequently the relaxation times for electrical and thermal resistance are similar and the Wiedemann–Franz law is well obeyed. The actual electron mean free path l_{ph} is inversely proportional to the phonon number. Since the lattice vibration energy at high temperature is $3Nk_BT$ and the phonon energy is constant, the phonon number is proportional to T. The electron scattering time τ_{ph} is therefore inversely proportional to T, and inserting this into Eq. (3.29) for the ideal resistivity we find

$$\rho_1(T) = \frac{m_e}{ne^2\tau_{ph}} = \frac{m_e v_F}{ne^2 l_{ph}} \propto T \qquad (T \gg \Theta_D)$$

(Fig. 3.10(a)). Similarly, inserting $\tau_{ph} \propto T^{-1}$ into Eq. (3.32) predicts that the thermal conductivity of metals is independent of temperature at high temperatures (Fig. 3.10(b)).

At low temperatures ($T \ll \Theta_D$) the average phonon energy is of order $k_B T$, and since the lattice vibration energy is proportional to T^4 (Eq. (2.47)) the phonon number is proportional to T^3. Phonons of energy $k_B T$ have just the energy required for the scattering events shown in Fig. 3.8(c) that are effective in producing thermal resistance; the corresponding mean free path l_{th} and the scattering time τ_{th} that determine the thermal conductivity are therefore just inversely proportional to the phonon number, so that

$$\tau_{th} \propto T^{-3}.$$

Thus from Eq. (3.32) we expect that the phonon-limited thermal conductivity will exhibit the temperature dependence $K \propto T^{-2}$ for $T \ll \Theta_D$ (Fig. 3.10(b)).

In the case of electrical resistance, however, it is the momentum of a typical phonon that is important. Since a phonon of energy $k_B\Theta_D$ has a momentum of order p_F, a phonon of energy $k_B T$ has momentum of order $(T/\Theta_D)p_F$ and for $T \ll \Theta_D$ this is too small to cause the large-momentum-change collisions of Fig. 3.8(b) that are effective in producing electrical resistance; large momentum changes of the electrons can only occur by the addition of many small changes. Because the initial and final electron states must be close to the Fermi surface, a small momentum change implies scattering through a small angle θ as shown in Fig. 3.11. The effectiveness of a collision in producing electrical resistance may be measured by the loss in momentum $\hbar\delta k$ of the electron along its original direction of motion, and from Fig. 3.11 this is $\hbar q^2/2k_F$ for small θ which is therefore proportional to T^2 for typical phonons at low temperature. Hence the scattering rate $1/\tau_{el}$ to be used in calculating the electrical resistivity must be

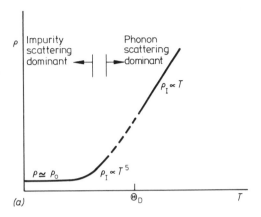

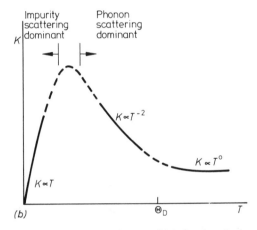

Fig. 3.10 Schematic temperature dependences of (a) the electrical resistivity and (b) the thermal conductivity of a metal

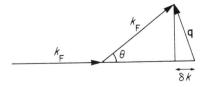

Fig. 3.11 Change in momentum of an electron when it is scattered through an angle θ by absorption of a phonon of wavevector **q**. Note that the magnitude of the momentum of the electron does not change, as it remains in a state close to the Fermi surface

weighted by an *effectiveness* factor of order $(T/\Theta_D)^2$ so that

$$\frac{1}{\tau_{el}} \sim \left(\frac{T}{\Theta_D}\right)^2 \frac{1}{\tau_{th}} \propto T^5$$

and, from Eq. (3.29),

$$\rho_i(T) \propto T^5. \tag{3.35}$$

In this temperature region therefore the Wiedemann–Franz law fails.

At the lowest temperatures electron–impurity collisions are dominant.† These collisions are elastic and thus unable to produce collisions like those shown in Fig. 3.8(c), but they are capable of producing large-momentum-change collisions like those in Fig. 3.8(b); the effective scattering times for electrical and thermal conductivity are therefore identical and the Wiedemann–Franz law is again obeyed. The electron mean free path due to electron–impurity scattering should be independent of temperature, leading to a temperature-independent electrical conductivity and a thermal conductivity proportional to T. The temperature dependences of the electrical resistivity and the thermal conductivity that we have deduced are summarized in Figs. 3.10(a) and 3.10(b); many metallic elements of high purity follow these dependences reasonably well.‡

The predicted T^5 dependence of the ideal resistivity at low temperatures (Eq. (3.35)) is not observed in some metals. This may be because of **phonon drag**; electron–phonon collisions cause the phonons to be dragged along with a drift velocity approaching that of the electrons and they are thus less effective in reducing the drift momentum of the electrons. Alternatively the T^5 dependence may be masked by the contribution to the resistivity of electron–electron scattering. Since momentum is conserved in the collision of two electrons it would appear at first as though such collisions do not lead to electrical resistivity; we advanced a similar argument in section 2.8.1 to suggest that phonon–phonon collisions could not limit the thermal conductivity of phonons. We saw there that the existence of a periodic lattice means that the momentum associated with a phonon cannot be unambiguously specified, and this permitted the existence of Umklapp scattering processes with an associated thermal resistance. When the effect of the periodic lattice potential is taken into account the electron momentum becomes ambiguous in the same way (section 4.3.3). The resulting electron–electron Umklapp scattering processes produce a small

---

† In very pure metals collisions of the electrons with the sample boundaries can become important and the effective electrical conductivity then depends on the shape and size of the sample; this is another example of a size effect (section 2.8.4). Size effects are most easily seen in thin films or fine wires where at least one of the sample dimensions is small.

‡ Interesting departures are sometimes observed when a non-magnetic metal contains a small amount of magnetic impurity, for example Fe impurity in Cu. The resistivity goes through a minimum with decreasing temperature at low temperatures. This is the **Kondo effect** (J. Kondo, *Solid State Phys.* **23**, 183 (1969)).

contribution to the electrical resistivity which has the same T^2 temperature dependence as the electron–electron scattering rate (section 13.5.4).†

3.3.5   The Hall effect

To this point we have been really quite successful in explaining the properties of metals using the free electron theory. This run of success comes to an abrupt halt in this section with the failure of the theory to obtain even the correct sign for the Hall effect in *some* metals. When a metal is placed in a magnetic field **B** and a current density **j** is passed through it, a transverse electric field E_H is set up given by

$$E_H = R_H B \times j. \tag{3.36}$$

This is the **Hall effect** and R_H is known as the Hall coefficient.

The geometry of the experiment is shown in Fig. 3.12. The origin of the effect is the Lorentz force $-e\mathbf{v} \times \mathbf{B}$ on the conduction electrons in the magnetic field. Fig. 3.12 shows the direction of the drift velocity **v** of electrons corresponding to a current **j** in the x direction. The Lorentz force tends to deflect the electrons downwards and this results in the rapid build up of a negative charge density on the lower surface of the metal. The consequent electric field E_H in the $-y$ direction causes the current to continue to flow in the x direction, as it must for a long rod with electrical connections at the ends.

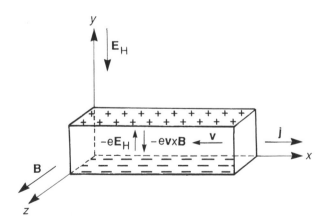

Fig. 3.12   Geometry of the Hall effect. The Lorentz force $-e\mathbf{v} \times \mathbf{B}$ on the electrons is just balanced by the force $-eE_H$ due to the Hall field E_H

---

† A recent review of the electrical and thermal conductivities of metals at low temperatures has been given by R. J. M. van Vacht *et al.*, *Rep. Prog. Phys.* **38**, 853 (1985).

Thus, in the steady state with $v_y = 0$, the x and y components of the equation of motion (3.23) in this geometry are

$$m_e v_x/\tau = -eE_x$$
$$0 = -e(E_y - v_x B). \tag{3.37}$$

The first of these equations just predicts that the electrical conductivity σ is unaffected by the presence of the magnetic field† (cf. Eq. (3.24)). From the second we obtain

$$E_y = v_x B = j_x B/(-ne), \tag{3.38}$$

so that, by comparison with Eq. (3.36),

$$R_H = -1/ne. \tag{3.39}$$

Thus the Hall coefficient should be negative and give a direct measurement of the free electron concentration. By combining Eqs. (3.27) and (3.39) we find

$$\mu = |R_H|\sigma \tag{3.40}$$

so that it should be possible to determine the electron mobility μ from measured values of R_H and σ.

TABLE 3.1   Hall coefficient of various metals. According to Eq. (3.39), $-1/(R_H Ne)$ should equal the number of conduction electrons per atom. A negative sign for this quantity indicates a positive value for R_H and thus that the charge carriers are positively charged particles!

Metal	Group	$-1/(R_H Ne)$
Na	I	+0.9
K		+1.1
Cu	IB	+1.3
Au		+1.5
Be	II	-0.2
Mg		+1.5
Cd	IIB	-2.2
Al	III	+3.5

(Data from the *American Institute of Physics Handbook*, 3rd edn, McGraw-Hill, New York (1972))

† In fact the electrical resistivity of metals often depends weakly on the magnetic field. The change in resistivity is called **magnetoresistivity** and can arise because of the failure in an anisotropic crystal of our assumption of a single relaxation time for all the electrons.

If N is the number of *atoms* per unit volume then the quantity $-1/(R_H Ne)$ should equal n/N and therefore give an estimate of the number of conduction electrons per atom; values of this quantity for various metals are shown in Table 3.1. The measured values for the group I and III elements appear reasonable, as does the value for the group II element magnesium. The values for the group II elements beryllium and cadmium, however, are negative, implying that *positively* charged particles are responsible for the conduction in these metals! To account for this surprising result it is necessary to consider the effect of the periodic lattice potential on the conduction electron states, and this is the subject of Chapter 4.

PROBLEMS 3

3.1 The work function of a metal is the minimum energy required to remove an electron from the metal and is typically 3 eV. Deduce a value for the 'penetration length' of the electron wavefunction outside the metal for electrons of the Fermi energy.

3.2 Metallic lithium has a body-centred cubic structure with unit cell side 3.5 Å. Calculate using the free electron model the width of the K emission band of soft x-rays from lithium. How would you expect the width of the emission to depend on temperature?

3.3 show that the kinetic energy of a free electron gas at absolute zero is

$$E = \tfrac{3}{5} N \varepsilon_F$$

where ε_F is the Fermi energy. Derive expressions for the pressure $p = -\partial E/\partial V$ and the bulk modulus $B = -V(\partial p/\partial V)$.

Estimate the contribution of the conduction electrons to B for potassium and compare your answer to the experimentally measured bulk modulus $0.37 \times 10^{10}\,\text{N m}^{-2}$.

3.4 Add the contribution of the conduction electrons to Eq. (2.59) to obtain the Helmholtz free energy of a metal. Hence generalize the calculation of section 2.7.1 to obtain a value for the thermal expansion coefficient of a metal.

3.5 Prove that the loss of momentum of the electron along its original direction of motion in the collision depicted in Fig. 3.11 is $\hbar q^2/2k_F$.

3.6 Estimate the Fermi temperatures of:
(a) liquid ^{3}He (density 81 kg m^{-3}), and
(b) the neutrons in a neutron star (density 10^{17} kg m^{-3}).

Everything has its beauty, but not everyone sees it.—*Confucius*

CHAPTER

The effect of the periodic lattice potential—energy bands

4.1 NEARLY FREE ELECTRON THEORY

Despite the success of the free electron theory in explaining many of the properties of metals, it does not explain why some materials are metals and others insulators. Nor does it explain why some metals have positive Hall coefficients, indicating the presence of mobile positively charged particles within them. We now attempt to improve the free electron model by taking into account the fact that the positive ions do not produce a uniform attractive potential but one with strong negative peaks at the lattice sites. An example of such a potential for a one-dimensional chain of identical equally spaced atoms is shown in Fig. 4.1(*a*); the potential is periodic with a period equal to the lattice spacing a. Note that we have chosen the origin of the x axis to be centred on one of the atoms.

We shall estimate the correction to the free electron energy by using the standard formula of first-order perturbation theory,

$$\Delta\varepsilon = \frac{\int \psi^* V \psi \, dx}{\int \psi^* \psi \, dx} \tag{4.1}$$

where V is the difference between the true potential and the constant potential assumed in the free electron calculation and ψ is the *unperturbed* wavefunction.

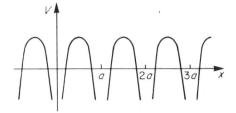

(a) One-dimensional periodic potential associated with a chain of identical atoms

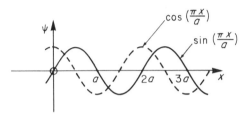

(b) Standing waves $\sin(\pi x/a)$ and $\cos(\pi x/a)$, with nodes and antinodes, respectively, at the lattice sites

Fig. 4.1

We are regarding the lattice potential as a small perturbation on the free electron potential so that this approach is known as the **nearly free electron theory**. Inspection of Fig. 4.1(a) suggests that the perturbation is really quite large so that at best we would expect the results of our perturbation theory calculation to be only qualitatively correct. Our confidence in the answers will be increased however in section 4.3 where we obtain the same qualitative results by adopting the extreme opposite viewpoint that the unperturbed state is one in which electrons are tightly bound to atoms by atomic potentials and the change in potential associated with the proximity of neighbouring atoms is the perturbation.

If we take the free electron potential to be the mean value of the true potential then the perturbation V for our one-dimensional example can be written as a Fourier series in the form

$$V = - \sum_{n=1}^{\infty} V_n \cos \left(\frac{2\pi n x}{a} \right) \qquad (4.2)$$

where we expect the V_n to be positive numbers for a potential with strong negative peaks at the lattice sites as in Fig. 4.1(a). If we use the potential of Eq. (4.2) and a running wave $\psi = e^{ikx}$ for the unperturbed wavefunction then, since $|\psi|^2 = 1$, Eq. (4.1) gives the non-informative answer: $\Delta \varepsilon = 0$ for all values of k! We must however be more careful than this since the states e^{ikx} and e^{-ikx} are degenerate. When doing first-order perturbation theory with two degenerate unperturbed states ψ_1 and ψ_2 the appropriate wavefunctions to insert in Eq.

(4.1) are the two *orthogonal* linear combinations, ϕ_1 and ϕ_2, of ψ_1 and ψ_2 that satisfy

$$\int \phi_1^* V \phi_2 \, dx = 0. \tag{4.3}$$

The only orthogonal linear combinations of e^{ikx} and e^{-ikx} that satisfy this condition for *all* values of k are $\sin(kx)$ and $\cos(kx)$ (see problem 4.1), and inserting these in Eq. (4.1) we find that $\Delta\varepsilon$ is now finite for certain special k values for which the periodic change density associated with the wavefunction is in synchronism with the periodicity of the lattice (Fig. 4.1(b)).

Thus for $\psi = \sin(kx)$ we calculate the perturbation as

$$\Delta\varepsilon = -\frac{\sum_{n=1}^{\infty} \int dx \, \sin^2(kx) V_n \cos(2\pi nx/a)}{\int dx \, \sin^2(kx)}$$

$$= -\frac{\sum_{n=1}^{\infty} \int dx \, [1 - \cos(2kx)] V_n \cos(2\pi nx/a)}{\int dx \, [1 - \cos(2kx)]}.$$

The integrand in the numerator oscillates about zero and the integral over all space vanishes unless the periodicity of $\cos(2kx)$ coincides with that of one of the Fourier coefficients $\cos(2\pi nx/a)$; thus $\Delta\varepsilon$ is only non-zero if $k = n\pi/a$, in which case

$$\Delta\varepsilon = \frac{\int dx \, V_n \cos^2(2\pi nx/a)}{\int dx} = \tfrac{1}{2} V_n, \tag{4.4}$$

all other terms in the series having integrated to zero. Similarly for $\psi = \cos(kx)$ and $k = n\pi/a$ we find

$$\Delta\varepsilon = -\tfrac{1}{2} V_n. \tag{4.5}$$

The physical reason for these results, for the case $n = 1$, becomes obvious on inspection of Fig. 4.1(b): $\psi = \sin(\pi x/a)$ has antinodes where the potential is repulsive, so its energy is raised; $\psi = \cos(\pi x/a)$ has antinodes where the potential is attractive, so its energy is lowered.

In Fig. 4.2 we show these perturbed energies in relation to the free electron parabola $\varepsilon(k) = \hbar^2 k^2/2m$. Since the stationary states for $k = n\pi/a$ are *standing* waves it follows that the group velocity and hence $d\varepsilon/dk$ must be zero at these points. To calculate other points on the $\varepsilon(k)$ curve requires a different approach (see problem 4.3), since the first-order perturbation result is zero. But, knowing ε and $d\varepsilon/dk$ at $k = n\pi/a$, we can guess that the form of the full $\varepsilon(k)$ curve is as indicated by the full curves on Fig. 4.2; this guess is confirmed by the calculation. The lattice potential thus has the dramatic effect on the $\varepsilon(k)$ curve of producing regions of energy in which there are no electron states; these regions, known as **band gaps** or **energy gaps**, are indicated in Fig. 4.2. The regions of ε in which states do exist are known as **energy bands** and these are also indicated. The

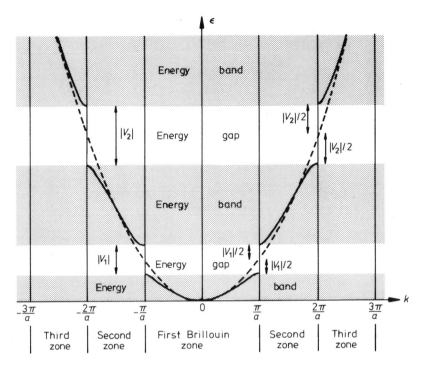

Fig. 4.2   The full curves show the effect of a periodic lattice potential on the parabolic free electron dispersion relation (broken curve). The bands of allowed energy levels are indicated by shading. The departures of the energy from the free electron values on the Brillouin zone boundaries are given by Eqs. (4.4) and (4.5)

grouping of the electron energies into bands has a very important influence on the behaviour of electrons in solids, as will become apparent to the reader in the remainder of this book.

We also see from Fig. 4.2 that the k axis is naturally split into a number of regions by the effect of the periodic lattice potential on the electron states; these regions are known as **Brillouin zones**. Thus, in our one-dimensional example the region $|k| < \pi/a$ (see Fig. 4.2) is the first Brillouin zone and states with k values in this range lie in the lowest energy band. The two regions with $\pi/a < |k| < 2\pi/a$ are the second Brillouin zone and states in this zone lie in the next-to-lowest energy band.

The alert reader may well have noticed that the boundaries of the first Brillouin zone at which the first energy gap appears and at which the electron group velocity vanishes occur at precisely the same k values, $k = \pm\pi/a$, at which the group velocity, $d\omega/dk$, vanishes (see Fig. 2.4) for the lattice vibration waves of the one-dimensional chain. As we pointed out in Chapter 2, wave motions with these k values satisfy the Bragg law (Eq. (1.3)) for diffraction by the one-dimensional lattice. Inspection of Eq. (2.12) shows that the Bragg diffraction condition in fact generates all the k values at which the energy gaps in the $\varepsilon(k)$ curve appear. The interference of an incident running wave with the resulting diffracted wave leads to the creation of standing waves at these k values and thus to the vanishing of the group velocity. In Chapter 11 we will introduce a beautiful formalism which allows the generalization of these ideas to crystals of arbitrary structure.

4.2 CLASSIFICATION OF CRYSTALLINE SOLIDS INTO METALS, INSULATORS AND SEMICONDUCTORS

We now consider how the occupation of the electron states in k-space is affected by the changes in the $\varepsilon(k)$ relation discussed in the previous section. If periodic boundary conditions are applied, the allowed k values for electron states in the one-dimensional crystal are exactly the same as those for lattice vibration waves (Eq. (2.11)). Thus $k = 2\pi p/L$, where p is an integer, $L = Na$ is the length of the chain and N is the number of *atoms* in the chain. As in the lattice vibration problem, therefore, there are N allowed values of k in the range $-\pi/a < k \leqslant \pi/a$ so that the first Brillouin zone and hence the lowest energy band of Fig. 4.2 contain exactly N k-states and can accommodate $2N$ electrons.

For monovalent atoms the electronic ground state corresponds to the states in the lowest energy band being filled up to the Fermi energy ε_F as indicated in Fig. 4.3(a). There are enough electrons exactly to half-fill the states in this band; that is the first Brillouin zone is half-filled by one electron per atom. Because there are vacant states immediately adjacent in energy to the occupied states it is possible to construct an electric-current-carrying state by shifting the whole electron distribution in k-space as indicated on Fig. 4.3(a), essentially as for free electrons (cf. Fig. 3.8(b)); we therefore expect metallic behaviour in this case. Indeed, since the $\varepsilon(k)$ relation near the Fermi energy is similar to that for free electrons, we might expect the free electron theory to work well for one-dimensional monovalent metals.

If on the other hand the atoms are divalent, then we have enough electrons to fill the first Brillouin zone exactly. Because of the energy gaps at $k = \pm \pi/a$, a current-carrying state can only be produced by expending a finite amount of energy V_1 for each electron shifted to k values greater than π/a (Fig. 4.3(b)). Since this energy is not available from a dc electric field such a material would be an insulator at absolute zero. At a finite temperature, if V_1 is sufficiently small, we would expect some electrons to be thermally excited into the second Brillouin zone; we will see in Chapter 5 that this leads to semiconducting behaviour. For larger V_1 the material would continue to act as an insulator at finite temperatures.

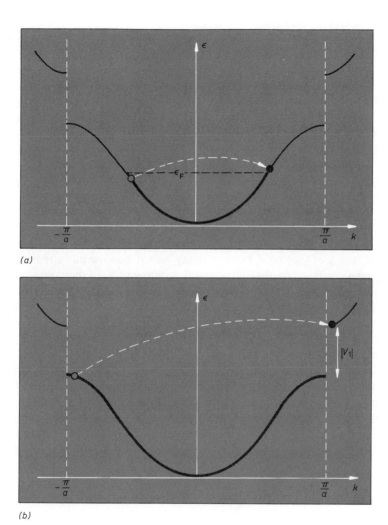

(a)

(b)

Fig. 4.3 (a) The occupied states for a one-dimensional chain of monovalent atoms are indicated by thickening of the $\varepsilon(k)$ curve. Transfer of electrons as indicated shifts the whole electron distribution in k-space and produces a current-carrying state. (b) For a one-dimensional chain of divalent atoms all the states in the first Brillouin zone are full. To produce a current-carrying state some electrons must be promoted to states in the second Brillouin zone as indicated, that is to a higher energy band

For trivalent atoms the *first* Brillouin zone will be filled and the *second* Brillouin zone half-filled; we thus expect metallic behaviour again from the electrons in the second zone. In general, we expect metallic behaviour from odd-valence atoms, and insulating or semiconducting behaviour from even-valence atoms. Actually it is the number of valence electrons per primitive unit cell rather than the number per atom that is important. This is because the positions of the Brillouin zone boundaries are determined by the periodicity of the lattice potential and hence by the size of the primitive unit cell. The number of k states in the first Brillouin zone is always equal to the number of primitive unit cells in the crystal. For even-valence atoms there will always be an even number of valence electrons in a primitive cell so that according to the above argument these should always give insulating crystals.

The problem with this elegant and simple picture is that divalent metals actually exist: the alkaline earths Ca, Sr and Ba, for example. The existence of divalent metals does not invalidate our conclusion because our argument applies only to one-dimensional crystals. To explain the existence of divalent metals we must generalize the idea of energy bands to crystals in more than one dimension. It will be sufficient to consider the simplest possible two-dimensional crystal: a square crystal of side $L \times L$ consisting of a simple square lattice of identical atoms of spacing a. For simplicity we will confine our attention to the energy gaps produced by the fundamental component V_1 of the Fourier series of the lattice potential. The energy gaps occur when the electron waves are in synchronism with the periodicity of the lattice. There will be synchronism with the fundamental periodicity in the x direction if $k_x = \pm\pi/a$, and with the periodicity in the y direction if $k_y = \pm\pi/a$. The boundaries of the first Brillouin zone in the two-dimensional $\mathbf{k}$-space are therefore the lines $k_x = \pm\pi/a$, $k_y = \pm\pi/a$, as shown in Fig. 4.4. Electron energies can be displayed on this diagram by means of contour lines of constant energy as a function of k_x and k_y. The concentric circles shown in the figure are the contour lines according to free electron theory, $\varepsilon = \hbar^2 k^2/2m = \hbar^2(k_x^2 + k_y^2)/2m$.

To find how these energy contours are perturbed by the lattice potential we note from Fig. 4.2 that the energy is depressed below the free electron value just

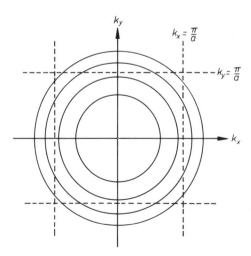

Fig. 4.4   Boundaries of the first Brillouin zone (broken lines) for a simple square lattice of atoms of spacing a. The circles are the free electron energy contours (at equal energy intervals)

inside the zone boundary and this has the effect of moving the energy contours out towards the boundary. Similarly the increased energy outside the zone boundary moves a constant energy contour in towards the boundary. Thus zone boundaries 'attract' energy contours and a perturbed energy contour is shown in Fig. 4.5(*a*). The perturbed contour meets the zone boundaries at right angles. This is because the component of the electron group velocity

$$\frac{\mathrm{d}\omega}{\mathrm{d}k} = \frac{1}{\hbar}\frac{\mathrm{d}\varepsilon}{\mathrm{d}k}$$

normal to the boundary vanishes; the *gradient* of ε in **k**-space is therefore *parallel* to the boundary and the contour of constant ε is *perpendicular* to the boundary. The vanishing of the group velocity component perpendicular to the boundary has the same physical interpretation in terms of standing waves produced by Bragg diffraction as in the one-dimensional crystal (see problem 4.2).

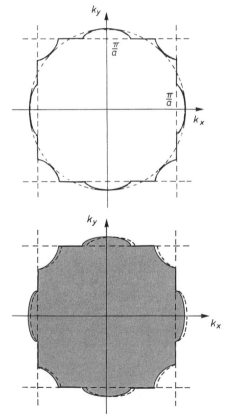

(*a*) Perturbation (full curves) of a free electron energy contour (broken circle) by a periodic lattice potential

(*b*) The shading indicates the occupied states at $T = 0$ for a two-dimensional divalent metal. The displacement of the electron distribution in k-space to give a current-carrying state is indicated by the broken curves

Fig. 4.5

The area of the first Brillouin zone is $(2\pi/a)^2$ and the density of running wave states in **k**-space is

$$\rho_R(\mathbf{k}) = L^2/(2\pi)^2. \tag{4.6}$$

This is the two-dimensional equivalent of the one- and three-dimensional densities of states for running waves, given by Eqs. (2.30) and (2.41) respectively. The first Brillouin zone therefore contains $\rho_R(\mathbf{k})(2\pi/a)^2 = (L/a)^2 = N$ **k**-states where N is the number of atoms in the crystal; this is precisely the same as the one-dimensional result and is also true in the three-dimensional case. The first Brillouin zone can therefore accommodate exactly $2N$ electrons.

To determine which states are actually occupied for a lattice of divalent atoms we must satisfy the following two conditions:

(i) we must fill an area of **k**-space equal to that of the first zone; and

(ii) we must fill all levels below some fixed energy ε_F.

These conditions are satisfied if the electrons occupy the shaded area in **k**-space shown in Fig. 4.5(b). The boundary between occupied and unoccupied states is still referred to as the Fermi surface although its shape is very different to the free electron circle. The Fermi surface in Fig. 4.5(b) does however have some free 'area', and can be slightly displaced as indicated by the broken curves to give a current-carrying state. We therefore have a two-dimensional divalent metal.

The reason for this behaviour is that for a free electron (Fig. 4.4) the energy at a zone corner is twice that at the centre of a zone edge because $|\mathbf{k}|$ is $\sqrt{2}$ times larger. Hence, if the perturbation due to the lattice potential is small, the lowest-energy states in the second zone are below the highest states in the first zone, and we have **overlapping energy bands**, the situation portrayed in Fig. 4.5. As the perturbation and hence the associated energy gap increases we reach a situation where the bands no longer overlap; all the states in the first zone have an energy lower than any state in the second. For divalent atoms the first zone is entirely occupied and the second zone completely empty. The occupied region of **k**-space is then completely bounded by the first Brillouin zone boundaries and there is no free area of Fermi surface. Electrons now have to be given a finite amount of energy to create a current-carrying state and we have an insulator.

There is a very strange consequence of the fact that an energy band filled with electrons does not conduct electricity and therefore appears to contain no mobile charge carriers. If a few electrons are removed from such a band to produce pockets of empty states such as those of the corners of the Brillouin zone in Fig. 4.5(b) then the removal of negative charge looks like the addition of positive charge and the empty states appear to conduct electricity like particles of charge $+e$. These 'carriers' are known as **holes** and are responsible for the positive Hall coefficients of some divalent metals. We discuss the properties of holes in more detail in the following chapter.

The situation in a three-dimensional crystal is much the same as in two dimensions; in Chapter 11 we give the general method for determining the positions of the Brillouin zone boundaries in three-dimensional k-space for an arbitrary crystal structure. The important conclusions are:

(i) an odd number of valence electrons in a primitive unit cell leads to metallic behaviour; and

(ii) an even number of valence electrons per primitive unit cell gives metallic behaviour if there is band overlap, a semiconductor if there is a small band gap (see Chapter 5) and an insulator if there is a large band gap.

Note the very important discovery that we have made in this section that the absence of conduction in a solid does not imply that the electrons are localized on the atoms. All the wavefunctions that we discuss in this chapter extend throughout the crystal.

4.3   THE TIGHT BINDING APPROACH

Although the nearly free electron theory has provided a solution to the problem of why some crystals conduct electricity and others are insulators, it has not given us much insight into the nature of the electron wavefunctions. It is not, for example, a suitable model for describing the covalent bonding by the valence electrons in the important semiconductor crystals, silicon and german-ium. The tight binding approach will give us the insight we require and at the same time serve to confirm the qualitative predictions of the nearly free electron theory. The tight binding approach is also important in that it forms the basis of many of the more advanced methods of energy band calculations in solids.

4.3.1   Coupled probability amplitudes

To introduce the tight binding method we again use the simplest possible crystal, namely the one-dimensional chain of identical atoms of separation a with the ends of the chain joined together so that periodic boundary conditions are appropriate. When the atoms are widely separated the wavefunctions of the valence electrons will be those of isolated atoms. We denote the wavefunction of the appropriate atomic state on the nth atom by ϕ_n; for sodium atoms, for example, this will be the 3s state. As the distance between the atoms decreases we might expect at some stage that the electrons begin to move from one atom to another. Provided the atoms are not too close we can picture this motion as the transfer of the electron from the state ϕ_n to the same state on a neighbouring atom, ϕ_{n-1} or ϕ_{n+1}. This picture suggests that we might be able to write, to a good approximation, the wavefunction of the electron as a linear combination of the states ϕ_n on the N atoms in the chain. Thus

$$\Psi = \sum_{n=1}^{N} a_n(t)\phi_n. \tag{4.7}$$

Using the time-dependent Schrödinger equation to calculate Ψ is a cumbersome procedure. It is simpler and physically more meaningful to use an alternative formulation of quantum mechanics which gives directly the probability $|c_n(t)|^2$ for finding the electron in the state ϕ_n at time t. Note that, since atomic states on different atoms are not orthogonal, the probability amplitude $c_n(t)$ is not equal to the coefficient $a_n(t)$ in Eq. (4.7); however, we wish to find the stationary states of the system† and for these the time dependences of $a_n(t)$ and $c_n(t)$ are the same, namely

$$a_n \sim c_n \sim \exp\left(-Et/\hbar\right) \tag{4.8}$$

for all n, where E is the energy of the state. The formal justification of this alternative formulation of quantum mechanics is given in appendix A and it leads to the following set of N coupled equations (one for each atom in the chain) for the probability amplitudes c_n:

$$i\hbar\frac{dc_n}{dt} = Bc_n - Ac_{n-1} - Ac_{n+1}. \tag{4.9}$$

We can given a simple physical interpretation to the terms Ac_{n-1} and Ac_{n+1}. They represent the changes in c_n (and hence in the probability of the electron being on the nth atom) associated with the transfer of the electron to and from the neighbouring atoms. The parameter A therefore measures the strength of the coupling between the states on neighbouring atoms. In writing Eqs. (4.9) we have ignored the possibility that transfer of electrons might occur between second nearest neighbours.

It is instructive to consider first the case of widely separated atoms so that the coupling between neighbours represented by the parameter A can be ignored. Eqs. (4.9) then reduce to N uncoupled equations

$$i\hbar\frac{dc_n}{dt} = Bc_n \tag{4.10}$$

with solutions

$$c_n(t) = c_n(0)\,e^{-iBt/\hbar}. \tag{4.11}$$

If we suppose that an electron is definitely in the state ϕ on atom m at $t = 0$ then we have

$$c_n(0) = \begin{cases} 1 & \text{for } n = m, \\ 0 & \text{otherwise}, \end{cases} \tag{4.12}$$

so that at time t the wavefunction is $\Psi = \phi_m \exp\left(-iBt/\hbar\right)$, thus confirming that the electron remains in the state ϕ on the mth atom and identifying the parameter B in this limit as the energy of the state ϕ.

---

† For stationary states the probability $|c_n(t)|^2$ of finding an electron on atom n is independent of time.

Before proceeding to apply Eqs. (4.9) to the one-dimensional chain we will digress slightly and show in section 4.3.2 that these equations are capable of describing covalent bonding by applying them to the simplest possible covalently bonded molecule, the H_2^+ ion. The reader anxious to learn the answer for the chain can skip the next section the first time around but is recommended to read it eventually since a chain of two atoms provides a useful and informative interpolation point between a single atom and a macroscopic crystal.

4.3.2   The H_2^+ ion—covalent bonding

The H_2^+ ion consists of two protons bound together by a single electron. Applying Eqs. (4.9) to a chain of two protons (labelled 1 and 2) we obtain†

$$i\hbar \frac{dc_1}{dt} = Bc_1 - Ac_2$$

$$i\hbar \frac{dc_2}{dt} = Bc_2 - Ac_1. \qquad (4.13)$$

B and A are functions of the separation between the protons. To calculate these energies requires more detailed quantum mechanics. The choice of the minus sign in front of A makes it a positive quantity (see below).

We must now seek to find the stationary state solutions of Eqs. (4.13); such solutions must possess the property (4.8). The solutions are easily found by noticing that Eqs. (4.13) look rather like the coupled oscillator equations that one gets, for example, for two identical pendulums coupled together; the main difference is that the differential operator is $i\,d/dt$ instead of d^2/dt^2. Finding solutions of the form of Eq. (4.8) is equivalent to determining the normal modes of the coupled pendulums and the same technique can be used in both cases, namely to note that by taking the sum and difference of Eqs. (4.13) two *uncoupled* equations are obtained for the new variables: $c_+ = c_1 + c_2$ and $c_- = c_1 - c_2$. The uncoupled equations are

$$i\hbar \frac{dc_+}{dt} = (B - A)c_+$$

$$i\hbar \frac{dc_-}{dt} = (B + A)c_- \qquad (4.14)$$

with stationary state solutions $c_+ \propto \exp[-i(B - A)t/\hbar]$ and $c_- \propto \exp[-i(B + A)t/\hbar]$.

To find the corresponding wavefunctions we must determine the values of $a_1(t)$ and $a_2(t)$ in Eq. (4.7); we first note that if the electron is definitely in the state represented by c_+ then $|c_-| = 0$ so that $c_1 = c_2$. Because ϕ_1 and ϕ_2 are the

---

† Note that connecting the ends of the chain to form a ring is not appropriate in this case.

same function centred on different points, Eq. (A3) of appendix A shows that $a_1 = a_2$ if $c_1 = c_2$ and hence the wavefunction corresponding to c_+ is

$$\Psi_+ \propto [\phi_1 + \phi_2] \exp [-i(B - A)t/\hbar] \qquad (4.15)$$

with an energy $E_+ = B - A$. Similarly the amplitude c_- refers to the wavefunction

$$\Psi_- = [\phi_1 + \phi_2] \exp [-i(B + A)t/\hbar] \qquad (4.16)$$

with energy $E_- = B + A$. For hydrogen atoms the state ϕ is the 1s state, shown in Fig. 4.6(a) for two isolated atoms. The variation of the wavefunctions Ψ_+ and Ψ_- along the internuclear line is sketched in Figs. 4.6(b) and (c) respectively.

Knowing these wavefunctions enables us to establish the sign of A and hence which state has the lower energy. The electron energy consists of two contributions: electrostatic potential energy of the electron in the field of the two

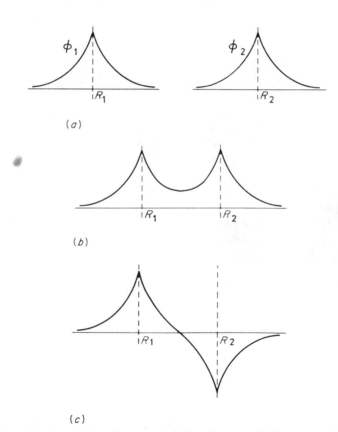

(a)

(b)

(c)

Fig. 4.6   (a) The 1s wavefunctions on two isolated atoms. (b) The symmetric wavefunction $(\phi_1 + \phi_2)$. (c) The antisymmetric wavefunction $(\phi_1 - \phi_2)$

protons, and kinetic energy proportional to $|\nabla\Psi|^2$. Comparison of Figs. 4.6(b) and (c) shows that both these contributions are lower for the symmetric state $\phi_1 + \phi_2$ of Fig. 4.6(b); the potential energy is low because $|\Psi|^2$ is large in the region in which the electrons 'see' the attractive potential of both protons and the kinetic energy is low because $|\nabla\Psi|^2$ is lower between the nuclei and much the same elsewhere. The energy A is therefore positive.

Our wavefunctions are not exact because we have not included atomic excited states in Eq. (4.7). But in spite of this we can obtain some exact energies for the limit where the separation, R, of the protons goes to zero so that the nuclear potential becomes that of the helium atom. The wavefunctions must go smoothly to states of the He^+ ion in this process. The symmetric state in Fig. 4.6(b) is nodeless, like an atomic 1s state; we therefore expect this state to become the 1s state of the He^+ ion with an electronic energy of -4 Rydberg (1 Rydberg is the ground state binding energy of the hydrogen atom, 13.6 eV). On the other hand the antisymmetric state shown in Fig. 4.6(c) has a single nodal plane perpendicular to the internuclear line. This is the symmetry of an atomic 2p state and we therefore expect that in the limit $R \to 0$ it will become the 2p state of He^+, with an energy of -1 Rydberg. In the opposite limit $R \to \infty$ both states tend to the ground state energy of atomic hydrogen, -1 Rydberg.

These limiting energies are seen to be correct in Fig. 4.7(a), which shows the results of exact calculations of the energies of the two states as functions of R. Fig. 4.7(a) shows the electronic energy only. To obtain the total energy we must add the internuclear Coulomb repulsion $e^2/4\pi\varepsilon_0 R$ to obtain the effective internuclear potential energy curves of Fig. 4.7(b). We see that only the symmetric combination Ψ_+ gives a minimum in the potential curve, representing the covalent bond. For this reason the symmetric wavefunction is called the **bonding orbital** and the antisymmetric wavefunction is known as the **antibonding orbital**. A wavefunction of the form of Eq. (4.15) to describe covalent bonding is referred to as a **linear combination of atomic orbitals**. Fig. 4.7 shows an embryonic band structure; the degeneracy of the atomic states of two isolated atoms has been lifted by the approach of the two nuclei; the number of levels in the resulting 'band' is equal to the number of nuclei in the molecule.

The binding of two protons serves to illustrate the principle of covalent bonding, namely the reduction in energy by concentration of electrons near the internuclear line, but it is not a typical example since the resulting molecular ion H_2^+ is charged. When neutral atoms are bound by equally shared electrons this necessarily involves taking one electron from each to form the bond. The simplest example of a bond of this type is that in the neutral hydrogen molecule, H_2, and it will be instructive for us to discuss briefly the electron wavefunctions in this case. If we neglect the mutual Coulomb repulsion between the two electrons, then they can be treated independently and our results for the H_2^+ ion are immediately applicable. The lowest-energy state is obtained by putting both electrons in the bonding orbital, and this is allowed by the exclusion principle

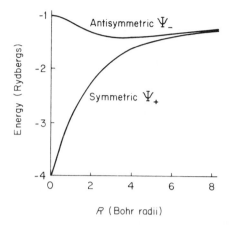

(a) Electron energy as a function of the proton separation, R

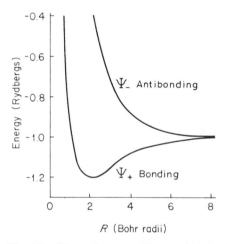

(b) Total energy including the nuclear repulsion, $e^2/4\pi\varepsilon_0 R$.

Fig. 4.7   (From *Quantum Theory of Molecules and Solids*, Vol. 1, by J. C. Slater © 1963 McGraw-Hill Book Company Inc.)

provided that the electrons have opposite spin. This situation is represented by a simple product wavefunction

$$\Psi(\mathbf{r}_a, \mathbf{r}_b) = \Psi_+(\mathbf{r}_a)\Psi_+(\mathbf{r}_b) \qquad (4.17)$$

which is symmetric under interchange of the two electron coordinates, $\mathbf{r}_a$ and $\mathbf{r}_b$; the opposite spins imply an antisymmetric spin wavefunction so that the total wavefunction is antisymmetric as required.

This independent particle approach is the one that we will adopt in the following section for a chain of atoms, not so much because it is a good approximation (although this turns out to be the case) as because it is much more difficult to do otherwise. It is therefore instructive in the simple case of the hydrogen molecule to examine the sort of errors to which it gives rise. To do this we write out in full the wavefunction of Eq. (4.17), using our approximate form of Ψ_+ of Eq. (4.15);

$$\Psi(\mathbf{r}_a, \mathbf{r}_b) = \phi_1(a)\phi_1(b) + \phi_2(a)\phi_2(b) + \phi_1(a)\phi_2(b) + \phi_2(a)\phi_1(b). \tag{4.18}$$

Here we use notation such that, for example, $\phi_1(a)$ corresponds to electron a in the state ϕ on atom 1. We have omitted the time dependence of $\Psi(\mathbf{r}_a, \mathbf{r}_b)$ which just tells us that in the absence of electron–electron interactions the energy of the two electrons is $2(B - A)$. If we note that ϕ_1 is large near nucleus 1 and ϕ_2 large near nucleus 2 then we see that the first two terms in the wavefunction (4.18) have a rather different physical interpretation to the last two. The first term has a large amplitude when both electrons are near nucleus 1, the second when both are near nucleus 2; in contrast the last two terms have a large amplitude when one electron is near each nucleus. Thus at large separations the last two terms give the probability amplitude for finding two neutral hydrogen atoms, whereas the first two give the probability amplitude for finding a bare proton H^+ and a negative hydrogen ion H^-. Within the independent particle approximation these states have the same energy because an electron has a binding energy of 1 Rydberg to a proton irrespective of whether another electron is already bound. The electron–electron interaction, however, reduces the binding energy of the second electron to only 0.05 Rydberg and thus makes a clear distiction between the states $H + H$ and $H^+ + H^-$; the higher energy of the latter means that we are almost certain to find one electron near each proton in the large separation limit. In contrast Eq. (4.18) gives equal amplitudes to the states $H + H$ and $H^+ + H^-$.

At smaller internuclear separations where ϕ_1 and ϕ_2 overlap considerably the clear distinction between the two types of state can no longer be made and it is no longer obvious that the independent particle approximation is bad; we cannot deduce from this however that it is good and we should always treat it with caution. One important deduction that we can make from the above argument is that when the atoms come together to form a solid we should not expect to obtain electron wavefunctions extending throughout the crystal when the atoms are widely separated. Extended states imply a finite probability that some atoms will have more valence electrons than others and, because of electron–electron interactions, such states have a higher energy than those in which the valence electrons are localized on their atoms. We might expect the transition to extended states and hence to possible metallic behaviour to occur at some critical finite separation and this does indeed seem to be the case. We discuss this possibility further in section 13.5.6.

4.3.3   Electron states on a one-dimensional chain

We will now try to find stationary state solutions of the form of Eq. (4.8) to the coupled probability equations (4.9). From the previous section we know that the approach that we are adopting is essentially a generalization to a macroscopic chain of the 'linear combination of atomic orbitals' technique for finding the electronic states in molecules. We find solutions of Eqs. (4.9) by noticing that there are very similar to Eqs. (2.7) for the lattice vibrations of the chain of atoms; the only essential difference is that the time derivatives are $i\,d/dt$ instead of d^2/dt^2 because Eqs. (4.9) are obtained from Schrödinger's equation rather than a classical equation of motion. Finding solutions of Eqs. (4.9) of the form of Eq. (4.8) corresponds to finding the normal modes of vibration of the chain; since these are running waves (Eq. (2.8)) when periodic boundary conditions are applied, we will look for running wave solutions of the same form to Eqs. (4.9),

$$c_n = C \exp\left[i(kx_n^0 - \omega t)\right], \tag{4.19}$$

where $x_n^0 = na$ is the equilibrium position of the nth atom in the chain. Substituting into Eqs. (4.9) gives

$$\hbar\omega\, e^{i(kna - \omega t)} = B\, e^{i(kna - \omega t)} - A\, e^{i[k(n-1)a - \omega t]} - A\, e^{i[k(n+1)a - \omega t]},$$

or, on cancelling factors $e^{i(kna - \omega t)}$, we find that Eq. (4.19) is a solution of Eqs. (4.9) provided that the energy ε of the state is related to the wavenumber k by

$$\varepsilon = \hbar\omega = B - A\, e^{-ika} - A\, e^{ika} = B - 2A\cos(ka). \tag{4.20}$$

The coefficients $a_n(t)$ in the electron wavefunction (Eq. (4.7)) also have the wavelike form of Eq. (4.19); the electron states of the one-dimensional chain are thus wavelike and extend throughout the crystal.† The dispersion relation for the waves (Eq. (4.20)) is plotted in Fig. 4.8. We see that the degeneracy of the states on the isolated atoms has been broken by the coupling to produce an energy band of width $4A$. As for lattice vibrations, ω is a periodic function of k with period $2\pi/a$. Also as there, our assumption of coupling to nearest neighbours only in Eqs. (4.9) is not essential to this periodicity; terms involving c_{n-2} and c_{n+2} in Eq. (4.9) would just give rise to a term in $\cos(2ka)$ in Eq. (4.20). More complicated coupling therefore only adds harmonics to the $\varepsilon(k)$ curve; the fundamental periodicity remains at $2\pi/a$, depending only on the lattice spacing.

By an argument identical to that used for lattice vibrations we can show that the wavenumber range $-\pi/a < k < \pi/a$, that is the first Brillouin zone, describes all possible physical situations. If we take the ordinate on Fig. 2.5 as representing c_n rather than the atomic displacement u_n, then Fig. 2.5(b) shows how c_n values given by a wave with $|k| > \pi/a$ (point A on Fig. 4.8) can equivalently be represented by a wave with $|k| < \pi/a$ (points B and C on Fig. 4.8). Points A and

---

† There is a general result known as **Bloch's theorem**, which shows that electron states in a periodic potential are always wavelike. We discuss this theorem in section 11.3.

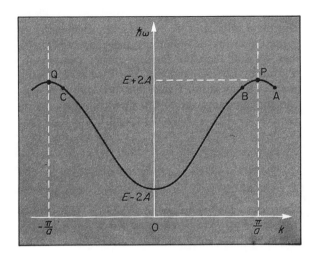

Fig. 4.8   Tight binding result for the energy as a function of wavenumber for electrons on a one-dimensional chain of atoms

C represent states with a negative group velocity $d\omega/dk$, and thus represent electron wavepackets moving to the left. A and C are thus completely equivalent; point B represents an otherwise similar wavepacket moving to the right. The application of periodic boundary conditions, as before (sections 2.3.1 and 4.2), determines that the number of allowed values of k in each Brillouin zone is equal to the number of atoms N in the chain.

Fig. 4.8 provides more insight into why a full energy band cannot carry an electric current. The filled band corresponds to all states between Q and P being occupied. An electric field causes a shift of the distribution in k-space (Eq. (3.34)), to the states between C and A say. The effect of the field is therefore to cause electrons in states in QC to shift to states in PA. We have shown above however that the states in PA are exactly the same as those in QC so that the electric field does not change the electron distribution at all and produces no current. A current-carrying state can only be produced if some of the electrons are promoted to a higher energy band as in Fig. 4.3(b).

The freedom to describe the same state by values of k differing by integral multiples of $2\pi/a$ allows us to choose which Brillouin zone we will use to plot the $\varepsilon(k)$ relation. This freedom allows us to make contact with the $\varepsilon(k)$ relation given by the nearly free electron theory (Fig. 4.2). We must first suppose that higher-energy atomic states will give rise to energy bands at higher energies as shown on Fig. 4.9(a); this method of plotting in which the dispersion relation for each band is continued periodically through the whole of **k**-space is known as a **repeated zone scheme**. Note that we have assumed that A changes sign for successive bands so that the minimum energy is alternately at $k = 0$ and $k = \pi/a$.

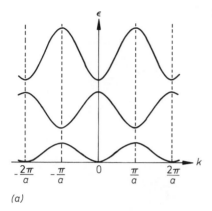

(a)

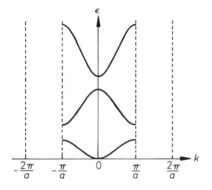

(a) Repeated zone scheme

(b) Reduced zone scheme

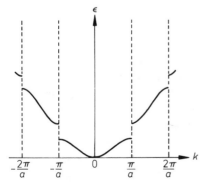

(c) Extended zone scheme

Fig. 4.9   Different ways of plotting the $\varepsilon(k)$ curves for electrons in different energy bands

This assumption will produce a dispersion relation with most similarity to that of the nearly free electron theory; our argument concerning the sign of A in section 4.3.2 depended on the use of 1s wavefunctions that are positive everywhere. In Fig. 4.9(b) the dispersion relations of Fig,. 4.9(a) are plotted according to the **reduced zone scheme** in which only the first Brillouin zone is used since this contains all the physically distinct solutions. Fig. 4.9(c) shows the dispersion relations in the **extended zone scheme** in which successively higher energy bands are plotted in successively higher Brillouin zones; it is this plot in which the qualitative similarities to the nearly free electron approach are most apparent. The tight binding approach enables us to label the energy bands by the atomic levels from which they arise. Thus the valence electrons in solid sodium are to be found in the 3s energy band.

In Fig. 4.10 we illustrate the use of the repeated zone scheme to plot the Fermi surface of the two-dimensional metal of Fig. 4.5(b). Two diagrams are needed, one for each of the overlapping energy bands.

We have so far ignored the possibility that the electron states of the ion cores might couple together and give rise to an energy band. Normally the coupling is so weak that the degeneracy of the energy levels is not significantly broken and the core electron states are localized. This is not the case however when the outermost core states are not completely filled, as is the case with the 3d states of the transition metals. It is by no means obvious whether these electrons are best regarded as localized in atomic states or part of a band of mobile electrons. The closer packing of the atoms and the larger binding energy of the transition metals can be attributed to the covalent bonding associated with the overlap of the incompletely filled 3d shells on neighbouring ions.

As an illustration of what tight binding wavefunctions would look like in one-dimensional sodium metal we have used the *hydrogen* 3s orbital† (Fig. 4.11(a)) for ϕ_n in Eq. (4.7). $a_n(t)$ takes the same form, Eq. (4.19), as $c_n(t)$ and we thus obtain the *real part* of the wavefunction as shown in Fig. 4.11(b); we took $a = 3.66$ Å (the nearest neighbour distance in real sodium metal) and $k = \pi/3a$ (equivalent to $\lambda = 6a$). The broken curve is proportional to the real part, $\cos (kx)$, of the free electron wavefunction for the same k. The large spikes on the tight binding wavefunction near each atom are from the 3s orbital on that atom; the 3s character of the wavefunction near each nucleus serves to make the wavefunctions approximately orthogonal to the lower lying 1s, 2s and 2p atomic states of the ion core. We might expect that the tight binding wavefunction would have a lower energy than the free electron wavefunction because the spikes increase the probability of finding the electron in the neighbourhood of the nuclei where the potential energy is lowest. This decrease in potential energy is however largely compensated by the increase in kinetic energy associated with

---

† The 3s orbital in sodium will be qualitatively similar and have the same asymptotic dependence on position at large distances from the nucleus.

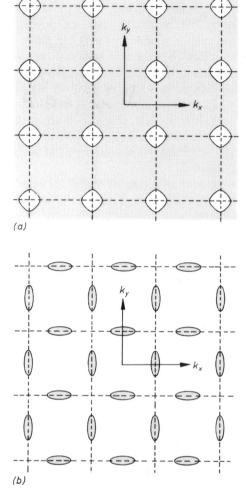

(a)

(b)

Fig. 4.10   The Fermi surface of the two-dimensional divalent metal of Fig. 4.5(b) plotted in a repeated zone scheme; occupied states are shaded. (a) Pockets of *unoccupied* states obtained by repeating periodically in k-space the pattern inside the first Brillouin zone in Fig. 4.5. (b) pockets of *occupied* states obtained by repeating periodically the pattern from the second Brillouin zone in Fig. 4.5(b)

the rapid spatial variation of the 3s wavefunction near the nucleus; thus, despite the difference in the wavefunctions, the two theories give similar dispersion relations.

The compensation of kinetic and potential energies we have just discussed can be seen already in an isolated atom. The potential of a Na^+ ion is that of a single positive charge at large distances, but that of a greater charge at short distances because the nuclear charge is not fully screened by the inner shell electrons. We might therefore expect that the 3s electron in sodium would be more strongly bound than the 1s electron in hydrogen. Yet the ionization potential of

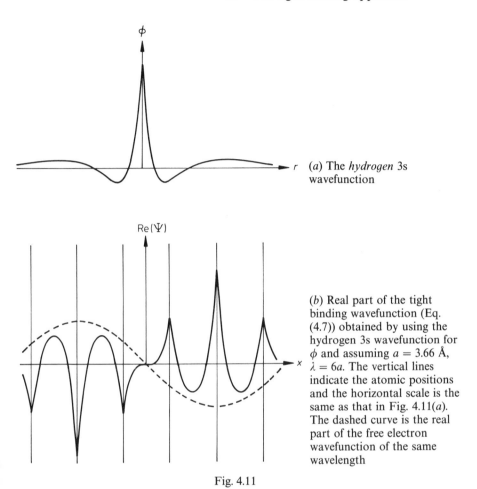

(a) The *hydrogen* 3s wavefunction

(b) Real part of the tight binding wavefunction (Eq. (4.7)) obtained by using the hydrogen 3s wavefunction for ϕ and assuming $a = 3.66$ Å, $\lambda = 6a$. The vertical lines indicate the atomic positions and the horizontal scale is the same as that in Fig. 4.11(a). The dashed curve is the real part of the free electron wavefunction of the same wavelength

Fig. 4.11

hydrogen is 13.6 eV and that of sodium is 5.1 eV. The reason for this is that the 3s wavefunction has more wiggles than the 1s wavefunction and the consequent kinetic energy largely compensates the attractive potential. Thus the ionization potential of the free atom is a better guide to the *effective* periodic potential in the solid than the Coulomb potential of the free ion.

4.3.4   Electron states in diamond, silicon and germanium

The electron configuration of an isolated carbon atom is $1s^2 2s^2 2p^2$. In a crystal of diamond (Fig. 1.15(a)) there are two carbon atoms and hence eight valence electrons in a primitive unit cell. To construct approximate wavefunctions for these electrons using the tight binding approach we must generalize Eq.

(4.7) (with c_n and hence a_n of the form of Eq. (4.19)) to a three-dimensional crystal containing more than one atom in a primitive unit cell. The appropriate generalization is

$$\Psi = \sum_n C \, e^{i(\mathbf{k} \cdot \mathbf{r}_n - \omega t)} \phi_n \tag{4.21}$$

where the sum is *not* over the atoms but over the lattice points at positions $\mathbf{r}_n$ given by Eq. (1.2). Correspondingly the values of $\mathbf{k}$ at which standing waves and hence energy gaps occur are determined by the periodicity of the lattice, and the wavefunction ϕ_n must be an orbital appropriate to the basis of two atoms associated with each lattice point rather than with an isolated atom.

The function ϕ must reflect the tetrahedral arrangement of the nearest neighbours in the diamond structure. The 2s and 2p states of the isolated atom do not have the appropriate symmetry but it is possible to form linear combinations of these states which do (see problem 4.5). This procedure is also used in explaining the tetrahedral bonding of carbon in organic molecules, such as methane; in this case the mixing of the s and p wavefunctions to generate wavefunctions with tetrahedral symmetry is known as sp^3 **hybridization** since all three p wavefunctions are involved. Four different linear combinations of the wavefunctions can be obtained, each one corresponding to a large electron concentration in a lobe along one of the four tetrahedral directions (Fig. 4.12).†

Once atomic orbitals of the appropriate tetrahedral symmetry have been obtained by the sp^3 hybridization process, a suitable 'molecular' orbital ϕ for the diatomic basis can be obtained by linear combination of the atomic orbitals on neighbouring atoms. The obvious way of doing this is to use the bonding and antibonding orbitals (see section 4.3.2) associated with the overlap of the appropriately directed tetrahedral lobe on one atom with that of its neighbour; the enhanced electron concentration in the region between the nearest neighbours associated with the bonding combination (Fig. 4.12(c)) can be identified with covalent bonding between atoms. The four bonding combinations and the four antibonding combinations provide eight possible molecular orbitals to insert in Eq. (4.21) and thus lead to eight energy bands, sufficient to accommodate $16N$ electrons where N is the number of primitive unit cells in the crystal. As there are $8N$ valence electrons and diamond is an insulator we can deduce that four of the bands are completely full (those associated with the covalent bonding) and the other four are separated from the filled bands by an energy gap.

The situation for silicon and germanium at absolute zero is similar to that for carbon except that it is the 3s, 3p and 4s, 4p orbitals respectively that are involved in the hybridization process. Also the energy gap (see the following chapter) between the occupied and unoccupied bands steadily decreases as the

---

† sp^2 and sp^1 hybridization also occur. The former produces orbitals with a symmetry appropriate to the layers of carbon atoms in the graphite structure (see problem 4.6).

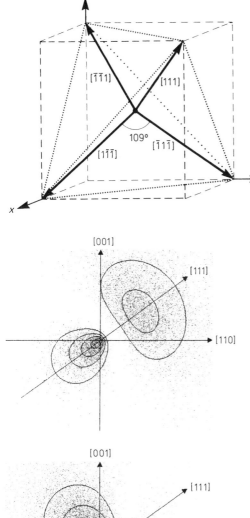

(a) The tetrahedral directions of the nearest neighbours in the diamond structure are the [1 1 1], [$\bar{1}$ $\bar{1}$ 1], [$\bar{1}$ 1 $\bar{1}$] and [1 $\bar{1}$ $\bar{1}$] directions towards the corners of a cube

(b) Section through a sp^3 hybridized orbital directed along [1 1 1]. The spot concentration and contour lines indicate the electron probability density. The nucleus is at the origin. The figure was generated using hydrogen 2s and 2p wavefunctions

(c) Bonding orbital formed by overlap of sp^3 hybridized orbitals (like that of (b)) on neighbouring atoms; the spot concentration and contour lines indicate the electron probability density. Insertion of an orbital of this type into Eq. (4.21) generates approximate wavefunctions for the valence electrons involved in the covalent bonding in diamond. The nuclear positions are indicated by circles; the distance between the nuclei has been chosen to maximize the overlap. Note that the tetrahedron of hybridized orbitals on one atoms is oppositely directed to that on its neighbour

Fig. 4.12

atomic number increases so that at finite temperatures sufficient electrons are excited across it in silicon and germanium to produce semiconducting behaviour. The semiconducting compounds such as InSb and GaAs are formed by combining a group III element (such as In or Ga) and a group V element (such as Sb or As). The electron states are similar to those in germanium and silicon but three of the valence electrons responsible for the covalent bonding come from the group III element and five from the group V element.

4.4 BAND STRUCTURE EFFECTIVE MASSES

In this section we justify the claim made in section 3.3.1 that the effect of the periodic lattice potential on the dynamics of the conduction electron wavepackets can be taken into account by using an effective mass m_e in the equations of

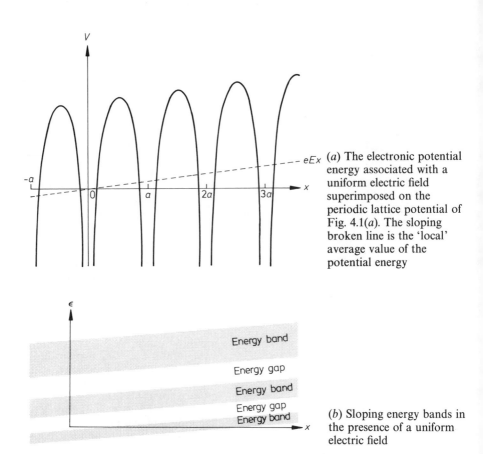

(a) The electronic potential energy associated with a uniform electric field superimposed on the periodic lattice potential of Fig. 4.1(a). The sloping broken line is the 'local' average value of the potential energy

(b) Sloping energy bands in the presence of a uniform electric field

Fig. 4.13

motion rather than the bare mass m. We will consider the effect of a uniform static electric field E applied to a one-dimensional chain. The electrostatic potential energy eEx of the electrons in the presence of the field superimposes a uniform gradient on top of the periodic lattice potential as shown in Fig. 4.13(a) and the local average potential energy thus acquires a gradient eE as indicated. For the fields that can normally be applied the change in the average potential energy on an atomic length scale is small and in these circumstances the $\varepsilon(k)$ relation for the electrons is the same as in the absence of the field except that ε has to be measured with respect to the local average potential. We thus obtain 'sloping' energy bands also with a gradient eE as indicated in Fig. 4.13(b).

We will calculate the motion of an electron wavepacket constructed by superimposing states from one of the energy bands. Let us suppose that at a particular time t the wavepacket is constructed from states centred on energy ε and wavenumber k. We wish to calculate the motion of the wavepacket in the subsequent time interval δt. To do this we make the following assumptions:

(i) The velocity of the wavepacket is the group velocity

$$v = \frac{\mathrm{d}\omega}{\mathrm{d}k} = \frac{1}{\hbar}\frac{\mathrm{d}\varepsilon}{\mathrm{d}k}.$$ (4.22)

(ii) The motion of the wavepacket resembles that of a classical particle in that its *total* energy remains constant. We see from Fig. 4.14 that if the wavepacket moves a distance δx in time δt then the change $\delta\varepsilon$ in kinetic energy is given by

$$\delta\varepsilon = -eE\delta x.$$ (4.23)

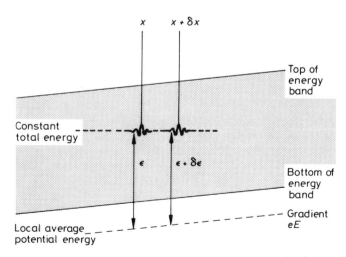

Fig. 4.14   Motion of an electron wavepacket at constant energy in the presence of an electric field. The positions of the wavepacket at times t and $t + \delta t$ are x and $x + \delta x$ respectively

We will use Eqs. (4.22) and (4.23) to derive the equation of motion in two useful forms. From Eq. (4.22), the change δk in k in time δt is

$$\delta k = \frac{dk}{d\varepsilon}\,\delta\varepsilon = \frac{1}{\hbar v}\,\delta\varepsilon.$$

Hence, using Eq. (4.23) and $v = dx/dt$,

$$\frac{dk}{dt} = \frac{1}{\hbar v}\frac{d\varepsilon}{dt} = -\frac{eE}{\hbar v}\frac{dx}{dt} = -\frac{eE}{\hbar}$$

or

$$\hbar\frac{dk}{dt} = -eE. \tag{4.24}$$

In this form the equation of motion is just a statement that the rate of change of momentum is equal to the applied force. The change in momentum cannot be attributed to the electron only, since, as we will show in section 13.3.1, some of the momentum is associated with the crystal lattice as a whole. For this reason $\hbar k$ is referred to as the **crystal momentum** of the electron.

The use of the effective-mass concept makes it possible to write the equation of motion in an alternative familiar form and to use it without having to worry about the subtle difference between the true momentum and crystal momentum of the electron. To do this we take the time derivative of Eq. (4.22) and use Eq. (4.24). Thus

$$\frac{dv}{dt} = \frac{1}{\hbar}\frac{d}{dt}\left(\frac{d\varepsilon}{dk}\right) = \frac{1}{\hbar}\frac{d^2\varepsilon}{dk^2}\frac{dk}{dt} = -\frac{1}{\hbar^2}\frac{d^2\varepsilon}{dk^2}\,eE, \tag{4.25}$$

which can be written

$$m_e\frac{dv}{dt} = -eE. \tag{4.26}$$

This is just Newton's second law for a particle of charge $-e$ and mass m_e. Comparison of Eqs. (4.25) and (4.26) identifies the effective mass m_e as

$$m_e = \hbar^2 \left/ \left(\frac{d^2\varepsilon}{dk^2}\right)\right. . \tag{4.27}$$

We have justified the use of an effective mass in the electron dynamics only for a dc electric field applied to a one-dimensional crystal and in the absence of collisions, but we hope that the reader will accept that appropriate generalization of Eq. (4.26) is indeed Eq. (3.23), as we have already assumed. The three-dimensional equivalent of Eq. (4.27) for the effective mass is given in section 13.3.1.

Readers should check for themselves that Eq. (4.27) yields $m_e = m$ for the free electron dispersion relation $\varepsilon = \hbar^2 k^2 / 2m$. For the dispersion relation within a typical energy band as shown in Fig. 4.15(a) (cf. Figs. 4.2 and 4.8), $\mathrm{d}^2 \varepsilon / \mathrm{d}k^2$ is positive at small k, zero at the value of k at which the curvature of the dispersion vanishes and negative for values of k greater than this. The effective mass therefore varies with k as shown in Fig. 4.15(b), and is negative for states close to the top of the energy band. Note that the effective mass varies very slowly with k near the top and bottom of the energy band; this is because the $\varepsilon(k)$ curve is approximately parabolic near a maximum or a minimum and therefore resembles in form the dispersion relation for free particles (see problem 4.7). Physicists find the idea of a particle with a negative effective mass rather indigestible and use an alternative approach to describe the motion of electrons in states near

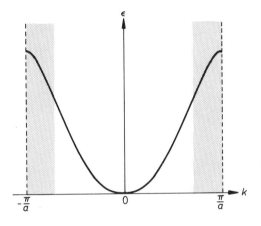

(a) Electron dispersion relation for a typical energy band. $\mathrm{d}^2 \varepsilon / \mathrm{d}k^2$ is negative in the shaded regions of k-space

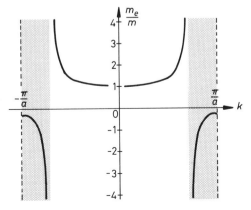

(b) Ratio of the effective mass m_e of Eq. (4.27) to the bare mass of the electron for the dispersion relation of (a). m_e / m is negative in the shaded region of k-space

Fig. 4.15

the top of an energy band. This approach uses the fact, explained in the following chapter, that the behaviour of the electrons in states near the top of an energy band is equivalent to the *unoccupied* states behaving like positively charged particles with positive masses. These fictitious particles are called holes and they play an important role in explaining the properties of semiconductors.

We should hasten to reassure the reader worried about the implications of a negative effective mass for the heat capacity (see section 3.2.3) that a different effective mass $m*$ is appropriate for this property. From our discussion of section 3.2.3 it is clear that the heat capacity is determined by the value, at the Fermi energy, of the density of states per unit energy range $g(\varepsilon)$ and this is always positive. To calculate $m*$ for a one-dimensional metal we first calculate $g(\varepsilon)$. Proceeding as in section 3.2 and using the density of running wave states $\rho_R(k)$ of Eq. (2.30), we find

$$g(\varepsilon) = 4\rho_R(k)\left|\frac{dk}{d\varepsilon}\right| = \frac{2L}{\pi}\left|\frac{dk}{d\varepsilon}\right| \tag{4.28}$$

for a one-dimensional crystal of length L. The factor 4 allows for both the spin degeneracy and the two regions of k-space, symmetrically disposed around $k = 0$, in which the energy range $d\varepsilon$ occurs (in one of these regions $d\varepsilon/dk > 0$ and in the other $d\varepsilon/dk < 0$, but both regions make the same positive contribution to the density of states; hence it is the modulus of $d\varepsilon/dk$ that matters). From Eqs. (4.28) and (3.5) we deduce that for free electrons

$$g(\varepsilon_F) = \frac{2Lm}{\pi\hbar^2 k_F}, \tag{4.29}$$

whereas in general

$$g(\varepsilon_F) = \frac{2L}{\pi}\left|\frac{dk}{d\varepsilon}\right|_F = \frac{2L}{\pi\hbar v_F}. \tag{4.30}$$

The subscript F denotes the value at the Fermi surface. Comparison of Eqs. (4.29) and (4.30) shows that in determining the specific heat the electrons behave like particles of effective mass

$$m* = \hbar^2 k_F\left|\frac{dk}{d\varepsilon}\right|_F = \frac{\hbar k_F}{v_F}. \tag{4.31}$$

For the dispersion relation of Fig. 4.15(a), $m*$, as calculated from this equation, becomes infinite if the Fermi 'surface' is at $k = \pi/a$,. This situation corresponds however to a filled band and insulating behaviour; despite the infinite density of states, the electronic heat capacity then vanishes at low temperatures because there are no low lying vacant states into which the electrons can be thermally excited.

PROBLEMS 4

4.1 Prove that $\phi_1 = \sin(kx)$ and $\phi_2 = \cos(kx)$ are the only orthogonal linear combinations of e^{ikx} and e^{-ikx} that satisfy

$$\int \phi_1^* V \phi_2 \, dx = 0$$

for all values of k, where V is the periodic lattice potential of Eq. (4.2).

4.2 Show that the plane wave $e^{i\mathbf{k}\cdot\mathbf{r}}$ satisfies the Bragg condition (Eq. (1.3)) for diffraction by a simple square lattice of identical atoms of spacing a provided that

$$k_x = \pm \pi/a \qquad \text{or} \quad k_y = \pm \pi/a.$$

This identifies the boundaries of the first Brillouin zone in **k**-space.

4.3 Near the boundary $k = \pi/a$ of the first Brillouin zone for the one-dimensional chain of identical atoms of spacing a, the nearly free electron theory predicts that the only important term in the lattice potential of Eq. (4.2) is $V_1 \cos(2\pi x/a)$ and that the wavefunction is approximately

$$\psi = \alpha \, e^{ikx} + \beta \, e^{i(k - 2\pi/a)x}.$$

Subsitute this wavefunction into the Schrödinger equation

$$-\frac{\hbar^2}{2m}\frac{d^2\psi}{dx^2} + V\psi = \varepsilon\psi.$$

Multiply the resulting equation (*a*) by e^{-ikx} and integrate over all space, and (*b*) by $e^{-i(k - 2\pi/a)x}$ and integrate over all space. By requiring that the resulting two equations have a non-trivial solution for α and β, show that the energy ε associated with the above wavefunction can be written

$$\varepsilon = \frac{\hbar^2 k^2}{2m} + \frac{\hbar^2 \pi}{ma}\left\{\left(\frac{\pi}{a} - k\right) \pm \left[\left(\frac{\pi}{a} - k\right)^2 + \left(\frac{amV_1}{2\pi\hbar^2}\right)^2\right]^{1/2}\right\}.$$

This confirms our guess for the form of the $\varepsilon(k)$ curve away from the Brillouin zone boundary in Fig. 4.2. You can check that it gives the correct answer at $k = \pi/a$ and that it reduces to the free electron result for values of k well away from the zone boundary.

4.4 A hypothetical monovalent metal consists of a simple cubic lattice of atoms of spacing a. Use the free electron theory to calculate the radius of the Fermi sphere. Calculate the distance of closest approach of this sphere to the Brillouin zone boundary. Is the sphere completely contained within the first Brillouin zone?

How would you expect this Fermi surface to be modified by the periodic lattice potential?

4.5 A set of normalized and mutually orthogonal p-state wavefunctions for an atom can be written in the form:

$$p_x = xf(r), \qquad p_y = yf(r), \qquad p_z = zf(r).$$

Consider the linear combination

$$\psi = a_x p_x + a_y p_y + a_z p_z.$$

Find four sets of coefficients (a_x, a_y, a_z) that give normalized p-state wavefunctions with positive lobes pointing towards the corners of a regular tetrahedron. (Remember that four of the corners of a cube are corners of an inscribed regular tetrahedron.)

Consider the linear combination

$$\phi = bs + c\psi$$

where ψ is one of the four wavefunctions calculated above and s is an s-state wavefunction, normalized and orthogonal to p_x, p_y and p_z. Find values of b and c which make the four resulting ϕ wavefunctions orthogonal to each other and normalized. Write out these four wavefunctions in terms of p_x, p_y, p_z and s. (These are the sp^3 hybrid wavefunctions.)

4.6 The sp^2 hybrid wavefunctions involved in the bonding of the two-dimensional layers of carbon atoms in graphite are of the form

$$\chi = \alpha s + \beta p_x + \gamma p_y$$

where s, p_x and p_y are s and p wavefunctions as defined in the previous question. Find values for α, β and γ that give three normalized mutually orthogonal wavefunctions with positive lobes directed at 120° with respect to each other in the xy plane.

4.7 Calculate the variation with wavenumber of the electron effective mass m_e for the tight binding dispersion relation of Eq. (4.20). Show that the value obtained at $k = \pi/a$ agrees with that obtained by expanding ε to second order in $k - \pi/a$ about $k = \pi/a$.

The unreasonable man persists in trying to adapt the world to himself. Therefore all progress depends on the unreasonable man.—*George Bernard Shaw*

CHAPTER

Semiconductors

5.1 INTRODUCTION

The important semiconductor materials silicon and germanium form covalently bonded crystals with the diamond structure (Fig. 1.15); the semiconductor compounds such as GaAs and InSb form the analogous zincblende structure. At absolute zero the highest occupied energy band is completely full; this is known as the **valence band** since it contains the electrons responsible for the covalent bonding (see section 4.3.4). The lowest unoccupied energy band is known as the **conduction band** and is typically separated from the valence band by a gap of order 1 eV. The behaviour of the semiconductor is dominated by electrons in states close to the top of the valence band and the bottom of the conduction band for which the energy dispersion relations $\varepsilon(k)$ are shown in Fig. 5.1.

Since we are dealing with states close to a maximum or minimum of energy we can take the dispersion curve $\varepsilon(k)$ to be parabolic to a good approximation and write:

$$\text{conduction band} \qquad \varepsilon = E_G + \frac{\hbar^2 k^2}{2m_e} \qquad (5.1)$$

$$\text{valence band} \qquad \varepsilon = -\frac{\hbar^2 k^2}{2m_h} \qquad (5.2)$$

where E_G is the energy gap and we have taken the top of the valence band as the zero of potential energy. Electrons near the bottom of the conduction band

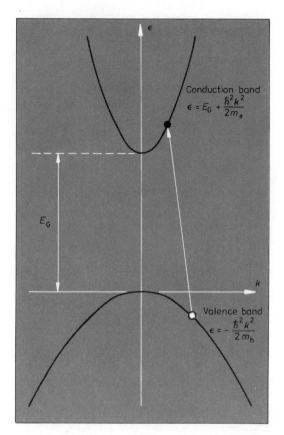

Fig. 5.1   Dispersion relations for electrons near the top of the valence band and the bottom of the conduction band (cf. Fig. 4.9(b)). The transition of an electron from the valence band to the conduction band, as indicated, creates a hole in the valence band

therefore behave like free particles of positive mass m_e. Those in states near the top of the valence band, however, appear to have a negative effective mass $-m_h$, although electrons in states lower down in the valence band do have positive effective masses (see Fig. 4.15(b)). At first sight it would seem to be a very difficult problem to calculate the properties of a nearly full valence band, but there is a simple and elegant approach which avoids the complexities. The behaviour of a nearly full valence band can be calculated by ignoring the filled states completely and regarding each empty state as being occupied by a particle of positive charge $|e|$, positive mass m_h† and energy $+\hbar^2 k^2/2m_h$ (as shown on Fig. 5.2). These fictitious particles are referred to as **holes**.

† Since there are only empty states near the top of the band the appropriate effective mass for all of these is m_h.

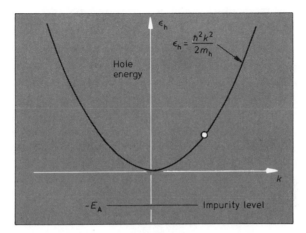

Fig. 5.2   The dispersion relation for holes in the valence band is the negative of that for electrons. The impurity level is the energy of a hole bound to an acceptor impurity

5.2   HOLES

We will consider the properties of the valence band with just one electron missing, from the state **k** say. We identify the energy and momentum of this hole by asking how much of each must be added to the crystal in order to create it by transfer of the electron from state **k** in the valence band to state **k₁** in the conduction band as indicated in Fig. 5.1. From Eqs. (5.1) and (5.2) the energy required is

$$\Delta \varepsilon = E_G + \frac{\hbar^2 k_1^2}{2m_e} - \left(- \frac{\hbar^2 k^2}{2m_h} \right)$$

$$= \left(E_G + \frac{\hbar^2 k_1^2}{2m_e} \right) + \frac{\hbar^2 k^2}{2m_h}.$$

The first term, in brackets, is clearly the energy of the electron in the conduction band. The second term is therefore identified as the energy required to create the hole and it is positive. Thus the energy of the hole in state **k** is

$$\varepsilon_h = \frac{\hbar^2 k^2}{2m_h} \tag{5.3}$$

which is plotted in Fig. 5.2. The hole dispersion relation is therefore obtained by inverting that for the valence band electrons. By a similar argument, the removal of an electron of momentum (strictly crystal momentum) $\hbar \mathbf{k}$ from the valence band corresponds to the addition of momentum $(-\hbar \mathbf{k})$ to the valence band, so that the (crystal) momentum of the hole in state $\mathbf{k}$ is

$$\mathbf{p_h} = -\hbar\mathbf{k}. \tag{5.4}$$

To establish the equation of motion of a hole we consider the effect of a dc electric field in the absence of collisions. We know from Eq. (4.24) that the effect of the field is to cause the $\mathbf{k}$ vector of all the electrons to move at the same uniform rate through $\mathbf{k}$-space. Figs. 5.3(a) and (b) show the electron distribution at two successive times. We see that the hole in the distribution is 'swept' along by the electrons in the occupied states. The important conclusion is therefore that the equation of motion of the hole is exactly the same as that of an electron in the same state; using results from Chapter 4 we can write this in either of the forms (4.24) or (4.26), i.e.

$$\hbar\, d\mathbf{k}/dt = -e\mathbf{E}$$

or

$$-m_h\, d\mathbf{v_h}/dt = -e\mathbf{E}$$

where Eq. (4.27) has been used with Eq. (5.2) to establish that the effective mass of electrons near the top of the valence band is indeed $-m_h$. Here $\mathbf{v_h}$ is the

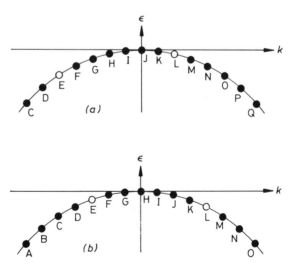

Fig. 5.3   Parts (a) and (b) show the electron distribution in k-space for the valence band at two successive times, illustrating the way in which a dc electric field, in the absence of collisions, causes the electrons to move through k-space at a steady rate $dk/dt = -eE/\hbar$. The unoccupied states (E and L) are swept along in this process so that the motion of a hole is the same as that of an electron in the same state

velocity of a hole wavepacket (which can be formed by Fourier synthesis from different valence band states in each of which the one missing electron is from a different $\mathbf{k}$ state). Because the hole motion is determined by that of the electrons in neighbouring states v_h is just the electron group velocity as given by Eq. (4.22), i.e.

$$v_h = \frac{1}{\hbar} \frac{d\varepsilon}{dk}. \tag{5.5}$$

The two forms of the equation of motion given above can be rewritten:

$$\frac{d\mathbf{p_h}}{dt} = e\mathbf{E}, \tag{5.6}$$

$$m_h \frac{d\mathbf{v_h}}{dt} = e\mathbf{E}, \tag{5.7}$$

where we have used Eq. (5.4). In these forms the equation of motion looks like that of a particle of positive charge $+e$ and positive mass m_h. Note also that the velocity of the hole can be written

$$\mathbf{v_h} = \frac{d\varepsilon_h}{d\mathbf{p_h}} \tag{5.8}$$

appropriate to a particle of energy ε_h and momentum $\mathbf{p_h}$. We hope that the reader will accept that the appropriate generalization of Eq. (5.7) to allow for the existence of collisions and for a magnetic field is

$$m_h\left(\frac{d\mathbf{v_h}}{dt} + \frac{\mathbf{v_h}}{\tau_h}\right) = e(\mathbf{E} + \mathbf{v_h} \times \mathbf{B}) \tag{5.9}$$

(cf. Eq. (3.23)). The scattering of a hole from state $\mathbf{k}_1$ to state $\mathbf{k}_2$ corresponds to the scattering of an electron from state $\mathbf{k}_2$ to state $\mathbf{k}_1$ so that the scattering time for holes τ_h is directly related to that for electrons.

To complete our demonstration that the behaviour of a nearly full valence band can be explained by considering only the vacant states, we must show that the current carried by the band as a whole can be expressed as a hole current. The crucial step is to use the fact that a full band carries no current. Removing an electron from the state $\mathbf{k}$ therefore causes a total current $\mathbf{j}$ which is minus the current carried by the electron. Hence $\mathbf{j} = -(-e)\mathbf{v}$, where $\mathbf{v}$, the group velocity of the electron, is the same as that of a hole in the same state. The current is thus $+e\mathbf{v_h}$ and is therefore that naturally associated with a hole in state $\mathbf{k}$. Since the total current can be written as the sum of the contributions from the electrons in the conduction band and holes in the valence band, these are referred to as the **charge carriers** in the semiconductor.

5.3   METHODS OF PROVIDING ELECTRONS AND HOLES

5.3.1   Donor and acceptor impurities

If atoms from group V of the periodic table (such as phosphorus or arsenic) are added to molten silicon or germanium they crystallize when the melt is cooled into a position normally occupied by a silicon or germanium atom. It is important that the impurity takes up a **substitutional** rather than an **interstitial**† position because this means that after forming the four covalent bonds demanded by the structure there is an extra valence electron left over which can occupy one of the states in the conduction band.

Escape of the electron to large distances leaves the impurity atom with a net positive charge; at finite separations the positive charge exerts an attractive force on the electron and leads to the existence of a bound state for the electron. The 'charged impurity plus electron' system is analogous to the 'proton plus electron' system and we can therefore estimate the strength of this binding by adapting the standard result for the energy levels of the hydrogen atom to allow for the fact that the electron is moving through a crystal rather than a vacuum. Thus we use m_e for the electron mass and assume that the crystal has a dielectric constant (relative permittivity) ε to obtain

$$E_n = -\frac{m_e e^4}{2\varepsilon^2 \hbar^2 n^2 (4\pi\varepsilon_0)^2}. \tag{5.10}$$

To estimate the spatial extent of the bound state wavefunctions we use the radii of the corresponding orbits as given by the Bohr theory,

$$r_n = \frac{\varepsilon n^2 \hbar^2}{m_e e^2} 4\pi\varepsilon_0. \tag{5.11}$$

The effective mass of electrons in germanium is 0.2 m; and the dielectric constant is 15.8. Using these values in Eqs. (5.10) and (5.11) gives an estimate

$$E_1 = -\left(\frac{m_e}{m\varepsilon^2}\right) \times 13.6\,\text{eV} \approx -0.01\,\text{eV} \tag{5.12}$$

for the ground state binding energy of the extra electron and

$$r_1 = \left(\frac{\varepsilon m}{m_e}\right) \times 0.53\,\text{Å} \approx 40\,\text{Å} \tag{5.13}$$

for the radius of the corresponding orbit (-13.6 eV and 0.53 Å are the corresponding values for hydrogen). Thus the combination of small effective mass and large dielectric constant gives very weak binding of the extra electron

---

† That is as an additional atom inserted in a gap between the atoms on the crystallographic sites.

to the impurity and a very extended wavefunction for the bound state. Since the bound state wavefunction extends over many atomic diameters, our approximation of using an effective mass and a macroscopic dielectric constant should work reasonably well. Note that our estimate of the binding energy is less than $k_B T$ at room temperature (0.026 eV), so we would expect most of the impurity atoms to be ionized at this temperature, with the extra electrons free to move through the crystal; the degree of ionization will be discussed more fully in the next section. Measured values of donor ionization energies for silicon and germanium are given in Table 5.1.

Eq. (5.10) gives an infinite series of bound states, but this applies only to the idealized case of a single impurity in an infinite crystal. In practice we can expect Eq. (5.10) to apply only if the mean separation between impurities is large compared to the size of the bound state wavefunction. Since the size of the wavefunction, according to Eq. (5.11), goes as the square of the quantum number n, the highest concentration for which we expect Eq. (5.10) to apply is proportional to $1/n^6$. The limiting concentration is thus of order: 10^{24} m^{-3} for $n = 1$, 10^{22} m^{-3} for $n = 2$ and 10^{21} m^{-3} for $n = 3$. In practice, therefore, the notion of hydrogenic bound states applies only to the lowest few levels. The important parameter in determining the number of ionized impurities is the energy difference between the lowest bound state and the lowest mobile state at the bottom of the conduction band. The presence of a few more bound states much closer to the bottom of the conduction band has very little effect and is usually ignored.

The electron energy level diagram of Fig. 5.1 is therefore modified near the bottom of the conduction band as shown in Fig. 5.4. The bound state is referred to as a **donor impurity level** because it is capable of giving an electron to the conduction band. The donor level is shown as a horizontal line in Fig. 5.4 to indicate that, since it is localized in space, the Fourier synthesis of the wavefunction of the bound state would require **k** states from a finite range of **k**. At absolute zero the extra electron associated with the donor impurity will

TABLE 5.1   Band gaps, donor and acceptor energies, relative permittivities and intrinsic carrier concentrations for Si and Ge

	E_G (eV) at $T = 0$ K	E_D (eV) P	As	E_A (eV) B	Al	ε	Intrinsic carrier concentration (m^{-3}) at 300 K
Si	1.08	0.045	0.049	0.045	0.057	11.7	2×10^{16}
Ge	0.66	0.012	0.013	0.010	0.010	15.8	2×10^{19}

(Data from the *American Institute of Physics Handbook*, 3rd edn, McGraw-Hill, New York (1972))

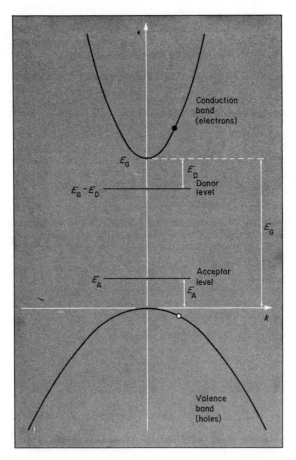

Fig. 5.4   Electron energy states of Fig. 5.1 modified to allow for the bound states near
donor and acceptor impurities

occupy the donor impurity level, but a relatively small energy is required to
ionize it into the conduction band.

Everything we have said about group V impurities and electrons applies
equally to group III impurities and holes. A group III element, such as boron or
aluminium, has one too few electrons to form the four covalent bonds demanded
by the diamond structure of silicon or germanium. This missing electron
represents a hole in the valence band with a tendency to be bound to the B^- or
Al^- ion. This results in the appearance of an **acceptor impurity level** (so called
because it accepts electrons from the valence band) for a hole about 0.01 eV
below the bottom of the hole valence band as shown in Fig. 5.2. Fig. 5.4 shows
the equivalent electron energy level; since the hole has a lower energy when it is

bound to the acceptor impurity, the occupation of the acceptor level by an electron represents a state of higher energy for the electron than a state in the valence band. Measured values of acceptor ionization energies for silicon and germanium can be found in Table 5.1.

Note that adding an impurity does not alter the total number of electron energy levels. Rather, levels are detached from the conduction and valence bands to form the donor and acceptor impurity levels. Therefore, although a crystal containing an acceptor atom has one too few electrons to form all the covalent bonds, the valence band has lost one energy level to form the localized acceptor level. There are thus just enough electrons to fill the valence band at $T = 0$. Similarly, the conduction band is completely empty at $T = 0$.

5.3.2 Thermal excitation of carriers

To calculate the number of charge carriers at any temperature T we use the electron energy level diagram of Fig. 5.4. This represents a simplification of the actual situation in silicon and germanium since in these materials the conduction band minimum is not at $k = 0$; this makes no difference for our present purposes (but see sections 5.4 and 5.5.3). The probability of occupation of a state of energy ε is given by the Fermi distribution function (Eq. (3.14))

$$f(\varepsilon) = \frac{1}{e^{(\varepsilon - \mu)/k_B T} + 1}.$$

in which the value of the chemical potential μ at each temperature has to be adjusted to obtain the correct total number of particles. The energy level $\varepsilon = \mu(T)$ is called the **Fermi level** and in semiconductor textbooks it is often denoted by the symbol ε_F. This can be very confusing since in Chapter 3 we followed the normal convention in statistical mechanics and reserved this notation for the value of μ at $T = 0$. For self-consistency we will continue this practice but the reader should be careful to distinguish between our use of the term 'Fermi energy' to refer to $\mu(0)$ and 'Fermi level' to refer to $\mu(T)$.

The other factor that determines the occupancy of states in the conduction and valence bands is the density of states per unit energy range. This can be calculated in the same way as in section 3.2 for free electrons, except that we must use the dispersion relations, Eqs. (5.1) and (5.2) respectively, for the conduction and valence bands rather than the free electron dispersion relation, Eq. (3.5). The resulting density of states (cf. Eq. (3.7)) is:

$$\text{conduction band} \qquad g(\varepsilon) = \frac{V}{2\pi^2 \hbar^3}(2m_e)^{3/2}(\varepsilon - E_G)^{1/2}, \qquad (5.14a)$$

$$\text{valence band} \qquad g(\varepsilon) = \frac{V}{2\pi^2 \hbar^3}(2m_h)^{3/2}(-\varepsilon)^{1/2}. \qquad (5.14b)$$

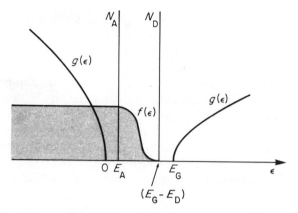

Fig. 5.5   Density of states $g(\varepsilon)$ and Fermi function $f(\varepsilon)$ for a semiconductor with the energy level scheme of Fig. 5.4. The Fermi level μ is normally within the band gap and is sufficiently far from the band edges that the probabilities of occupation of valence and conduction band states by electrons are close to 1 and 0, respectively

The density of states given by Eqs. (5.14) is sketched in Fig. 5.5. In addition to the parabolic densities of states in the conduction and valence bands there are $N_D V$ states at energy $E_G - E_D$ and $N_A V$ states at energy E_A corresponding to the N_D donors and N_A acceptors per unit volume† respectively. The Fermi function is also shown in Fig. 5.5.

At room temperature $k_B T$ is much less than E_G so that provided the Fermi level is somewhere in the band gap, not too close to the band edge, as illustrated (this is almost always the case), the Fermi function is very close to unity in the valence band and very small in the conduction band. This enables us to approximate the Fermi function is such a way that we can obtain analytic expressions for the number of electrons in the conduction band and the number of holes in the valence band. Thus for an electron energy ε in the conduction band we have $\varepsilon - \mu \gg k_B T$, and hence

$$e^{(\varepsilon - \mu)/k_B T} \gg 1$$

and

$$f(\varepsilon) \approx e^{(\mu - \varepsilon)/k_B T}. \tag{5.15}$$

---

† Since we are assuming that all the donor levels are degenerate, the *density* of states at the donor level is in fact infinite. A function that is infinite at one point and zero everywhere else but which has a finite area underneath it (the area under the density-of-states curve is the number of states) is known as a δ-function. A δ-function of unit area at $x = 0$ is denoted by $\delta(x)$. We should therefore write the density of states for the donor levels as $N_D V \delta(\varepsilon - (E_G - E_D))$, which says that it is a δ-function at $\varepsilon = E_G - E_D$ with area $N_D V$ underneath it.

Thus the number of electrons per unit volume in the conduction band is given by

$$n = \frac{1}{V} \int_{E_G}^{\infty} f(\varepsilon) g(\varepsilon) \, d\varepsilon$$

$$\approx \frac{1}{2\pi^2 \hbar^3} \int_{E_G}^{\infty} (2m_e)^{3/2} (\varepsilon - E_G)^{1/2} e^{(\mu - \varepsilon)/k_B T} \, d\varepsilon$$

$$= \frac{(2m_e)^{3/2}}{2\pi^2 \hbar^3} e^{(\mu - E_G)/k_B T} \int_{0}^{\infty} (\varepsilon - E_G)^{1/2} e^{-(\varepsilon - E_G)/k_B T} \, d(\varepsilon - E_G)$$

$$= N_C e^{(\mu - E_G)/k_B T}, \qquad (5.16)$$

where

$$N_C = 2 \left(\frac{2\pi m_e k_B T}{h^2} \right)^{3/2}. \qquad (5.17)$$

By comparison of Eqs. (5.15) and (5.16) we see that N_C is the effective number of levels per unit volume in the conduction band if we imagine them concentrated at the bottom of the band, $\varepsilon = E_G$. The temperature dependence of N_C arises because they are not so concentrated.

The probability that a state in the valence band is occupied by a hole is $1 - f(\varepsilon)$, which can be written

$$1 - f(\varepsilon) = 1 - \frac{1}{e^{(\varepsilon - \mu)/k_B T} + 1} = \frac{e^{(\varepsilon - \mu)/k_B T}}{e^{(\varepsilon - \mu)/k_B T} + 1}$$

$$= \frac{1}{e^{(\mu - \varepsilon)/k_B T} + 1}. \qquad (5.18)$$

For $\mu - \varepsilon \gg k_B T$ this can be approximated in a similar manner to that used in deriving Eq. (5.15) to give

$$1 - f(\varepsilon) \approx e^{(\varepsilon - \mu)/k_B T}. \qquad (5.19)$$

The number of holes per unit volume in the valence band is therefore

$$p = \frac{1}{V} \int_{-\infty}^{0} [1 - f(\varepsilon)] g(\varepsilon) \, d\varepsilon$$

$$\approx \frac{(2m_h)^{3/2}}{2\pi^2 \hbar^3} \int_{-\infty}^{0} (-\varepsilon)^{1/2} e^{(\varepsilon - \mu)/k_B T} \, d\varepsilon$$

$$= \frac{(2m_h)^{3/2}}{2\pi^2 \hbar^3} e^{-\mu/k_B T} \int_{0}^{\infty} \varepsilon_h^{1/2} e^{-\varepsilon_h/k_B T} \, d\varepsilon_h$$

$$= N_V e^{-\mu/k_B T}, \qquad (5.20)$$

where

$$N_V = 2\left(\frac{2\pi m_h k_B T}{h^2}\right)^{3/2} \qquad (5.21)$$

is the effective number of states per unit volume in the valence band if they were all concentrated at the top of the band, $\varepsilon = 0$.

Note that Eq. (5.15) predicts that the probability for occupation of a state in the conduction band by an electron is given by the classical Boltzmann distribution $f(\varepsilon) \propto \exp\left(-\varepsilon/k_B T\right)$. We thus expect the results of classical statistical mechanics to be valid for the conduction band electrons provided the approximation leading to Eq. (5.15) is justified. For example, the electron energy of Eq. (5.1) depends quadratically on k so that equipartition of energy predicts a mean thermal kinetic energy of $3k_B T/2$ for conduction band electrons just as for the particles in a classical gas. Inserting $\varepsilon_h = -\varepsilon$ in Eq. (5.19) shows that holes also obey a classical Boltzmann distribution.

The effect of doping a pure semiconductor with donor or acceptor impurities is to shift the chemical potential, as we will discuss shortly, and hence alter the carrier concentrations. But let us first note the following important result, obtained by multiplying together Eqs. (5.16) and (5.20):

$$np = N_C N_V e^{-E_G/k_B T}. \qquad (5.22)$$

The product of hole and electron concentrations is therefore independent of μ and hence of impurity concentration, although it does depend on temperature. Eq. (5.22) can be viewed as an example of the **law of mass action** used in the theory of chemical reactions; it is analogous to the constant product of hydrogen and hydroxyl ion concentrations in different aqueous solutions at the same temperature. The 'chemical reaction' in the semiconductor is the equilibrium of the electons and holes with thermal energy in the form of lattice vibrations and black-body radiation.

5.3.3 Intrinsic behaviour

In a pure semiconductor the electron and hole concentrations are equal since a hole in the valence band can only be created by excitation of an electron into the conduction band (Fig. 5.1). From Eq. (5.22) therefore

$$n_i = p_i = (N_C N_V)^{1/2} e^{-E_G/2k_B T}. \qquad (5.23)$$

The subscript i identifies these as the **intrinsic carrier concentrations**, so called because they are an intrinsic property of the pure semiconductor. The electrical conductivity that they give rise to is likewise called the **intrinsic conductivity**. The intrinsic carrier concentrations in silicon and germanium at room temperature are 2×10^{16} and 2×10^{19} m^{-3} respectively. Equating the values of n and p from

Eqs. (5.16) and (5.20) enables us to deduce the chemical potential of an intrinsic semiconductor. Thus

$$e^{(2\mu - E_G)/k_B T} = N_V/N_C$$

and

$$\mu = \tfrac{1}{2}E_G + \tfrac{1}{2}k_B T \ln (N_V/N_C) = \tfrac{1}{2}E_G + \tfrac{3}{4}k_B T \ln (m_h/m_e). \qquad (5.24)$$

Since $k_B T \ll E_G$ the second term is small and the Fermi level is essentially in the middle of the band gap. Note that, using Eqs. (5.22) and (5.23), the product of electron and hole concentrations in *any* semiconductor can conveniently be written

$$np = n_i^2(T), \qquad (5.25)$$

where $n_i(T)$ is the intrinsic carrier concentration at the same temperature.

To find the circumstances in which we might expect to observe intrinsic behaviour experimentally, let us consider what happens when a small amount of donor impurity is added. At room temperature small amounts of impurity alter the chemical potential only slightly so that it remains close to the centre of the band gap. The donor level remains in the high-energy tail of the Fermi function as in Fig. 5.5 and consequently almost all the electrons from the donor atoms will be in the conduction band. The electron concentration will therefore be seriously affected unless the donor concentration is small compared to the intrinsic carrier concentration given by Eq. (5.23). Thus the requirement for intrinsic behaviour in silicon at room temperature is $N_D < 2 \times 10^{16}$ m^{-3}. Since the atomic concentration is 5×10^{28} m^{-3} the impurity content needs to be less than 1 in 10^{12}.

Although the requirement for germanium is less stringent (< 1 in 10^9), the technical problem of making crystals of such extraordinarily high purity is one reason why semiconductors were not widely used earlier. The breakthrough that made modern developments possible was the discovery of **zone refining**. This depends on the fact that impurities are more soluble in the liquid than in the solid, so that if a molten zone is moved along a solid the impurities are swept along with it. In practice the crystal is held just below its melting point in a furnace and a small zone is melted by an auxiliary induction heater; by pulling the sample through the induction heater the molten zone is moved along it. The process is repeated until the required purity is achieved. The purified material is subsequently **doped** with controlled amounts of donors or acceptors to give the desired properties.

5.3.4   Extrinsic behaviour

When acceptors and donors are present, the chemical potential is determined by the requirement that the total number of electrons is correct. A convenient way of ensuring this is to require that μ be chosen so that the crystal is electrically neutral; it was formed from electrically neutral atoms so that a charge that appears at any point must be accompanied by an equal and opposite charge. The charges that concern us are electrons in the conduction band, holes in the valence band and ionized donor and acceptor impurities. The condition for electrical neutrality is that the densities of negative and positive charge associated with these should be equal. That is

$$n + N_A^- = p + N_D^+ \qquad (5.26)$$

where N_A^- and N_D^+ are the concentrations of ionized acceptors and donors, given in terms of the Fermi function by†

$$N_D^+ = N_D[1 - f(E_G - E_D)] \qquad (5.27)$$

and

$$N_A^- = N_A f(E_A). \qquad (5.28)$$

By inserting Eqs. (5.16) and (5.20) for n and p into Eq. (5.26) it is possible to determine μ and hence n and p. Analytic solutions can only be found in special limiting cases and we now proceed to discuss some of these.

The commonest situation is that in which impurities of both types are present. Let us suppose that the number of donors exceeds the number of acceptors. Since the acceptor levels have a lower energy they will be fully occupied at absolute zero by electrons from the donor impurities, leaving $N_D - N_A$ donor levels un-ionized. Since only the Fermi level can be partly occupied at $T = 0$ we have $\mu = E_G - E_D$. At very low temperatures, $k_B T \ll E_D$, where the number of ionized donors has not changed much, the Fermi level remains close to this value, so that the electron concentration in the conduction band is, from Eq. (5.16),

$$n \approx N_C\, e^{-E_D/k_B T} \qquad \text{for } k_B T \ll E_D. \qquad (5.29)$$

Because $E_D \ll E_G$ the electron concentration is much greater than the intrinsic concentration given by Eq. (5.23); consequently, because of the law of mass action, Eq. (5.22), the hole concentration is very much less than its intrinsic value. Material of this type, in which donor impurities are dominant and the number of electrons exceeds the number of holes, is known as **n-type material**; the electrons are referred to as the **majority carrier** and the holes as the **minority carrier**. In an exactly similar way excess acceptors give **p-type material** in which holes are the majority carrier. In p-type material containing more acceptors than donors the Fermi level at absolute zero is at the acceptor level, $\mu = E_A$, and

---

† These expressions for N_D^+ and N_A^- are not strictly correct. They do not allow for the fact that the bound impurity states can be occupied by carriers of either up (↑) or down (↓) spin, but that the probability of dual occupation is zero (because of electron–electron interactions). Using the correct expressions (C. Kittel and H. Kroemer, *Thermal Physics*, 2nd edn, W. H. Freeman, San Fransisco (1980), p. 143) does not make much difference.

the result corresponding to Eq. (5.29) is

$$p \approx N_V \, e^{-E_A/k_B T} \qquad \text{for } k_B T \ll E_A. \tag{5.30}$$

Returning to the n-type material, with increasing temperature the number of ionized donors becomes comparable to the total number of donors; because the donor level must then be in the tail of the Fermi distribution function, the Fermi level lies below the donor level. We thus have a range of temperature in which essentially all the donors and acceptors are ionized and the electron density, from Eq. (5.26), is

$$n = N_D - N_A. \tag{5.31}$$

The Fermi level, obtained by equating the values of n of Eqs. (5.16) and (5.31), is

$$\mu = E_G - k_B T \ln \left(\frac{N_C}{N_D - N_A} \right). \tag{5.32}$$

The doping levels in most semiconductor devices are such as to bring them into the region where all the impurities are ionized at room temperature;† the resulting temperature independence of the majority carrier concentration is often important for the successful operation of the device. The temperature-dependent minority carrier concentration in this region can be obtained by using the law of mass action, Eq. (5.22).

On increasing the temperature of an n-type semiconductor still further the hole concentration increases towards the electron concentration, the Fermi level falls towards the centre of the gap and eventually intrinsic behaviour is observed. The temperature dependences of the chemical potential and carrier concentrations in an n-type semiconductor are thus as shown in Fig. 5.6. In the region where the concentrations are determined by the impurities the behaviour of the semiconductor is said to be **extrinsic**. For a p-type semiconductor the results corresponding to Eqs. (5.31) and (5.32) for the hole concentration and chemical potential in the temperature range in which all the acceptors and donors are ionized are

$$p = N_A - N_D \tag{5.33}$$

$$\mu = k_B T \ln \left(\frac{N_V}{N_A - N_D} \right). \tag{5.34}$$

The very low-temperature behaviour is different if only one type of impurity is present. If only donors are present, the donor level is completely full at absolute zero and almost so at very low temperatures. The Fermi level must then lie between the donor level and the conduction band; the situation is similar to

---

† Although the donor and acceptor ionization energies are comparable to $k_B T$ (Table 5.1), all the donors and acceptors are ionized because the effective densities of states in the conduction and valence bands, N_C and N_V, are very much larger than the impurity concentrations.

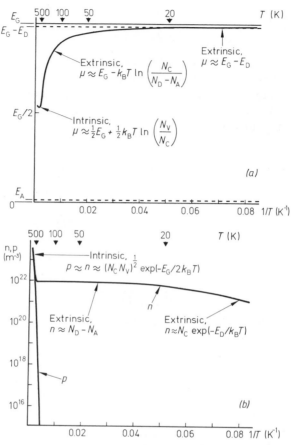

Fig. 5.6   Variations of (a) the Fermi level μ and (b) the electron and hole concentrations (note the logarithmic scale) with $1/T$ for an n-type semiconductor containing a significant number of acceptor impurities. The figure was calculated for a germanium semiconductor with $N_D = 10^{22}$ m^{-3}, $E_D = 0.012$ eV, $N_A = 10^{21}$ m^{-3} and $E_A = 0.010$ eV; the scale at the top shows temperature values for this case

intrinsic material, except that the donor level takes the place of the valence band Thus, approximating Eq. (5.27) in the way we used to obtain Eq. (5.19), we find

$$N_D^+ \approx N_D \, e^{(E_G - E_D - \mu)/k_B T}. \tag{5.35}$$

The hole concentration can be ignored so that electrical neutrality (Eq. (5.26)) requires $n = N_D^+$. Therefore, using Eqs. (5.16) and (5.35),

$$n = (N_C N_D)^{1/2} \, e^{-E_D/2k_B T}. \tag{5.36}$$

The exponents of Eqs. (5.29) and (5.36) differ by a factor of 2, so appreciable amounts of minority impurity have a noticeable effect at very low temperatures.

5.4   ABSORPTION OF ELECTROMAGNETIC RADIATION

Fig. 5.7 shows the absorption coefficient for electromagnetic radiation as a function of photon energy (frequency) for germanium at two different temperatures. At $T = 77$ K there is an onset of absorption as the photon energy increases through about 0.73 eV and a further sharp increase in absorption at an energy of 0.87 eV. The increases in absorption occur when the photons have sufficient energy to excite a valence band electron into the conduction band, creating an electron–hole pair in the process. The photon energies at the onset of absorption therefore provide a measure of the energy gap in semiconductors.

To understand why there are apparently two energy gaps in germanium it is necessary to know that germanium (like silicon) is an **indirect (band) gap semiconductor** in that the maximum of the valence band and the minimum of the conduction band occur at different values of **k**. The valence band maximum is at **k** = 0 but the minimum in the conduction band is at a **k** vector on the Brillouin zone boundary in the [1 1 1] direction. This is illustrated in Fig. 5.8(*a*), which shows electron dispersion relations for germanium for **k** in the [1 1 1] direction. The lower onset energy for photon absorption corresponds to a good approximation to the minimum energy difference, E_G, between the two bands as indicated in the figure. The higher onset energy corresponds to the minimum energy at which an electron can be promoted to the conduction band with no change in wavenumber; this is denoted E'_G in the figure. When an electron

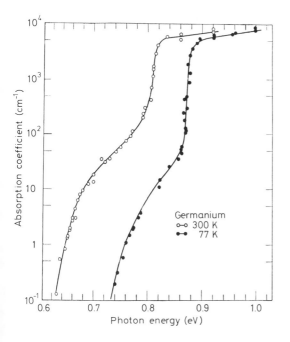

Fig. 5.7   Absorption coefficient for electromagnetic radiation of germanium versus photon energy at temperatures of 77 and 300 K. (Reproduced with permission from W. C. Dash and R. Newman, *Phys. Rev.* **99**, 1151 (1955))

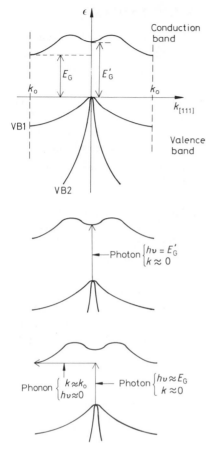

(a) Dispersion relations for the
conduction and valence bands of
germanium for **k** in the [1 1 1]
direction. E_G is the minimum energy
difference between the two bands. E'_G is
the minimum vertical distance. The two
valence bands VB1 and VB2 are
degenerate at their maximum energy
and lead to holes with two different
effective masses

(b) Vertical transition of an electron
from the valence band to the
conduction band caused by absorption
of a photon of energy E'_G; the
momentum change produced by such a
photon is very small

(c) Simultaneous absorption of a
photon and a phonon provides the
energy E_G and momentum $\hbar k_o$,
respectively, to enable the electron to
make a transition from the valence
band maximum to the conduction band
minimum

Fig. 5.8

absorbs a photon, energy and momentum must be conserved. The change in
energy $\Delta\varepsilon$ and (crystal) momentum $\hbar\Delta\mathbf{k}$ of the electron are therefore

$$\Delta\varepsilon = \hbar\omega_{ph} \qquad \text{and} \qquad \hbar\Delta\mathbf{k} = \hbar\mathbf{k}_{ph}$$

where ω_{ph} and $\mathbf{k}_{ph}$ are the angular frequency and wavevector of the photon. For
a photon of energy 1 eV, the wavelength is about 10^{-6} m and thus much greater
than the interatomic spacing R. The momentum h/λ of the photon is thus much
less than h/R, which is the order of the momentum of a conduction band
electron. Photons of energy 1 eV cannot therefore cause significant momentum
changes of the electrons and transitions are effectively vertical as shown in Fig.
5.8(b). The onset energy for such transitions is the minimum vertical difference

between valence and conduction bands and this explains the higher of the two onset energies of Fig. 5.7.

Why then are photons absorbed at energies between E_G and E'_G? Clearly there must be some source (or sink) of momentum within the crystal. The quanta of lattice vibrations, phonons, provide this; in contrast to photons, phonons have large momenta and small energy. A phonon of momentum h/R has an energy of order $k_B \Theta_D$ where Θ_D is the Debye temperature (section 2.6.4), and this is usually very small compared to 1 eV. To explain the lower onset energy of Fig. 5.7 we envisage a process like that shown in Fig. 5.8(c) in which the electron simultaneously absorbs a photon and emits (or absorbs) a phonon. The photon provides the energy change of the electron, the phonon the momentum change. In time-dependent perturbation theory such a process would appear in second order whereas a process involving a photon only is present in first order, and this explains why the onset of absorption at $\hbar \omega_{ph} = E_G$ is less dramatic than that at $\hbar \omega_{ph} = E'_G$. In the **direct (band) gap semiconductor**, InSb, in which the maximum of the valence band and minimum of the conduction band occur at the same **k** value, only one onset energy is observed.

The absorption spectrum of a semiconductor thus provides a direct method for measuring the energy gap. It is apparent from Fig. 5.7 that the band gap and hence the electron energies are dependent on temperature. The temperature dependence has two causes: thermal expansion changes the interatomic spacing and hence the lattice potential; also the electron energies are changed by the presence of thermally excited lattice vibrations.

5.5   TRANSPORT PROPERTIES

To describe the motion of the carriers in the presence of electric and magnetic fields we use Eq. (3.23) for the electrons and Eq. (5.9) for the holes. For easy reference we write these equations again here and renumber them.

$$m_e\left(\frac{d\mathbf{v}_e}{dt} + \frac{\mathbf{v}_e}{\tau_e}\right) = -e\mathbf{E} - e\mathbf{v}_e \times \mathbf{B} \qquad (5.37a)$$

$$m_h\left(\frac{d\mathbf{v}_h}{dt} + \frac{\mathbf{v}_h}{\tau_h}\right) = +e\mathbf{E} + e\mathbf{v}_h \times \mathbf{B}. \qquad (5.37b)$$

5.5.1   Electrical conductivity

When only a dc electric field is present the solutions of Eqs. (5.37) are

$$\mathbf{v}_e = -\frac{e\tau_e}{m_e}\mathbf{E} = -\mu_e\mathbf{E} \qquad (5.38a)$$

$$\mathbf{v}_h = \frac{e\tau_h}{m_h}\mathbf{E} = \mu_h\mathbf{E} \qquad (5.38b)$$

(cf. Eq. (3.24)), which gives the electron and hole mobilities as

$$\mu_e = \frac{e\tau_e}{m_e} \quad \text{and} \quad \mu_h = \frac{e\tau_h}{m_h}. \tag{5.39}$$

The resulting electric current density, obtained by summing the electron and hole contributions, is

$$\mathbf{j} = -ne\mathbf{v}_e + pe\mathbf{v}_h = \left(\frac{ne^2\tau_e}{m_e} + \frac{pe^2\tau_h}{m_h}\right)\mathbf{E}$$

$$= (ne\mu_e + pe\mu_h)\mathbf{E} = \sigma\mathbf{E}, \tag{5.40}$$

which is Ohm's law with an electrical conductivity σ given by

$$\sigma = ne\mu_e + pe\mu_h. \tag{5.41}$$

Since the electron and hole mobilities are usually comparable the relative carrier densities determine the relative contributions of the electrons and holes to the conductivity. In the intrinsic region the two contributions are usually similar but in the extrinsic region the conductivity is normally dominated by the majority carrier.

Measured conductivities for arsenic-doped n-type germanium are plotted logarithmically against $1/T$ in Fig. 5.9. The steep increase in σ at high temperatures given by the broken line on the left, which was observed in the purest specimen only, represents the increase in n and p associated with the transition to intrinsic behaviour (Eq. (5.23)). The fall in σ at low temperatures on the right of the figure for the two purest specimens is associated with the 'freezing' out of electrons on the donor levels; since the slopes of the two curves differ approximately by a factor of 2, it is tempting to associate the upper curve with Eq. (5.36) and the lower curve with Eq. (5.29) (see problem 5.4). At intermediate temperatures the donors are fully ionized and the observed decrease in conductivity with rising temperature results from the decrease in electron mobility caused by increased scattering from thermally excited lattice vibrations.

Qualitatively different behaviour is displayed by the most impure specimen; the conductivity is independent of temperature at the lowest temperatures just as for metals. As we discussed in section 5.3.1, the impurity concentration in this specimen is that at which the wavefunctions of the lowest bound state on neighbouring donor atoms begin to overlap. This results in the formation of an 'impurity' energy band of mobile electron states (section 13.5.6); the electrons in the donor levels can then conduct electricity, leading to a temperature-independent carrier concentration at low temperatures. The formation of an impurity energy band associated with spatially disordered donor atoms shows that crystalline order is not essential for the presence of mobile electron states;

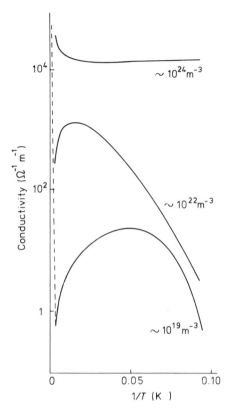

Fig. 5.9   Temperature dependence of the conductivity of three samples of germanium containing arsenic donor impurities with the approximate concentrations indicated. (Reproduced with permission from P. P. Debye and E. M. Conwell, *Phys. Rev.* **93**, 693 (1954))

the existence of *liquid* metals is an even more convincing demonstration of this fact.

We can account roughly for the order of magnitude of the conductivities shown on Fig. 5.9. We take a moderately low temperature (50 K) and the specimen of intermediate purity so that we might expect the scattering to be largely due to the impurities, which are almost 100% ionized at this temperature. The experimental conductivity is therefore of order $10^3 \, \Omega^{-1} \, m^{-1}$ for $n = 10^{22} \, m^{-3}$. From Eq. (5.40), with $e \approx 10^{-19} \, C$, a typical mobility is thus $1 \, m^2 \, V^{-1} \, s^{-1}$, and from Eq. (5.39) with $m_e = 10^{-31}$ kg a typical collision time is $\tau_e = 10^{-12}$ s. We can relate this to the collision cross section A of an impurity by using the elementary kinetic theory expression,

$$l = 1/N_D A, \tag{5.42}$$

for the mean free path l (Flowers and Mendoza,[1] section 6.4.1). We have taken the impurity concentration as the donor concentration. The mean free path l is related to the scattering time by

$$\tau = l/\bar{v}_e$$

where $\bar{v}_e$ is the mean group velocity of the electrons; we take this to be $(3k_BT/2m_e)^{1/2}$ since, as we have already shown, equipartition of energy can be used for the conduction band electrons. From the experimental value of τ_e we therefore deduce

$$A = \frac{1}{N_D l} = \frac{1}{N_D \bar{v}_e \tau_e} = \frac{1}{N_D \tau_e} \left(\frac{2m_e}{3k_B T}\right)^{1/2} \approx 10^{-15} \text{m}^2$$

for $N_D = 10^{22} \text{ m}^{-3}$, $m_e = 10^{-31}$ kg and $T = 50$ K.

An impurity thus has a collision diameter of 3×10^{-8} m ($= 300$ Å) which is large on an atomic scale. Since the donors are ionized it is Coulomb scattering that is responsible for this cross section. The Rutherford scattering formula can be used to obtain an accurate value of A, but for our purpose we can make a rough estimate by saying that the radius inside which scattering is appreciable is that at which the Coulomb potential energy $(e^2/4\pi\varepsilon_0 \varepsilon r)$ is equal to the kinetic energy $(3k_B T/2)$ of the incident electron. Thus

$$r \approx \frac{e^2}{4\pi\varepsilon_0 \varepsilon k_B T} \approx 10^{-8} \text{ m} = 100 \text{ Å}$$

for $\varepsilon \approx 10$, so that the effective collision diameter is about 200 Å, in agreement with our rough estimate from the conductivity. Note that, in contrast to the situation in a metal, because the electron velocity and scattering cross section are temperature-dependent, the contribution of impurity scattering to the total resistivity is temperature-dependent in a semiconductor.

5.5.2   Hall effect

We have already defined the Hall coefficient R_H in section 3.3.5 as determining the electric field generated transverse to an applied current flow $\mathbf{j}$ and magnetic field $\mathbf{B}$ via Eq. (3.36), i.e.

$$\mathbf{E}_H = R_H \mathbf{B} \times \mathbf{j}.$$

We also calculated its value for a free electron metal. In a semiconductor, where there are two types of carrier, the calculation is more complicated. Often however the density of one type of carrier is much greater than that of the other and we can then use the calculation of section 3.3.5 directly to obtain†

$$R_H = -\frac{1}{ne} \qquad \text{in n-type semiconductor } (n \gg p),$$

$$R_H = +\frac{1}{pe} \qquad \text{in p-type semiconductor } (p \gg n).$$

(5.43)

---

† These results depend on our assumption of a single relaxation time, independent of carrier velocity, for each type of carrier; failure of this assumption can modify Eqs. (5.43) by factors of order unity, and give rise to a change with magnetic field in the electrical resistance measured along the current direction, a phenomenon referred to as **magnetoresistance**.

The sign of the Hall effect is thus determined by the sign of the majority charge carrier, and measurement of R_H enables the carrier concentration to be determined. By combining Eqs. (5.41) and (5.43) we find that simultaneous measurements of σ and R_H enable the carrier mobility to be determined from

$$\mu = |R_H|\sigma. \tag{5.44}$$

We have assumed that the conductivity is also dominated by the majority carrier.

For the example considered in the previous section

$$R_H = \frac{1}{ne} \approx \frac{1}{10^{22} \times 10^{-19}} = 10^{-3}\ \Omega\,\text{m}\,\text{T}^{-1}$$

compared with a resistivity $1/\sigma = 10^{-3}\ \Omega\,\text{m}$; the ohmic and Hall electric fields are therefore equal in a field of 1 T. At this magnetic field the total electric field is at $45°$ to the current flow, that is the **Hall angle** is $45°$. This magnetic field, the **Hall field** B_0, is a useful measure of the strength of the Hall effect (the smaller B_0 the larger the Hall effect); from the above calculation we see that it is given by

$$B_0 = \frac{1}{|R_H|\sigma} = \frac{1}{\mu}.$$

The smaller carrier density means that the Hall effect is larger in semiconductors than metals (see problem 5.5) and hence makes semiconductors useful for the construction of **Hall probes** for measuring magnetic fields. Fairly lightly doped material gives the largest sensitivity although it is desirable that the carrier concentration remain independent of temperature (all impurities ionized and negligible intrinsic carriers) throughout the range of operation.

When both electrons and holes are present in significant numbers, in the intrinsic region for example, the calculation of the Hall coefficient is more complicated. The steady state solutions of Eqs. (5.37) are

$$\mathbf{v}_e = -\frac{e\tau_e}{m_e}(\mathbf{E} + \mathbf{v}_e \times \mathbf{B}) = -\mu_e(\mathbf{E} + \mathbf{v}_e \times \mathbf{B})$$

$$\mathbf{v}_h = \frac{e\tau_h}{m_h}(\mathbf{E} + \mathbf{v}_h \times \mathbf{B}) = \mu_h(\mathbf{E} + \mathbf{v}_h \times \mathbf{B}) \tag{5.45}$$

so that the current density $\mathbf{j}$ is

$$\mathbf{j} = -ne\mathbf{v}_e + pe\mathbf{v}_h = ne\mu_e(\mathbf{E} + \mathbf{v}_e \times \mathbf{B}) + pe\mu_h(\mathbf{E} + \mathbf{v}_h \times \mathbf{B}). \tag{5.46}$$

We assume the experimental geometry of Fig. 3.12 in which the current flows along x, the magnetic field is applied along z and the Hall field appears along y.

The boundary conditions in this geometry require $j_y = 0$, but this does not mean that the hole and electron currents in this direction are individually zero, merely that they are equal and opposite.

For simplicity we will restrict our calculation to small magnetic fields. If we note that the y components of the carrier velocities are linear in B and keep only terms up to first order in B, then from Eq. (5.46)

$$j_x = eE_x(n\mu_e + p\mu_h) \tag{5.47a}$$

and

$$j_y = 0 = eE_y(n\mu_e + p\mu_h) - eB_z(n\mu_e v_{ex} + p\mu_h v_{hx})$$
$$= eE_y(n\mu_e + p\mu_h) + eB_z E_x(n\mu_e^2 - p\mu_h^2) \tag{5.47b}$$

where we have used Eq. (5.45) to obtain the last line. Elimination of E_x between these two equations gives

$$E_y = -\frac{j_x B_z(n\mu_e^2 - p\mu_h^2)}{e(n\mu_e + p\mu_h)^2}. \tag{5.48}$$

From the definition of the Hall coefficient (Eq. (3.36)) therefore

$$R_H = \frac{(p\mu_h^2 - n\mu_e^2)}{e(p\mu_h + n\mu_e)^2}. \tag{5.49}$$

Thus a minority carrier can determine the sign of the Hall coefficient if its mobility is high enough.

It is interesting to work out the magnitude of the equal and opposite electron and hole currents in the y direction. Thus, using Eqs. (5.47) and (5.48),

$$j_{ey} = -j_{hy} = n\mu_e e(E_y + \mu_e E_x B_z)$$

$$= n\mu_e eE_x B_z\left(\frac{p\mu_h^2 - n\mu_e^2}{p\mu_h + n\mu_e} + \mu_e\right)$$

$$= j_x B_z(\mu_e + \mu_h)\frac{\sigma_e \sigma_h}{(\sigma_e + \sigma_h)^2}, \tag{5.50}$$

where $\sigma_e (= ne\mu_e)$ and $\sigma_h (= pe\mu_h)$ are the electron and hole contributions to the conductivity. Eq. (5.50) shows that there is a steady flow of electrons and holes in the negative y direction, which is largest when the electron and hole conductivities are comparable. This implies the creation of electron–hole pairs on one side of the sample, absorbing energy, and their mutual annihilation at the other side, releasing energy. Consequently a transverse temperature gradient, known as the **Ettinghausen effect**, develops; such an effect is indicative of two types of carrier.

If terms of higher order in B are retained in the calculation then the resistance measured along the direction of current flow is found to be dependent on the magnetic field. The magnetoresistance depends quadratically on the field in small fields (Smith,[18] p. 114).

5.5.3 Cyclotron resonance

A cyclotron is a machine used to accelerate charged particles; its operation depends critically on the fact that in a constant dc magnetic field non-relativistic particles of mass m move in circular orbits at an angular frequency $\omega_C = eB/m$, independent of their energy, and can thus absorb energy from a suitably phased ac electric field at this frequency. In semiconductors the same principle applies and, by measuring the cyclotron frequencies at which absorption of energy from ac electric fields occurs, it is possible to determine the effective masses of the carriers. For one type of carrier, holes say, we can analyse the situation by writing Eq. (5.37b) in component form for B in the z direction and E in the xy plane:

$$\frac{dv_{hx}}{dt} + \frac{v_{hx}}{\tau_h} = \frac{e}{m_h}(E_x + v_y B)$$

$$\frac{dv_{hy}}{dt} + \frac{v_{hy}}{\tau_h} = \frac{e}{m_h}(E_y - v_x B).$$

The symmetry of these two equations enables us to reduce them to one equation by adding i times the second to the first and writing $u = v_{hx} + iv_{hy}$ and $\mathscr{E} = E_x + iE_y$.† This gives

$$\frac{du}{dt} + \frac{u}{\tau_h} = \frac{e}{m_h}(\mathscr{E} - iuB). \tag{5.51}$$

We assume an ac electric field of the form, $\mathscr{E} = \mathscr{E}_0\, e^{i\omega t}$; note that this corresponds to a circularly polarized wave since $E_x = \mathscr{E}_0 \cos(\omega t)$ and $E_y = \mathscr{E}_0 \sin(\omega t)$. The magnetic field of the radiation is negligible at the frequencies that concern us and we will ignore it. The solution for u will be of the form $u = u_0\, e^{i\omega t}$ where u_0 is in general complex. Inserting this solution into Eq. (5.51) we find

$$\left(\frac{1}{\tau_h} + i\omega\right)u_0 = \frac{e}{m_h}(\mathscr{E}_0 - iu_0 B)$$

or

$$u_0 = \frac{-ie\mathscr{E}_0}{m_h(\omega - \omega_C - i/\tau_h)} \tag{5.52}$$

where $\omega_C = -eB/m_h$ is the cyclotron frequency of the holes. The minus sign in ω_C defines the direction of rotation of the holes around the field as shown in Fig. 5.10 and hence the direction of rotation of the circularly polarized electric field

---

† We are using the general rule that two-dimensional vectors can be represented by complex numbers on an Argand diagram.

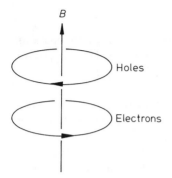

B

Holes

Electrons

Fig. 5.10   The directions of the
cyclotron orbits of electrons and holes

that will couple to them. With the sign convention that we have adopted the cyclotron frequency of electrons is positive.

Eq. (5.52) gives the response of the holes to an applied electric field. To interpret this equation we note that it is of the same form as the response of a series L-C-R circuit to an applied ac voltage (Grant and Phillips,[3] p. 268). If the current in the circuit of Fig. 5.11 is $I = I_0 e^{i\omega t}$, then, for frequencies close to the resonant frequency $\omega_R (= 1/(LC)^{1/2})$ and in the limit of a large quality factor, I_0 can be written

$$I_0 = \frac{-iV_0}{2L(\omega - \omega_R - i/\tau)} \tag{5.53}$$

where $\tau = 2L/R$ is the time constant for the decay of free oscillations in the circuit. The L-C-R circuit exhibits a maximum power absorption at the resonant frequency, and comparison of Eqs. (5.52) and (5.53) therefore suggests that a maximum in the absorption of electromagnetic energy by the semiconductor should occur at the cyclotron frequency.

For a sharp resonance the damping must be small and in the case of the L-C-R circuit this requires $\omega_R \tau \gg 1$. The corresponding condition for the semiconductor is $|\omega_C| \tau_h \gg 1$, which is just the condition that the hole should complete several orbits between collisions. This is not an easy condition to

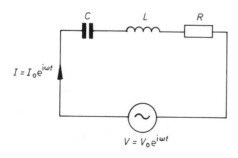

C L R

$I = I_0 e^{i\omega t}$

$V = V_0 e^{i\omega t}$ Fig. 5.11   Series L-C-R circuit achieve; cyclotron resonance experiments typically operate at frequencies of about 20 GHz and thus

$$\omega_C \tau \geqslant 1 \quad \text{when} \quad \tau \geqslant 10^{-11}\,\text{s}.$$

Collision times of this order are only achieved in pure semiconductors at low temperatures where scattering by lattice vibrations is small. Under such conditions the conductivity is usually small enough that the electromagnetic skin effect does not prevent penetration of the sample by the electric field but the number of carriers in thermal equilibrium is so small that the resonance can only be observed easily if the number is increased, for example, by shining light on the semiconductor with a frequency greater than E_G/h.

The resonance is often observed using plane polarized radiation; since this can be considered as a coherent sum of left and right circularly polarized light, it does not distinguish between electrons and holes. Typical results for silicon are shown in Fig. 5.12. We see that there are four absorption peaks, two corresponding to holes with different effective masses and two to electrons with different effective masses. The two effective masses for holes reflect the fact that the maximum valence band energy in silicon is actually a degenerate level in two different energy bands with electron dispersion relations of different curvature (germanium also exhibits this property as can be seen from Fig. 5.8(a)). The two effective masses for electrons reflect the fact that the conduction band dispersion relation in silicon is anisotropic; electrons travelling in different directions have different effective masses.

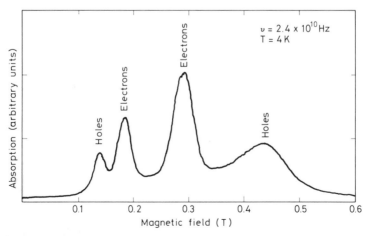

Fig. 5.12   Typical cyclotron resonance signal from silicon. The field lies in the (1 1 0) plane and is at 30° to the [0 0 1] axis. (Reproduced with permission from G. Dresselhaus, A. F. Kip and C. Kittel, *Phys. Rev.* **98**, 368 (1955))

To extend our theory of cyclotron resonance to cover this possibility we need to know that the conduction band minimum in silicon occurs at six degenerate points in k-space situated on the coordinate axes as shown on Fig. 5.13. The constant energy surfaces around each point are ellipsoids of revolution as shown. Thus if we consider the electrons situated close to $(0, 0, k_0)$ on the $+k_z$ axis the dispersion relation is of the form

$$\varepsilon = \frac{\hbar^2}{2}\left(\frac{k_x^2 + k_y^2}{m_T} + \frac{(k_z - k_0)^2}{m_L}\right), \tag{5.54}$$

where the subscripts T and L refer to transverse and longitudinal to the axis of the ellipsoid. We cannot use Eq. (5.37a) to calculate the dynamics of the electrons since this assumes a single isotropic effective mass. Instead we use the three-dimensional generalization of Eq. (4.24). Thus we take

$$\hbar\frac{d\mathbf{k}}{dt} = -e\mathbf{v} \times \mathbf{B}, \tag{5.55}$$

in which we have ignored collisions and also the driving electric field for the cyclotron motions; this is self-consistent since, in the absence of damping by collisions, the cyclotron motions will not decay and so should be steady state solutions of Eq. (5.55).

The electron velocity to insert in Eq. (5.55) is obtained from the three-dimensional equivalent of Eq. (4.22),

$$\mathbf{v} = \frac{1}{\hbar}\nabla_k\varepsilon, \tag{5.56}$$

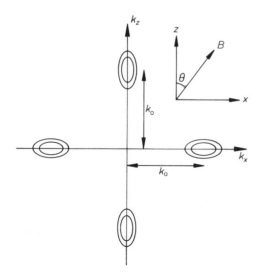

Fig. 5.13   Energy contours in k-space near the conduction band minimum in silicon. Two more pockets of conduction band electrons are found on the $\pm k_y$ axis. In the text we calculate the cyclotron resonance of the electrons on the $\pm k_z$ axis for a magnetic field B in the direction indicated

which says that $\mathbf{v}$ is determined by the gradient of ε in $\mathbf{k}$-space. From Eqs. (5.54) and (5.56) we deduce therefore

$$v_x = \frac{\hbar k_x}{m_{\mathrm{T}}}, \qquad v_y = \frac{\hbar k_y}{m_{\mathrm{T}}}, \qquad v_z = \frac{\hbar(k_z - k_0)}{m_{\mathrm{L}}}, \qquad (5.57)$$

for electrons near to the conduction band minimum at $(0, 0, k_0)$. The most general situation for electrons in this region of $\mathbf{k}$-space is a magnetic field at an angle θ to the axis of the ellipsoid as shown in Fig. 5.13. For convenience we take B to be in the xz plane. Using Eqs. (5.57) we find that the components of Eq. (5.55) can be written

$$\hbar \frac{dk_x}{dt} = -\frac{e\hbar B k_y}{m_{\mathrm{T}}} \cos \theta,$$

$$\hbar \frac{dk_y}{dt} = \frac{e\hbar B k_x}{m_{\mathrm{T}}} \cos \theta - \frac{e\hbar B(k_z - k_0)}{m_{\mathrm{L}}} \sin \theta,$$

$$\hbar \frac{dk_z}{dt} = \frac{e\hbar B k_y}{m_{\mathrm{T}}} \sin \theta.$$

If we look for an oscillatory solution of the form

$$k_x = k_1 e^{i\omega t}, \qquad k_y = k_2 e^{i\omega t}, \qquad k_z = k_0 + k_3 e^{i\omega t},$$

then we readily find that such a solution can exist only at the frequency

$$\omega_{\mathrm{C}} = eB\left(\frac{\sin^2 \theta}{m_{\mathrm{L}} m_{\mathrm{T}}} + \frac{\cos^2 \theta}{m_{\mathrm{T}}^2}\right)^{1/2}. \qquad (5.58)$$

The cyclotron frequency therefore depends on the orientation of the field; the frequency is often quoted in terms of a **cyclotron effective mass** defined by

$$m_{\mathrm{C}} = eB/\omega_{\mathrm{C}}. \qquad (5.59)$$

In the two extreme limits, $\theta = 0°$ and $\theta = 90°$, we see that $m_{\mathrm{C}} = m_{\mathrm{T}}$ and $m_{\mathrm{C}} = (m_{\mathrm{L}} m_{\mathrm{T}})^{1/2}$ respectively. The two electron resonances on Fig. 5.12 represent contributions from two of the different **pockets** of conduction band electrons on Fig. 5.13 (see problem 5.7). Analysis of cyclotron resonance experiments for fields at different angles therefore enables the details of the electron dispersion relation to be deduced.

5.6   NON-EQUILIBRIUM CARRIER DENSITIES

In the operation of most semiconductor devices the carrier concentrations are disturbed from their thermal equilibrium values. To understand the behaviour of such devices we must derive the equations that describe the variation in space and time of the disturbances. We write the carrier densities as

$$n(x, t) = n_0 + n'(x, t)$$
$$p(x, t) = p_0 + p'(x, t)$$
(5.60)

where p_0 and n_0 are the thermal equilibrium concentrations, which in a uniformly doped homogeneous semiconductor are independent of position. We take the departures n' and p' of the concentrations to be dependent on one spatial coordinate only, although the generalization to dependence on more than one is straightforward. We will also assume that the disturbances n' and p' are small compared to the *majority* carrier concentration; thus in n-type semiconductor we would have

$$n', p' \ll n_0.$$
(5.61)

5.6.1   The continuity equations

We consider the processes that can change the carrier concentrations in the region of semiconductor between x and $x + \delta x$ (Fig. 5.14):

(1) *Recombination.* An electron in the conduction band can fall into an empty state in the valence band, resulting in the loss of an electron–hole pair. As well as by this direct process, recombination can also occur through traps or at surfaces. Traps are localized states with energies near the middle of the band gap (gold impurity atoms in silicon are a source of such levels) into which the electron falls before dropping at some later time into the valence band. Whatever the mechanism, recombination results in the disappearance of holes and electrons at the same rate, which we denote by r per unit volume.

(2) *Generation.* This is the opposite process to recombination; an electron in the valence band receives enough energy from some source to promote it to the conduction band, creating an electron–hole pair in the process. We denote the generation rate by g per unit volume and note that both types of carrier are

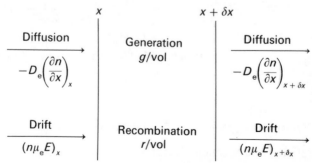

Fig. 5.14   The ways in which the concentration of electrons in the region between x and $x + \delta x$ can change

generated at the same rate. The internal source of energy for generation is the photons associated with thermal equilibrium radiation within the crystal.

(3) *Diffusion*. If the carrier concentration depends on position then diffusion will occur. We assume that the resulting *number* current densities, J_e and J_h, for electrons and holes are given by **Fick's law** as

$$J_e = -D_e \frac{\partial n}{\partial x} \qquad \text{and} \qquad J_h = -D_h \frac{\partial p}{\partial x}$$

where D_e and D_h are the respective diffusion constants. The net rate per unit area at which electrons enter the region between x and $x + \delta x$ as a result of diffusion is given by the difference in the values of J_e at x and $x + \delta x$ and can thus be written

$$-\frac{\partial J_e}{\partial x} \, dx = D_e \frac{\partial^2 n}{\partial x^2} \, dx,$$

with a similar expression for holes.

(4) *Motion in an electric field*. In the presence of an electric field (assumed to be in the x direction) the electric current densities of electrons and holes are $ne\mu_e E$ and $pe\mu_h E$ respectively (Eq. (5.40)). The corresponding number current densities are

$$J_e = -n\mu_e E \qquad \text{and} \qquad J_h = p\mu_h E.$$

As with diffusion currents the net rate per unit area at which electrons enter the region between x and $x + \delta x$ is determined by the difference in the values of J_e at x and $x + \delta x$ and is given by

$$-\frac{\partial J_e}{\partial x} \, dx = \mu_e \frac{\partial(nE)}{\partial x} \, dx.$$

By adding together the contributions listed above we deduce that the net rate of change of the electron density in the region between x and $x + \delta x$ is

$$\frac{\partial n}{\partial t} = g - r + D_e \frac{\partial^2 n}{\partial x^2} + \mu_e \frac{\partial(nE)}{\partial x} \qquad (5.62a)$$

where we have cancelled a factor dx from every term. Eq. (5.62a) is known as the **continuity equation** for electrons. The continuity equation for holes is likewise determined to be

$$\frac{\partial p}{\partial t} = g - r + D_h \frac{\partial^2 p}{\partial x^2} - \mu_h \frac{\partial(pE)}{\partial x}. \qquad (5.62b)$$

5.6.2   Electrical neutrality

Eqs. (5.62) give the impression that the electrons and holes move independently, but this is misleading since the electric field that appears in both equations contains a contribution from the disturbed carrier densities, which couples the motion of electrons and holes very strongly. The effect of this coupling is that any momentary departure from electrical neutrality in a homogeneously doped semiconductor disappears on a very short time scale. Alternatively, if a departure from electrical neutrality is maintained at a particular point in the semiconductor, electrical neutrality is attained a very short distance away from that point. In both cases electrical neutrality is achieved by the redistribution of the majority carrier resulting from the electric field associated with the charged region.

The charge density at a point in the semiconductor is

$$\rho = e(p + N_D^+ - n - N_A^-). \tag{5.63}$$

Inserting Eqs. (5.60) and recalling that the semiconductor is electrically neutral in thermal equilibrium (Eq. (5.26)) we obtain

$$\rho = e(p' - n'). \tag{5.64}$$

Thus we see that electrical neutrality is attained when the disturbances of the electron and hole densities are equal. The electric field generated by the departure from electrical neutrality is given by Gauss' law, $\text{div } \mathbf{E} = \rho/\varepsilon\varepsilon_0$. For variation only with x therefore

$$\frac{dE}{dx} = \frac{e(p' - n')}{\varepsilon\varepsilon_0}. \tag{5.65}$$

We wish to calculate the effect of this electric field on the distribution of the majority carrier, which we take to be electrons. We substitute Eq. (5.65) into the continuity equation for electrons (Eq. (5.62a)) to obtain

$$\frac{\partial n'}{\partial t} = g - r + D_e \frac{\partial^2 n'}{\partial x^2} + \mu_e E \frac{\partial n'}{\partial x} + \mu_e n \frac{e(p' - n')}{\varepsilon\varepsilon_0}$$

where we have used the fact that n_0 is constant to replace derivatives of n by derivatives of n'. From Eq. (5.65) we see that (at least in the absence of an externally applied electrical field) E is first order in departures from equilibrium; the term $\mu_e E \, \partial n'/\partial x$ is therefore second order and we will ignore it. We shall see that recombination and generation represent slow processes when compared with the remaining terms in the equation and we will ignore them also. The resulting equation can be written in the form

$$\tau_D \frac{\partial n'}{\partial t} = (p' - n') + \lambda_D^2 \frac{\partial^2 n'}{\partial x^2} \tag{5.66}$$

where $\tau_D = \varepsilon\varepsilon_0/ne\mu_e = \varepsilon\varepsilon_0/\sigma_e$ is known as the **dielectric relaxation time** and $\lambda_D = (\varepsilon\varepsilon_0 D_e/ne\mu_e)^{1/2}$ is called the **Debye length**.

We will not need to consider solutions of Eq. (5.66) in detail but just note their general behaviour (see problem 5.8) that a finite value of $p' - n'$ at any point in space or time disappears by a redistribution of the electrons on a time scale τ_D and a length scale λ_D. Clearly the order of magnitude of these quantities is important. To estimate τ_D we take a typical value of $100\ \Omega^{-1}\ m^{-1}$ for σ_e, the contribution of the electrons to the conductivity, and obtain

$$\tau_D = \frac{\varepsilon\varepsilon_0}{\sigma_e} \approx \frac{10 \times 10^{-11}}{100} = 10^{-12}\ \text{s.}$$

To calculate λ_D we can use the Einstein relation (proved in section 6.2),

$$D_e = \frac{k_B T}{e}\mu_e, \tag{5.67}$$

to obtain

$$\lambda_D = \left(\frac{k_B T \varepsilon \varepsilon_0}{ne^2}\right)^{1/2} \approx 400\ \text{Å} \tag{5.68}$$

for $T = 300$ K, $\varepsilon = 10$ and $n = 10^{22}\ m^{-3}$. These estimates tell us that τ_D and λ_D are very much shorter than the time scales and length scales, respectively, of most phenomena in semiconductors; in calculating such phenomena it is then possible to ignore departures from electrical neutrality.

In homogeneous semiconductors therefore the motion of the majority carrier is such as to make $n' = p'$ everywhere and it is only necessary to solve the continuity equation for the *minority* carriers:

$$\frac{\partial n'}{\partial t} = g - r + D_e \frac{\partial^2 n'}{\partial x^2} + \mu_e E \frac{\partial n'}{\partial x} \qquad \text{in p-type semiconductor,} \tag{5.69a}$$

$$\frac{\partial p'}{\partial t} = g - r + D_h \frac{\partial^2 p'}{\partial x^2} - \mu_h E \frac{\partial p'}{\partial x} \qquad \text{in n-type semiconductor.} \tag{5.69b}$$

E is now an applied spatially constant electric field; Eq. (5.65) and electrical neutrality ensure that the gradient of E vanishes.

5.6.3 Generation and recombination

We now consider the recombination and generation terms in Eqs. (5.69). In thermal equilibrium the rates of recombination and generation are equal and we denote the value of each by $g_0(T)$ to indicate that it is temperature-dependent. The rate g at which valence band electrons make transitions to the conduction band depends on the number of electrons in the valence band and the probability that at any one time one of them acquires enough energy to make the transition. Neither of these factors is significantly affected by small disturbances in the carrier concentrations (recall that the concentration of electrons in the valence band is much greater than n' and p'), and consequently the thermal generation rate remains at the value $g_0(T)$ when the carrier concentrations are disturbed from their equilibrium values by small amounts. In contrast the recombination rate for direct recombination depends on the carrier concentrations. Since each recombination process involves the interaction of an electron in the conduction band with a hole in the valence band we might expect the recombination rate to depend linearly on both the electron and hole concentrations so that $r = k(T)np$, where $k(T)$ depends on temperature but not on carrier concentration. Thus we can write

$$g - r = g_0(T) - k(T)np = g_0(T) - k(T)(n_0 + n')(p_0 + p')$$
$$= -k(T)(n_0 p' + p_0 n' + n'p')$$

where to obtain the last line we have used $g_0(T) = k(T)n_0 p_0$, which is just the condition that the recombination and generation rates should be equal in thermal equilibrium. One of the three remaining terms is much bigger than the other two; if we recall that electrical neutrality requires $n' = p'$ and that both n' and p' are much smaller than the majority carrier concentration, then the largest term is $-k(T)n_0 p'$ in n-type material and $-k(T)p_0 n'$ in p-type material. These terms can be written as $-p'/\tau_n$ and $-n'/\tau_p$ respectively, where τ_n and τ_p are known as the **minority carrier lifetimes** in n- and p-type material respectively. Do not confuse the carrier lifetimes, τ_n and τ_p, with the scattering times, τ_h and τ_e. The continuity equations for the minority carriers can therefore be written

$$\frac{\partial n'}{\partial t} = -\frac{n'}{\tau_p} + D_e \frac{\partial^2 n'}{\partial x^2} + \mu_e E \frac{\partial n'}{\partial x} \qquad (5.70a)$$

for electrons in p-type semiconductor, and

$$\frac{\partial p'}{\partial t} = -\frac{p'}{\tau_n} + D_h \frac{\partial^2 p'}{\partial x^2} - \mu_h E \frac{\partial p'}{\partial x} \qquad (5.70b)$$

for holes in n-type semiconductor.

Although we have deduced the form of $g - r$ for the case of direct recombination in extrinsic semiconductor, the form that we have determined is more generally valid. The expressions for the carrier lifetimes will be different for other types of recombination. The minority carrier lifetime is typically 10^{-7} s; this is much longer than the dielectric relaxation time τ_D, thus justifying our neglect of recombination and generation processes in section 5.6.2.

We will now apply Eqs. (5.70) to two important situations, which will provide the reader with a physical picture of the processes that occur.

5.6.4   Injection of minority carriers at a steady rate

We consider a long thin rod of p-type semiconductors of cross-sectional area A as in Fig. 5.15(a) and assume that electrons are injected at a steady rate N per unit area per second at one end. In a steady state ($\partial n'/\partial t = 0$) and in the absence of an applied electric field, Eq. (5.70a) becomes

$$\frac{\partial^2 n'}{\partial x^2} = \frac{n'}{L_e^2} \qquad (5.71)$$

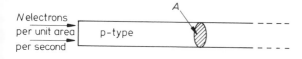

(a) Injection of electrons at one end of a long rod of p-type semiconductor

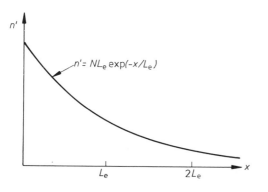

(b) Excess concentration of electrons as a function of position within the rod

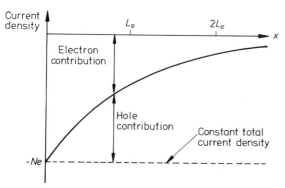

(c) The total electric current density is constant but the relative contribution of the electrons and holes varies with position

Fig. 5.15

where

$$L_e = (D_e \tau_p)^{1/2} \qquad (5.72)$$

is known as the **diffusion length** of the electrons and is typically about 20 μm; the significance of this quantity will become clear shortly.

The solution of Eq. (5.71) is

$$n' = C_1 e^{-x/L_e} + C_2 e^{+x/L_e} \qquad (5.73)$$

where C_1 and C_2 are constants of integration. If the rod occupies the region $x \geqslant 0$ then the second term predicts an electron concentration which increases exponentially with x at large x; this is unphysical in a long rod and we therefore set $C_2 = 0$ in this case. C_1 is then determined from the number current density at $x = 0$. Thus

$$N = -D_e \left[\frac{\partial n'}{\partial x} \right]_{x=0} = \frac{C_1}{L_e}$$

and the excess electron concentration is given by

$$n' = N L_e e^{-x/L_e}. \qquad (5.74)$$

The electron concentration is shown in Fig. 5.15(b) and we see that it decays exponentially with increasing x with a decay length L_e. The diffusion length is thus the mean distance that an electron diffuses before recombination. Fig. 5.15(c) shows the corresponding electron diffusion current, $D_e \, \partial n'/\partial x$, which also decays exponentially with decay length L_e. To preserve electrical neutrality the total electric current density must remain constant. A hole current that increases with x enables this to be achieved as indicated in Fig. 5.15(c). The holes are flowing towards $x = 0$ and they replace those lost by recombination with the incoming electrons. We could of course have chosen the steady flow of holes into an n-type semiconductor to illustrate the same principles.

5.6.5   Injection of a pulse of minority carriers

We consider the simple circuit shown in Fig. 5.16(a), which forms the basis of the Haynes–Shockley experiment (*Phys. Rev.* **75**, 691 (1949)); the purpose of the circuit is to inject a pulse of the minority carrier, holes in this case, at a point on a long thin rod of semiconductor. Suppose that the switch is closed for a short period of time. The effect of this is to inject a pulse of holes at point A on the rod and a pulse of electrons at point B. Immediately after the pulse the excess carrier distributions are as shown on Fig. 5.16(b) and thus electrical neutrality is violated; majority carriers therefore flow until neutrality is effectively achieved after a few dielectric relaxation times.† The excess carrier distributions are then

---

† In a typical experiment the switch is closed for about 1 μs; as this is much longer than τ_D, electrical neutrality is maintained, to a good approximation, throughout the injection process.

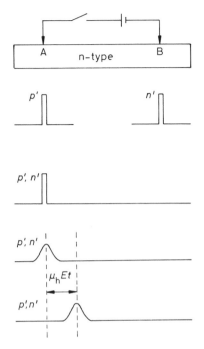

(a) Closing the switch for a short period of time injects a pulse of holes at point A and a pulse of electrons at point B

(b) The excess carrier densities immediately after the injection of the pulses

(c) The excess carrier concentrations after a few dielectric relaxation times. Electrical neutrality has been achieved by the flow of majority carriers (electrons)

(d) On a time scale comparable to the minority carrier lifetime the pulse of holes has broadened due to diffusion and the area underneath has decreased due to recombination

(e) In the presence of an additional steady electric field the pulse of holes also drifts at a velocity $\mu_h E$

Fig. 5.16

as shown in Fig. 5.16(c); there are no excess carriers left at point B and equal hole and electron pulses at point A. Subsequently the much slower diffusion and recombination processes occur; these result in a broadening of the pulse and a decrease in the area underneath it (the area determines the number of excess holes left in the semiconductor) as shown in Fig. 5.16(d).

The solution of the continuity equation (Eq. (5.70b)) that describes this behaviour is well known from the theory of diffusion (problem 5.9), and is

$$p'(x, t) = \frac{P}{(4\pi D_h t)^{1/2}} \exp\left(-\frac{t}{\tau_n} - \frac{x^2}{4D_h t}\right) \tag{5.75}$$

where P is the initial number of holes (per unit area of cross section) injected into the semiconductor at $x = 0$ and $t = 0$. Thus the area underneath the pulse decreases as $\exp(-t/\tau_n)$ as the holes recombine, and the width of the pulse at time t is of order $(D_h t)^{1/2}$ as would be expected for a diffusion process. In the Haynes–Shockley experiment the holes were also subject to a dc electric field and the effect of this was to cause the pulse to move steadily along the rod at a velocity determined by the hole mobility; thus after a time t the whole pulse had shifted to $x = \mu_h t E$ as shown in Fig. 5.16(e). Observations of the shape and position of the pulse after a time t therefore provide a direct method for

168   Semiconductors   Chap. 5

measuring the mobility, diffusion constant and lifetime of the holes. The Haynes–Shockley experiment demonstrated the importance of the minority carrier in determining the behaviour of an extrinsic semiconductor.

PROBLEMS 5

5.1 The electron energy near the top of the valence band in a semiconductor is given by

$$\varepsilon = -10^{-37}k^2 \text{ J},$$

where $\mathbf{k}$ is the wavevector. An electron is removed from the state

$$\mathbf{k} = 10^9 \hat{\mathbf{k}}_x \text{ m}^{-1}$$

where $\hat{\mathbf{k}}_x$ is a unit vector along the x axis. Calculate (a) the effective mass, (b) the energy, (c) the momentum and (d) the velocity of the resulting hole. (Each answer must include the sign (or direction).)

5.2 A sample of silicon is purified until it contains only 10^{18} donors m^{-3}. Below what temperature does it cease to show intrinsic behaviour? ($E_G = 1.1$ eV and the intrinsic carrier concentration at 300 K is 2×10^{16} m^{-3}.)

5.3 Indium antimonide has dielectric constant $\varepsilon = 17$ and electron effective mass $m_e = 0.014m$. Calculate:
(a) the donor ionization energy,
(b) the radius of the ground state orbit, and
(c) the donor concentration at which orbits around adjacent impurities begin to overlap. What effects occur at about this concentration, and why?

5.4 Use the data of Fig. 5.9 to estimate a value of the donor ionization energy for arsenic impurities in germanium.

5.5 Calculate values for the Hall coefficients of sodium and intrinsic indium antimonide at 300 K. Sodium has a bcc structure with unit cell side 4.28 Å. Indium antimonide has $E_G = 0.15$ eV, $m_e = 0.014m$, $m_h = 0.18m$ and the electrons are the only effective carrier (why?).

Estimate the Hall voltages generated in each case across the 5 mm width of a sample of 1 mm thickness when a current of 100 mA is passed along it and the perpendicular field is 0.1 T.

5.6 A sample of germanium is doped with a single type of impurity. Outline the measurements you would make to determine the sign and concentration of the carriers, their mobility and effective mass.

If there are 10^{20} donors m^{-3} what are the conditions necessary for satisfactory observation of cyclotron resonance? (Collision diameter of donor = 300 Å, $m_e = 10^{-31}$ kg.)

5.7 Deduce values of m_L and m_T in Eq. (5.54) from the data of Fig. 5.12. (The direction of the magnetic field is given in the figure caption.)

5.8 Show that solutions of Eq. (5.66) have the qualitative behaviour indicated in section 5.6.2 by considering the two important limits:
(a) no time dependence;
(b) no position dependence.

5.9 Show that Eq. (5.75) is a solution of the continuity equation for holes in n-type semiconductor in the absence of an electric field. Confirm that the total number of holes remaining at time t decreases as $\exp(-t/\tau_n)$.

I find television very educating. Every time somebody turns on
the set I go into the other room and read a book.—*Groucho Marx*

CHAPTER

Semiconductor devices

6.1 INTRODUCTION

The use of semiconductors in electronics is probably the biggest contribution
of solid state physics to twentieth century technology. To understand the great
majority of semiconductor devices it is necessary to consider the behaviour of
charge carriers near a surface or interface. Of particular importance are the
boundary between an n-type region and a p-type region, the boundary between
a semiconductor and an insulator, and the boundary between two different
semiconductors. In this chapter we consider devices using these three possibilities. Our emphasis will be on understanding the physics of the devices, not the
technical applications.

6.2 THE p–n JUNCTION WITH ZERO APPLIED BIAS

A boundary between p and n regions can be produced in a number of ways.
The deposition of a thin layer of donor impurities on the surface of a p-type
semiconductor, followed by a controlled period of time at a high temperature to
allow diffusion of the donors into the substrate, creates an n-type region near the
surface where the donors outnumber the original acceptors. A p–n junction can
also be produced by epitaxial growth (section 6.6) on a p-type substrate of
material containing donor impurities. Both of the above methods can of course
alternatively create an acceptor-rich region in contact with an n-type substrate.

Other methods also exist for producing a region of semiconductor in which a changeover from p-type to n-type behaviour occurs on a short length scale.

The useful behaviour of a p–n junction results from the effect on the electron energy levels in the region of the junction as shown in Fig. 6.1. The energy levels are shown as a function of position only, and no distinction is made between different **k** values; electrons in the conduction band are indicated schematically by full circles and holes in the valence band by open circles. The factor controlling the relative positions of the levels on the two sides of the junction is the necessity for a uniform chemical potential; this is the condition for thermal equilibrium between two or more systems when particles can move freely between them (see Mandl,[2] chapter 8). In the case of a p–n junction the equilibrium is achieved by a small transfer of electrons from the n region to the p region, where they annihilate with holes, leaving a region with very few free carriers† near the junction, known therefore as the **depletion layer** (Fig. 6.2(b)). The positively charged ionized donors in the n region of the depletion layer and the negatively charged ionized acceptors in the p region leave the n region positively charged and the p region negatively charged (Fig. 6.2(c)). This results in the lowering of electron energy levels on the n side and the raising on the p side, shown in Fig. 6.1, which causes the chemical potential to be position-independent as required; remember that Fig. 6.1 is a diagram of *electron* energy levels so that a region of low energy is a region of high electrostatic potential.

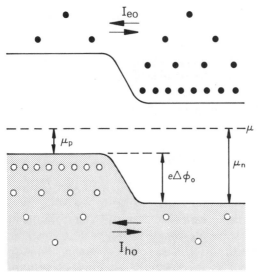

Fig. 6.1   A p–n junction in equilibrium with electrons (full circles) and holes (open circles) indicated. The relationship $e\Delta\phi_o = \mu_n - \mu_p$ (Eq. (6.1)) is obtained by inspection of the figure. The arrows indicate the equal and opposite electron currents, I_{e0}, and hole currents, I_{h0}, that are discussed in section 6.3

---

† Note however that the electron and hole concentrations must still satisfy the law of mass action (Eq. (5.22)) everywhere.

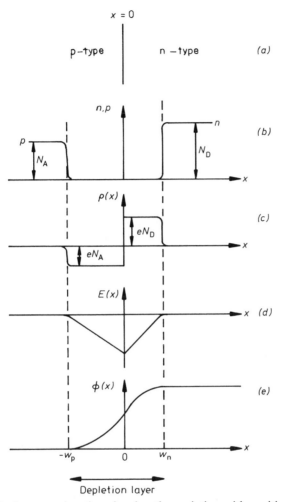

Fig. 6.2   (*a*) A sharp p-n junction showing the variation with position of: (*b*) carrier concentrations, *n* and *p*; (*c*) net charge density, ρ; (*d*) electric field, *E*; and (*e*) electrostatic potential, ϕ

The total potential difference $\Delta\phi_0$ required to produce a uniform chemical potential can be deduced from Eqs. (5.32) and (5.34). Thus if we ignore the acceptor concentration on the n side of the junction and denote the donor concentration there by N_D, then the chemical potential μ_n relative to the valence band edge on the n side at large distances from the junction is given by Eq. (5.32) as

$$\mu_n = E_G - k_B T \ln\left(\frac{N_C}{N_D}\right).$$

Similarly for a p region containing N_A acceptors per unit volume and no donors, the chemical potential μ_p relative to the valence band edge on the p side at large distances from the junction as given by Eq. (5.34) is

$$\mu_p = k_B T \ln \left(\frac{N_V}{N_A} \right).$$

Hence, for equality of the chemical potential, we see from Fig. 6.1 that the valence band edges on the two sides of the junction must differ by

$$e\Delta\phi_0 = \mu_n - \mu_p = E_G + k_B T \ln \left(\frac{N_D N_A}{N_C N_V} \right) \tag{6.1}$$

which, by the use of Eq. (5.23), may be written

$$\Delta\phi_0 = \frac{k_B T}{e} \ln \left(\frac{N_D N_A}{n_i^2} \right) \tag{6.2}$$

where n_i is the electron (or hole) concentration in an intrinsic sample of the semiconductor at the same temperature. It is important to remember that $\Delta\phi_0$ is in the nature of a contact potential; in any complete circuit there will be compensating potential differences at the other junctions so that no current flows in thermal equilibrium. Typical values of $\Delta\phi_0$ for Si and Ge of 0.3 V and 0.7 V respectively are obtained at $T = 300$ K by substituting $N_A = N_D = 10^{22}$ m^{-3} in Eq. (6.2) (n_i is given in Table 5.1); $\Delta\phi_0$ is only weakly dependent on the impurity concentration as readers can check for themselves by substituting different values for N_A and N_D in Eq. (6.2).

The width of the depletion layer and the variation of the electrostatic potential $\phi(x)$ within it may be calculated to a good approximation by making two simplifying assumptions:

(1) The boundary between the n and p regions is sharp as shown in Fig. 6.2(a),† where the boundary is taken to be at $x = 0$.

(2) The majority carrier concentrations decrease very rapidly from their 'bulk' values at the edges of the depletion layer, which we take to be at $x = -w_p$ on the p side and $x = w_n$ on the n side; this very rapid decrease is depicted in Fig. 6.2(b) and is necessary for the edges of the depletion layer to be well defined.

If these assumptions are made then the charge density near the junction is well approximated by

$$\rho(x) = \begin{cases} -N_A e & -w_p < x < 0, \\ +N_D e & 0 < x < w_n, \\ 0 & \text{elsewhere,} \end{cases} \tag{6.3}$$

as shown on Fig. 6.2(c).

---

† Such a junction is described as **abrupt**. Problem 6.3 calculates $\phi(x)$ for a junction with no sharp boundary between the p and n regions.

The electrostatic potential is related to the charge density by Poisson's equation

$$\frac{d^2\phi}{dx^2} = -\frac{\rho(x)}{\varepsilon\varepsilon_0}, \tag{6.4}$$

the first integral of which gives the electric field within the depletion layer as

$$E = -\frac{d\phi}{dx} = \begin{cases} -\dfrac{N_A e}{\varepsilon\varepsilon_0}(x + w_p) & -w_p < x < 0, \\[2ex] \dfrac{N_D e}{\varepsilon\varepsilon_0}(x - w_n) & 0 < x < w_n. \end{cases} \tag{6.5}$$

The linear variation of E with x follows from integrating Eq. (6.4) with the constant charge density of Eq. (6.3). The electric field must vanish in the bulk semiconducting regions outside the depletion layer and the integration constants in Eq. (6.5) have been chosen to ensure that E is continuous at the boundaries of the depletion layer. Using Eq. (6.5) and requiring that E should be continuous at $x = 0$ provides the relation

$$N_A e w_p = N_D e w_n \tag{6.6}$$

which is just the statement of overall electrical neutrality, that the number of ionized acceptors on the p side in the depletion layer equals the number of ionized donors on the n side. The electric field of Eq. (6.5) is shown in Fig. 6.2(d).

Integrating the electric field (Eq. (6.5)) gives

$$\phi(x) = \begin{cases} \dfrac{eN_A}{2\varepsilon\varepsilon_0}(x + w_p)^2 & -w_p < x < 0 \\[2ex] \Delta\phi_0 - \dfrac{eN_D}{2\varepsilon\varepsilon_0}(x - w_n)^2 & 0 < x < w_n \end{cases} \tag{6.7}$$

for the potential; the linear variation of E with x in Eq. (6.5) leads to the quadratic dependence of ϕ on x in Eq. (6.7). The constants of integration have been chosen so that the potential of the p region outside the depletion layer is zero (this defines the zero of potential) and the total potential difference across the junction is $\Delta\phi_0$ as given by Eq. (6.1) or (6.2). The potential $\phi(x)$ must be continuous at $x = 0$ and this provides the relationship

$$\Delta\phi_0 = \frac{e}{2\varepsilon\varepsilon_0}(N_A w_p^2 + N_D w_n^2). \tag{6.8}$$

The potential is shown as a function of position in Fig. 6.2(e).

Finally, solving Eqs. (6.6) and (6.8) simultaneously gives the widths of the depletion layer on the two sides of the junction as

$$w_n = \left(\frac{2\varepsilon\varepsilon_0 N_A \Delta\phi_0}{eN_D(N_A + N_D)}\right)^{1/2}, \qquad w_p = \left(\frac{2\varepsilon\varepsilon_0 N_D \Delta\phi_0}{eN_A(N_A + N_D)}\right)^{1/2}. \tag{6.9}$$

It follows that the depletion layer is wider in more lightly doped junctions (smaller N_A and N_D); the width $w_n + w_p$ is about 1 μm for $N_A \sim N_D \sim 10^{21}$ m^{-3} and about 0.1 μm for $N_A \sim N_D \sim 10^{23}$ m^{-3}. For our assumption of a sharp junction to be valid it is necessary that the change in doping from p-type to n-type should occur on a length scale much more rapid than this.

It remains for us to investigate the assumption that the majority carrier concentration falls off very rapidly at the edges of the depletion layer. As the variation of $\phi(x)$ is slow on an atomic length scale our calculation of the carrier concentrations, given in section 5.3.2, remains valid, the values of n and p at any point being determined, through Eqs. (5.16) and (5.20), by the position of the chemical potential relative to the conduction and valence band edges respectively. The chemical potential is constant and the spatial variation of the band edges is given by $e\phi(x)$. We can therefore conveniently write the carrier concentrations as

$$n(x) = n_0 \exp\left[e\phi(x)/k_B T\right]$$
$$p(x) = p_0 \exp\left[-e\phi(x)/k_B T\right], \tag{6.10}$$

where n_0 and p_0 are the concentrations of electrons and holes at points where $\phi(x)$ is zero; in the present situation this is all points on the p side of the junction outside the depletion layer. The rapid fall-off in majority carrier concentrations at the edges of the depletion layer occurs because $k_B T$ is small compared to the total energy difference $e\Delta\phi_0$ across the junction; a small change in $\phi(x)$ therefore produces a large change in the carrier concentration (see problem 6.1).

Eqs. (6.10) are valid in other situations in which there is a position-dependent electrostatic potential and they can be used to deduce the very useful **Einstein relations** between the diffusion constants and mobilities of the carriers. In thermal equilibrium the current density of electrons must vanish everywhere, and in a region of semiconductor where there is an electric field such as a p–n junction this means that the contributions due to diffusion and to the electric field must cancel each other. Thus

$$eD_e \frac{\partial n}{\partial x} + n\, e\mu_e E = 0. \tag{6.11}$$

Alternatively by differentiating Eq. (6.10) we obtain

$$\frac{\partial n}{\partial x} = \frac{e}{k_B T} \frac{\partial \phi}{\partial x} n(x). \tag{6.12}$$

By recalling that $E = -\partial\phi/\partial x$ we see that Eqs. (6.11) and (6.12) are consistent only if

$$\frac{D_e}{\mu_e} = \frac{k_B T}{e} \tag{6.13}$$

which is the Einstein relation for electrons, which we have already used in section 5.6.2. Similarly by requiring that the hole current should vanish in thermal equilibrium, we can deduce the Einstein relation for holes,

$$\frac{D_h}{\mu_h} = \frac{k_B T}{e}. \tag{6.14}$$

These Einstein relations remain valid as long as the electron and hole concentrations are given by the Boltzmann distributions of Eqs. (6.10).

6.3   THE p–n JUNCTION WITH AN APPLIED BIAS

The application of an additional potential difference V across a p–n junction causes an electric current to flow through it. If the positive side of the potential is attached to the p region, the junction is said to be **forward biased** and V is taken to be positive; if the positive side is attached to the n region then the junction is **reverse biased** and V is negative. Because of the low carrier density in the depletion layer this region has a high resistivity compared to the neutral semiconducting regions on either side and consequently the applied voltage appears across this layer. Hence, using the sign conventions that we have adopted, the total potential difference across the depletion layer is

$$\Delta\phi = \phi_n - \phi_p = \Delta\phi_0 - V \tag{6.15}$$

where $\Delta\phi_0$ is the potential drop in the absence of applied bias as given by Eqs. (6.1) and (6.2). Forward bias therefore reduces the total potential difference whereas reverse bias increases it (see Fig. 6.3). To be consistent with our sign convention for V we must take the current as positive when it flows from p to n.

The variation of the potential within the depletion layer is obtained by solving Poisson's equation (Eq. (6.4)) as in the previous section. The width of the depletion layer is thus obtained by replacing $\Delta\phi_0$ in Eqs. (6.9) by $\Delta\phi$ to give

$$w_n = \left(\frac{2\varepsilon\varepsilon_0 N_A(\Delta\phi_0 - V)}{eN_D(N_A + N_D)}\right)^{1/2}, \qquad w_p = \left(\frac{2\varepsilon\varepsilon_0 N_D(\Delta\phi_0 - V)}{eN_A(N_A + N_D)}\right)^{1/2}. \tag{6.16}$$

We see that the width of the depletion layer is decreased by forward bias and increased by reverse bias. Because the charged ionized impurity atoms within the depletion layer are not compensated by carriers, equal and opposite charge densities of magnitude $\sigma = eN_D w_n = eN_A w_p$ per unit area are located at a p–n junction (see Fig. 6.2(c)). The junction therefore behaves as though it has a capacitance

$$C = \frac{\sigma}{V} = \frac{eN_D w_n}{V} = \frac{1}{V}\left(\frac{2\varepsilon\varepsilon_0 eN_A N_D(\Delta\phi_0 - V)}{(N_A + N_D)}\right)^{1/2} \tag{6.17}$$

per unit area of junction, where we have used the value of w_n from Eqs. (6.16).

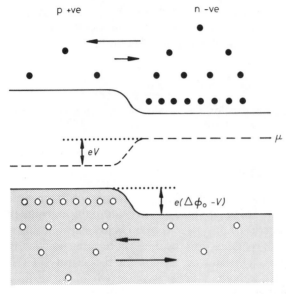

(a) A p–n junction with forward bias. The electron energy levels in the p region are lowered relative to those in n. Because the energy barrier is reduced, the flow of electrons from n to p increases as indicated by the longer arrow. The flow of holes from p to n is similarly increased so that both carriers contribute to the net electric current from p to n

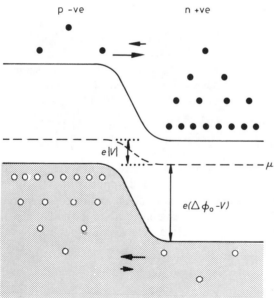

(b) A p–n junction with reverse bias. Note that V is negative in this case. As indicated by the arrows, the flow of electrons from n to p and the flow of holes from p to n are both reduced because of the increase in the potential barrier. There is a very small net electric current from n to p

Fig. 6.3

Although this capacitance can be a problem in electronic circuits where speed is important, it is put to good use in circuits where a voltage variable capacitance is required (see problem 6.2); p–n junctions intended for this use are known as **varactor diodes** or **varicaps**. They can be used only under conditions of reverse bias where the current flow is very small.

We now proceed to calculate the electric current through a p–n junction produced by the applied bias. In the absence of the bias the shift in energy levels associated with the potential difference $\Delta\phi_0$ can be regarded as a potential barrier limiting the flow of conduction band electrons from n to p; from Fig. 6.1 we see that only those few electrons with sufficient energy to overcome this barrier can make the transition and we suppose that these electrons give rise to an *electron* current I_{e0} from n to p. In the absence of bias the total electron current must vanish and to achieve this there must be an equal and opposite electron current from p to n; this counter-current arises because, although there are very few electrons on the p side, any that diffuse to the edge of the depletion layer are swept across the junction by the favourable electric field within this layer. These equal and opposite currents are indicated by arrows on Fig. 6.1.

It is necessary for us to calculate I_{e0} and it is more straightforward to calculate the flow from p to n. If the lifetime of the n_p electrons per unit volume on the p side of the junction is τ_p then the recombination and generation rates for electrons on this side of the junction are both equal to n_p/τ_p per unit volume (section 5.6.3). On average a newly generated electron moves a distance of one diffusion length L_e (see section 5.6.4) before recombination. Hence we can estimate I_{e0} by saying that only those electrons generated within one diffusion length of the depletion layer edge are likely to diffuse to this edge and cross to the n region before recombination. Thus

$$I_{e0} \approx e \times (\text{generation rate/volume}) \times (\text{volume within } L_e \text{ of depletion layer})$$

$$= e\left(\frac{n_p}{\tau_p}\right)(L_e A)$$

where A is the area of the junction. By assuming that all the acceptors on the p side are ionized, so that $n_p = n_i^2/p_p = n_i^2/N_A$ (from Eq. (5.25)), we can write I_{e0} as

$$I_{e0} = \frac{eD_e n_i^2}{L_e N_A} A \tag{6.18}$$

where we have also used $L_e = (D_e \tau_p)^{1/2}$ (Eq. (5.72)).

The effect of forward bias V is to reduce the potential barrier by an amount eV as shown on Fig. 6.3(a). Since the occupancy of electron states within the conduction band is given by a Boltzmann distribution (Eq. (5.15)), this leads to an increase by a factor $\exp(eV/k_B T)$ in the number of electrons on the n side with sufficient energy to overcome the barrier. The electron current from n to p therefore increases to $I_{e0} \exp(eV/k_B T)$ but the flow from p to n is not changed since there is no potential barrier for motion in this direction. The imbalance in the electron currents is indicated in Fig. 6.3(*a*) and the net *electron* current through the junction is given by

$$I_e = I_{e0}(e^{eV/k_B T} - 1). \tag{6.19}$$

Eq. (6.19) is also valid for reverse bias; in this case the flow of electrons from n to p is reduced by a factor $\exp(eV/k_B T)$† because of the increase in the potential barrier by an amount $e|V|$ as shown in Fig. 6.3(*b*).

We leave it as an exercise for the reader to fill in the details of the corresponding argument for the hole contribution to the current. In this case $e\Delta\phi_0$ acts as a potential barrier preventing the flow of holes from p to n; the barrier height is again decreased by forward bias and increased by reverse bias giving a net hole current

$$I_h = I_{h0}(e^{eV/k_B T} - 1) \tag{6.20}$$

where, by comparison with Eq. (6.18),

$$I_{h0} \approx \frac{eD_h n_i^2}{L_h N_D} A \tag{6.21}$$

is the value of the equal and opposite hole currents through the junction in the absence of an applied potential difference.

The total current is obtained by summing the electron and hole contributions to give

$$I = I_e + I_h = I_0(e^{eV/k_B T} - 1) \tag{6.22}$$

where

$$I_0 = I_{e0} + I_{h0} = en_i^2 A\left(\frac{D_e}{L_e N_A} + \frac{D_h}{L_h N_D}\right). \tag{6.23}$$

Since the diffusion constants and diffusion lengths depend only weakly on temperature, the temperature dependence of I_0 is dominated by the factor $\exp(-E_G/k_B T)$ that appears in n_i^2 (Eq. (5.23)). The current–voltage relation of Eq. (6.22) is plotted in Fig. 6.4. Because of the rectifying action apparent from this characteristic, the **p–n junction diode** has largely replaced the vacuum diode in electronic circuits where such action is required. Eq. (6.21) gives the best rectifier characteristic that can be obtained with carriers of charge e. For an interesting discussion showing that the mechanical analogue of such a rectifier (a ratchet) cannot violate the second law of thermodynamics, consult Feynman,[6] vol. 1, chapter 46.

The current–voltage relations of real p–n junctions differ from Eq. (6.22) in a number of ways and we discuss three of these here. The rapid increase in current

---

†This factor is less than unity because V is negative for reverse bias.

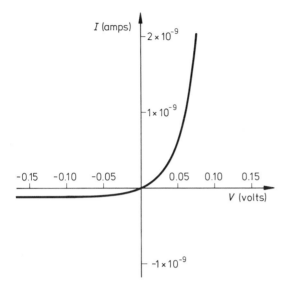

Fig. 6.4   The current–voltage relation for a p–n junctions as predicted by Eq. (6.22). We have taken $I_0 = 10^{-10}$ A. The resistance of the diode is much larger for reverse bias than forward bias

with increasing forward bias that is apparent in Fig. 6.4 corresponds to a decrease in the effective resistance of the depletion layer; when this resistance becomes comparable to that of the semiconductor regions on either side of this layer, it is no longer possible to ignore the potential drop across these regions and this leads to a less rapid increase in current with voltage than that predicted by Eq. (6.22). The second correction to Eq. (6.22) arises because the calculation leading to it ignores the recombination and generation of carriers within the depletion layer itself; taking these into account gives an additional contribution to the current for *forward bias* of the form

$$I = I_0' \, e^{eV/2k_{\mathrm{B}}T} \tag{6.24}$$

(see van der Ziel,[19] p. 321). Eq. (6.24) is valid only for a sufficiently large bias that $I \gg I_0'$; the main difference to Eq. (6.22) is the factor 2 in the exponent, which means that the contribution of Eq. (6.24) tends to dominate at lower voltages. Also, since $I_0' \propto n_i$ whereas $I_0 \propto n_i^2$, the current of Eq. (6.24) is more important in silicon diodes than germanium diodes because n_i is much smaller in the former.

The most obvious deviation from Eq. (6.22) is the phenomenon of **reverse breakdown**; this is the sudden increase of current that occurs when the reverse bias increases through some critical value, ≈ -3 V for the diode characteristic shown in Fig. 6.5(*a*). At values of reverse bias of this order the top of the valence band on the p side of the junction lies above the bottom of the conduction band on the n side as shown in Fig. 6.5(*b*). A valence band electron on the p side therefore has states available to it at the same energy in the conduction band on

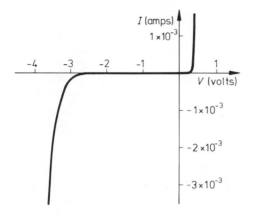

(a) Reverse breakdown of a p-n junction diode at $V \approx -3$ V. Note that because of the very different scale on the current axis as compared to Fig. 6.4, the reverse saturation current I_0 is too small to be seen.

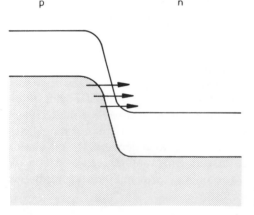

(b) A p-n junction with large reverse bias. Zener breakdown occurs when electrons in the valence band on the p side can tunnel into the conduction band on the n side as indicated by the arrows

Fig. 6.5

the n side; there is a finite probability that such an electron can cross the junction by quantum mechanical tunnelling through the potential barrier consisting of the central region of the depletion layer where no states of the appropriate energy are available.† The tunnelling current increases exponentially with decreasing barrier thickness and this leads to a rapid increase in tunnelling current when the reverse bias increases through some critical value. This type of reverse breakdown is known as **Zener breakdown** and it is the normal mechanism for breakdown in more heavily doped p-n junctions where the depletion layer and hence the potential barrier is narrower (Eq. (6.16)).

---

† For an account of quantum mechanical tunnelling through potential barriers see French and Taylor,[4] p. 383.

With decreasing levels of doping the threshold voltage increases and eventually the mechanism of reverse breakdown changes; for silicon diodes the change in behaviour occurs when the breakdown voltage reaches about 5 V. As the breakdown voltage increases, the electric field within the depletion layer at breakdown becomes larger and eventually reaches a value at which carriers within the depletion layer gain enough energy from the field between collisions to excite an electron from the valence band to the conduction band, thus creating an electron-hole pair, the members of which can themselves create new pairs in the same way. This breakdown mechanism is known as **avalanche breakdown** because of the amplification of the current that it implies.

The rapid increase in current with only small increases in voltage in the breakdown region (whatever the mechanism of breakdown) leads to the use of reverse biased p-n junctions as voltage references in electronic circuits. Junctions constructed to exploit this possibility are known as **Zener diodes**. By changing the doping levels it is possible to tune the breakdown voltage from a value of around 2 V up to values above 1000 V.

In very heavily doped p-n junctions the Fermi level can lie in the valence band on the p side and in the conduction band on the n side as depicted in Fig. 6.6(a); the bottom of the conduction band on the n side is then below the top of the valence band on the p side even with zero applied bias. In these circumstances the depletion layer is very narrow and a large tunnelling current is observed at small forward bias as shown in Fig. 6.5(b). With increasing forward bias the overlap in energy between the conduction band on the n side and the valence band on the p side eventually disappears (point A on Fig. 6.6(b)); tunnelling can no longer occur and the current decreases to the lower value appropriate to the current-voltage characteristic of a conventional diode (Fig. 6.4). Devices with a current-voltage characteristic like that of Fig. 6.6(b) are known as **tunnel diodes**. Their use in electronic circuits results from the region of the characteristic (between points B and A on Fig. 6.6(b)) in which they have a negative differential resistance, $\mathrm{d}V/\mathrm{d}I < 0$.

6.4   OTHER DEVICES BASED ON THE p-n JUNCTION

In this section we describe briefly some of the many semiconductor devices that contain p-n junctions; we choose devices that are interesting from a physical viewpoint.

6.4.1   Light emitting diodes and lasers

A p-n junction with forward bias corresponds to a situation of minority carrier injection; electrons from n flow into the p region and holes from p into the n region. As we discussed in section 5.6.4, excess minority carriers recombine

p   n

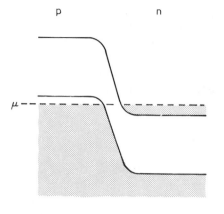

(a) Heavily doped p–n junction with zero bias. The constant chemical potential lies within the valence band on the p side and within the conduction band on the n side. Tunnelling of electrons from the valence band on the p side to the conduction band on the n side can occur in the neighbourhood of zero bias. The Fermi–Dirac distribution function, not the Boltzmann distribution function, must be used to describe the degenerate carriers

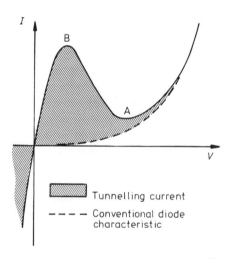

Tunnelling current

– – – – Conventional diode characteristic

(b) Current–voltage characteristic of the p–n junction tunnel diode of (a)

Fig. 6.6

with majority carriers within about one diffusion length of the depletion layer edge. If the dominant recombination process is a direct one in which an electron from the conduction band falls into a vacant state in the valence band, emitting a photon to conserve energy, then the diode acts as a source of light. **Light emitting diodes** (LEDs) are made from direct band gap semiconductors, where direct processes are more likely because momentum conservation can be achieved in the recombination process without the involvement of lattice vibrations (see section 5.4). The dominant photon energy is close to the energy gap E_G because the excess carriers come into thermal equilibrium (energies close to the band edges) prior to recombination; thermal equilibrium can be achieved for example through collisions with lattice vibrations (phonons) and the mean time between such collisions is normally much shorter than the carrier lifetime at room temperature. By using semiconductors of different band gap it is possible to produce light of different colours. The efficiency of an LED is usually small since the large refractive index of semiconductor materials means that many of the emitted photons are totally internally reflected at the surface and reabsorbed within the semiconductor.

In very heavily doped junctions the Fermi level lies within the energy bands outside the depletion layer (see Fig. 6.6(a)). Consequently states near the top of the valence band on the p side are empty. Applying a forward bias that approximately cancels the thermal equilibrium potential difference $\Delta\phi_0$ results in the injection of a large number of electrons into states near the conduction band edge on the p side, and it is possible to obtain a situation of **population inversion** as shown in Fig. 6.7 in which there are more electrons near the conduction band edge than near the valence band edge on the p side of the junction. Photons are generated in this region by electron–hole recombination processes. Because of the population inversion such a photon is more likely to cause recombination of a conduction band electron with the associated **stimulated emission** of a second coherent photon than to be absorbed by a valence band electron. If the p-type region is sandwiched between mirrors (the total internal reflection mentioned above can be used for this purpose) then escape of the photons created by stimulated emission is inhibited and the coherent photon amplitude increases. The energy of the photons emitted by this **p–n junction laser** is close to the energy gap of the semiconductor (see Smith and Thomson,[5] chapter 17, for a fuller account of the physics of lasers). To achieve population inversion in a simple p–n junction, a very high current is required; in section 6.6 we will show how the current can be reduced by using a three-layer heterojunction structure.

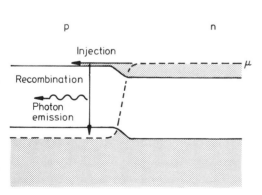

p   n

Injection

Recombination

Photon emission

μ

Fig. 6.7   The application of a forward bias to the p–n junction of Fig. 6.6(a) that almost cancels the thermal equilibrium potential difference $\Delta\phi_0$ results in a large injection of electrons into states near the conduction band edge on the p side. A population inversion can result in which the concentration of electrons in states at the bottom of the conduction band exceeds that in states at the top of the valence band. Recombination by stimulated emission of a photon can then result in laser action

6.4.2 Solar cells

The use of p–n junctions as solar cells exploits the inverse effect to that
discussed in the previous section. When photons with an energy exceeding the
energy gap are incident on a p–n junction, electron–hole pairs are produced. The
electrons produced within the depletion layer and many of those produced
within about a diffusion length of it on the p side are swept to the n side by the
electric field within this layer. Similarly holes generated in the depletion layer
and within about a diffusion length of it on the n side can be swept into the p
region. The junction therefore behaves as a current source, with the current
proportional to the intensity of the incident radiation.

6.4.3 The junction transistor

A p–n–p junction transistor consists of a thin ($<1\,\mu$m) layer of n-type
semiconductor sandwiched between two more heavily doped p-type layers as
shown schematically in Fig. 6.8(a). The three layers are referred to as the
collector, base and emitter as indicated, and electrical contact is made to each
layer by metallic contacts. We can obtain a semi-quantitative understanding of
the behaviour of the transistor by regarding it as two p–n junctions back-to-
back. An n–p–n junction transistor has a layer of p-type semiconductor
sandwiched between two more heavily doped n-type layers; its behaviour is
analogous to that of a p–n–p transistor with the roles of electrons and holes
interchanged and the signs of all the currents and voltages reversed.

When the transistor is being used as an amplifier the base–emitter junction is
forward biased and the base–collector junction is reverse biased, as shown in
Fig. 6.8(b); this produces the energy level diagram for electrons shown in Fig.
6.8(c). The emitter current I_E is just that of a forward biased p–n junction and
can be estimated using Eq. (6.22)†

$$I_E = I_0\, e^{eV_{EB}/k_B T} \qquad (6.25)$$

where V_{EB} is the potential difference across the junction and we have assumed
that the contribution $-I_0$ to the current in Eq. (6.22) is negligible (true under
normal operating conditions). We will suppose that a fraction f_1 of I_E is carried
by electrons and a fraction $1 - f_1$ is carried by holes as shown in Fig. 6.8(d);
because of the higher doping level in the emitter the current is predominantly
due to holes flowing from the emitter to the base, so that $f_1 \ll 1$. From Eq. (6.25)
we deduce that I_E is a very rapid function of V_{EB}; inserting $I_0 = 1$ nA we find
that I_E increases by a factor of 10 from 1 to 10 mA as V_{EB} increases by only 17%
from 0.36 to 0.42 V. This property of a p–n junction can also be seen in Fig.
6.5(a).

---

† Because the width w_B of the base region is small compared to the diffusion length L_h for holes,
our method of estimating I_0 in section 6.3 is no longer valid; a better estimate of I_0 in Eq. (6.25) is
obtained by replacing L_h in Eq. (6.23) by w_B (see problem 6.5).

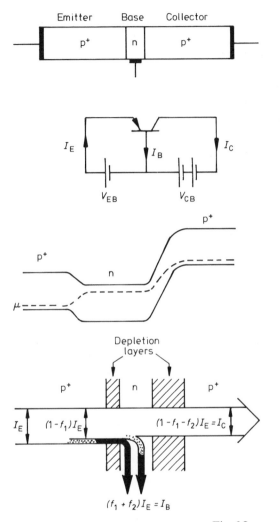

(a) A p–n–p junction transistor; the superscript on p^+ indicates that the p region is more heavily doped

(b) Circuit symbol and biasing arrangement for the transistor; I_E, I_B and I_C denote the emitter, base and collector currents respectively (note that our sign convention differs from that normally used by electronics engineers)

(c) Effect of biasing on the electron energy levels

(d) Current flows within the transistor with hole and electron contributions indicated by unshaded and shaded arrow shafts respectively; regions within which recombination occurs are dotted. The electron arrows are in the direction of conventional current flow, that is in the opposite direction to the actual flow of electrons

Fig. 6.8

The injection of the holes into the base region from the emitter is a situation of minority carrier injection similar to that discussed in section 5.6.4. The important difference in the transistor is that the width of the base is narrow compared to the diffusion length of the holes and the majority of the holes therefore diffuse to the edge of the depletion layer of the base–collector junction prior to recombination; these are immediately accelerated into the collector by the electric field within the depletion layer (recall that hole energies are the negative of electron energies so that, by reference to Fig. 6.8(c), we see that the collector is a region of low energy for holes).

These holes flowing into the collector are the only significant contribution to the collector current I_C since the current intrinsic to the reverse biased base–collector junction can usually be ignored in comparison. The collector and emitter currents are therefore almost equal. The small difference between them is the base current I_B; I_B provides the electron contribution to the base–emitter current $f_1 I_E$, and also electrons to replace those annihilated by the small fraction of injected holes that fail to diffuse to the collector–base junction prior to recombination. This latter contribution is proportional to the hole current and thus to the total emitter current; we write it as $f_2 I_E$ where $f_2 \ll 1$ (see Fig. 6.8(d)). Hence

$$I_B = (f_1 + f_2)I_E \ll I_E$$
$$I_C = I_E - I_B = (1 - f_1 - f_2)I_E \approx I_E. \tag{6.26}$$

The ratio I_E/I_B is denoted by β and is called the **current gain** of the transistor. From Eqs. (6.26) we see that

$$\beta = \frac{1}{(f_1 + f_2)} \gg 1.$$

A typical value for β is 100. It follows from Eqs. (6.26) that, since a small variation i_B in base current is associated with a much bigger variation $(\beta - 1)i_B$ in collector current, the transistor can be pictured as a current amplifier of gain $\beta - 1$.

Ideally a current amplifier should have zero input impedance and infinite output impedance. The input impedance, $\partial V_{EB}/\partial I_B$, seen at the base-emitter junction, is (for $T = 300$ K, $I_E = 5$ mA and $\beta = 100$)

$$\frac{\partial V_{EB}}{\partial I_B} = \frac{\partial V_{EB}}{\partial I_E}\frac{\partial I_E}{\partial I_B} = \frac{k_B T \beta}{e I_E} = 500 \ \Omega$$

where we have used Eqs. (6.25) and (6.26). The output impedance is infinite according to our simple model since the emitter and collector currents are independent of the collector–emitter voltage V_{CE} provided that the reverse bias at this junction is maintained. In practice an increase in V_{CE} causes the depletion layer at the base–collector junction to expand and hence the width of the base region to decrease; this causes the value of I_0 in Eq. (6.25) to increase (see the footnote at the bottom of p. 184) and thus I_E and I_C to increase slightly. The resulting output impedance is typically around 5 kΩ. The transistor can be converted to a voltage amplifier by allowing the collector current to flow through a suitable load resistance; the changes in collector current caused by variations in V_{EB} through Eq. (6.25) are then converted to variations in voltage difference across this resistance.

6.4.4   The junction–gate field-effect transistor

Fig. 6.9(*a*) is a schematic diagram of an n-channel junction–gate field-effect transistor (JUGFET). The source and drain connections are linked by a continuous conducting channel of n-type semiconductor; flow of current through this channel can be controlled by varying the reverse bias applied to the p-type gate. Figs. 6.9(*b*) and (*c*) show the conventional circuit symbol and the normal biasing arrangement for this type of JUGFET. Comparison of Figs. 6.9(*a*) and (*c*) shows that the width of the depletion layer between the p and n regions is increased by the reverse bias and thus the area for current flow through the n channel is reduced. Variations in the gate potential cause changes in the thickness of the depletion layer and, since this is a region of high electrical resistivity, there are corresponding changes in the resistance of the channel

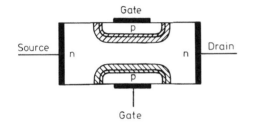

(*a*)An n-channel JUGFET; the gate and p region form a collar encircling the n channel. The shading indicates the width of the depletion layer appropriate to zero bias

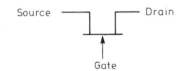

(*b*) Circuit symbol of an n-channel JUGFET

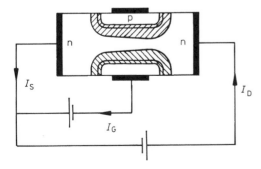

(*c*) Normal biasing arrangement for the n-channel JUGFET and its effect on the width of the depletion layer; the gate current I_G is small so that the drain current I_D and source current I_S are very nearly equal

Fig. 6.9

between the drain and source. A sufficiently negative gate voltage, or alternatively a large enough positive drain or source voltage, can cause the depletion layer to extend across the whole of the channel, a situation referred to as **pinch-off**. When the JUGFET is used as a voltage amplifier the input voltage is fed to the gate; the resulting variations in channel resistance cause changes in the drain–source current and hence voltage variations across a load resistance in series with the channel. The main advantage of a JUGFET voltage amplifier is that the input resistance is that of a reverse biased p–n junction and is consequently very large; the gate current is therefore very small.

6.5   METAL–OXIDE–SEMICONDUCTOR TECHNOLOGY AND THE MOSFET

Metal–oxide–semiconductor (MOS) technology derives its name from the basic structure shown in Fig. 6.10(a), in which a metal electrode (referred to as a gate) is separated from a semiconductor substrate by a thin layer of insulating material. In the most common MOS devices the substrate is silicon and the insulator is silicon dioxide. The structure shown in Fig. 6.10(a) is useful because it is possible to alter the properties of the semiconductor close to the surface by applying a potential to the gate.

To illustrate this we suppose that the substrate is weakly p–type and that, in the absence of a potential on the gate, the energy level diagram for electrons is as shown in Fig. 6.10(b); the filled circles and open circles are used to indicate electrons and holes respectively. When a positive potential is applied to the gate as in Fig. 6.10(c), holes are repelled from the surface towards the interior of the semiconductor and electrons are attracted towards the surface. For small applied potentials the latter, being the minority carrier, are small in number and, as a result, a region with very few carriers is generated close to the surface; this region is called a **depletion layer** and it has similar properties to the depletion layer at a p–n junction (see section 6.2) such as a high electrical resistivity.

There is a net negative charge density within the depletion layer because the charge density of the ionized acceptor impurities is no longer compensated by that of the holes. As in the p–n junction this charge density $\rho(x)$ is associated with a spatially varying electrostatic potential $\phi(x)$ through Poisson's equation (Eq. (6.4)). Since $\rho(x)$ depends on the concentration of electrons and holes and these in turn depend on $\phi(x)$ via Eqs. (6.10) we obtain a non-linear equation for $\phi(x)$ which has in general to be solved using numerical methods. The spatially dependent potential obtained by solving Eq. (6.4) causes bending of the electron energy bands near the surface just as it does in the depletion layer of a p–n junction; we thus obtain the electron energy level diagram of Fig. 6.10(c). Since there is no net flow of electrons or holes the chemical potential must be constant as shown.

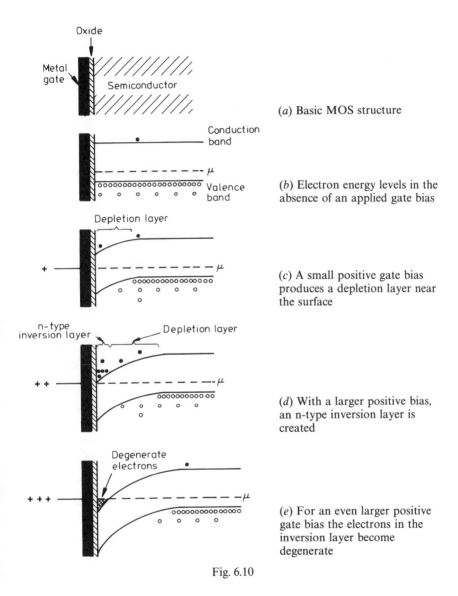

(a) Basic MOS structure

(b) Electron energy levels in the absence of an applied gate bias

(c) A small positive gate bias produces a depletion layer near the surface

(d) With a larger positive bias, an n-type inversion layer is created

(e) For an even larger positive gate bias the electrons in the inversion layer become degenerate

Fig. 6.10

Increasing the positive bias on the gate increases the bending of the energy bands until the conduction band edge at the surface approaches the chemical potential as shown in Fig. 6.10(d). The conduction band electron density then becomes significant near the surface as indicated in the figure. There is thus a layer near the surface in which the electrical conductivity is enhanced. This layer is known as an **inversion layer** since the behaviour has been essentially changed from p-type to n-type within it. The high-resistivity depletion layer, in which there are very few carriers, has been displaced towards the interior of the semiconductor where it separates the n-type inversion layer from the p-type substrate; the analogy with the p–n junction is now even stronger.

Because the electrons in the inversion layer are restricted to moving in a thin layer close to the surface, their behaviour is often similar to a two-dimensional electron gas; we explain further the implications of this in Chapter 14. The density of electrons within the inversion layer and hence the conductivity of this layer can be increased by increasing the positive potential on the gate. For sufficiently high potentials the chemical potential lies within the conduction band near the surface as in Fig. 6.10(e) and the two-dimensional electron gas then becomes degenerate with the occupancy of the states being given by the Fermi function (Eq. (3.14)) rather than the Boltzmann distribution (Eq. (6.10)).

The modulation of the conductance of the electrons in the inversion layer by the gate potential explains the operation of the metal–oxide–semiconductor field-effect transistor (MOSFET). A section through a MOSFET is shown in Fig. 6.11; the width shown for the depletion layer is appropriate to the bias voltages indicated. The structure of Fig. 6.10(a) can be seen in the centre of the top face in Fig. 6.11. The n-type regions near the drain and source contacts are more heavily doped than the substrate and are consequently denoted by n^+ in the figure; they provide a means for making contact with the n-type inversion layer in order to pass a current through it. Note that the depletion layers at the p–n^+ junctions link up with the depletion layer created by the gate bias so that the n-type conducting path is completely insulated from the p-type substrate. The negative bias on the substrate electrode provides additional control of this layer; this electrode takes very little current provided both p–n^+ junctions remain reverse biased.

As the similarity in the naming of the contacts suggests, the JUGFET (section 6.4.4) and MOSFET behave very similarly in electronic circuits. The more

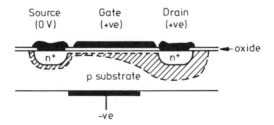

Fig. 6.11   Section through an n-channel MOSFET. The shading shows the depletion layer appropriate to the bias voltages indicated. The n-type inversion channel can be seen, sandwiched between the depletion layer and the oxide layer underneath the gate electrode. Electrical connections to the metal electrodes can be made through metal strips evaporated onto the surface

widespread use of the latter in integrated circuits is due to the relative ease with which many MOSFETs can be manufactured within a small area on the surface of a thin slice of silicon. Other circuit components, such as junction transistors, resistors and capacitors, can be fabricated on the surface using essentially the same manufacturing methods, and it is thus possible to build integrated circuits to fulfil many different functions. It is currently possible to construct circuits with about 10^5 individual components per square millimetre and this density is likely to increase further.

The basic MOS structure of Fig. 6.10(a) will only operate in the way we have outlined above if the behaviour of the surface layer is controlled by the gate potential rather than by charges resident on the surface. Unwanted surface charge density can result from electrons in states localized near the surface; these states often have energies within the band gap of the semiconductor. Also the large number of unsatisfied covalent bonds on a bare semiconductor surface makes the surface very reactive chemically and liable to contamination by impurity atoms (section 12.6); these chemical impurities in general have an associated surface charge density. The scale of the problem is indicated by noting that there are about 10^{13} semiconductor atoms per square millimetre of surface (and hence a similar number of unsatisfied covalent bonds) whereas a charge density of 10^{11} electronic charges per square millimetre is sufficient to prevent the gate from determining the surface behaviour.

It is also important that the surface should be flat and free from defects so that the mobility of the carriers in the inversion layer should not be severely reduced by increased scattering. The following surface preparation procedure can be used. A thin slice of silicon is cut with a diamond saw and the surface is polished to produce a *mirror* finish. The surface is then etched to remove the strained material produced by the sawing and polishing processes. Finally the surface is sealed by depositing a protective silicon dioxide layer on it. The unwanted charge density can be reduced in this way to about 10^8 electronic charges per square millimetre. To construct a MOSFET on this virgin silicon surface the sequence of operations described in Fig. 6.12 can be used.

The MOSFET described above is referred to as an **n-channel enhancement MOSFET** because it is necessary to apply a positive potential to the gate in order to create an n-type channel of enhanced conductivity. It is possible also to prepare a semiconducting surface with a controlled amount of surface charge density. The right amount of positive charge density on the surface of the semiconductor in the MOS structure of Fig. 6.10(a) can cause sufficient downward bending of the energy bands that an n-type inversion layer exists in the absence of an applied gate potential. In this case the application of a negative gate potential removes the inversion layer and hence the n-type conduction channel. A device exploiting this possibility is referred to as an **n-channel depletion MOSFET**. The reader will not be surprised to learn that p-channel MOS devices also exist in which the roles of electrons and holes are reversed.

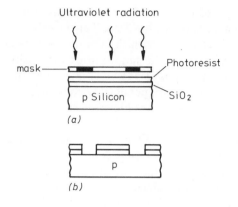

(a) The oxide surface is coated with a layer of photoresist varnish. The ultraviolet radiation hardens the photoresist except in the areas protected by the opaque regions of the mask

(b) The unhardened photoresist is dissolved in a suitable solvent and the exposed SiO_2 is etched away with HF

(c) The remaining photoresist is dissolved. Suitable donor atoms are diffused into the regions of Si from which the oxide layer has been removed; this creates the n^+ gate and source regions

(d) A thin metal film is evaporated over the whole surface and this is covered with another layer of photoresist. Exposure to ultraviolet radiation through a suitable mask enables regions of the metal film to be selectively removed by steps similar to those in (a) and (b). The electrode pattern of Fig. 6.11 is thus established

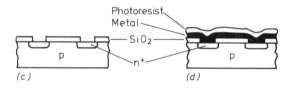

Fig. 6.12   Processes used to create an n-channel MOSFET on the surface of a thin slice of Si

6.6 MOLECULAR BEAM EPITAXY AND SEMICONDUCTOR HETEROJUNCTIONS

The development of the **molecular beam epitaxy** (MBE) method for the epitaxial growth† of fresh semiconductor material on a substrate has made it possible to manufacture semiconductor structures in which the chemical composition varies on an atomic length scale. In this method the new material arrives at the substrate in the form of a beam of atoms; individual sources are

---

† An epitaxial growth process is one in which the arrangement of atoms in the newly added material is a continuation of the ordered crystal structure of the substrate. In **vapour phase** and **liquid phase epitaxy** (VPE and LPE) the new material is deposited from the gaseous and liquid phases respectively. MBE provides a more sophisticated method of achieving epitaxial growth.

used for each element in the beam so that it is possible to vary the relative amounts of the different elements arriving at the substrate. In this way the chemical composition of the new material can be controlled and varied; growth rates are usually in the range 1 to 10 Å s^{-1}. The temperature of the substrate must be high enough for good epitaxial growth but low enough to prevent diffusion of the deposited atoms over significant distances. An example of the type of structure that can be created using this technique is shown in Fig. 6.13; a layer of GaAs is sandwiched between two layers of $Ga_{1-x}Al_xAs$ $(0 < x < 1,$ typically $x \sim 0.3)$. The interfaces (known as **heterojunctions** since they are between two different semiconductors) are sharp on an atomic length scale and layers as thin as a few atomic spacings can be made.

We now discuss the motivation for making such a structure. GaAs has the zincblende (ZnS) crystal structure (Fig. 1.15) and the structure of $Ga_{1-x}Al_xAs$ differs only in that a fraction x of the Ga atoms have been directly substituted by Al atoms. Since both Ga and Al are in group III of the periodic table, $Ga_{1-x}Al_xAs$ is also a III–V semiconductor, but with an energy gap that varies from 1.42 to 2.16 eV as x varies from 0 to 1; the end points in the range correspond to unsubstituted GaAs and AlAs respectively. The larger energy gap in $Ga_{1-x}Al_xAs$ explains the energy level diagram for electrons shown in Fig. 6.13.

This energy level diagram provides many interesting possibilities. The conduction band edge provides a finite square potential well for electrons; the valence band edge provides a similar well for holes. Most elementary courses on quantum mechanics contain calculations of the bound states of particles in one-dimensional square potential wells and the semiconductor heterojunction structure of Fig. 6.13 provides an experimental system in which the results of these calculations can be demonstrated. Of course the heterojunction structure is not strictly one-dimensional in that the electrons and holes are free to move in the plane of the GaAs layer; this difference can however be simply allowed for as

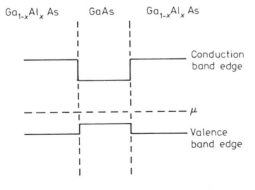

$Ga_{1-x}Al_x As$ GaAs $Ga_{1-x}Al_x As$

Conduction band edge

μ

Valence band edge

Fig. 6.13   Heterojunction structure consisting of a thin layer of GaAs sandwiched between two regions of $Ga_{1-x}Al_xAs$. The electron energy levels are also shown. Some transfer of charge between the GaAs and $Ga_{1-x}Al_xAs$ is in general necessary to cause alignment of the chemical potential μ but for very weakly doped layers the resulting charge density does not cause significant bending of the band edges on the length scale shown (see however Fig. 6.14)

we shall demonstrate in Chapter 14. Because the carriers in the bound states are restricted to moving parallel to the GaAs layer their behaviour is essentially two-dimensional if this layer is thin enough. The heterojunction structure of Fig. 6.13 thus provides an alternative to the MOSFET for the study of two-dimensional electron systems and we discuss this possibility further in Chapter 14.

More complicated structures than that illustrated in Fig. 6.13 can be manufactured by the MBE technique. One possibility is to create a regular array of alternate layers of GaAs and $Ga_{1-x}Al_xAs$. The result of this is to add a periodicity to a solid state system in addition to the underlying crystal structure. Systems containing such an additional periodicity are often referred to as **superlattices**. One interesting consequence of the existence of a superlattice is the appearance of band gaps in the ε/k dispersion relations of the electrons at values of k such that the electron wavefunction has the same periodicity as that of the superlattice. These band gaps are analogous to those that appear at the Brillouin zone boundaries when the electron wavefunction has the same periodicity as the crystal lattice (see section 4.1). Since the periodicity of the superlattice is greater than that of the crystal lattice the superlattice-induced band gaps occur at smaller values of k than those due to the crystal lattice.

Because the mobility of conduction band electrons in GaAs is very high, this material is often used in semiconductor devices where high-frequency operation or fast switching is required. One problem is that if the electrons are provided by donor atoms then the carrier mobility is reduced by scattering of the electrons by the ionized donors. This problem can be overcome by the use of a heterojunction structure as in the **high-electron-mobility transistor** (HEMT), the crucial part of which is shown in Fig. 6.14. A thin (~ 500 Å) layer of n-type

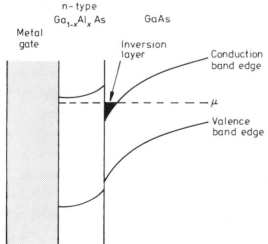

Fig. 6.14   Electron energy levels in an HEMT showing the formation of an n-channel inversion layer. Contact is made to the channel through drain and source electrodes as in the MOSFET of Fig. 6.11

$Ga_{1-x}Al_xAs$ is sandwiched between a metal gate electrode and a thick layer of pure GaAs; the donors are introduced into the $Ga_{1-x}Al_xAs$ in the MBE process by including the dopant atoms in the atomic beam. The behaviour of the structure of Fig. 6.14 is analogous to that of the MOS structure shown in Fig. 6.10; the potential applied to the metal gate determines the conductance of an n-type inversion layer at the surface. Inspection of the electron energy diagram of Fig. 6.14 shows that, although the electrons in the inversion layer are provided by the donors in the $Ga_{1-x}Al_xAs$, the inversion layer itself is situated in the GaAs; this is because there are conduction band states of lower energy within the GaAs. Since the carriers are physically separated from the donor atoms, their mobility is not decreased by donor scattering.

The electrostatic potential that gives rise to the curvature of the band edges in Fig. 6.14 can be calculated from the charge density using Poisson's equation (Eq. (6.4)) just as for the p-n junction (section 6.2). The charge density in the $Ga_{1-x}Al_xAs$ is predominantly that of the positive ionized donors; in the GaAs the charge density is due to the electrons in the inversion layer and to the negative ionized acceptors (we take the GaAs to be weakly p-type). The curvature of the bands is therefore in opposite directions in the two regions. Since the electron concentration depends on the potential it is necessary to ensure that the electron concentration inserted into Poisson's equation is the appropriate one for the potential obtained by solving the equation. This self-consistency requirement is often quite tricky to achieve due to the small length scale of the potential variation in the HEMT; because of this, Eqs. (6.10) do not adequately relate the electron concentration to the potential. Instead the electron charge density has to be obtained by solving Schrödinger's equation for the electron wavefunctions in the inversion layer.

The HEMT is now widely used in situations where low-noise amplification at high frequencies is required. The performance can be significantly improved by cooling the transistor to low temperatures.

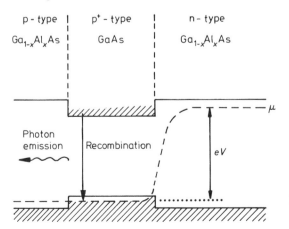

Fig. 6.15   Three-layer heterojunction laser

A more satisfactory p–n junction laser than that illustrated in Fig. 6.7 can be obtained by using the three-layer heterojunction structure of Fig. 6.15. The p–n junction is formed between n-type $Ga_{1-x}Al_xAs$ and p-type GaAs. The electrons injected into the p region are confined to a small region in the neighbourhood of the junction by the potential barrier provided by a layer of *p-type* $Ga_{1-x}Al_xAs$. Because of the confinement, population inversion and laser action are achieved at much lower currents than in a simple p–n junction. A further advantage of the structure shown in Fig. 6.15 is that $Ga_{1-x}Al_xAs$ has a lower refractive index than GaAs so that total internal reflection helps to confine the photons to the GaAs layer where the laser action occurs.†

PROBLEMS 6

6.1 By assuming that the potential in the depletion layer of a p–n junction is given by Eqs. (6.7) with w_n and w_p given by Eqs. (6.9), estimate the distance from the edge of the depletion layer (as a fraction of the total width, $w_n + w_p$) in which the majority carrier concentrations decrease by a factor of 2. (Assume $N_A = N_D$.)

6.2 The capacitance at a silicon p–n junction diode is used with a 100 μH inductance to create a resonant circuit. Calculate the change in resonant frequency when the bias applied to the junction changes from -1 to -10 V. (Take $N_A = N_D = 10^{22}$ m^{-3}, $T = 300$ K, $\varepsilon = 12$, $n_i = 2 \times 10^{16}$ m^{-3} and assume that the area of the junction is 10^{-6} m^2.)

6.3 Calculate the potential variation $\phi(x)$ and the depletion layer width for a graded p–n junction, that is one in which the doping level varies linearly with position: $N_D - N_A = kx$. You should assume that this linear variation is valid throughout the depletion layer.

6.4 A current of 5 μA flows through a p–n junction diode at room temperature when it is reverse biased with 0.15 V. Calculate the current flow when it is forward biased with the same voltage.

6.5 Show that the following alternative method for deriving the current–voltage relation for a p–n junction also leads to Eq. (6.22). Calculate the electron contribution to the current by using Eq. (5.74) to deduce the diffusive flow of electrons injected into the p region by a forward bias. The excess electron concentration at the edge of the depletion layer on the p side can be deduced by assuming that Eqs. (6.10) remain valid *within the depletion layer* even in the non-equilibrium situation where a bias is present.

Use this method to justify the statements in the footnote on p. 184 concerning the value of I_{h0} appropriate to the base–emitter junction of a p–n–p transistor. You will have to generalize Eq. (5.74) to the case of minority carrier injection into a semiconductor of length much shorter than a diffusion length. You can assume that the excess hole concentration vanishes at the edge of the depletion layer at the base–collector junction.

6.6 Use the generalization of Eqs. (6.5)–(6.9) to a situation of finite reverse bias to calculate the maximum electric field within the depletion layer.

---

† For more information on semiconductor heterojunctions, see M. Jaros, *Physics and Application of Semiconductor Microstructures*, Oxford University Press, 1989.

Estimate the width T of the potential barrier through which electrons involved in the Zener breakdown process must tunnel. By assuming that the tunnelling probability contains a factor of the form $\exp(-2\alpha T)$ (why?) show that the tunnelling current–voltage relation contains a factor $\exp(-b/|V|^{1/2})$ for $|V| \gg \Delta\phi_0$.

Given a time between collisions for carriers ($m_e = 0.1m$) in the depletion layer of 10^{-12} s, estimate the critical electric field for avalanche breakdown. Hence deduce the doping level ($N_A = N_D$) required to achieve a breakdown voltage of 100 V in a silicon diode ($E_G = 1.1$ eV, $\varepsilon = 12$).

Quantum Mechanics is the key to understanding magnetism. When one enters the first room with this key there are unexpected rooms beyond, but it is always the master key that unlocks each door.—*J. H. Van Vleck (Nobel prize address, 1977)*

CHAPTER 7

Diamagnetism and paramagnetism

7.1 INTRODUCTION

Increasing from zero the magnetic field applied to any material causes an induced emf (Faraday's law), which accelerates electrons within the material. According to Lenz's law the resulting electric current is in such a direction as to reduce (screen) the applied field. For reasons that we will explain in section 7.3 the current persists when the applied field is maintained at a constant value (even though the induced emf is then zero) and the material thus acquires a magnetization **M** (magnetic moment per unit volume) in the direction opposite to the field. This phenomenon is known as **diamagnetism**. The strength of magnetic effects of materials is quantified by the **magnetic susceptibility** χ, which is defined as the dimensionless constant of proportionality between **M** and **H**, where **H** is the macroscopic magnetic field within the material (see appendix B). Thus

$$\mathbf{M} = \chi\mathbf{H}. \qquad (7.1)$$

From our discussion above we deduce that χ is negative for diamagnetic materials; it is normally also very small, the value -8.1×10^{-6} for ice being typical. The diamagnetic susceptibility normally depends only very weakly on temperature at temperatures of order room temperature and below.

In materials in which some or all of the atoms possess a permanent magnetic dipole moment (i.e. a dipole moment that is non-zero in the absence of an applied field), the diamagnetic effects are normally small in comparison to the **paramagnetic** effects associated with the permanent moments. The magnetization of paramagnetic materials vanishes in zero applied field and satisfies Eq. (7.1) in small fields. We show in section 7.2 that this behaviour can be understood by assuming that the permanent dipoles behave independently of each other. In zero field the atomic moments are randomly oriented and there is no *net* magnetization; in small fields there is competition between the aligning effect of the field and thermal disorder, but on average there are more moments with components parallel to the field than antiparallel to it. The paramagnetic susceptibility is therefore positive and decreases with increasing temperature as the thermal disorder increases. A typical value of χ for a paramagnetic solid at room temperature is 3.8×10^{-4} for $CuSO_4$.

At low temperature the interactions between the permanent dipole moments can no longer be ignored. Thermal energy is insufficient to cause the dipoles to point in random directions in zero applied field. Their directions are correlated in such a way as to minimize the interaction energy. Thus the permanent dipoles in all paramagnetic materials make a transition to some kind of ordered state as the temperature is decreased in zero applied field. The ordering temperature varies widely; it is as high as 1388 K in cobalt but is below 1 K in some ionic salts in which the magnetic ions are widely separated. Magnetic ordering is the subject of Chapter 8.

In this book we shall consider only magnetic properties arising from the electrons in solids. The magnetic effects due to the nuclei are in general weaker by a factor of order the electron mass divided by the proton mass ($\sim 1/2000$). The reader should not however assume that nuclear magnetism is unimportant. The technique of **nuclear magnetic resonance** (NMR), for example, provides an excellent tool for investigating the properties of solids.

We conclude this introduction with a warning to the reader of problems that can be encountered when looking up values of the magnetic susceptibility in tables. The susceptibility defined by Eq. (7.1) is dimensionless although it is, perhaps misleadingly, often referred to as the susceptibility per unit volume or volume susceptibility. The reader is likely to encounter also the susceptibility per unit mass (or mass susceptibility) and the susceptibility per mole (or molar susceptibility). These are given, in terms of the dimensionless susceptibility we have defined, by χ/ρ and χV_m, respectively, where ρ is the mass density and V_m the molar volume; unfortunately the notation χ is used frequently to refer to all three quantities. Another problem the reader may encounter is that, although the volume susceptibility is dimensionless, the value appropriate to the SI units used throughout this book differs from that appropriate to the cgs system, which is still often encountered in tabulations of χ: the relationship is

$$\chi_{SI} = 4\pi\chi_{cgs}.$$ (7.2)

7.2 PARAMAGNETISM

7.2.1 The origin of permanent dipole moments

To investigate the origin of atomic dipole moments we consider the simple classical picture of an atom, i.e. an electron undergoing circular motion of radius r at velocity v about the nucleus (Fig. 7.1). The period τ of the orbit is $2\pi r/v$ and the orbiting electron is therefore equivalent to an electric current $i = (-e)/\tau = -ev/2\pi r$: the minus sign indicates that the electron is moving in the opposite direction to the current. It is a principle of electromagnetism (Ampère's law) that such a current loop has a magnetic dipole moment

$$\mathbf{\mu} = i\mathbf{a} \tag{7.3}$$

where $\mathbf{a}$ is the 'area' vector of the loop, directed so that the current is in a clockwise sense when looking along $\mathbf{a}$. Thus

$$\mathbf{\mu} = -\frac{ev}{2\pi r}\mathbf{a} = -\frac{e\hbar}{2m}\mathbf{l}, \tag{7.4}$$

where $\hbar\mathbf{l}$ is the angular momentum vector of the orbiting electron ($|\hbar\mathbf{l}| = mvr$) and we have used $|\mathbf{a}| = \pi r^2$. We write the angular momentum as $\hbar\mathbf{l}$ since $\hbar$ is the natural unit for the orbital angular momentum of atoms. It thus follows from Eq. (7.4) that the natural unit for the magnetic moment is the Bohr magneton μ_B, where

$$\mu_B = \frac{e\hbar}{2m} = 9.27 \times 10^{-24}\,\text{J T}^{-1}. \tag{7.5}$$

Eq. (7.4) indicates that there will be a contribution to the magnetic moment of an atom from the orbital angular momentum of the electrons within it. Eq. (7.4) remains valid in a quantum mechanical treatment provided that $\hbar\mathbf{l}$ is regarded as the angular momentum operator of the electron.

There is also a magnetic moment,

$$\mathbf{\mu} = -g_0\mu_B\mathbf{s}, \tag{7.6}$$

associated with the intrinsic (spin) angular momentum $\hbar\mathbf{s}$ of the electron; to a very good approximation (certainly good enough for our purposes) $g_0 = 2$.

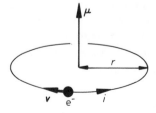

Fig. 7.1   An electron in a circular orbit is equivalent to a current loop and hence to a magnetic moment

Comparison of Eqs. (7.4) and (7.6) then shows that spin angular momentum is twice as effective in generating magnetic moment as orbital angular momentum. The eigenvalues of the z component, $\hbar s_z$, of $\hbar s$ are $\pm\frac{1}{2}\hbar$ so that, from Eq. (7.6), the z component of the intrinsic magnetic moment can take the values $\pm\mu_B$.

The total magnetic dipole moment of the atom is obtained by summing Eqs. (7.4) and (7.6) over all the electrons within it. Thus

$$\boldsymbol{\mu} = -\mu_B(\mathbf{L} + 2\mathbf{S}),\tag{7.7}$$

where $\hbar\mathbf{L} = \hbar\sum\mathbf{l}$ and $\hbar\mathbf{S} = \hbar\sum\mathbf{s}$, the sums being over the electrons in the atom; $\hbar\mathbf{L}$ and $\hbar\mathbf{S}$ are the total orbital and spin angular momenta of the atom respectively. The contribution of a closed shell of electrons to $\mathbf{L}$ or $\mathbf{S}$ is zero, so that permanent dipole moments only occur in atoms or ions with incomplete shells such as those of the transition metals and rare earths, which have incomplete 3d and 4f shells respectively; transition metal and rare-earth ions thus exhibit paramagnetism and we will use them to illustrate this phenomenon. In weak applied fields the angular momenta $\hbar\mathbf{L}$, $\hbar\mathbf{S}$ and $\hbar\mathbf{J}$ ($= \hbar\mathbf{L} + \hbar\mathbf{S}$) associated with the incomplete shells in *isolated* ions of these elements are determined by the **Russell–Saunders coupling scheme**, sometimes called the **L–S coupling**. According to this scheme the stationary states of the shell are eigenstates of $\mathbf{L}^2$, $\mathbf{S}^2$, and $\mathbf{J}^2$, with eigenvalues $L(L + 1)$, $S(S + 1)$ and $J(J + 1)$ respectively.

The values of L, S and J for the state of lowest energy are given by **Hund's rules**, which, in the order they must be obeyed, are:

(1) S takes the maximum value allowed by the exclusion principle—as many as possible of the electrons must have parallel spins;

(2) L takes the maximum value consistent with this value of S—the electrons have their orbital angular momenta as well aligned as possible;

(3) $J = |L - S|$ for a shell less than half-full and $J = L + S$ for a shell more than half-full.

In Fig. 7.2 we illustrate the use of these rules to calculate the values of L, S and J for the ground states of the transition metal ions V^{3+} and Fe^{2+}. The spectroscopic notation (see Fig. 7.2) is used in Tables 7.1 and 7.2 to show the values of L, S and J predicted by Hund's rules for all the rare-earth and transition metal ions. Note that Hund's rules 1 and 2, which determine the values of L and S, are associated with the Coulomb forces between electrons; since these are much larger than magnetic forces† the application of a magnetic field does not interfere with these. The third rule, which determines the value of J, is associated with the spin–orbit interaction, that is with the magnetic field generated by the motion of the electrons within the atom; this is of order 10 T (see problem 7.1) so there is a possibility that this rule can be disrupted by an applied field of this order. Note also that the splitting of the levels corresponding

---

† See Chapter 8 for justification of this statement and of Hund's first rule.

	V^{3+}		Fe^{2+}
Number of 3d electrons	2		6

		l_z	
Occupancy of states as given by Hund's rules	—↑————	2	—↑——↓—
	—↑————	1	—↑————
	————	0	—↑————
	————	−1	—↑————
	————	−2	—↑————

	V^{3+}	Fe^{2+}
Hund's rule 1: determination of S. Make as many spins parallel as possibble	$S = \frac{1}{2} + \frac{1}{2}$ $= 1$	$S = \frac{1}{2} + \frac{1}{2} + \frac{1}{2} + \frac{1}{2} + \frac{1}{2} - \frac{1}{2}$ $= 2$
Hund's rule 2: determination of L. Make $\sum l_z$ as large as possible	$L = \sum l_z$ $= 2 + 1 = 3$	$L = 2 + 1 + 0 - 1 - 2 + 2$ $= 2$
Hund's rule 3: determination of J.	Shell less than half-full $J = |L - S| = 2$	Shell more than half-full $J = L + S = 4$
Spectroscopic notation, $^{2S+1}L_J$	$^3F_2$	$^5D_4$

Fig. 7.2   Use of Hund's rules to calculate the quantum numbers S, L and J of the ground states of the V^{3+} and Fe^{2+} ions. The 3d shell has $l = 2$ so there are $2l + 1$ sublevels corresponding to $l_z = -2, -1, 0, 1, 2$ as indicated. In the spectroscopic notation, values of L of 0, 1, 2, 3, 4, 5, 6,... are indicated by letters S, P, D, F, G, H, I,..

to different J values can be comparable to $k_B T$ at room temperature so that levels other than the ground state may be occupied in thermal equilibrium. In solids, Hund's third rule can also fail because of the effect of the *electric* field of the neighbouring ions.

7.2.2   The interaction of a permanent dipole moment with an applied magnetic field

The magnetic field $\mathbf{B}$ is conventionally taken to be along the z axis. The alignment of an atomic dipole by the field occurs because of a term

$$H_P = -\mathbf{\mu} \cdot \mathbf{B} = -\mu_z B \qquad (7.8)$$

in the energy of the atomic electrons, where μ is the dipole moment given by Eq. (7.7). Although H_P causes the reorientation of the magnetic moment by a magnetic field, it is *incorrect* to interpret it as the potential energy of interaction of an 'atomic bar magnet' with the field. We show for example in appendix C that the term $\mu_B \mathbf{L} \cdot \mathbf{B}$ in H_P is part of the kinetic energy of the electrons in a magnetic field. The reader is recommended to consult Mandl[2] for further discussion of magnetic energies.

The simplest approach to calculating the effect of the alignment energy H_P is to assume that it is small and to use first-order perturbation theory. This involves calculating the expectation value of H_P for the *unperturbed* ground state of the ion as given by Hund's rules. The calculation requires a competence in quantum mechanics above that assumed in this book, so we will not reproduce it here but refer the interested reader to Ashcroft and Mermin,[11] appendix P (see problem 7.3 for a simpler geometrical calculation which gives the same answer). The result of the calculation has a simple interpretation which we now present; the effect of the alignment energy (7.8) is to break the degeneracy of the ground state of the ion associated with the $2J + 1$ different values of J_z. The energies of the states are

$$E_P = \langle H_P \rangle = g\mu_B J_z B, \qquad (7.9)$$

where $J_z = -J, \ldots -1, 0, 1, \ldots, J$ and the **Landé g-factor** is

$$g = \frac{3}{2} - \frac{L(L+1) - S(S+1)}{2J(J+1)}. \qquad (7.10)$$

Eq. (7.9) corresponds to $2J + 1$ equally spaced levels and comparison of Eqs. (7.8) and (7.9) shows that the ion behaves as though it had an *effective* magnetic moment

$$\mu_{eff} = -g\mu_B \mathbf{J}. \qquad (7.11)$$

The Landé g-factor gives the number of Bohr magnetons associated with the effective moment, and in the simple geometrical model of problem 7.3 it arises because μ ($\propto (\mathbf{L} + 2\mathbf{S})$) is not parallel to μ_{eff} ($\propto \mathbf{J}$ ($= \mathbf{L} + \mathbf{S}$)). Fig. 7.3 shows the energy levels predicted by Eq. (7.9) for V^{3+} and Fe^{2+} ions in a field B. The lowest energy corresponds to the maximum possible alignment of μ_{eff} with $\mathbf{B}$ ($J_z = -J$).

7.2.3   Calculation of the magnetization of paramagnetic ions

If the permanent dipoles in a solid behave independently of each other, then, at a temperature T, the relative occupation of the energy levels of Eq. (7.9) will be given by a Boltzmann factor

$$\exp\left(-E_P/k_B T\right) = \exp\left(+\mu_{eff} \cdot \mathbf{B}/k_B T\right) = \exp\left(-g\mu_B B J_z/k_B T\right).$$

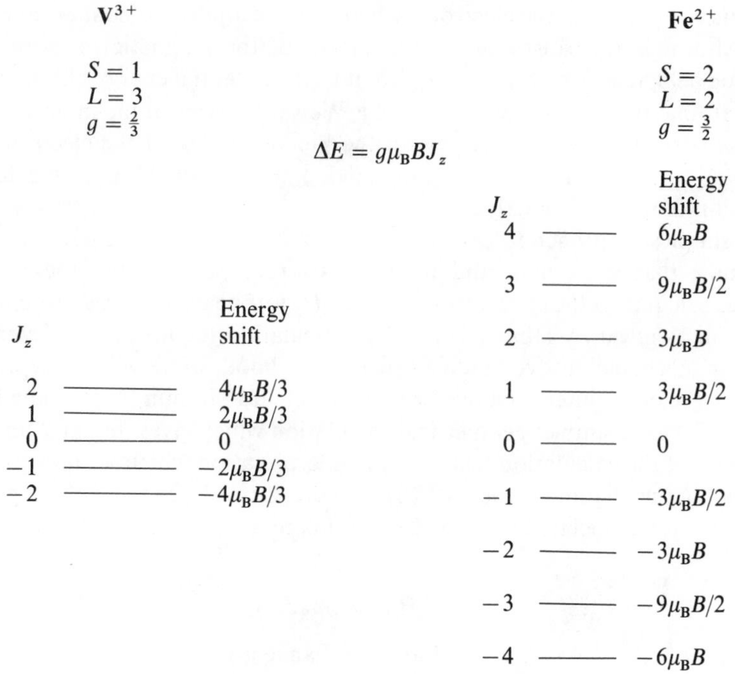

$\mathbf{V}^{3+}$

$\mathbf{Fe}^{2+}$

$S = 1$
$L = 3$
$g = \frac{2}{3}$

$S = 2$
$L = 2$
$g = \frac{3}{2}$

$$\Delta E = g\mu_B B J_z$$

J_z	Energy shift
4	$6\mu_B B$
3	$9\mu_B B/2$
2	$3\mu_B B$
1	$3\mu_B B/2$
0	0
-1	$-3\mu_B B/2$
-2	$-3\mu_B B$
-3	$-9\mu_B B/2$
-4	$-6\mu_B B$

J_z	Energy shift
2	$4\mu_B B/3$
1	$2\mu_B B/3$
0	0
-1	$-2\mu_B B/3$
-2	$-4\mu_B B/3$

Fig. 7.3   Splitting of the ground state degeneracy of the V^{3+} and Fe^{2+} ions by a magnetic field B, as given by Eq. (7.9)

From Eq. (7.11) the contribution of an atom to the z component of the magnetization is $-g\mu_B J_z$. The net magnetization for N moments per unit volume is therefore given by

$$M = N \sum_{J_z = -J}^{+J} -g\mu_B J_z \exp\left(-g\mu_B B J_z/k_B T\right) \bigg/ \sum_{J_z = -J}^{+J} \exp\left(-g\mu_B B J_z/k_B T\right).$$

This can be evaluated in a manner similar to that used in section 2.6.1 for evaluating the thermal energy of a simple harmonic oscillator by noting that it can be written

$$M = -\frac{Nk_B T^2}{B}\frac{1}{Z}\left(\frac{\partial Z}{\partial T}\right)_B = -\frac{Nk_B T^2}{B}\left(\frac{\partial \ln Z}{\partial T}\right)_B, \qquad (7.12)$$

where

$$Z = \sum_{J_z = -J}^{+J} \exp\left(-g\mu_B B J_z/k_B T\right) \qquad (7.13)$$

is the partition function of the dipole. Z is readily calculated by noting that it is a geometric series of $2J + 1$ terms with first term $\exp(g\mu_B BJ/k_B T)$ and ratio of successive terms $\exp(-g\mu_B B/k_B T)$. Thus

$$Z = \frac{e^x(1 - e^{-(2J+1)x/J})}{1 - e^{-x/J}} = \sinh\left[\left(\frac{2J+1}{2J}\right)x\right] \Bigg/ \sinh\left(\frac{x}{2J}\right) \qquad (7.14)$$

where $x = g\mu_B BJ/k_B T$ is essentially a dimensionless measure of the magnetic field.

Thus, using Eq. (7.14) in Eq. (7.12), we obtain, after some manipulation,†

$$M = -\frac{Nk_B T^2}{B}\left(\frac{d\ln Z}{dx}\right)\left(\frac{\partial x}{\partial T}\right)_B = Ng\mu_B J B_J(x), \qquad (7.15)$$

where

$$B_J(x) = \frac{2J+1}{2J}\coth\left[\left(\frac{2J+1}{2J}\right)x\right] - \frac{1}{2J}\coth\left(\frac{x}{2J}\right) \qquad (7.16)$$

is known as the Brillouin function. The behaviour of this function is qualitatively similar for all values of J. It increases linearly with x for small x but saturates at 1 for large x; as a result the magnetization increases linearly with field at small fields but approaches the **saturation magnetization** $Ng\mu_B J$ at large fields, corresponding to maximum possible alignment of the dipoles with the field (all the ions in the $J_z = -J$ state). The changeover in behaviour occurs when $x \sim J$, that is when $g\mu_B B/k_B T \sim 1$. For $g = 1$ this corresponds to a field of order 450 T at $T = 300$ K and hence to a field of order 1.5 T at $T = 1$ K. Fig. 7.4 shows that the experimentally observed dependence of the magnetization on B and T for Gd^{3+}, Fe^{3+} and Cr^{3+} ions is well described by the Brillouin function.

For small x, $B_J(x) \approx (J+1)x/3J$ so that in weak fields

$$M = \frac{Ng^2\mu_B^2 J(J+1)B}{3k_B T}. \qquad (7.17)$$

Before using this result to deduce a value for the susceptibility χ we must recall that B in Eq. (7.8) and subsequent equations is the *local* magnetic field at the ion, which will in general differ from the applied field because of the contribution from the magnetic moments of neighbouring ions. For small χ, $\mu_0 M \ll B$ and the difference between the local and applied fields as well as that between B and $\mu_0 H$ is unimportant. We then obtain, from Eqs. (7.17) and (7.1),

$$\chi = \frac{M}{H} = \mu_0 \frac{M}{B} = \frac{Np^2\mu_B^2\mu_0}{3k_B T} \qquad (7.18)$$

---

† Readers unwilling or unable to do this manipulation are advised to do problem 7.4 which contains the same calculation for the much simpler (but very important) case where $J = S = \frac{1}{2}$, $g = 2$.

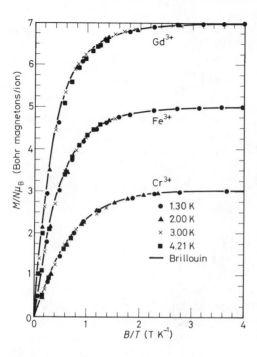

Fig. 7.4   Magnetization curves for Cr^{3+} ions in potassium chromium alum, Fe^{3+} ions in iron ammonium alum and Gd^{3+} ions in gadolinium sulphate octahydrate. Note that the values of $M/N\mu_B$ at different temperatures fall on the same curve, indicating that the magnetization is a function of B/T in agreement with Eq. (7.15). The full curves are obtained from Eq. (7.15) using $g = 2$ and values for J of $\frac{3}{2}$ for Cr^{3+}, $\frac{5}{2}$ for Fe^{3+} and $\frac{7}{2}$ for Gd^{3+}. Comparison with tables 7.1 and 7.2 shows agreement of these values with the Hund's rules predictions for Fe^{3+} and Gd^{3+} but disagreement for Cr^{3+}. The values needed to obtain agreement for Cr^{3+} indicate that the orbital angular momentum is quenched by the crystal field so that $L \approx 0$ and consequently $J \approx S$. (Reproduced with permission from W. E. Henry, *Phys. Rev.* **88**, 559 (1952))

where $p = g[J(J + 1)]^{1/2}$. This notation is used because Eq. (7.18) is the result of a classical calculation by Langevin of the paramagnetic susceptibility for ions of magnetic moment $p\mu_B$ (see section 9.1.3 for the Langevin calculation for electric dipoles). Our quantum calculation relates the value of p to the properties of the ion. Eq. (7.18) explains the **Curie law** for a paramagnet, that the susceptibility is inversely proportional to the absolute temperature; if it is written in the form $\chi = C/T$ then

$$C = \frac{Np^2\mu_B^2\mu_0}{3k_B} \tag{7.19}$$

is known as the **Curie constant.**

Inserting typical numbers in Eq. (7.18) ($N \approx 10^{28}$ m^{-3}, $p^2 \approx 3$) shows that χ becomes of order unity at a temperature of order 0.1 K; thus the distinctions between the applied field and the local field and between B and $\mu_0 H$ only become important at temperatures below about 1 K. In appendix B we discuss the relationship of the local and applied fields when the difference cannot be ignored; the relationship depends on the shape of the sample and the symmetry of the atomic arrangement. We show in appendix B that the contribution of the other dipoles to the local field vanishes for a spherically shaped sample in which each ion either has an environment of cubic symmetry or is surrounded by dipoles arranged at random (as in a liquid or gas); for this case the local and applied fields are equal.

Using Eq. (7.18) it is possible to deduce values for p^2 from measured values of χ. In tables 7.1 and 7.2, values obtained in this way are compared with theoretical values for isolated ions deduced using Hund's rules. Agreement is generally good for the rare-earth ions but it can be seen from Table 7.2 that for most of the transition metals better agreement is obtained by using S rather than J to calculate the value of p in Eq. (7.18). This seems to suggest that the transition metal ions are behaving as though they had $L = 0$; their orbital angular momentum is said to be **quenched**. Quenching is also apparent from the value of the saturation magnetization observed for the Cr^{3+} ion in Fig. 7.4. Quenching occurs because in a solid the electronic wavefunctions are affected by the electric field produced by neighbouring ions; this field is referred to as the **crystal field**. The crystal field is usually insufficient to disrupt the first two Hund's rules so that L and S remain good quantum numbers for each ion. It does however compete with the spin–orbit interaction responsible for Hund's third rule to decide which linear combinations of the $(2L+1)(2S+1)$ substates in the L, S manifold constitute the states of lowest energy.

If the crystal field effect is dominant, as it usually is for the transition metal ions, then the symmetry of the environment is often such that the eigenstates determined by the crystal field have $\langle L_x \rangle = \langle L_y \rangle = \langle L_z \rangle = 0$, so that the

TABLE 7.1   Values of p deduced, using Eq. (7.18), from measured values of the Curie constant for rare-earth ions compared with the value, $g[J(J+1)]^{1/2}$, predicted for the Hund's rule ground state of the ion. The discrepancies between theory and experiment for Sm and Eu can be accounted for by the existence of levels with different J values within $\sim k_B T$ of the ground state

Ion	Number of 4f electrons	Hund's rule ground state	Measured p	Calculated p
La^{3+}	0	$^1S_0$	0	0
Ce^{3+}	1	$^2F_{5/2}$	2.4	2.54
Pr^{3+}	2	$^3H_4$	3.5	3.58
Nd^{3+}	3	$^4I_{9/2}$	3.5	3.62
Pm^{3+}	4	$^5I_4$	–	2.68
Sm^{3+}	5	$^6H_{5/2}$	1.5	0.84
Eu^{3+}	6	$^7F_0$	3.4	0.00
Gd^{3+}	7	$^8S_{7/2}$	8.0	7.94
Tb^{3+}	8	$^7F_6$	9.5	9.72
Dy^{3+}	9	$^6H_{15/2}$	10.6	10.63
Ho^{3+}	10	$^5I_8$	10.4	10.60
Er^{3+}	11	$^4I_{15/2}$	9.5	9.59
Tm^{3+}	12	$^3H_6$	7.3	7.57
Yb^{3+}	13	$^2F_{7/2}$	4.5	4.54
Lu^{3+}	14	$^1S_0$	0	0

(The experimental values are from R. Kubo and T. Nagamiya, *Solid State Physics*, McGraw-Hill, New York, 2nd edn (1968), p. 451)

TABLE 7.2   Comparison of experimental and theoretical values of p for transition metal ions. The experimental values agree much better with $2[S(S + 1)]$ than with $g[J(J + 1)]$, indicating that the orbital angular momentum is quenched

Ion	Number of 3d electrons	Hund's rule ground state	Measured p	$g[J(J + 1)]$	$2[S(S + 1)]$
K^+	0	$^1S_0$	0	0	0
V^{4+}	1	$^2D_{3/2}$	1.8	1.55	1.73
V^{3+}	2	$^3F_2$	2.8	1.63	2.83
V^{2+}	3	$^4F_{3/2}$	3.8	0.77	3.87
Cr^{3+}	3	$^4F_{3/2}$	3.7	0.77	3.87
Mn^{4+}	3	$^4F_{3/2}$	4.0	0.77	3.87
Cr^{2+}	4	$^5D_0$	4.8	0	4.90
Mn^{3+}	4	$^5D_0$	5.0	0	4.90
Mn^{2+}	5	$^6S_{5/2}$	5.9	5.92	5.92
Fe^{3+}	5	$^6S_{5/2}$	5.9	5.92	5.92
Fe^{2+}	6	$^5D_4$	5.4	6.70	4.90
Co^{2+}	7	$^4F_{9/2}$	4.8	6.54	3.87
Ni^{2+}	8	$^3F_4$	3.2	5.59	2.83
Cu^{2+}	9	$^2D_{5/2}$	1.9	3.55	1.73

(The experimental values are from R. Kubo and T. Nagamiya, *Solid State Physics*, McGraw-Hill, New York, 2nd edn (1968), p. 453)

contribution of the orbital motion to the magnetic moment vanishes; only S is free to respond to an applied magnetic field in this case. On the other hand, Hund's third rule works for the rare-earth ions because the incomplete 4f shell is much closer to the nucleus and not so strongly affected by the crystal field. In this respect it is important to realize that, because the electron density is symmetrically disposed with respect to the nucleus, it is the electric field gradient rather than the field itself that is responsible for the effect on the wavefunction; the change in the field across a 3d wavefunction is much bigger than that across a 4f wavefunction.

7.2.4   Conduction electron paramagnetism

The theory of section 7.2.3 does not apply when the permanent moments within the solid are those associated with the conduction electron spins in a metal. This is because the behaviour of the conduction electrons is dominated by their indistinguishability and hence by the Pauli principle; we were able to consider the paramagnetic ions as distinguishable particles obeying a classical Boltzmann distribution because they were located in definite positions within the crystal. The Pauli principle restricts the states that a conduction electron can occupy and hinders spin alignment, thereby reducing the susceptibility below the Curie law value.

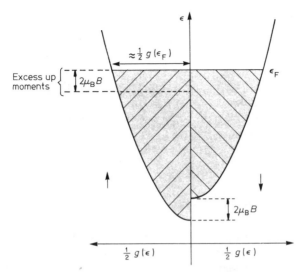

Fig. 7.5   Occupied states for conduction electrons in a magnetic field at $T = 0$, showing an excess of electrons with dipole moments parallel to the field ($\uparrow$) over those with moments antiparallel to the field ($\downarrow$). Note that the spin angular momentum is in the *opposite* direction to the dipole moment

The effect may be calculated by reference to Fig. 7.5, which shows the conduction electron density-of-states curve (see Fig. 3.2.(*b*)) split into two halves, one for each of the two spin states of the electrons. The component of the spin magnetic moment can take values $\pm\mu_B$ parallel to an applied field. The energies of the electrons with their magnetic moment parallel to the field B (labelled $\uparrow$) are lowered by $\mu_B B$ and the energies of electrons with moments antiparallel to the field (labelled $\downarrow$) are raised by $\mu_B B$, giving the density-of-states curves as drawn. The Fermi energy ε_F must be the same for both spin states in thermal equilibrium. For fields that can be applied in the laboratory the shift in energy levels $\mu_B B$ is usually much smaller than the Fermi energy ε_F (see problem 7.5), and if this is the case there is an excess of parallel moments over antiparallel moments Δn equal to the approximately rectangular area indicated in Fig. 7.5. Thus

$$\Delta n = \tfrac{1}{2} g(\varepsilon_F) \times 2\mu_B B, \qquad (7.20)$$

so that the magnetization is

$$M = \mu_B \Delta n = \mu_B^2 g(\varepsilon_F) B, \qquad (7.21)$$

where now $g(\varepsilon_F)$ is the density of states at the Fermi surface per *unit volume* of metal. Assuming that the magnetic effects are weak we take $B = \mu_0 H$ and deduce a value for the susceptibility,

$$\chi_P = \frac{M}{H} = \mu_0 \mu_B^2 g(\varepsilon_F).$$ (7.22)

The effect we have just considered is known as the **Pauli spin paramagnetism** of the conduction electrons.

For N free electrons per unit volume we can write $g(\varepsilon_F) = 3N/2\varepsilon_F$ (Eq. (3.8)) and Eq. (7.22) becomes

$$\chi_P = \frac{3N\mu_0\mu_B^2}{2\varepsilon_F}.$$ (7.23)

It is interesting to compare this result with the classical Curie law prediction $\chi_{cl} = N\mu_0\mu_B^2/k_B T$ obtained from Eq. (7.18) by putting $J = S = \frac{1}{2}$, $g = 2$. Since $\varepsilon_F = k_B T_F$ we see that the spin susceptibility of a degenerate Fermi gas, like the heat capacity (section 3.2.3), is reduced by a factor of order T/T_F below its classical value. The classical value would only be achieved at temperatures $T \gg T_F$, where the electrons become non-degenerate; all metals vaporize before reaching such a high temperature!

Because it is reduced by such a large factor from the classical value, the Pauli paramagnetic susceptibility is comparable to the diamagnetic susceptibility of conduction electrons. Indeed Landau calculated a value

$$\chi_L = -\tfrac{1}{3}\chi_P$$ (7.24)

for the diamagnetic susceptibility of a *free electron* metal. The net susceptibility for *free* electrons is thus positive and equal to $\frac{2}{3}\chi_P$. Band structure effects and electron–electron interactions modify this result but it remains correct in order of magnitude. Comparison between theoretical and experimental values of χ_P for the alkali metals is given in table 7.3; there is agreement within a factor of order 2. The prediction of a temperature-independent χ is well satisfied in practice for many metals over a wide temperature range.

The magnetic susceptibility of some transition metals is larger and this is due to the contribution of the 3d electrons. If these are pictured as inhabiting an energy band associated with the overlap of the 3d wavefunctions on neighbouring atoms, then the enhancement to the susceptibility can be attributed to the contribution of the 3d band to $g(\varepsilon_F)$ in Eq. (7.22); the overlap of the 3d wavefunctions is small and this means that the band is narrow and has a high density of states. In order that the 3d band should contribute to $g(\varepsilon_F)$ it is necessary that this band should be only partly filled so that the Fermi energy lies within it. In the rare-earth metals the 4f electrons contribute to the susceptibility; in this case the overlap of the wavefunctions on neighbouring atoms is negligibly small and the contribution is better calculated by assuming that these electrons are localized in the atomic states with a susceptibility given by

TABLE 7.3 Comparison of the free electron theory prediction (Eq. (7.23)) with the measured Pauli spin susceptibility for the alkali metals. Note that the experimental value is just the contribution from the conduction electron spin, *not* the total susceptibility

	Li	Na	K	Rb	Cs
$10^5 \times \chi_P$ (experiment)	2.5	1.4	1.1	1.0	1.0
$10^5 \times \chi_P$ (theory)	1.01	0.83	0.67	0.63	0.58

(The data in the table were obtained by converting the values from Ashcroft and Mermin,[11] p. 664, into SI units)

Eq. (7.18). The temperature below which many transition metals and rare-earth metals exhibit magnetic ordering is high and this is an indication that the magnetic moments interact strongly in these metals.

7.3 DIAMAGNETISM

We have already described how diamagnetism arises because of the induced currents set up on applying a magnetic field; these currents tend to screen the applied field from the interior of the material. According to classical mechanics the currents are destroyed by the interactions that maintain thermal equilibrium within the material. We shall see, however, that quantum mechanics gives a certain stability to the screening currents, resulting in a tendency (usually very weak) for all matter to exclude applied fields. To understand this effect we will need to consider in more detail what happens when a charged particle is accelerated by a changing magnetic field.

7.3.1 Momentum in a magnetic field

Consider a stationary particle of mass M and charge q at a distance r from a point on the axis of a long solenoid carrying a current i as in Fig. 7.6 For simplicity we suppose that the solenoid is superconducting so that, if the ends are connected together as shown to form a complete circuit, it will carry the current indefinitely without an external power supply. If the solenoid is heated above its superconducting transition temperature the current and hence the magnetic field decay. The resulting induced emf around the circle C accelerates the particle, giving it a momentum $M\mathbf{v}$. Where has this momentum come from? We certainly applied no force to the system in heating the coil to cause the current to disappear. This paradox can be resolved by arguing that the particle

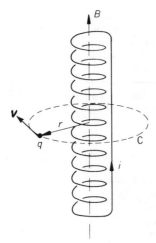

Fig. 7.6 A charged particle accelerated by a decaying magnetic field. The motion indicated is appropriate to a *negatively* charged particle.

possessed the momentum throughout; we demonstrate below that to recover the law of conservation of momentum it is necessary to write the momentum of the particle as

$$\mathbf{p} = M\mathbf{v} + q\mathbf{A}, \tag{7.25}$$

where $\mathbf{A}$ is the magnetic vector potential ($\mathbf{A}$ determines the magnetic field through $\mathbf{B} = \mathrm{curl}\ \mathbf{A}$). The momentum $q\mathbf{A}$ possessed by a charged particle *at rest* is known as **electromagnetic momentum**; in the experiment described above the momentum is initially all electromagnetic and in the final state it is all in the kinetic form $M\mathbf{v}$.

To establish that the momentum defined by Eq. (7.25) is conserved in the above process we assume that the current decays in a short time so that the motion of the particle during the decay period may be ignored. We do this only to simplify the calculation; the result we obtain is a general one. The kinetic momentum acquired by the particle is the impulse of the force on it. That is

$$M\mathbf{v} = \int q\mathbf{E}\ \mathrm{d}t. \tag{7.26}$$

The electric field $\mathbf{E}$ at the position of the particle is given by Faraday's law

$$\oint_{\mathrm{C}} \mathbf{E} . \mathrm{d}\mathbf{l} = -\frac{\mathrm{d}\Phi}{\mathrm{d}t} \tag{7.27}$$

where Φ is the magnetic flux through C. We will relate the electric field to the magnetic vector potential $\mathbf{A}$. If the solenoid in Fig.7.6 is long the field $\mathbf{B}$ is vanishingly small at the position of the particle; $\mathbf{A}$ must however be finite there in order to satisfy the requirement

$$\Phi = \iint \mathbf{B} \cdot d\mathbf{S} = \iint \text{curl } \mathbf{A} \cdot d\mathbf{S} = \oint_C \mathbf{A} \cdot d\mathbf{l}, \tag{7.28}$$

where the surface integrals are over the surface of circle C. Inserting this value for Φ in Eq. (7.27) we obtain

$$\oint_C \mathbf{E} \cdot d\mathbf{l} = -\oint_C \frac{d\mathbf{A}}{dt} \cdot d\mathbf{l}. \tag{7.29}$$

From the symmetry of Fig. 7.6, **E** must be constant around C and we can also choose **A** to have this property.† In this case we deduce from Eq. (7.29) that

$$\mathbf{E} = -\frac{d\mathbf{A}}{dt}, \tag{7.30}$$

so that Eq. (7.26) can be written

$$M\mathbf{v} = -\int \frac{d\mathbf{A}}{dt} dt = -q \int d\mathbf{A} = -q\Delta\mathbf{A} = -q\mathbf{A_0}. \tag{7.31}$$

$\mathbf{A_0}$ is the value of **A** prior to the decay; the value of **A** is zero at the end. Eq. (7.31) establishes that the momentum defined by Eq. (7.25) is indeed conserved.

Our discussion so far has been classical; the transition to quantum mechanics is made by replacing the momentum **p** of Eq. (7.25) by the operator $-i\hbar\nabla$. This is discussed further in appendix C. In the remainder of this chapter we apply Eq. (7.25) to electrons by substituting $q \to -e$ and $M \to m$.

7.3.2 Screening by induced currents

Classically the velocity **v** of an electron averages to zero in thermal equilibrium so there are no induced currents; from Eq. (7.25) the average momentum $\langle \mathbf{p} \rangle$ is therefore $-e\mathbf{A}$ and changes as the magnetic field changes. Quantum mechanically the identification of **p** with $-i\hbar\nabla$ means that **p** is determined by the geometry of the wavefunction. The confinement of atomic electrons by the Coulomb attraction of the nucleus means that there are only discrete electronic states with orthogonal, qualitatively different, wavefunctions (differing for example in the number of nodes); the states are separated in energy by typically 1 eV. Atomic wavefunctions thus possess a measure of rigidity so that they are

---

† **E** is a measurable quantity and its cylindrical symmetry is thus ensured by the symmetry of the apparatus. **B** = curl **A** is also measurable and must be cylindrically symmetric. **A** cannot be measured and the addition of $\nabla\theta$ to it, where θ is *any* scalar function of position, gives an equally acceptable vector potential (it gives the same **B** because curl $(\nabla\theta) = 0$). $\nabla\theta$ and hence **A** need not be cylindrically symmetric. We can force **A** to be cylindrically symmetric in the geometry of Fig. 7.6 by imposing the additional condition div **A** = 0. Equations tend to be be simpler if this condition is imposed on **A**.

perturbed only slightly by a weak magnetic field; using the language of perturbation theory we can say that the wavefunction is unchanged to first order in the perturbation. It is a reasonable guess that an atomic wavefunction cannot change much until the applied field is large enough to make free electron cyclotron orbits (see section 5.5.3) of atomic dimensions.

Because of this rigidity of the atomic wavefunctions we might expect that it is $\langle \mathbf{p} \rangle$ that remains constant and $\langle \mathbf{v} \rangle$ that changes when a field is applied. If this is so then, from Eq. (7.25), the induced velocity is $\mathbf{v} = -q\mathbf{A}/M = e\mathbf{A}/m$ and the resulting induced current density is

$$\mathbf{j} = n(-e)\mathbf{v} = -\frac{ne^2}{m}\mathbf{A}. \tag{7.32}$$

Since the electron density is independent of time, we must have div $\mathbf{j} = 0$. Thus Eq. (7.32) is only valid if $\mathbf{A}$ is chosen so div $\mathbf{A} = 0$.†

To show that Eq. (7.32) implies screening of the applied magnetic field by the electrons, we use $\mathbf{B} = $ curl $\mathbf{A}$ and Maxwell's equation (for static fields) curl $\mathbf{B} = \mu_0\mathbf{j}$ to obtain‡

$$\text{curl curl } \mathbf{A} = \text{curl } \mathbf{B} = \mu_0\mathbf{j} = -\frac{\mu_0 ne^2}{m}\mathbf{A}.$$

On using the vector identity curl curl $\mathbf{A} = $ grad(div $\mathbf{A}$) $- \nabla^2\mathbf{A}$ and recalling that div $\mathbf{A} = 0$, this becomes

$$\nabla^2\mathbf{A} = \frac{\mu_0 ne^2}{m}\mathbf{A} = \frac{1}{\lambda^2}\mathbf{A} \tag{7.33}$$

where $\lambda = (m/\mu_0 ne^2)^{1/2}$ is a characteristic length; the solutions of Eq. (7.33) have the property (see problem 7.7) that $\mathbf{A}$, and hence $\mathbf{B}$, decays exponentially as we go into the interior of a region containing electrons, such as an atom.

The characteristic length of this decay is λ, and if we compare this length with atomic dimensions we will see how effective atomic electrons are in screening an applied magnetic field. The typical electron density in an atom is 1 electron/Å^3 so that $n \sim 10^{30}$ m^{-3}. Hence

$$\lambda = \left(\frac{m}{\mu_0 ne^2}\right)^{1/2} \sim \left(\frac{10^{-30}}{10^{-6} \times 10^{30} \times 10^{-38}}\right)^{1/2} = 10^{-8}\text{ m} = 100\text{ Å}, \tag{7.34}$$

which is much larger than an atomic size. Magnetic fields are therefore screened only slightly from an atom and in the following section we will calculate the diamagnetic moment of the atom by using Eq. (7.32) with $\mathbf{A}$ the vector potential

---

† For any other choice of $\mathbf{A}$ it is no longer true that the change in $\langle \mathbf{p} \rangle$ can be ignored.

‡ In this section the response of the medium to the magnetic field is described by the current density $\mathbf{j}$ rather than the equivalent magnetization; we can therefore set $\mathbf{B} = \mu_0\mathbf{H}$. In deriving Eq. (7.33) we have ignored the spatial dependence of n and assumed that it can be replaced by its average value; since our intention is to determine only the qualitative behaviour this is an acceptable assumption.

of the *applied* field. Thus, even if the atomic wavefunctions are completely unperturbed by a magnetic field, the resulting diamagnetism is very weak. On an atomic scale, the inertia of the electrons is too great for them to provide effective screening currents.

When electronic wavefunctions extend over a larger region, as for the conduction electrons in a metal, they are usually strongly perturbed by a magnetic field; the change in $\langle \mathbf{p} \rangle$ is then of order $-e\mathbf{A}$, the screening currents are small and the diamagnetism remains weak. Landau's calculation of the diamagnetism for free electron metals (see section 7.2.4) confirms this. The weak **Landau diamagnetism** of normal metals contrasts sharply with the situation in superconductors; we shall see in Chapter 10 that superconducting electron wavefunctions are only slightly perturbed by a magnetic field, so that the decay of field predicted by Eq. (7.33) occurs on a macroscopic scale.

7.3.3   Calculation of the diamagnetic susceptibility

We will use Eq. (7.32) to calculate the magnetic moment induced on an atom by an applied field and hence to calculate the diamagnetic susceptibility. We use a cylindrical system of coordinates with origin at the centre of the atom and z axis parallel to $\mathbf{B}$ as in Fig. 7.7. By using Eq. (7.28) it is possible to show that the uniform field can be described by a vector potential (problem 7.6)

$$\mathbf{A} = \hat{\mathbf{c}}\, B\rho/2, \tag{7.35}$$

where $\hat{\mathbf{c}}$ is a unit vector directed tangential to the circular ring of radius ρ in Fig. 7.7. This vector potential also satisfies div $\mathbf{A} = 0$ and so it is the appropriate vector potential to use in Eq. (7.32).

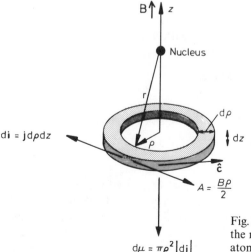

Fig. 7.7   Current element for calculating the magnetic dipole moment μ of an atom due to induced screening currents

From Eq. (7.32), the current density is

$$\mathbf{j} = - \frac{ne^2}{m} \mathbf{A} = - \hat{\mathbf{c}} \frac{ne^2 B \rho}{2m}. \tag{7.36}$$

If we make the simplifying assumption that n is constant within the circular ring in Fig. 7.7, then the contribution $d\mu$ of this ring to the magnetic moment of the atom is that of a loop of current

$$d\mathbf{i} = \mathbf{j} \, d\rho \, dz = - \hat{\mathbf{c}} ne^2 B \rho \, d\rho \, dz / 2m.$$

According to Ampere's law, Eq. (7.3), therefore

$$d\boldsymbol{\mu} = - \hat{\mathbf{z}} |d\mathbf{i}| \pi \rho^2 = - \mathbf{B} \frac{ne^2 \rho}{2m} \pi \rho^2 d\rho \, dz,$$

and the total induced moment on the atom is

$$\boldsymbol{\mu} = - \mathbf{B} \frac{e^2}{4m} \iint n\rho^2 2\pi\rho \, d\rho \, dz = - \mathbf{B} \frac{e^2}{4m} \int n\rho^2 \, dV,$$

where $dV = 2\pi\rho \, d\rho \, dz$ is the volume of the ring. Since $\int n \, dV = Z$, where Z is the total number of electrons in the atom, we can write $\int n\rho^2 \, dV = Z \langle \rho^2 \rangle$ where $\langle \rho^2 \rangle$ is the mean square distance of the electrons from the z axis. Thus

$$\boldsymbol{\mu} = - \frac{Ze^2}{4m} \langle \rho^2 \rangle \, \mathbf{B}. \tag{7.37}$$

Note that for a spherically symmetric distribution of electrons $\langle \rho^2 \rangle = 2\langle r^2 \rangle / 3$ where r is the mean square distance of the electrons from the nucleus (see Fig. 7.7); the appearance of Eq. (7.37) with a 6 rather than a 4 in the denominator indicates that this substitution has been made.

Our calculation is appropriate to an isolated atom, but, since diamagnetic effects are weak, we can ignore the effect on the field at an atom of the induced moments of its neighbours and take $\mathbf{B} = \mu_0 \mathbf{H}$ with $\mathbf{B}$ the externally applied field. Thus the magnetization for N identical atoms per unit volume is

$$\mathbf{M} = N\boldsymbol{\mu} = - \frac{NZe^2}{4m} \langle \rho^2 \rangle \, \mathbf{B}. \tag{7.38}$$

By comparison with Eq. (7.1), we identify the magnetic susceptibility as

$$\chi = - \frac{NZe^2\mu_0}{4m} \langle \rho^2 \rangle = - \frac{\langle \rho^2 \rangle}{4\lambda^2} \sim 10^{-5}. \tag{7.39}$$

To obtain an order of magnitude value for χ, we have used the estimate of the screening length λ calculated in the previous section and taken $\langle \rho^2 \rangle \sim a_0^2$, where $a_0 = 0.53$ Å is the Bohr radius. More precisely we might expect from Eq. (7.39) that the quantity $4m\chi/\mu_0 NZe^2 a_0^2$ is of order unity; we see that this is true in table 7.4, which gives values of this quantity and of the molar susceptibility for the inert gases.

TABLE 7.4   Values of the molar susceptibility χ_M (in m^3 mol^{-1}) for the inert-gas atoms. The molar susceptibility is given by Eq. (7.39) with N replaced by Avogadro's number N_A. The quantity $-4m\chi_M/N_A\mu_0 Ze^2 a_0^2$ should therefore be of order unity. The diamagnetic properties of an inert-gas atom are independent of the state (gas, liquid or solid) of the atom to a good approximation

	He	Ne	Ar	Kr	Xe
χ_M (10^{-11} m^3 mol^{-1})	-2.36	-8.47	-24.6	-36.2	-55.2
Z	2	10	18	36	54
$-\dfrac{4m\chi_M}{N_A\mu_0 Ze^2 a_0^2}$	0.79	0.57	0.91	0.67	0.68

(The data in the table were obtained by converting values from the *Handbook of Chemistry and Physics* (61st edn. Copyright CRC Press Inc., Boca Raton, Florida. Reprinted with permission) into SI units (multiply cgs value in cm^3 mol^{-1} by $4\pi \times 10^{-6}$)

PROBLEMS 7

7.1 For an electron in the lowest energy state of the Bohr model of the hydrogen atom, calculate in tesla the magnetic field at the nucleus resulting from the motion of the electron around it.

(Note however that a correct quantum mechanical calculation shows the electron to have zero angular momentum in its ground state and thus to produce zero magnetic field.)

7.2 Apply Hund's rules to a 4f shell containing n electrons to show that;

$$S = \begin{cases} n/2 & \text{for } n \leq 7 \\ (14-n)/2 & \text{for } n \geq 7 \end{cases}$$

$$L = \begin{cases} n(7-n)/2 & \text{for } n \leq 7 \\ (14-n)(n-7)/2 & \text{for } n \geq 7 \end{cases}$$

$$J = \begin{cases} n(6-n)/2 & \text{for } n \leq 6 \\ (14-n)(n-6)/2 & \text{for } n \geq 6. \end{cases}$$

Use these results to check the values of S, L and J given for the rare-earth elements in Table 7.1

7.3 The coupling of **L** and **S** to give **J** may be represented on the vector diagram

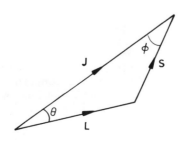

The magnetic moment is $\boldsymbol{\mu} = -\mu_B(\mathbf{L} + 2\mathbf{S})$ (Eq. (7.7)). Unlike $\mathbf{J}^2$, $\boldsymbol{\mu}^2$ is not a good quantum number so that only the component of $\boldsymbol{\mu}$ along $\mathbf{J}$ contributes to the magnetic properties (the component perpendicular to $\mathbf{J}$ is sometimes pictured as precessing rapidly around $\mathbf{J}$ and thus averaging to zero). Show that the effective moment can be written

$$\boldsymbol{\mu} = -\frac{\mathbf{J}}{|\mathbf{J}|}\mu_B(|\mathbf{L}|\cos\theta + 2|\mathbf{S}|\cos\phi),$$

and hence evaluate the Landé g-factor of Eq. (7.10).

7.4 The most important contribution to the paramagnetism of $CuSO_4$ comes from the Cu^{2+} ions for which the magnetic moment is due to a single unpaired spin ($L = 0$, $J = S = \frac{1}{2}$, $g = 2$). Write down the probabilities at temperature T that the moment lies parallel and antiparallel to the field. Hence show that the magnetization for N ions per unit volume in a field B is

$$M = N\mu_B \tanh{(\mu_B B/k_B T)}.$$

Write down the internal energy and hence calculate the magnetic heat capacity C_B of the ions in a constant field B. Deduce the limiting form of the specific heat at high and low temperatures. Sketch C_B as a function of temperature. What is a high temperature if $B = 0.5$ T?

7.5 In deriving the Pauli spin susceptibility for free electrons in a metal we made the assumption that $\mu_B B \ll \varepsilon_F$. Show that the *exact* relation between the spin magnetization M and the field B for a free electron gas at absolute zero can be written

$$\frac{2\mu_B B}{\varepsilon_F} = \left(1 + \frac{M}{M_s}\right)^{2/3} - \left(1 - \frac{M}{M_s}\right)^{2/3}$$

provided $M < M_s$, where ε_F is the Fermi energy and M_s is the saturation magnetization. (You will need to choose the energy ($\neq \varepsilon_F$) of the maximum occupied level in Fig. 7.5 so that the shaded area is equal to the number of electrons.)

Estimate, for potassium ($N = 1.4 \times 10^{28}\ m^{-3}$), the magnetic field required to saturate the spin magnetization. Estimate the field required at $T = 0$ to saturate the nuclear spin magnetization of the ^{3}He atoms present at a concentration of 0.1 % in a liquid mixture of ^{3}He and ^{4}He. (The ^{3}He behaves like a degenerate Fermi gas and the nuclear moment on each atom is $2.1 \times 10^{-26}\ J\ T^{-1}$; the total density of the liquid is 130 kg m^{-3}.)

7.6 Use Eq. (7.28) to calculate the magnetic vector potential at the position of the particle when there is a flux Φ through circle C in Fig. 7.5. Show that the vector potential satisfies div $\mathbf{A} = 0$.

Show that Eq. (7.35) gives the vector potential of a uniform field that satisfies div $\mathbf{A} = 0$.

7.7 Show that $\mathbf{A} = A_0\hat{\mathbf{y}}\exp{(-x/\lambda)}$ is a solution of Eq. (7.33) for the penetration of the vector potential into the interior of a region containing electrons with rigid wavefunctions if the region occupies the space $x > 0$. Write down the magnetic field $\mathbf{B}$ and the induced current density $\mathbf{j}$ for this solution.

7.8 In benzene the carbon atoms form a regular hexagon of side 1.4 Å. One outer electron from each atom has a wavefunction that extends round the whole ring of atoms (the other three outer electrons from each atom are in sp^2 atomic orbitals). Estimate roughly the contribution of these electrons to the diamagnetic susceptibility of liquid benzene (density = 880 kg m^{-3}, molecular weight = 78 (C_6H_6)). The experimental value of χ for benzene is -7.7×10^{-6}.

It is well to observe the force and virtue and consequence of discoveries, and these are to be seen nowhere more conspicuously than in those three which were unknown to the ancients, and of which the origin, though recent, is obscure and inglorious; namely, printing, gunpowder and the magnet (i.e. Mariner's Needle). For these three have changed the whole face of things throughout the world.—*Francis Bacon (1561–1626)*

Chapter

Magnetic order

8.1 INTRODUCTION

At low temperatures it is observed that many paramagnetic materials possess a finite magnetization in the absence of an applied magnetic field. This **spontaneous magnetization** is due to alignment of the permanent dipole moments and indicates that each dipole is aware of the direction in which other dipoles are pointing. This awareness results from the interactions between the moments that we ignored in our calculation of paramagnetism in the previous chapter. The transition to a state in which the dipoles are aligned represents an increase in the degree of order within the solid and thus a decrease in entropy. The simplest type of magnetic order is **ferromagnetic order** (section 8.3) in which all the moments contribute equally to the spontaneous magnetization. The ordering in **antiferromagnets** (section 8.4) is such that there is no spontaneous magnetization because half the dipoles are aligned in one direction and the other half in the opposite direction. In **ferrimagnets** (section 8.6.1) there are oppositely directed moments which do not cancel and thus there is a net spontaneous magnetization.

The magnetic interaction between the dipoles is *too small* to be responsible for magnetic ordering. To demonstrate this we estimate the magnetic interaction between two moments of magnitude μ_B a distance $r = 3$ Å apart; the field B at one moment due to the other is of order $\mu_0 \mu_B / 4\pi r^3$ so that the interaction energy can be estimated as

$$\Delta E \sim \mu_B B \sim \frac{\mu_0 \mu_B^2}{4\pi r^3} \sim \frac{10^{-7} \times 10^{-46}}{3 \times 10^{-29}} \sim 3 \times 10^{-25} \text{ J} \sim 2 \times 10^{-6} \text{ eV}.$$

This energy is equal to $k_B T$ at a temperature of order 0.03 K. Random thermal disorder would be sufficient to destroy alignment of magnetic moments by this mechanism above this temperature. Many ferromagnets retain a spontaneous magnetization at temperatures of order 1000 K, indicating a much stronger interaction. The only possibility is that the interaction results from the electro-static interactions of the electrons with each other and with the nuclei in the solid; **exchange** provides a mechanism whereby the electrostatic interaction energy of two electrons can depend on the relative orientation of their magnetic moments.

8.2   THE EXCHANGE INTERACTION

We give here a qualitative explanation of the exchange interaction; a fuller discussion is given in appendix D. The interaction is a consequence of the fact that the wavefunction of two electrons must be antisymmetric under the exchange of all electron coordinates, space and spin:

$$\psi(\mathbf{r}_1, \mathbf{s}_1 : \mathbf{r}_2, \mathbf{s}_2) = -\psi(\mathbf{r}_2, \mathbf{s}_2 : \mathbf{r}_1, \mathbf{s}_1).$$

It follows that the wavefunction vanishes when the coordinates of both electrons are identical: $\mathbf{r}_1 = \mathbf{r}_2, \mathbf{s}_1 = \mathbf{s}_2$. There is thus zero probability of finding two electrons of the same spin at the same point in space. The antisymmetry of the wavefunction therefore tends to keep electrons of parallel spin apart so that the expectation value of the Coulomb repulsion energy $e^2/4\pi\varepsilon_0|\mathbf{r}_1 - \mathbf{r}_2|$ is smaller for parallel spins than for antiparallel spins. This is the exchange interaction and it can be represented in the form $-2\mathscr{J}\mathbf{s}_1 . \mathbf{s}_2$, corresponding to the Coulomb energy of the parallel spin state being $2\mathscr{J}$ less than that of the antiparallel spin state (see appendix D and problem 8.1).

The argument we have given suggests that $\mathscr{J}$ is positive and that ferromag-netic (parallel) alignment of spins is preferred; exchange interactions between electrons on the same atom explain Hund's first rule (section 7.2.1). The Coulomb interaction between two electrons on different atoms also depends on their relative spin orientation because of the antisymmetry of the wavefunction but the exchange energy $\mathscr{J}$ falls off rapidly with increasing distance between the atoms; the region $\mathbf{r}_1 = \mathbf{r}_2$ is no longer so important so that our argument that $\mathscr{J}$ is positive no longer applies. A negative value for $\mathscr{J}$ for nearest neighbours favours antiparallel spins and thus antiferromagnetic ordering.

The type of exchange that we have described above is known as **direct exchange**. Direct exchange cannot explain magnetic ordering in the rare-earth metals because there is little overlap of the 4f wavefunctions on neighbouring atoms. Other types of exchange exist and the important mechanism in the rare earths is believed to be the **indirect exchange** process illustrated in Fig. 8.1. Indirect exchange also leads to a coupling betwen spins of the form $-2\mathscr{J}\mathbf{s}_1 . \mathbf{s}_2$ where $\mathscr{J}$ alternates in sign as well as decreasing in magnitude as the distance between the atoms increases.

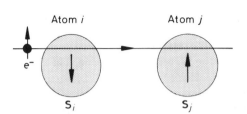

Fig. 8.1   The indirect exchange interaction—the direction of polarization of a conduction electron spin is affected by its direct exchange interaction with the magnetic moment of atom i; atom j then senses the polarization of the conduction electron and thereby interacts indirectly with atom i

In practice there are many atoms in the solid and usually more than one magnetic electron on each atom. It is a difficult task, involving some dubious assumptions, to proceed from an exchange interaction of the above form to the **Heisenberg Hamiltonian**

$$H = -\sum_i \sum_{j \neq i} \mathscr{J}_{ij} \mathbf{S}_i \cdot \mathbf{S}_j \tag{8.1}$$

for the exchange energy of the whole solid. Here $-2\mathscr{J}_{ij}\mathbf{S}_i \cdot \mathbf{S}_j$ is the contribution from atoms i and j,† and $\hbar\mathbf{S}_i$ and $\hbar\mathbf{S}_j$ are the *total* angular momenta of the electrons on atoms i and j. Despite this, convention dictates that they are referred to as spins and denoted by $\mathbf{S}$, in contradiction with the notation $\mathbf{J}$ normally used in discussing the paramagnetism of ions (section 7.2). It is possible to create even more confusion by taking $\mathbf{S}$ to represent an angular momentum *parallel* to the magnetic moment. We will not do this because this approach can easily lead to incorrect signs being obtained in the equations describing the dynamics of spins (sections 8.5); instead we will continue to regard electrons as negatively charged particles with magnetic moment *anti-parallel* to their angular momentum. The Heisenberg Hamiltonian is the starting point for many of the calculations of the properties of magnetically ordered materials.

8.3   FERROMAGNETISM

8.3.1   The Weiss molecular field

Before the advent of quantum mechanics, Weiss had suggested that the spontaneous magnetization of iron was due to the alignment of the atomic magnetic moments and had postulated the existence of a **molecular field** proportional to the magnetization to explain this alignment. He proposed that the effective magnetic field acting on any moment was

$$\mathbf{B}_{\mathrm{eff}} = \mathbf{B}_{\mathrm{loc}} + \lambda\mu_0\mathbf{M} \tag{8.2}$$

---

† The factor 2 that appears in $-2\mathscr{J}\mathbf{S}_1 \cdot \mathbf{S}_2$ does not appear in Eq. (8.1). This avoids counting each interaction twice; we regard the exchange energy as being equally shared between the two atoms.

where $\mathbf{B}_{loc}$ is the real magnetic field at the atom and $\lambda \mu_0 \mathbf{M}$ is the Weiss molecular field. Inserting this value of $\mathbf{B}_{eff}$ into Eq. (7.8) leads to the additional term $-\lambda \mu_0 \boldsymbol{\mu} . \mathbf{M}$ in the energy of the dipole moment.

By approximating the Heisenberg Hamiltonian (Eq. (8.1)), we show below how exchange interactions can give rise to the Weiss molecular field. The nature of the approximation is to assume that the effect on a spin $\mathbf{S}_i$ of its exchange interaction with another spin $\mathbf{S}_j$ can be calculated by replacing $\mathbf{S}_j$ by its average value $\langle \mathbf{S} \rangle$. Note that in a ferromagnetic material $\langle \mathbf{S} \rangle$ is the same for all spins and is related to the magnetization by (see Eq. (7.11))†

$$\mathbf{M} = -Ng\mu_B \langle \mathbf{S} \rangle, \tag{8.3}$$

where N is the number of spins per unit volume. Making this approximation to Eq. (8.1) leads to the following value for the dimensionless constant λ in Eq. (8.2) that determines the molecular field acting on the ith moment:

$$\lambda = \frac{2\sum_{j \neq i} \mathscr{J}_{ij}}{N\mu_0 g^2 \mu_B^2}. \tag{8.4}$$

Therefore λ is proportional to the sum of the exchange energies of a spin with all the other spins in the solid. We shall see that $\lambda \gg 1$, reflecting the fact that the electrostatic interactions giving rise to $\mathscr{J}_{ij}$ are much stronger than magnetic interactions between atoms.

The systematic way of replacing the spin operators in the Heisenberg Hamiltonian by their average values in order to obtain the Weiss molecular field approximation is to insert

$$\mathbf{S}_i = \langle \mathbf{S} \rangle + (\mathbf{S}_i - \langle \mathbf{S} \rangle)$$
$$\mathbf{S}_j = \langle \mathbf{S} \rangle + (\mathbf{S}_j - \langle \mathbf{S} \rangle) \tag{8.5}$$

in Eq. (8.1) and to ignore the term $(\mathbf{S}_i - \langle \mathbf{S} \rangle)(\mathbf{S}_j - \langle \mathbf{S} \rangle)$, quadratic in the difference between the operators and their average value; note that Eqs. (8.5) are identities so no approximation is involved in writing them. Proceeding in this way simplifies the Hamiltonian considerably because it no longer contains the products of operators, which make more exact calculations very difficult. Eq. (8.1) becomes (see problem 8.2 for the details)

$$H \approx \tfrac{1}{2}\lambda\mu_0 \mathbf{M}^2 - \sum_i \lambda\mu_0 \boldsymbol{\mu}_i . \mathbf{M} \tag{8.6}$$

where $\boldsymbol{\mu}_i = -g\mu_B \mathbf{S}_i$ is the operator corresponding to the magnetic moment of the atom and λ is given by Eq. (8.4). The second term in H represents the interaction of $\boldsymbol{\mu}_i$ with the Weiss molecular field.

---

† Since $\mathbf{S}$ contains in general a contribution from both orbital and spin angular momentum (it is really $\mathbf{J}$) we retain the Landé g-factor.

The replacement of spins by their average value means that fluctuations about the average value are ignored; this type of approach is used for phase transitions other than that to a ferromagnetic state and is referred to generally as a **mean field theory**. We shall discuss some of the failures of the mean field theory of ferromagnetism later, but for the moment we note that our model is approximate and proceed to analyse it. Fortunately the model is simple enough for a complete solution to be obtained.

8.3.2 Calculation of ferromagnetic properties using the mean field theory

Our procedure is identical to that used for paramagnetism in section 7.2 except that we replace the real magnetic field $\mathbf{B}$ by the effective field $\mathbf{B}_{eff}$ of Eq. (8.2). Thus the magnetization is given by Eq. (7.15). For simplicity we consider the case where the magnetic moment on each atom is due to a single electron spin so that we put $L = 0$, $J = S = \frac{1}{2}$, $g = 2$ in Eq. (7.15), which becomes (see problem 7.4)

$$M = N\mu_B \tanh\left(\frac{\mu_B B_{eff}}{k_B T}\right). \tag{8.7}$$

Consider first the high-temperature limit, $\mu_B B_{eff}/k_B T \ll 1$, in which the approximation $\tanh x \approx x$ can be used to write Eq. (8.7) as

$$M = \frac{N\mu_B^2}{k_B T} B_{eff} = \frac{N\mu_B^2}{k_B T}(B_{loc} + \lambda\mu_0 M), \tag{8.8}$$

where we have used Eq. (8.2) for B_{eff}. In this limit M is proportional to B_{loc} so there is no spontaneous magnetization. To identify the resulting paramagnetic susceptibility we must relate B_{loc} to the macroscopic field $\mu_0 H$ in the material; since these fields differ by an amount of order $\mu_0 M$,† the difference can be incorporated into the molecular field where it is dwarfed by the exchange interaction and we will consequently ignore it. Thus substituting $\mu_0 H$ for B_{loc} into Eq. (8.8) and solving for M we can identify the susceptibility using Eq. (7.1) as

$$\chi = \frac{M}{H} = \frac{C}{T - T_C} \tag{8.9}$$

---

† In ferromagnetic materials it is not in general possible to ignore the differences between the various fields and to take $\mathbf{B}_{loc} = \mu_0\mathbf{H}$ = applied field. The relationship between these fields is discussed in appendix B. We show there that in the long rod samples often used in experiments it is $\mu_0\mathbf{H}$ rather than the macroscopic $\mathbf{B}$ field ($= \mu_0(\mathbf{H} + \mathbf{M})$) that equals the applied field (Eq. (B5)). The relationship of $\mathbf{B}_{loc}$ to $\mu_0\mathbf{H}$ depends on the arrangement of atoms within the material; if the arrangement is random or has cubic symmetry then $\mathbf{B}_{loc} = \mu_0(\mathbf{H} + \mathbf{M}/3)$ (Eq. (B18)). Substituting $\mu_0\mathbf{H}$ for $\mathbf{B}_{loc}$ in Eq. (8.8) in this case therefore corresponds to adding $\frac{1}{3}$ to λ; since $\lambda \gg 1$ this change is unimportant.

where

$$C = N\mu_0 \mu_B^2/k_B \qquad \text{and} \qquad T_C = \lambda C. \qquad (8.10)$$

Eq. (8.9) is a modified Curie law (Eq. (7.18)), known as the **Curie–Weiss law**, which describes fairly well the susceptibility of ferromagnetic metals at high temperatures. At $T = T_C$ the susceptibility diverges, and below this temperature our assumption that $\mu_B B_{loc}/k_B T \ll 1$ is no longer true. We show below that a spontaneous magnetization is predicted by the Weiss model for $T < T_C$ so that T_C represents the upper temperature at which the material displays ferromagnetic properties; this temperature is referred to as the **Curie temperature**. To estimate the Curie constant C we substitute $N = 9 \times 10^{28}$ m^{-3} (the value for Fe) and obtain $C \approx 1$ K. Since T_C is typically 1000 K for ferromagnets we deduce that the Weiss molecular field constant $\lambda \sim 1000$. From T_C it is also possible to estimate the exchange energy $\mathcal{J}$. If we assume that only nearest exchange interactions are important, we can replace $\sum \mathcal{J}_{ij}$ in Eq. (8.4) by $z\mathcal{J}$, where $\mathcal{J}$ is the nearest neighbour exchange interaction and z the number of nearest neighbours. Then by using Eqs. (8.10), we see that $\mathcal{J} \approx 2k_B T_C/z$, confirming our supposition at the beginning of this chapter that ferromagnetism disappears when the thermal energy $k_B T$ is of the order of the interaction energy between spins. For $T_C \approx 1000$ K and $z = 8$, $\mathcal{J} \approx 0.03$ eV.

At temperatures below T_C, Eq. (8.7) (with B_{eff} given by Eq. (8.2)) cannot be solved analytically for the magnetization; a graphical solution is however possible. We wish to show that there is a non-zero magnetization in the absence of an applied field. Putting $B_{loc} = 0\dagger$ and introducing convenient dimensionless measures of the magnetization, $y = M/N\mu_B$, and the effective field, $x = \mu_B B_{eff}/k_B T$, enables us to write Eqs. (8.2) and (8.7), respectively, in the forms

$$x = \frac{T_C}{T} y \qquad (8.11)$$

$$y = \tanh x \qquad (8.12)$$

where we have used Eqs. (8.10). The simultaneous solution of these equations is obtained by plotting both on the same graph, and this is done in Fig. 8.2 for three different values of T/T_C. For $T/T_C > 1$ the only intersection is at the origin, which corresponds to zero magnetization, $y = 0$; this is the temperature region in which paramagnetic behaviour is predicted as discussed above. For $T/T_C = 1$ the only intersection is still at the origin but now both curves have the same slope there. For $T < T_C$ there are two intersections, one at the origin and the other at the point labelled A.

---

† When there is a spontaneous magnetization B_{loc} does not vanish in zero applied field, but is of order $\mu_0 M$. The term containing B_{loc} in Eq. (8.2) thus contributes a small correction to the Weiss molecular field just as did the difference between B_{loc} and $\mu_0 H$ in the calculation of the high-temperature susceptibility; here, as there, we ignore this contribution.

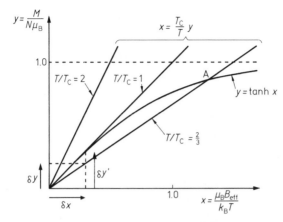

Fig. 8.2   Graphical solution of Eqs. (8.11) and (8.12). The figure also demonstrates that the solution at the origin is unstable for $T < T_C$

The solution at the origin corresponds to unstable equilibrium as the following argument demonstrates. Suppose that, by a fluctuation, a small magnetization appears, shown as δy on Fig. 8.2; this will cause an effective field given by Eq. (8.11) and this is indicated as δx on the figure. The magnetization that would be produced by this effective field is given by Eq. (8.12) and is indicated by $\delta y'$; we see that the original fluctuation produces an effective field that generates an even bigger magnetization. The fluctuation therefore grows and the solution $y = 0$ is unstable for $T < T_C$. A similar argument can be devised to show that point A represents a stable solution.

The spontaneous magnetization represented by point A increases from zero at T_C to the saturation value $N\mu_B$ at $T = 0$. Fig. 8.3 compares the temperature dependence of the spontaneous magnetization for nickel with that predicted by

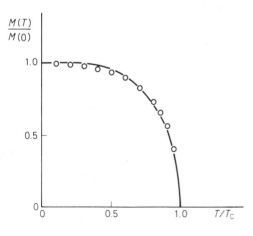

Fig. 8.3   Spontaneous magnetization of a ferromagnet, relative to its value at $T = 0$, as a function of T/T_C. The full curve is the theoretical mean field result for $J = S = \frac{1}{2}$, obtained by solving Eqs. (8.11) and (8.12). The circles show experimental values for Ni taken from the *American Institute of Physics Handbook*, 3rd edn, McGraw-Hill, New York (1972)

TABLE 8.1   Properties of ferromagnetic materials

Material	T_C (K)	Spontaneous magnetization (μ_B per atom) at $T = 0$	Theory		Critical exponents	
			gJ	$2S$	γ	β
Fe	1043	2.22	Fe^{3+} 5 Fe^{2+} 6	5 4	1.33 ± 0.02	0.34 ± 0.04
Co	1388	1.72	6	3	1.21 ± 0.04	–
Ni	627	0.61	5	2	1.35 ± 0.02	0.42 ± 0.07
Gd	292	7.63	7	–	1.30 ± 0.10	–
Dy	88	10.2	10	–	–	–
EuO	69	6.8	7	–	1.30 ± 0.01	0.36 ± 0.01

(Data taken from Kittel[7])

the theory; although there is qualitative agreement there are small but important discrepancies, which we discuss further below. In table 8.1 saturation values of the spontaneous magnetization for various ferromagnetic metals are compared with the values predicted by assuming that the moment on each ion is that predicted by Hund's rules. Good agreement is obtained for the rare-earth ferromagnets Gd and Dy but agreement is poor for the transition metals Fe, Ni and Co. In contrast to the situation for the paramagnetic salts of the transition metals, the agreement for the metals themselves is not improved by assuming that the orbital angular momentum is quenched.

As we have already indicated in section 7.2.4, the 3d electrons in these metals can be described as occupying a band of mobile electron states rather than being localized on the atoms. The mean field theory can be extended to cover this situation. The spontaneous magnetization then arises because the molecular field causes a relative displacement of the bands of up and down spins like that shown in Fig. 8.4. This figure also explains how a saturation magnetization for iron of 2.2 Bohr magnetons is produced in such a model. Confirmation that the spontaneous magnetization of iron is associated with electron spin only comes from the **Einstein–de Haas experiment**, in which a measurement is made of the angular momentum impulsively generated when an iron rod is suddenly magnetized; the ratio of angular momentum to magnetic moment corresponds to a Landé factor $g = 2$.

In the limits T just less than T_C and T close to zero, approximate analytic solutions of Eqs. (8.11) and (8.12) can be obtained. Near T_C, x is small and we can use $\tanh x \approx x - x^3/3$; solving for y then gives

$$M = N\mu_B y \approx \sqrt{3}\, N\mu_B \frac{T}{T_C}\left(1 - \frac{T}{T_C}\right)^{1/2}, \qquad (8.13)$$

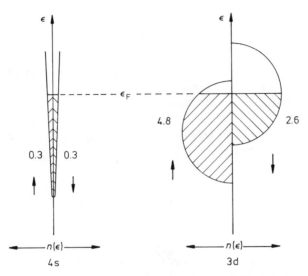

Fig. 8.4   Simplified band picture of ferromagnetism in iron. The vertical arrows indicate spin directions and the numbers indicate the division of the eight 3d and 4s electrons per atom between the 3d and 4s energy bands. A possible small polarization of the 4s band has been ignored in drawing the figure; the 4s electrons probably contribute to the exchange interaction between the 3d electrons through the indirect mechanism of Fig. 8.1

indicating that M goes to zero as $(1 - T/T_C)^{1/2}$ as the Curie temperature is approached from below. Near $T = 0$, x is large and $\tanh x \approx 1 - 2 \exp(-2x)$ so that, from Eqs. (8.11) and (8.12)

$$y \approx \left[1 - 2 \exp\left(-\frac{2T_C}{T} y \right) \right]$$

or

$$M \approx N\mu_B \left[1 - 2 \exp\left(-\frac{2T_C}{T} \right) \right] \tag{8.14}$$

where we have put $y = 1$ in the second term in brackets, which gives the small correction to perfect alignment of the moments.

Although mean field theory correctly describes the qualitative behaviour of ferromagnets, the temperature dependences predicted by Eqs. (8.9), (8.13) and (8.14) are not observed experimentally. The reason for the discrepancy near $T = 0$ is discussed in section 8.5. The failure of mean field theory near T_C is common to all second-order phase transitions.† It occurs because of the

---

† A second-order phase transition is one in which there is no entropy discontinuity and thus no latent heat (see problem 8.2). Other examples of such transitions are the superconducting transition in zero magnetic field and the liquid–gas transition at the critical temperature (but not at lower temperatures).

presence near T_C of large thermodynamic fluctuations (known as **critical fluctuations**) of the properties of the system around their average values; in magnetic systems the large fluctuations of an atomic dipole moment mean that it is no longer a good approximation to replace the moment by its average value.

The critical fluctuations dominate the behaviour of the susceptibility just above T_C and the spontaneous magnetization just below T_C. The susceptibility for iron is shown as a function of $T - T_C$ in Fig. 8.5; as the scales on both axes are logarithmic the **critical behaviour** of the susceptibility is described by the temperature dependence

$$\chi \propto (T - T_C)^{-\gamma}$$

with the **critical exponent** $\gamma = 1.33$. The spontaneous magnetization below T_C is found to vary as

$$M \propto (T - T_C)^\beta$$

where β is another critical exponent. Measured values of γ and β are given in table 8.1 for various ferromagnets. Similar values of γ (≈ 1.35) and β (≈ 0.35) are found for all materials; the mean field predictions are $\gamma = 1$ (Eq. (8.9)) and $\beta = 0.5$ (Eq. (8.13)).

Investigation of the critical behaviour has been the subject of intense research interest and one important conclusion is that the critical exponents for different second-order phase transitions appear to depend only on the nature of the

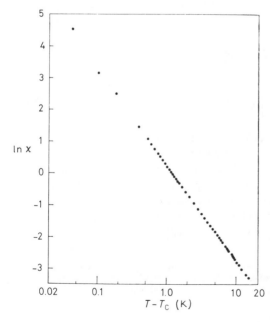

Fig. 8.5   Susceptibility of iron (containing 0.16 at % tungsten impurity) plotted as a function of $T - T_C$ using logarithmic scales on both axes. The straight line followed by the data indicates a relationship of the form $\chi = (T - T_C)^{-1.33}$. (Reproduced with permission from J. E. Noakes, N. E. Tornberg and A. Arrott, *J. Appl. Phys.* **37**, 1264 (1966))

symmetry change involved. Thus the critical exponents of a paramagnetic–ferro-magnetic transition described by the Heisenberg Hamiltonian are determined by the fact that the isotropic paramagnetic phase changes to one in which there is a special direction, that of the spontaneous magnetization; the ferromagnetic state, unlike the paramagnetic state, is not rotationally invariant. Note however that all the physically different ferromagnetic states produced by rotation of the spontaneous magnetization must have the same energy because the Heisenberg Hamiltonian itself is rotationally invariant.

To complete our discussion of the Weiss mean field theory of ferromagnetism we note that the theory appears to predict a hysteretic magnetization versus applied field curve, qualitatively similar to that observed in ferromagnetic particles consisting of a single domain (see section 8.7 for an explanation of domains). After the particle has been magnetized in one direction by a field in that direction, the magnetization reverses sign only when a field in the reverse direction exceeds a certain critical value. The hysteresis in our model is an artefact of our restriction of the magnetization to the z direction. In zero applied field there is nothing in the model to define a special direction for the magnetization, which thus rotates freely and reverses direction as the field decreases through zero. In a real crystal there will be a preferred direction for the magnetization, associated, for example, with the electrostatic fields of the neighbouring atoms; the hysteresis curve of single-domain ferromagnets results from this crystalline anisotropy.

8.4   THE NÉEL MODEL OF ANTIFERROMAGNETISM

If the exchange energy $\mathscr{J}_{ij}$ between nearest neighbours in Eq. (8.1) is negative then antiparallel alignment of their spins is preferred. In this case there is a tendency for ordering to occur into a state in which up and down spins alternate within the structure and there is no macroscopic magnetization in zero applied field. The lattice of magnetic atoms is divided into two identical sublattices, A and B, such that in the ordered state the mean magnetic moment on the sites of the A sublattice is antiparallel to that on the B sublattice. If the initial lattice is simple cubic then there is an obvious subdivision in which the A and B sites are the Na and the Cl sites of the NaCl structure.† There is also an obvious division of a bcc lattice in which the body-centred sites form one sublattice and the corner sites the other.‡ In both these cases the nearest neighbours of any B site

---

† We must be careful to distinguish the **chemical unit cell** (the original simple cubic cell) from the **magnetic unit cell** (fcc unit cell of the NaCl structure); since ↑ and ↓ spins have different scattering cross sections for a neutron of spin ↑, it is possible to deduce the magnetic unit cell from neutron diffraction experiments and thus to determine antiferromagnetic structures (see section 12.5.1).

‡ Note however that, in bcc solid ³He, the antiferromagnetic arrangement of the *nuclear* moments that occurs for $T < 1$ mK does not correspond to this 'obvious' subdivision.

are all A sites and vice versa. An fcc lattice cannot be subdivided into two sublattices with this property (see problem 8.4).

Néel generalized the Weiss molecular field approach to antiferromagnetism by supposing that the atoms on one sublattice experience a molecular field proportional to the magnetization of the other sublattice and opposite in direction to it. Thus instead of Eq. (8.2) we write the effective fields for the two sublattices as

$$\mathbf{B}_{\text{eff}}^{A} = \mu_0(\mathbf{H} - \lambda\mathbf{M}_B) \qquad \mathbf{B}_{\text{eff}}^{B} = \mu_0(\mathbf{H} - \lambda\mathbf{M}_A) \tag{8.15}$$

where $\mathbf{M}_A$ and $\mathbf{M}_B$ are the contributions of each sublattice to the total magnetization, $\mathbf{M} = \mathbf{M}_A + \mathbf{M}_B$. Here as in the previous section we assume that the differences between the local field $\mathbf{B}_{\text{loc}}$ and the macroscopic field $\mu_0 H$ can be incorporated as insignificant corrections to the molecular fields. It is also possible to include coupling of the sublattice to itself within the model; this adds complication without changing the major qualitative results, so we shall not do so (see problem 8.4).

Eqs. (8.15) can be derived from the Heisenberg Hamiltonian by assuming that only nearest neighbour interactions are important and that all the nearest neighbours of any A moment are from the B sublattice. A procedure analogous to that used in the ferromagnetic case identifies the molecular field constant λ in Eqs. (8.15) as

$$\lambda = -\frac{4\sum_{j \neq i}\mathcal{J}_{ij}}{N\mu_0 g^2\mu_B^2} = -\frac{4z\mathcal{J}}{N\mu_0 g^2\mu_B^2} \tag{8.16}$$

where $\mathcal{J}$ is the nearest neighbour exchange energy and z the number of nearest neighbours. Note that, since $\mathcal{J}$ is negative, λ is positive; Eq. (8.16) differs by a factor of 2 from Eq. (8.4) because each sublattice has only $N/2$ atoms.

As in the ferromagnetic case we will specialize for simplicity to the case $S = J = \frac{1}{2}, g = 2$, so that the sublattice magnetizations, as given by Eq. (8.7), are

$$M_A = \frac{N}{2}\mu_B \tanh\left(\frac{\mu_B B_{\text{eff}}^A}{k_B T}\right) \qquad M_B = \frac{N}{2}\mu_B \tanh\left(\frac{\mu_B B_{\text{eff}}^B}{k_B T}\right). \tag{8.17}$$

At high temperatures we can use $\tanh x \approx x$ as before and substitute the effective fields from Eqs. (8.15) to obtain, after rearrangement,

$$M_A + \frac{\lambda C}{2T}M_B = \frac{C}{2T}H \qquad \frac{\lambda C}{2T}M_A + M_B = \frac{C}{2T}H \tag{8.18}$$

where $C = N\mu_0\mu_B^2/k_B$ is the Curie constant (cf. Eq. (8.10)). Solving Eqs. (8.18) gives a magnetization

$$M = M_A + M_B = \frac{C}{T + \lambda C/2}H$$

corresponding to a susceptibility

$$\chi = \frac{M}{H} = \frac{C}{T + T_N} \tag{8.19}$$

where $T_N = \lambda C/2$. Thus the susceptibility is reduced below the Curie law value of Eq. (7.18) (see Fig. 8.6) and remains finite at all temperatures.

The onset of magnetic ordering occurs at the temperature at which Eqs. (8.18) have a non-zero solution in the absence of an applied field. This requires the determinant of coefficients on the left-hand side of these equations to vanish and thus

$$1 - \lambda^2 C^2/4T^2 = 0$$

or

$$T = \lambda C/2 = T_N. \tag{8.20}$$

The onset temperature for antiferromagnetism, known as the **Néel temperature**, is thus equal to the temperature that appears in the modified Curie law for the susceptibility (Eq. (8.19)); this feature is not shared by more advanced models or by real antiferromagnetic materials. The formation of an antiferromagnetic state is confirmed by substituting Eq. (8.20) into Eqs. (8.19) to find that the non-zero solution for $H = 0$ has the property $M_A = -M_B$. For temperatures below T_N Eqs. (8.15) and (8.17) continue to have an antiferromagnetic solution with $M_A = -M_B$ in zero applied field; the spontaneous magnetization of the A sublattice satisfies

$$M_A = \frac{N}{2} \mu_B \tanh \left(\frac{T_N}{T} \frac{2M_A}{N\mu_B} \right) \tag{8.21}$$

which is essentially the same equation (obtained by combining Eqs. (8.11) and (8.12)) as that which gives the spontaneous magnetization of a ferromagnet.

Since the total magnetization $(M_A + M_B)$ vanishes in zero field it is still possible to define a susceptibility for an antiferromagnet in its ordered state. The measured susceptibility is shown as a function of temperature for MnF_2 in Fig. 8.6. Below T_N there are two values depending on whether the applied field is parallel or perpendicular to the sublattice spontaneous magnetizations.† We will not calculate these susceptibilities (see problem 8.5) but the observed behaviour is qualitatively reasonable. When the field is perpendicular to the sublattice magnetizations the spins are fairly easily tilted as in Fig. 8.7 to give a magnetization parallel to the field with the approximately constant susceptibility $\chi_\perp$ shown in Fig. 8.6. Magnetization by a field parallel to the sublattice magnetizations is opposed by the full molecular field and does not occur at all to

---

† We must imagine that the direction of the sublattice magnetizations is determined by electrostatic fields within the crystal so that it does not change when a weak magnetic field is applied.

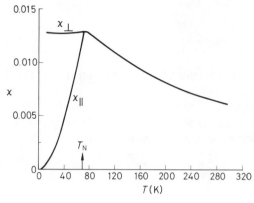

Fig. 8.6   Magnetic susceptibility of antiferromagnetic MnF_2. The two susceptibilities, $\chi_\parallel$ and $\chi_\perp$, measured below the Néel temperature of 67 K correspond, respectively, to fields applied parallel and perpendicular to the sublattice spontaneous magnetizations

Fig. 8.7   Magnetization of an antiferromagnet in a direction perpendicular to the sublattice spontaneous magnetizations can be achieved by slight tilting of the atomic moments as shown

first order in the applied field at $T = 0$ where both sublattices are perfectly aligned. It is only near T_N that the molecular field weakens and the two susceptibilities become comparable.

8.5   SPIN WAVES

8.5.1   Ferromagnets at low temperatures

The mean field theory of ferromagnetism fails at low temperatures because it does not predict correctly the low lying excited states. To see this consider the one-dimensional chain of ferromagnetically aligned spins shown in Fig. 8.8(a). If the spin at one end of the chain is rotated through 360°, then the other spins relax into the arrangement shown in Fig. 8.8(b) to minimize the exchange energy. This is a low lying excited state because neighbouring spins are very nearly parallel in the resulting structure so that very little exchange energy is lost. However the mean field theory attributes a very high energy to this state since the *average* magnetization and hence the molecular field vanish. The crucial point is that the energy of a spin should depend on the orientation of the other spins in its neighbourhood and not on the average magnetization of the sample.

(a) Chain of perfectly aligned spins

(b) Low lying excited state of the chain to which mean field theory incorrectly attributes a high energy. In real crystals crystalline anisotropy provides an 'easy' direction for the magnetization and the lowest lying excited states correspond to small departures of the spins from this preferred direction as in our spin wave calculation of section 8.5.2. Because of the crystalline anisotropy, the spin wave frequency tends to a finite value in the infinite-wavelength limit (see problem 8.6)

Fig. 8.8

8.5.2   Spin waves in a one-dimensional crystal

We will calculate the low lying excited states of the chain of aligned spins of Fig. 8.8(a) by assuming that the spins behave as classical angular momenta.† The method is similar to that already used to calculate the lattice vibrations of the chain in Chapter 2. For simplicity we consider only nearest neighbour exchange interactions so that, from Eq. (8.1), the exchange energy of the nth spin in the chain is

$$E_n = -2\mathscr{J}\, \mathbf{S}_n \cdot (\mathbf{S}_{n-1} + \mathbf{S}_{n+1}), \tag{8.22}$$

where $2\mathscr{J}$ is the nearest neighbour exchange interaction. Since the magnetic moment $\boldsymbol{\mu}_n$ of spin n is $-g\mu_B\mathbf{S}_n$, we can write E_n in the form $-\boldsymbol{\mu}_n \cdot \mathbf{B}_n$, where

$$\mathbf{B}_n = -\frac{2\mathscr{J}}{g\mu_B}(\mathbf{S}_{n-1} + \mathbf{S}_{n+1}) \tag{8.23}$$

is the effective field acting on the nth atom due to the exchange interaction. The equation of motion of the nth spin is obtained by equating its rate of change of angular momentum $\hbar\, d\mathbf{S}_n/dt$ to the torque $\boldsymbol{\mu}_n \times \mathbf{B}_n$ acting on it due to this field. Thus

$$\hbar\,\frac{d\mathbf{S}_n}{dt} = \boldsymbol{\mu}_n \times \mathbf{B}_n = 2\mathscr{J}\, \mathbf{S}_n \times (\mathbf{S}_{n-1} + \mathbf{S}_{n+1}). \tag{8.24}$$

---

† A classical approach is implicit in drawing pictures like those of Fig. 8.8, since exact specification of the orientation of a spin conflicts with the uncertainty principle.

Eqs. (8.24) are non-linear in the spins and therefore difficult to solve without approximation. Since we are interested in the low lying excited states, we may linearize the equations by writing

$$\mathbf{S}_n = -S\hat{\mathbf{z}} + \boldsymbol{\sigma}_n \qquad (8.25)$$

where $-S\hat{\mathbf{z}}$ is the *constant* value of $\mathbf{S}_n$ (and all the other spins) in the perfectly aligned state and $\boldsymbol{\sigma}_n$ is a small vector in the xy plane that represents the deviation of spin n from its perfect alignment.† This situation is illustrated in Fig. 8.9. Note that the magnetization direction $\hat{\mathbf{z}}$ need have no special relation to the direction of the chain. Substituting Eq. (8.25) in Eq. (8.24) and retaining only terms to first order in $\boldsymbol{\sigma}_n$ we obtain

$$\hbar \frac{d\boldsymbol{\sigma}_n}{dt} = -2\mathscr{J}S\hat{\mathbf{z}} \times (\boldsymbol{\sigma}_{n-1} + \boldsymbol{\sigma}_{n+1}) - 4\mathscr{J}S\boldsymbol{\sigma}_n \times \hat{\mathbf{z}}$$

$$= -2\mathscr{J}S\hat{\mathbf{z}} \times (\boldsymbol{\sigma}_{n-1} - 2\boldsymbol{\sigma}_n + \boldsymbol{\sigma}_{n+1}). \qquad (8.26)$$

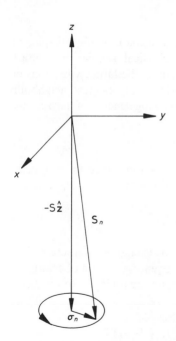

Fig. 8.9   Precession of a single spin in a classical spin wave

---

† The condition that $\boldsymbol{\sigma}_n$ is perpendicular to $\hat{\mathbf{z}}$ follows from the requirement that $|\mathbf{S}_n|^2$ remains constant and equal to S^2; to first order in $\boldsymbol{\sigma}_n$ this requires $S\hat{\mathbf{z}} \cdot \boldsymbol{\sigma}_n = 0$. We take the spins aligned along $-\hat{\mathbf{z}}$ because this corresponds to a magnetization along $+\hat{\mathbf{z}}$.

In component form Eq. (8.26) becomes

$$\hbar\left(\frac{d\sigma_n}{dt}\right)_x = 2\mathcal{J}S(\sigma_{n-1} - 2\sigma_n + \sigma_{n+1})_y$$

$$\hbar\left(\frac{d\sigma_n}{dt}\right)_y = -2\mathcal{J}S(\sigma_{n-1} - 2\sigma_n + \sigma_{n+1})_x.$$

(8.27)

If we multiply the first of these equations by i and add the second we obtain the single equation

$$i\hbar\frac{d\sigma_n^-}{dt} = -2\mathcal{J}S(\sigma_{n-1}^- - 2\sigma_n^- + \sigma_{n+1}^-)$$

(8.28)

for the complex variable

$$\sigma_n^- = \sigma_{nx} - i\sigma_{ny}.$$

(8.29)

Eqs. (8.28) are very similar in form to Eqs. (2.7) and (4.9), which describe, respectively, the lattice vibrations and the electron states of the one-dimensional chain; like these equations, Eqs. (8.28) have wavelike solutions. If we substitute

$$\sigma_n^- = Ae^{i(kna - \omega t)}$$

(8.30)

where a is the lattice spacing, then, on cancelling a factor $A \exp[i(kna - \omega t)]$, we obtain

$$\hbar\omega = -2\mathcal{J}S(e^{-ika} - 2 + e^{ika}) = 4\mathcal{J}S[1 - \cos(ka)]$$

(8.31)

which is the dispersion relation for the **spin waves**.

The dispersion relation is plotted in Fig. 8.10. As with the lattice vibration waves and the electron states the relation is periodic in k with period $2\pi/a$; also

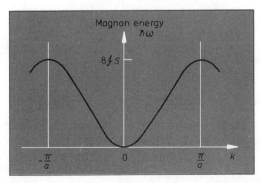

Fig. 8.10   Dispersion relation for spin waves in a one-dimensional ferromagnet. The first Brillouin zone, $-\pi/a < k < \pi/a$, contains all the physically distinct solutions

as in these cases the application of periodic boundary conditions determines that there are exactly N distinct modes of vibration. From Eqs. (8.29) and (8.30) we see that σ_{nx} and σ_{ny} are of the form

$$\sigma_{nx} \propto \cos (kna - \omega t) \qquad \sigma_{ny} \propto -\sin (kna - \omega t).$$

The $90°$ phase difference between σ_{nx} and σ_{ny} implies that (on a classical picture) each spin precesses about the z axis as shown in Fig. 8.9. Spin waves also propagate in a three-dimensional crystal with a dispersion relation similar to Eq. (8.31). As in the one-dimensional case the z direction of the magnetization need have no special relation to the propagation direction $\mathbf{k}$. Spin waves with $\mathbf{k}$ parallel and perpendicular to z are shown schematically in Fig. 8.11.

A quantum mechanical calculation of spin waves also leads to the dispersion relation Eq. (8.31), but, as we might expect from the analogy with lattice vibrations, the energy of a mode of wavenumber k is quantized and can only take the values

$$E = (n + \tfrac{1}{2})\hbar\omega(k) \tag{8.32}$$

appropriate to a simple harmonic oscillator. The quanta associated with the spin wave modes are called **magnons** and, like phonons, magnons are bosons. The quantum mechanical calculation also shows that each magnon within the crystal reduces the magnitude of the z component of angular momentum by $\hbar$ and thus decreases the magnetization by $g\mu_B$.

From Eq. (8.31) the magnon energy near $k = 0$ is

$$\varepsilon = \hbar\omega \approx 2\mathscr{J}Sa^2k^2. \tag{8.33}$$

The quadratic dependence of ε on k contrasts with the linear relation for phonons.[†] Comparison with the dispersion relation $\varepsilon = \hbar^2k^2/2m$ for free particles of mass m suggests that magnons of long wavelength behave like particles of effective mass

$$m^* = \frac{\hbar^2}{4\mathscr{J}Sa^2}. \tag{8.34}$$

If we take $S = \tfrac{1}{2}$, $a = 2.5$ Å and $\mathscr{J} = 0.03$ eV then we find that the effective mass of a magnon is approximately 20 electron masses.

8.5.3 Magnetization and heat capacity at low temperatures

The contribution of the spin waves to the heat capacity of ferromagnets at low temperatures can be calculated in a manner analogous to that used for calculating the heat capacity due to lattice vibrations in section 2.6. The number

---

[†] The dispersion relation for spin waves in antiferromagnetic materials is *linear* ($\omega \propto k$) at small k (see problem 8.8).

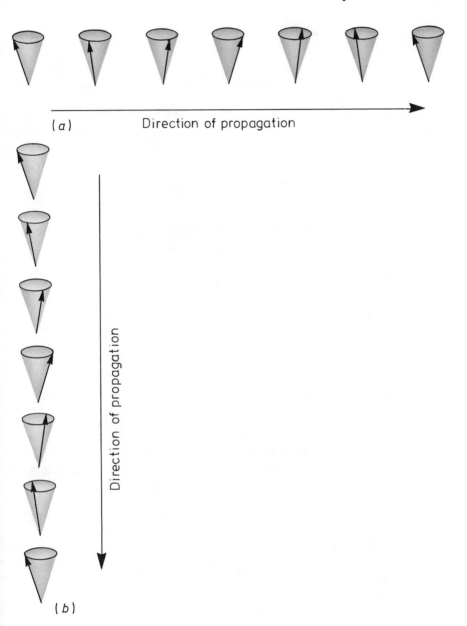

Fig. 8.11   Ferromagnetic spin waves propagating: (*a*) perpendicular to spontaneous magnetization; (*b*) parallel to spontaneous magnetization

of spin wave modes with wavenumber between k and $k + dk$ is given by Eq. (2.38) as

$$g(k)\, dk = \frac{Vk^2}{2\pi^2}\, dk.$$

At low temperatures only low-energy long-wavelength modes will be excited and for these we can use the limiting form of the dispersion relation, Eq. (8.33), to write the number of modes with frequency between ω and $\omega + d\omega$ as†

$$g(\omega)\, d\omega = g(k)\, dk = \frac{V}{4\pi^2}\left(\frac{\hbar}{2\mathscr{J}Sa^2}\right)^{3/2}\omega^{1/2}\, d\omega. \tag{8.35}$$

The energy associated with each mode is $(n + \frac{1}{2})\hbar\omega$, where the average number n of magnons at temperature T is given by the Bose–Einstein distribution function (Eq. (2.27))

$$n(\omega) = \frac{1}{e^{\hbar\omega/k_B T} - 1}.$$

Thus the contribution of the magnons to the energy is

$$E = E_0 + \int_0^\infty \hbar\omega n(\omega)g(\omega)\, d\omega$$

$$= E_0 + \frac{V}{4\pi^2}\left(\frac{\hbar}{2\mathscr{J}Sa^2}\right)^{3/2}\left(\frac{k_B T}{\hbar}\right)^{5/2}\int_0^\infty \frac{x^{3/2}}{e^x - 1}\, dx, \tag{8.36}$$

where E_0 is the zero point energy and the final line has been obtained by changing the variable to $x = \hbar\omega/k_B T$. The upper limit of the integral has been set to ∞ because $n(\omega)$ cuts off the integrand at small frequencies where Eq. (8.33) is still valid. The integral is just a number ($= 1.78$) so that the spin wave contribution to the heat capacity at low temperature is

$$C = \frac{dE}{dT} = \frac{5 \times 1.78 \times V}{2 \times 4\pi^2\hbar}\left(\frac{k_B}{2\mathscr{J}Sa^2}\right)^{3/2} k_B T^{3/2}. \tag{8.37}$$

The magnon heat capacity therefore varies as $T^{3/2}$ at low temperatures.

Since each magnon reduces the magnetic moment by $g\mu_B$, the magnetization at low temperatures is

$$M = M_s - g\mu_B N_{mag}$$

where $M_s = Ng\mu_B S$ is the saturation magnetization and

$$N_{mag} = \frac{1}{V}\int_0^\infty n(\omega)g(\omega)\, d\omega$$

---

† We assume that the dispersion relation is isotropic.

is the number of magnons per unit volume. Thus, proceeding as for the specific heat, we obtain

$$M = M_s \left[1 - \frac{1}{NS4\pi^2} \left(\frac{k_B T}{2 \mathscr{J} Sa^2} \right)^{3/2} \int_0^\infty \frac{x^{1/2}}{e^x - 1} \, dx \right].$$ (8.38)

The integral is again just a number (= 2.32); the decrease of the magnetization from its saturation value with a $T^{3/2}$ temperature dependence as predicted by Eq. (8.38) agrees much better with the experimentally observed behaviour than does the mean field theory prediction of Eq. (8.14).

★8.5.4   Ferromagnetic resonance and the experimental observation of spin waves

In the long-wavelength limit the lattice vibrations of a crystal are essentially sound waves, which can be excited by attaching suitable transducers. One can ask if there is anything similar for spin waves. To obtain sound waves from the equations of motion (2.7) for a one-dimensional crystal we note that $(u_{n+1} - 2u_n + u_{n-1})/a^2$ is the finite difference form for the second spatial derivative d^2u/dx^2, so that in the long-wavelength limit these equations can be written

$$M \frac{d^2 u}{dt^2} = Ka^2 \frac{d^2 u}{dx^2}$$ (8.39)

which is the wave equation with the velocity of sound correctly identified (Eq. (2.13)) as $a(K/M)^{1/2}$.

Let us suppose that there is a long-wavelength disturbance in the x direction of the spins in a ferromagnetic material. To describe the dynamics of such a disturbance it is convenient to rewrite Eq. (8.26) as an equation for the local magnetization $\mathbf{M}(x, t)$ defined, using Eq. (8.3), by

$$\mathbf{M}(x, t) = -Ng\mu_B \langle \mathbf{S}(x, t) \rangle$$ (8.40)

where $\langle \mathbf{S}(x, t) \rangle$ represents an average over the spins in the neighbourhood of point x. For small motions we can take, by analogy with Eq. (8.25),

$$\mathbf{M}(x, t) = \mathbf{M}_0 + \mathbf{M}_\perp(x, t),$$ (8.41)

where $\mathbf{M}_\perp$ is the small perpendicular deviation of $\mathbf{M}$ from its average value $\mathbf{M}_0$. Using Eqs. (8.40) and (8.41) enables us to write Eq. (8.26) in the form

$$\frac{\partial \mathbf{M}_\perp}{\partial t} = -\frac{2 \mathscr{J} Sa^2}{\hbar} \frac{\mathbf{M}_0}{|\mathbf{M}_0|} \times \frac{\partial^2 \mathbf{M}_\perp}{\partial x^2}$$ (8.42)

where the finite difference form of the derivative has been replaced by the derivative itself just as in Eq. (8.39). Proceeding as in section 8.5.2, it is easy to

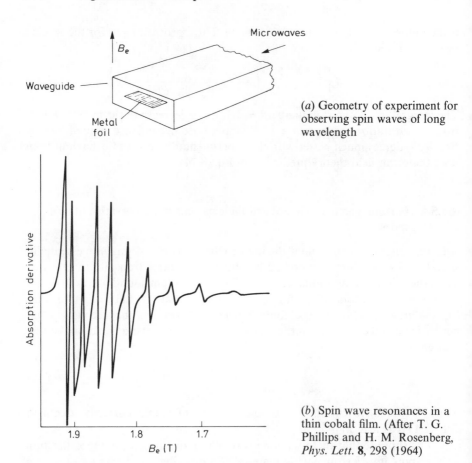

(a) Geometry of experiment for observing spin waves of long wavelength

(b) Spin wave resonances in a thin cobalt film. (After T. G. Phillips and H. M. Rosenberg, *Phys. Lett.* **8**, 298 (1964)

Fig. 8.12

show that Eq. (8.42) has wavelike solutions with $\mathbf{M}_\perp \propto \exp\left[i(kx - \omega t)\right]$ with ω and k related by the dispersion relation of Eq. (8.33) for spin waves of small k.

A typical experimental set-up for observing spin waves of long wavelength is shown in Fig. 8.12(a). A steady field $\mathbf{B}_e$ is applied perpendicular to a thin film of the ferromagnet; we take the field to be in the z direction and assume that the field orients the average magnetization $\mathbf{M}_0$ of the sample in this direction. Spin waves with $\mathbf{k}$ vector *perpendicular* to the film are then excited by applying a radiofrequency field in the plane of the film. To generalize Eq. (8.42) to allow for the existence of the external field we must add a term $\boldsymbol{\mu}_n \times \mathbf{B}_{loc}$ to the right-hand side of Eq. (8.24) where $\mathbf{B}_{loc}$ is the real local magnetic field at atom n. $\mathbf{B}_{loc}$ contains a contribution from the magnetization as well as from the external field, and we show in appendix B that in the thin film geometry under consideration

$$\mathbf{B}_{loc} = \mathbf{B}_e - \tfrac{2}{3}\mu_0\mathbf{M}_0 + \tfrac{1}{3}\mu_0\mathbf{M}_\perp. \tag{8.43}$$

The extra term in Eq. (8.24) leads to an extra term on the right-hand side of Eq. (8.42), which, for $\mathbf{M}_\perp$ varying in the z direction, becomes

$$\frac{\partial \mathbf{M}_\perp}{\partial t} = -\frac{2\mathscr{J}Sa^2}{\hbar}\frac{\mathbf{M}_0}{|\mathbf{M}_0|} \times \frac{\partial^2 \mathbf{M}_\perp}{\partial z^2} + \gamma\mathbf{M}_\perp \times (\mathbf{B}_e - \mu_0\mathbf{M}_0) \tag{8.44}$$

where $\gamma = -g\mu_B/\hbar = -ge/2m$ is the **gyromagnetic ratio** of the atom.

The spin wave dispersion relation (Eq. (8.33)) for waves of the form $\exp[\mathrm{i}(kz - \omega t)]$ is thus modified to

$$\omega = \frac{2\mathscr{J}Sa^2}{\hbar}k^2 - \gamma(B_e - \mu_0 M_0). \tag{8.45}$$

The presence of the additional term indicates a finite frequency for the precession of a spatially uniform magnetization ($k = 0$), and the absorption peak observed at the frequency $\omega = -\gamma(B_e - \mu_0 M_0)$ is referred to as **ferromagnetic resonance**. In general the frequency of ferromagnetic resonance depends on the sample shape.

The experimental results in Fig. 8.12(b) were obtained for a thin Co film in which the magnetization was pinned at the surface by strong local anisotropy so that $\mathbf{M}_\perp = 0$ there. The situation is analogous to a string stretched between two fixed points; only spin wave modes with an integral number of half-wavelengths in the thickness d of the film are excited. As the film is thin compared to the electromagnetic skin depth, the radiofrequency magnetic field is uniform across the width of the film and only the odd harmonics are excited ($k = (2n + 1)\pi/d$). Fig. 8.12(b) shows the absorption associated with harmonics corresponding to values of $2n + 1$ between 3 and 21. The value of $\mathscr{J}Sa^2$ and hence of the magnon effective mass m^* can be deduced from these measurements (problem 8.9).

★8.6   OTHER TYPES OF MAGNETIC ORDER

8.6.1   Ferrimagnetism

Ferrimagnetic ordering is intermediate between ferromagnetic and antiferromagnetic ordering; an important class of compounds that exhibit it are the ferrites, which have the general formula $MO.Fe_2O_3$, where M is a divalent cation such as Ni, Mn or Fe. Mankind's earliest experience of magnetism was almost certainly provided by magnetite (lodestone) $FeO.Fe_2O_3$, and we will use this material to illustrate the nature of ferrimagnetic ordering. Magnetite and the other ferrites have the **spinel** crystal structure with a cubic unit cell containing 32 oxygen ions in an approximately close-packed array; some of the interstices between the oxygen ions are filled by the (smaller) Fe^{2+} and Fe^{3+} ions. Eight of the Fe^{3+} ions in each cell are in **tetrahedral** (A) sites surrounded by four oxygen atoms; the remaining eight Fe^{3+} and the eight Fe^{2+} ions are in **octahedral** (B) sites surrounded by eight oxygen atoms.

The exchange interactions between neighbouring magnetic ions are all believed to be antiferromagnetic but the interactions between the neighbouring A and B sites are dominant, with the result that the spins on all the A sites are parallel and oppositely directed to all the spins on the B sites. Since Fe^{3+} ions inhabit the A and B sites equally there is no net magnetization from these. The Fe^{2+} ions are all on B sites however and give rise to a spontaneous magnetization. The low-temperature limiting value of the spontaneous magnetization confirms this picture as it corresponds to $32\mu_B$ per unit cell, just the value expected for eight Fe^{2+} ions in which the orbital angular momentum is quenched (hence $J = S = 2, g = 2$). The widespread use of ferrites in devices results from the combination of the high electrical resistivity of these materials with the high effective magnetic permeability associated with their spontaneous magnetization.

8.6.2   Spin density wave antiferromagnetism in chromium

Chromium crystallizes in the body-centred cubic structure. The earliest neutron scattering experiments below the Néel temperature of 311 K suggested a simple antiferromagnetic structure with the atoms at the corners of the unit cell having spins opposed to those at the body-centred positions. Since chromium is a transition metal we expect that a band picture of the 3d electrons will best describe the magnetic ordering. In such a picture antiferromagnetic ordering can be achieved by a periodic variation with position of the density $n_\uparrow$ of $\uparrow$ spin electrons, with the density $n_\downarrow$ of $\downarrow$ spin electrons varying in antiphase, so that the total electron density $(n_\uparrow + n_\downarrow)$ is constant but the spin density $(n_\uparrow - n_\downarrow)$ is oscillatory.

The result is a **spin density wave** (SDW). One possible way of creating the simple antiferromagnetic structure described above is to set up a SDW of wavevector $2\pi/a$ (wavelength a) in the [1 0 0] direction; Fig. 8.13(a) illustrates this situation with the arrows indicating the spin density at the lattice positions.†
This SDW is said to be **commensurate** since it has the same periodicity as the lattice; it is also described as **static** since the spin density does not vary with time.

Later neutron scattering experiments on Cr indicated that the SDW was in fact **incommensurate** with the lattice with a wavevector $\mathbf{Q}$ about 4 % less than

---

† SDWs of wavevector $2\pi/a$ in either the [0 1 0] or [0 0 1] directions give the same antiferromagnetic structure at the lattice sites, as indeed does the more complicated SDW produced by a sum of the three simple SDWs in the [1 0 0], [0 1 0] and [0 0 1] directions; this more complicated SDW reflects better the underlying bcc symmetry of the lattice.

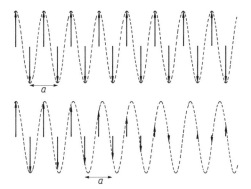

(a) Spin density wave commensurate with the bcc lattice. The wavevector is $2\pi/a$ in the x direction. The ↑ and ↓ arrows indicate the spin density at the corner atoms and body-centred atoms respectively

(b) Incommensurate spin density wave with a wavelength about $1.04a$. The arrows indicate the spin density at the lattice sites as in (a)

Fig. 8.13

$2\pi/a$ in the [1 0 0] direction. The resulting magnetic structure is shown in Fig. 8.13(b) and corresponds to a slow modulation in space of the spin density at the atomic sites. Note that, by symmetry, SDWs described by $\mathbf{Q}$ vectors of the same magnitude in the [0 1 0] and [0 0 1] directions must correspond to the same energy. The direction of the arrows in Fig. 8.13(b) is not necessarily meant to indicate the spin polarization direction relative to the wavevector $\mathbf{Q}$: neutron scattering experiments indicate that the spin polarization is perpendicular to $\mathbf{Q}$ for $T > 123$ K and parallel to $\mathbf{Q}$ below this temperature. Localized magnetic moments could not possibly produce such a magnetic structure; the existence of an incommensurate SDW is convincing evidence in favour of a band picture for the 3d electrons.

To demonstrate why such a structure might be energetically favoured we consider first a one-dimensional solid. We showed in Chapter 4 that the periodic lattice potential has a strong influence on the electron dispersion relation for wavevectors close to the Brillouin zone boundaries, $k = n\pi/a$, where a is the lattice spacing and n an integer; the energy is reduced below its free electron value for k just inside the boundary and increased above it for k just outside the boundary, as shown in Fig. 4.2. Fig. 4.1(b) shows that the reduction in energy for k just inside the boundary occurs because the wavefunction corresponds to a standing wave of electron density with maxima situated on the lattice sites where the potential is low; this situation could be described as a static commensurate *charge* density wave. Fig. 4.3(a) explains why the changes in the dispersion relation often have little effect on the properties of metallic solids, essentially because the Fermi wavenumber k_F is not close to the Brillouin zone boundary.

If however a periodic potential with period π/k_F were somehow to be introduced into the solid then this would produce energy gaps at the Fermi surface, that is at wavenumbers $k = \pm k_F$ as shown in Fig. 8.14. We see that the energy of the *occupied* states for $|k| < k_F$ is lowered by the potential and that of the *unoccupied* states with $|k| > k_F$ is raised, resulting in an overall decrease in

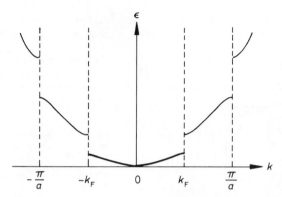

Fig. 8.14 A potential of period π/k_F introduces gaps in the electron dispersion relation at the Fermi surface, $k = \pm k_F$, analogous to those produced at $k = \pm \pi/a$ by a potential of period a. The occupied states are indicated by the thicker line

the energy of the electrons. It can therefore be advantageous for the electron gas to deform spontaneously to produce such a potential, and an incommensurate SDW of wavenumber $2k_F$ is a possible deformation of this kind; the decrease in energy associated with the changes in the dispersion relation more than compensates for any increase in energy associated with the deformation (from electron–electron interactions for example). An important consequence of the appearance of the energy gaps at the Fermi surface is to convert a metal into an insulator.†

This simple argument can be applied only to one-dimensional solids. In two and three dimensions a SDW cannot produce an energy gap at all points on the Fermi surface. The mechanism we have suggested is then only effective in reducing the energy if two different sections of the Fermi surface can be brought into near coincidence by a rigid translation in **k**-space. A section in the k_x,k_y plane through two pieces of the complicated Fermi surface of Cr is shown in Fig. 8.15(a); translation through the vector **Q** achieves the near coincidence shown in Fig. 8.15(b), a phenomenon that is described as **nesting**. A SDW of wavefactor **Q** produces energy gaps along the region of Fermi surface for which nesting is possible. The electrical resistivity of Cr increases just below the Néel temperature and this can be interpreted as arising because the formation of these energy gaps reduces the number of electrons that can partake in the conduction process.

---

† An incommensurate charge density wave can also lead to the appearance of energy gaps at the Fermi surface in a one-dimensional solid. Peierls showed that a one-dimensional metal would always be unstable against the formation of an incommensurate CDW with an associated lattice distortion; a one-dimensional metal cannot therefore exist. CDWs have been observed in the transition metal dichalcogenides in which there are chains of atoms that exhibit one-dimensional behaviour.

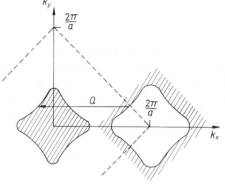

(*a*) Section through two pieces of the Fermi surface of Cr in the k_x, k_y plane in k-space. The two pieces occur in two different but overlapping energy bands; the shading indicates the occupied states. The broken lines indicate the first Brillouin zone boundaries, and the repeated zone scheme (see Figs. 4.9 and 4.10) has been used to extend the Fermi surface outside this zone

(*b*) A relative translation through the wavevector **Q** illustrates the nesting property of the two sections of Fermi surface. The periodic potential associated with a distortion of wavenumber Q can produce energy gaps in the 'overlapping' regions

Fig. 8.15

8.6.3   Magnetic ordering in rare-earth metals

The rare-earth metals crystallize either into the hexagonal close-packed structure (Fig. 1.11) or into very similar structures of hexagonal symmetry with a more complicated stacking sequence of close-packed layers. Neutron scattering experiments on single crystals have revealed a great variety of magnetic ordering arrangements in these metals. For example, dysprosium is a simple ferromagnet below 85 K with spontaneous magnetization along a direction parallel to the close-packed planes, that is perpendicular to the z axis. Between 85 and 179 K the ordering remains ferromagnetic within a close-packed layer, but on going from one layer to its neighbour the direction of the magnetization rotates through a temperature-dependent angle of order 30° about the z axis to produce a helical ordering of spins. Such ordering may be described as a transverse circularly polarized incommensurate spin density wave with a wavevector parallel to the z axis. Above 179 K dysprosium is paramagnetic. Most rare-earth metals show this helical ordering under some conditions. Even more complicated arrangements can occur in which helical ordering is accompanied by a component of magnetization along the z axis that may also show periodic behaviour.

8.7   FERROMAGNETIC DOMAINS

Despite the existence of a spontaneous magnetization, macroscopic samples of ferromagnetic materials often have a negligible total dipole moment in the absence of an applied field. This is because of the tendency of such materials to consist of a number of small regions, known as **domains**, in which the magnetization points in different directions. Fig. 8.16 shows the domains in a 50 μm single crystal or iron. In this body-centred cubic material, crystalline anisotropy favours directions for the magnetization parallel to one of the edges of the unit cell, that is in one of the six equivalent directions [1 0 0], [0 1 0], [0 0 1], [$\bar{1}$ 0 0], [0 $\bar{1}$ 0], [0 0 $\bar{1}$]; this explains the directions adopted by the magnetization within the domains in Fig. 8.16 as indicated by the arrows.

Although the boundaries between the domains in Fig. 8.16 appear sharp, there is in fact a narrow transition region, known as a **Bloch wall**, within which the magnetization changes smoothly from its value in one domain to that in another. Fig. 8.17 shows a Bloch wall in which the magnetization rotates smoothly by an angle 180° about an axis normal to the wall.

8.7.1   The energy and thickness of a Bloch wall

To calculate the energy of a Bloch wall it is necessary to take account of the change in exchange energy due to the variation of the magnetization with position and to allow for the fact that within the wall the magnetization is not pointing in a direction favoured by crystalline anisotropy. We illustrate the principles of the calculation by taking a simple example in which the spins lie on a simple cubic lattice of side a and there is a domain wall of the form of Fig. 8.17 in the yz plane. We take the spins to rotate about the x axis from the z direction to the $-z$ direction in a distance of N atomic spacings along the x axis. Such a domain wall is appropriate to a situation where $\pm z$ are the **easy directions** for the magnetization, favoured by the crystal anisotropy.†

The anisotropy energy density will be a function of the angle θ between the local direction of the magnetization and the z axis. In hexagonal close-packed cobalt, for example, the easy directions for the magnetization are $\pm z$ and the anisotropy energy density is of the form

$$E_{\text{anis}} = K_1 \sin^2 \theta + K_2 \sin^4 \theta \qquad (8.46)$$

where $K_1 = 4.1 \times 10^5 \, \text{J m}^{-3}$ and $K_2 = 1 \times 10^5 \, \text{J m}^{-3}$; in this material there-fore a state with spins parallel to the easy axis ($\theta = 0$) is lower in energy by $K = K_1 + K_2 = 5.1 \times 10^5 \, \text{J m}^{-3}$ than a state with spins perpendicular to this axis ($\theta = 90°$). Because of the anisotropy the magnetization will spiral most rapidly past the direction ($\theta = 90°$ for the above example) at which the energy is

---

† We ignore the inconsistency in our model of assuming uniaxial anisotropy in a crystal of cubic symmetry.

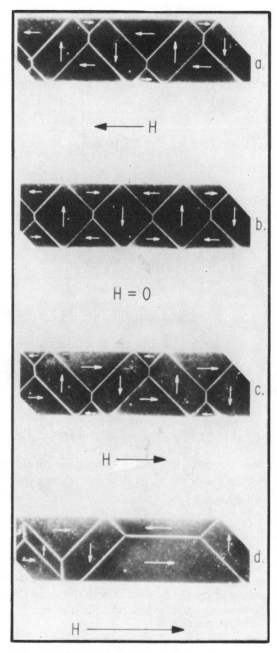

Fig. 8.16   Photographs showing the effect of an applied magnetic field on the domains in a 50 μm iron whisker: (a) and (c) show reversible domain wall motion associated with small applied fields; (d) shows that a stronger field causes disappearance of domains, a process that is not perfectly reversible. (Reproduced with permission from R. W. DeBlois and C. D. Graham, *J. Appl. Phys.* **29**, 931 (1958))

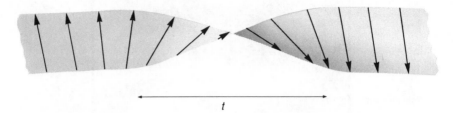

Fig. 8.17   Schematic diagram of a Bloch wall between two domains

greatest, but we will ignore this effect and assume a uniform rotation angle of π/N radians between neighbouring spins in the x direction. We will also ignore the details of the anisotropy energy as represented for example by Eq. (8.46) and obtain a crude estimate of the increased anisotropy energy by assuming that effectively half of the spins in the wall are pointing in the unfavourable direction. The anisotropy energy per unit area of wall is then

$$\sigma_{\text{anis}} \approx \tfrac{1}{2} K N a \tag{8.47}$$

where Na is the volume of wall per unit area and, from above, we expect $K \approx 5 \times 10^5 \text{ J m}^{-3}$.

To estimate the increase in exchange energy associated with the formation of the wall we take into account only nearest neighbour exchange interactions. Because the spins lie on a simple cubic lattice a spin in the wall has four parallel nearest neighbours and two at an angle π/N relative to it; from Eq. (8.1) its contribution to the exchange energy is therefore

$$- [4\mathscr{J}S^2 + 2\mathscr{J}S^2 \cos^2 (\pi/N)] \approx - 6\mathscr{J}S^2 + \mathscr{J}S^2\pi^2/N^2,$$

where we have used the small-angle expansion of the cosine function. The first term on the right-hand side gives the exchange energy per unit volume of a uniformly magnetized crystal as

$$- W = - \frac{6\mathscr{J}S^2}{a^3} \approx -5 \times 10^8 \text{ J m}^{-3}, \tag{8.48}$$

where $1/a^3$ is the number of atoms per unit volume and we have used $\mathscr{J} = 0.03$ eV, $S = \tfrac{1}{2}$ and $a = 2.5$ Å to obtain the numerical estimate. The second term gives the change in energy associated with the variation of magnetization with position and leads to a contribution to the wall energy per unit area

$$\sigma_{\text{exch}} = \frac{\mathscr{J}S^2\pi^2}{N^2} \times \frac{N}{a^2} = \frac{\pi^2 W a}{6N} \tag{8.49}$$

where N/a^2 is the number of spins per unit area of wall and we have used Eq. (8.48) to obtain the final expression.

The total wall energy per unit area obtained by adding Eqs. (8.47) and (8.49) is

$$\sigma = \tfrac{1}{2}KNa + \frac{\pi^2 Wa}{6N}.$$

As N (and hence the width of the wall) increases, the anisotropy energy increases but the exchange energy decreases. The optimum width is that for which σ is minimized and this leads to

$$\frac{\mathrm{d}\sigma}{\mathrm{d}N} = \tfrac{1}{2}Ka - \frac{\pi^2 Wa}{6N^2} \qquad \text{or} \qquad N = \left(\frac{\pi^2 W}{3K}\right)^{1/2},$$

corresponding to a wall thickness

$$t = Na = \pi a \left(\frac{W}{3K}\right)^{1/2} \tag{8.50}$$

and a wall energy per unit area

$$\sigma = \pi a \left(\frac{WK}{3}\right)^{1/2} \tag{8.51}$$

According to our above estimates the exchange energy W is much greater than the anisotropy energy K and this means that a Bloch wall is many (≈ 100) atoms thick. Since our calculation shows that a Bloch wall is higher in energy than a region of uniform magnetization, we must explain why such walls appear, and this we do in the following section.

8.7.2   Why do domains occur?

If a ferromagnetic crystal consists of a single domain, as in Fig. 8.18(a), there is a considerable contribution to the total energy ($B^2/2\mu_0$ per unit volume) from

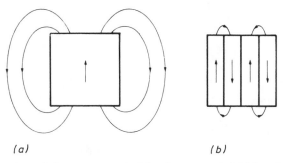

(a)   (b)

Fig. 8.18   Reduction of the energy stored in the external field by the formation of domains

the magnetic field outside the crystal, and it is to reduce this energy that all but the smallest samples find it preferable to split up into domains, as in Fig. 8.18(b). The optimum size of domains is determined by minimizing the sum of the external field energy and the energy of the Bloch walls. If the domains are larger than the optimum size then the energy stored in the field is the dominant contribution, and if they are smaller than the optimum size the energy of the domain walls is very large.

The situation is different in cubic crystals such as iron where there are several equivalent easy directions for the magnetization. The external field can be reduced almost to zero by the formation of **closure domains**, like those shown in Fig. 8.19(a) (see also Fig. 8.16(b)). One may wonder why the closure domains take the form shown in Fig. 8.19(a) rather than that in Fig. 8.19(b), which would appear to have a significantly smaller area of Bloch wall. The reason is **magnetostriction**; a magnetized crystal tends to expand or contract along the magnetization direction. The distortions of the different domains in Fig. 8.19(b) are incompatible, resulting in a large positive stress energy. The elastic stress energy is reduced by having the smaller closure domains shown in Fig. 8.19(a); thus the optimum size of the domains in this case is determined by a balance between Bloch wall energy and magnetoelastic energy.

8.7.3   Magnetization curves of ferromagnets

Although the total dipole moment of a macroscopic ferromagnetic sample may be small in zero applied field because of domain formation, a large dipole moment is often produced by a modest applied field. In this case the material has a high effective magnetic permeability. The magnetization process is illustrated in its simplest form, for a single-crystal iron whisker, by the domain patterns shown in Fig. 8.16. For the small fields shown in Figs. 8.16(a)–(c) magnetization occurs by the almost reversible movement of domain walls, but for larger fields (Fig. 8.16(d)) irreversible changes of magnetization occur by the disappearance of unfavourable domains. Eventually a situation is reached in which the crystal

(a)   (b)

Fig. 8.19   The formation of closure domains leads to a vanishingly small external field: (a) and (b) share this property but (b) has more magnetoelastic energy

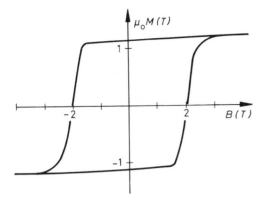

Fig. 8.20   Hysteresis curve for $Nd_2Fe_{14}B$ containing small amounts of Dy. NdFeB alloys are used for making high-performance permanent magnets. (Reproduced by permission of H. R. Kirchmayer)

consists of a single domain with the magnetization along the nearest easy direction to that of the applied field. Very strong fields will overcome the crystalline anisotropy and cause the magnetization to rotate from the easy direction towards the applied field if these are not already parallel.

The magnetization processes are essentially the same, though less distinct, in bulk polycrystalline material. Domain boundary motion may then be much more irreversible because of inhomogeneities. Indeed an important technique in making high-coercive-force materials for permanent magnets is the deliberate introduction of inhomogeneities to inhibit domain boundary motion, for example by introducing a two-phase alloy system. An example of the kind of hysteresis curve that can be obtained in this way is shown in Fig. 8.20. In contrast it is possible to prepare ferromagnetic materials with a very high effective permeability and almost reversible magnetization curves; such materials are used in transformer cores and for magnetic screening.

PROBLEMS 8

8.1 By using the result

$$S^2 = (s_1 + s_2)^2 = s_1^2 + s_2^2 + 2s_1 . s_2,$$

calculate the eigenvalues of $s_1 . s_2$ for the spin singlet and spin triplet states of two electrons. Hence show that an exchange interaction of the form $-2\mathcal{J}s_1 . s_2$ implies that the energy of the singlet state exceeds that of the triplet state by $2\mathcal{J}$.

8.2 State the assumptions of the Weiss molecular field theory of ferromagnetism. Derive Eq. (8.4) from Eq. (8.1).

Use the Hamiltonian (8.6) to determine the internal energy and hence show that the magnetic contribution to the heat capacity of a ferromagnet in zero applied field in the mean field approximation is $C_m = -\frac{1}{2}\lambda\mu_0 \, d(M^2)/dT$. Give the limiting forms of C_m for $T \to 0$ and $T \to T_C$. What is the thermodynamic nature of the transition at the Curie point in zero applied field?

8.3 Show that Eq. (8.7) can be written

$$m = \frac{M}{N\mu_B} = \frac{h + \tanh(m/t)}{1 + h\tanh(m/t)}$$

where $h = \tanh(\mu_B B/k_B T)$. Hence use the small-angle expansion $\tanh x \approx x - x^3/3$ to show that the Weiss model predicts that the magnetization M varies as $B^{1/3}$ in small applied fields at $T = T_C$. Would you expect this relation to hold in practice?

8.4 For an fcc lattice of magnetic spins it is impossible to find an antiferromagnetic arrangement in which all the nearest neighbours of any spin are antiparallel to it. The best that can be achieved, for example by having spins in alternate (2 0 0) planes ↑ and ↓, is eight antiparallel and four parallel neighbours. (Miller indices are referred to the conventional cubic unit cell.)

Develop a Néel theory appropriate to the case where a spin has z_a antiparallel nearest neighbours and z_p parallel nearest neighbours and only nearest neighbour exchange interactions are important; the effective field acting on an ion on the A sublattice would then be $B_{eff}^A = B_{loc} - \lambda_a M_B - \lambda_p M_A$ with a similar expression for B_{eff}^B. Show that $\lambda_a/\lambda_p = z_a/z_p$ and that the high-temperature susceptibility is of the form $\chi = C/(T + \theta)$, where θ is related to the Néel temperature T_N by

$$\frac{\theta}{T_N} = \frac{\lambda_a + \lambda_p}{\lambda_a - \lambda_p} = \frac{z_a + z_p}{z_a - z_p}.$$

Hence show that $\theta/T_N = 3$ for the fcc structure mentioned above.

8.5 Calculate according to the Néel model the parallel and perpendicular magnetic susceptibilities of an antiferromagnet below the Néel temperature.

8.6 Fig. 12.10 shows the measured magnon dispersion relation for different directions in an fcc colbalt alloy (92 % Co, 8 % Fe). Deduce a value for $\mathscr{J}S$ and suggest a reason for the finite value of ω at $k = 0$.

8.7 The low-temperature heat capacity of $Y_3Fe_5O_{12}$ (yttrium iron garnet) gives a straight line graph when $C/T^{3/2}$ is plotted against $T^{3/2}$. What information can be obtained from the slope and intercept of this line?

8.8 By analogy with the approach used in section 8.5.2 for a ferromagnet, develop a theory for spin waves in a one-dimensional antiferromagnet. Include only nearest neighbour exchange interactions and assume that the spins on the ↑ and ↓ sublattices may be written as

$$S_n^\uparrow = S\hat{z} + \sigma_n^\uparrow \qquad \text{and} \qquad S_{n+1}^\downarrow = -S\hat{z} + \sigma_{n+1}^\downarrow.$$

Write down equations of motion for ↑ and ↓ spins and assume solutions of the form

$$\sigma_n^{\uparrow-} = \sigma_{n,x}^\uparrow - i\sigma_{n,y}^\uparrow = u\, e^{i(nka - \omega t)}$$
$$\sigma_{n+1}^\downarrow = \sigma_{n+1,x}^\downarrow - i\sigma_{n+1,y}^\downarrow = \alpha u\, e^{i(nka - \omega t)}.$$

(Compare with Eqs. (2.8) and (2.17).) Calculate the dispersion relation for the spin waves and show that it is linear ($\omega \propto k$) for small k. What is the implication of this for the heat capacity at low temperature?

8.9 The experimental results illustrated in Fig. 8.12 were obtained at 9.7 GHz on a Co film about 600 nm thick. Estimate the effective mass of magnons in Co. (Remember that the graph shows the absorption *derivative*.)

There is no plea which will justify the use of high tension and alternating currents, either in a scientific or a commercial sense. They are employed solely to reduce investment in copper wire and real estate.—*Thomas Edison (1889)*

CHAPTER

Electric properties of insulators

9.1 DIELECTRICS

9.1.1 Dielectric constant and susceptibility

Dielectric materials are *electrical insulators*† for which the response to a weak static or low-frequency electric field is given by

$$\mathbf{P} = \varepsilon_0 \chi \mathbf{E} \tag{9.1}$$

where $\mathbf{P}$ is the electric polarization (dipole moment per unit volume) and $\mathbf{E}$ is conventionally the *macroscopic* electric field inside the material (see appendix B for a discussion of the electric fields in matter). Comparison with Eq. (7.1) shows that dielectrics are analogous to diamagnets and paramagnets; however, as the dimensionless static susceptibility χ is always positive for dielectrics, it would be more consistent with magnetism if they were called paraelectrics. **Pyroelectric** materials (of which **ferroelectric** materials are a subset) possess a spontaneous electric polarization in the absence of an applied field; we discuss them further in section 9.2

The electric susceptibility of a dielectric is normally much larger than the magnetic susceptibility and is typically of order unity or larger at room

---

† The response of a conducting material to a low-frequency electric field is dominated by the mobile charge carriers (see section 13.6).

temperature. Eq. (9.1) defines the SI susceptibility; as in the magnetic case, the reader is likely to encounter tabulations of cgs susceptibilities related by (Eq. (7.2))

$$\chi_{SI} = 4\pi\chi_{cgs}.$$

We consider only dielectrics for which $\mathbf{P}$ and $\mathbf{E}$ are parallel so that χ is a scalar quantity; crystals of cubic symmetry possess this property. The relative permittivity (dielectric constant) ε is related to the susceptibility by

$$\chi = \varepsilon - 1. \tag{9.2}$$

A theoretical value for the susceptibility can be obtained by first calculating the dipole moment of an isolated atom

$$\mathbf{p} = \alpha\mathbf{E} \tag{9.3}$$

produced by an applied field $\mathbf{E}$; α is known as the **polarizability** of the atom. If the behaviour of atoms is not greatly affected by their incorporation in a solid then the polarization of the solid is obtained by summing Eq. (9.3) over the N atoms in unit volume,

$$\mathbf{P} = \sum_{i=1}^{N} \mathbf{p}_i = \sum_{i=1}^{N} \alpha_i \mathbf{E}_{Li}. \tag{9.4}$$

where the subscript i refers to the ith atom and $\mathbf{E}_{Li}$ is the *local* electric field at the atom. To deduce the susceptibility it is necessary to evaluate the relation between the local and macroscopic electric fields, and this depends on the arrangement of atoms within the crystal. The most straightforward situation is for an atom in a position of cubic symmetry in a crystal of atoms with point-like time-independent dipole moments. The local field at the centre of the atom is then given by the **Lorentz relation** (appendix B, Eq. (B31))

$$\mathbf{E}_L = \mathbf{E} + \frac{\mathbf{P}}{3\varepsilon_0} \tag{9.5}$$

and is thus the same at all such atoms. Using Eqs. (9.1), (9.2), (9.4) and (9.5), we deduce the following relationships between ε, χ and the atomic polarizabilities:

$$\frac{\chi}{\chi + 3} = \frac{\varepsilon - 1}{\varepsilon + 2} = \frac{1}{3\varepsilon_0} \sum_{i=1}^{N} \alpha_i. \tag{9.6}$$

This is the **Clausius–Mossotti** relation. The Lorentz relation (Eq. (9.5)) also gives the average local field at an atom or molecule in a random arrangement; the Clausius–Mossotti result can therefore also be applied to gases, liquids and amorphous solids. The relation works well for gases but, as we will find in section 9.1.3, it must be used more cautiously in systems of higher density.

In the following sections we discuss three mechanisms that lead to polarization of a solid by an electric field; the relative displacement of electrons and nuclei in individual atoms; the orientation of the permanent dipole moments of molecules in molecular solids; and the relative displacement of positive and negative ions in ionic solids.

Eqs. (9.1) to (9.6) can also be used to describe the response of a dielectric to an alternating electric field, $E_0 e^{i\omega t}$, but the dielectric constant and susceptibility depend on the frequency ω and are generally complex; the imaginary part of ε indicates the existence of dissipation as can be seen by writing

$$\varepsilon = \varepsilon' + i\varepsilon''. \tag{9.7}$$

The resulting Maxwell displacement current is

$$\dot{D} = \varepsilon\varepsilon_0 \dot{E} = \varepsilon_0(\varepsilon' + i\varepsilon'')\dot{E} = \varepsilon_0\varepsilon' i\omega E - \varepsilon_0\varepsilon''\omega E. \tag{9.8}$$

The first term is 90° out of phase with the applied field and is thus reactive; the second term is in phase with the field and therefore resistive. The average dissipation associated with the resistive term is

$$\langle \dot{D} . E \rangle = -\varepsilon_0\varepsilon''\omega E_0^2 \langle \cos^2(\omega t) \rangle = -\tfrac{1}{2}\varepsilon_0\varepsilon''\omega E_0^2$$

per unit volume. The dissipation must be positive and hence ε'' is negative. The two terms in Eq. (9.8) can be represented on a phasor diagram as shown in Fig. 9.1: the quality of the dielectric is expressed by the **loss tangent**, which is the ratio of the dissipative term to the reactive term and is thus the tangent of the angle δ in Fig. 9.1. Hence

$$\text{loss tangent} = \tan \delta = -\varepsilon''/\varepsilon'. \tag{9.9}$$

The frequency-dependent refractive index $n(\omega)$ for the passage of an electromagnetic wave through a solid is defined in terms of the dispersion relation by

$$\frac{\omega}{k} = \frac{c}{n(\omega)}, \tag{9.10}$$

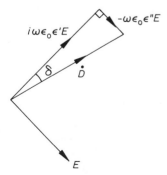

Fig. 9.1 Phase relationship with respect to E of the reactive and resistive contributions to the displacement current $\dot{D}$. Note that ε'' is negative

where $n(\omega)$ and $\varepsilon(\omega)$ are related by† (problem 9.1)

$$n(\omega) = [\varepsilon(\omega)]^{1/2};\qquad(9.11)$$

if $\varepsilon(\omega)$ is complex, so is $n(\omega)$.

In the very strong electric fields that occur in a focused laser beam $(\approx 10^{10}\ \text{V m}^{-1})$, Eq. (9.1) is inadequate and must be replaced by

$$\mathbf{P} = a\mathbf{E} + b\mathbf{E}^2 + c\mathbf{E}^3 + \cdots.\qquad(9.12)$$

For crystals with a centre of symmetry the coefficient b vanishes since equal and opposite fields $\mathbf{E}$ and $-\mathbf{E}$ must produce equal and opposite polarizations. The science of **non-linear optics** depends on the existence of the higher-order terms in Eq. (9.12). We can obtain an insight into the type of novel phenomena caused by these terms by noting that the quadratic term implies that an electric field of frequency ω will generate a contribution to the polarization varying at a frequency 2ω. The crystal re-radiates at this frequency and thus acts as a source of radiation at a frequency double that of the original laser.‡

9.1.2   Polarization due to relative motion of electrons and nuclei

An order of magnitude estimate of the polarizability of an atom can be obtained by assuming that the Z electrons in it are distributed with uniform density in a sphere of radius r. A displacement of the nucleus from the centre of the sphere by a distance x means that it is subject to a restoring electric field generated by electrons within a sphere of radius x (Fig. 9.2); the charge within the sphere is $-Ze(x/r)^3$ so that the field is

$$E = -\frac{Ze(x/r)^3}{4\pi\varepsilon_0 x^2} = -\frac{Zex}{4\pi\varepsilon_0 r^3}.\qquad(9.13)$$

An applied field $\mathbf{E}_{\text{L}}$ causes a relative displacement of electrons and nucleus until the restoring field just balances the applied field, $\mathbf{E} + \mathbf{E}_{\text{L}} = 0$. Eq. (9.13) gives the resulting electric dipole moment of the atom

$$\mathbf{p} = Ze\mathbf{x} = 4\pi\varepsilon_0 r^3 \mathbf{E}_{\text{L}}.\qquad(9.14)$$

Comparison with Eq. (9.3) then identifies the atomic polarizability as§

$$\alpha/4\pi\varepsilon_0 = r^3 \approx 10^{-30}\ \text{m}^{-3},\qquad(9.15)$$

---

† We assume that the solid is non-magnetic, $\mu = 1$.

‡ Further information on the application of solid state physics to non-linear optics can be found in Dalven.[13]

§ Exactly the same order of magnitude estimate of the polarizability is obtained if the atom is regarded as behaving like a perfectly conducting sphere of radius r. We have divided by $4\pi\varepsilon_0$ because this gives values for the polarizability in m^3 that can be compared directly with cgs values quoted in cm^3.

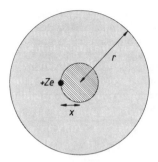

Fig. 9.2   Model of an atom in which the Z electrons are uniformly distributed in a sphere of radius r. Displacement of the nucleus by a distance x as shown means that it is subject to a restoring force provided by the electrons inside the sphere of radius x indicated. The electrons outside this sphere do not contribute because the electric field inside a uniformly charged spherical shell is zero

where we have obtained the numerical estimate by assuming $r \approx 1$ Å. Table 9.1 shows measured values of $\alpha/4\pi\varepsilon_0$ for various atoms and molecules, together with the value of r deduced using Eq. (9.15); agreement with our order of magnitude estimate is reasonable. A more rigorous quantum mechanical calculation of α can be made using second-order perturbation theory.

When both the atomic polarizabilities and the relationship between the macroscopic and local electric fields are known, it is possible to calculate the dielectric constant of liquids and solids. For the random arrangement of atoms or molecules in a liquid the Lorentz local field relation (Eq. (9.5)) and hence the Clausius–Mossotti result (Eq. (9.6)) can be used. The validity of this approach is checked in Table 9.1, which shows that there is good agreement between experimental and predicted values of the dielectric constant of various liquids at their normal boiling points. The molecular polarizabilities used in the calculation were obtained from gas phase measurements where the local field correction is very small.

TABLE 9.1   Comparison of observed values of the electric susceptibilities of various liquids at their normal boiling points with values predicted by the Clausius–Mossotti equation (9.6). The atomic or molecular polarizability used to obtain the predicted value was deduced from gas phase measurements of the susceptibility

Atom or molecule	Polarizability, $\alpha/4\pi\varepsilon_0$ $(10^{-30}$ m$^3)$	Effective radius $(\alpha/4\pi\varepsilon_0)^{1/3}$ (Å)	Number density in liquid $(10^{28}$ m$^{-3})$	Dielectric constant of liquid	
				Predicted	Observed
He	0.206	0.59	1.880	1.049	1.048
Ar	1.64	1.18	2.128	1.514	1.538
H_2	0.807	0.93	2.114	1.231	1.228
O_2	1.57	1.16	2.143	1.492	1.507
N_2	1.74	1.20	1.737	1.435	1.454

(Data taken with permission from the *Handbook of Chemistry and Physics*, 61st edn. Copyright CRC Press Inc., Boca Raton, Florida)

TABLE 9.2   Comparison of measured (upper value) and predicted (lower value) dielectric constants of the alkali halide crystals. The ionic polarizabilities ($\alpha/4\pi\varepsilon_0$) used in the calculation are indicated after the chemical symbol and are chosen to obtain as good a fit between measured and predicted values as possible. The observed values were measured at the frequency of the D lines in the spectrum of atomic sodium

	Anions ($\alpha/4\pi\varepsilon_0$)			
Cations ($\alpha/4\pi\varepsilon_0$)	F^- (0.644 Å^3)	Cl^- (2.960 Å^3)	Br^- (4.158 Å^3)	I^- (6.431 Å^3)
Li^+	0.920	2.980	4.159	6.248
(0.029 Å^3)	0.673	2.989	4.187	6.459
Na^+	1.186	3.360	4.560	6.721
(0.408 Å^3)	1.053	3.368	4.566	6.839
K^+	1.966	4.272	5.508	7.790
(1.334 Å)	1.981	4.297	5.495	7.767
Rb^+	2.572	4.856	6.147	8.532
(1.979 Å^3)	2.623	4.939	6.137	8.409
Cs^+	3.664	6.419	7.497	9.952
(3.335 Å^3)	3.979	6.295	7.493	9.765

(From J. R. Tessmann, A. H. Kahn and W. Shockley, *Phys. Rev.* **92**, 890 (1953))

The Clausius–Mossotti result can also be applied when the atoms are in positions of cubic symmetry; Table 9.2 compares experimental and predicted values of the dielectric constant of the cubic alkali halide crystals. The experimental values of ε are at the frequency ($\nu = 5 \times 10^{14}$ Hz) of the D lines in the spectrum of atomic sodium; at this frequency the contribution to ε from ionic motion (section 9.1.4) is negligible. The values used for the ionic polarizabilities are indicated in the table and were chosen to obtain a good fit between the experimental and calculated values of ε; except for the fluorides the fit is better than about 3 %.

The small differences between the experimental and fitted values arise partly because the polarizability of an ion depends on the arrangement of the electrons in it and hence on its environment; this is a relatively small effect for ionic crystals, because the electronic arrangement is little affected by the formation of the solid. When a covalently bonded crystal is formed the effect on the electrons is much larger and using the polarizability of an isolated atom does not give a good value for the dielectric constant of the solid. It is the outermost electrons that are most affected by the formation of the covalent bonds, and because of the r^3 dependence in Eq. (9.15) it is these that have the largest influence on the polarizability. The large dielectric constants of diamond, silicon and germanium indicate the existence of a significant density of electrons at some distance from the nuclei and this can be attributed to the electrons involved in the formation of the covalent bonds. A proper calculation of the dielectric constant for these materials must take band structure effects into account.

We can use the simple atomic model of Fig. 9.2 to investigate the frequency dependence of the atomic polarizability arising from the relative displacement of electrons and nuclei. The model predicts (Eq. (9.13)) a restoring force proportional to the displacement and this suggests that the response of the electrons to an alternating field $\mathbf{E_L} = \mathbf{E_0}e^{i\omega t}$ should be described by the driven simple harmonic oscillator equation,

$$Zm\frac{d^2\mathbf{x}}{dt^2} + Zm\gamma\,\frac{d\mathbf{x}}{dt} + \frac{Z^2e^2}{4\pi\varepsilon_0 r^3}\mathbf{x} = Ze\mathbf{E_0}\,e^{i\omega t}. \tag{9.16}$$

The expression for the restoring force has been obtained from Eq. (9.13). We have added a dissipative term $Zm\gamma\,d\mathbf{x}/dt$ to the equation and we discuss the origin of this later.

The solution of Eq. (9.16) is

$$\mathbf{x} = \frac{e}{m(\omega_0^2 - \omega^2 + i\omega\gamma)}\,\mathbf{E_0}e^{i\omega t} \tag{9.17}$$

where

$$\omega_0 = \left(\frac{Ze^2}{4\pi\varepsilon_0 r^3 m}\right)^{1/2} \tag{9.18}$$

is the 'natural' resonant frequency of our model atom. The dipole moment $Ze\mathbf{x}$ and the frequency-dependent polarizability can be deduced by proceeding in the same way as for a static field (Eq. (9.14) and (9.15)). For weak damping the response of the atom is a sharp resonance at frequency ω_0. Fig. 9.3 shows the resulting behaviour of the dielectric constant, deduced using the Clausius–Mossotti relation (Eq. (9.6)). At low frequencies, $\omega \ll \omega_0$, the real part ε' tends to the static value which follows from Eq. (9.15), and at high frequencies it tends to unity from below as

$$\varepsilon' = 1 - \frac{NZe^2}{\varepsilon_0 m\omega^2}, \tag{9.19}$$

where N is the number of atoms per unit volume. We show in section 13.6 that the dielectric constant of free electrons at high frequencies is also given by Eq. (9.19), and this indicates that at high frequencies the electrons in all materials behave as free particles. The sharp peak in the imaginary part ε'' of the dielectric constant reflects the resonant absorption of energy that occurs in the vicinity of ω_0.

The physical interpretation of the resonant frequency ω_0 becomes apparent if we note from Table 9.1 that r is of order $2a_0$, where a_0 is the Bohr radius. The energy $\hbar\omega_0$ is thus found to be close to the binding energy of a single-electron

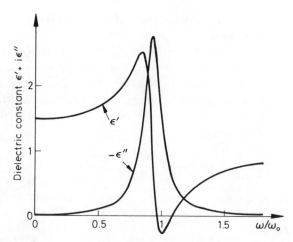

Fig. 9.3   Variation of the real and imaginary parts of the dielectric constant, ε' and ε'', with frequency as predicted by Eqs. (9.14) and (9.17). The Clausius–Mossotti relation (Eq. (9.6)) was used to relate the atomic polarizability to the dielectric constant. The local field correction embodied in this relation causes the peak in the absorption (ε'') to occur at a frequency just below ω_0. In the region of frequency where ε' is negative, waves incident on the crystal are totally externally reflected (see section 9.1.4)

atom with a nucleus of charge Ze (problem 9.3); this identifies the frequency ω_0 with electron energy levels in the atom. A quantum mechanical calculation of the polarizability of an atom predicts several resonances, each at a frequency corresponding to a transition between electronic energy levels of the atom; the qualitative behaviour of the dielectric constant at each resonance is similar to that shown in Fig. 9.3. The limiting behaviour at high frequencies given by Eq. (9.19) is only attained above the highest of the resonant frequencies, which is usually in the long-wavelength x-ray region of the spectrum. The dissipation peak associated with each resonance corresponds to resonant absorption of photons by electrons undergoing the corresponding transition.

9.1.3   Orientation of permanent dipole moments

Molecules without a centre of symmetry possess a permanent electric dipole moment and are said to be **polar**. The interaction energy,

$$E = -\mathbf{p}\cdot\mathbf{E}_{\mathrm{L}}, \qquad (9.20)$$

of the dipole moment $\mathbf{p}$ with an electric field $\mathbf{E}_{\mathrm{L}}$ tends to orient the moment parallel to the field. If the dipoles behave independently of each other the resulting alignment is limited by thermal disorder and the probability $p(\theta)$ of finding a dipole at an angle θ to the field $\mathbf{E_L}$ is given by the Boltzmann factor

$$p(\theta) \propto \exp\left(-E/k_B T\right) = \exp\left(+\mathbf{p} \cdot \mathbf{E_L}/k_B T\right) = \exp\left[(pE_L \cos \theta)/k_B T\right]. \quad (9.21)$$

Eq. (9.21) describes polar liquids and gases reasonably well. In solids, however, it is not possible to ignore the potential energy of interaction between a dipole and its neighbours. This can hinder the free rotation of molecules and sometimes suppress it completely (see section 9.2); where rotation is possible it is often between one favoured orientation and another. To get a rough idea of what to expect we ignore this and assume that the average component of dipole moment parallel to the field $\bar{p}_{\parallel}$ can be obtained using the Boltzmann factor (Eq. (9.21)). Thus

$$\bar{p}_{\parallel} = \langle p \cos \theta \rangle = p \langle \cos \theta \rangle$$

$$= p \frac{\int_0^\pi \cos \theta \exp[(pE_L \cos \theta)/k_B T] 2\pi \sin \theta \, d\theta}{\int_0^\pi \exp[(pE_L \cos \theta)/k_B T] 2\pi \sin \theta \, d\theta}.$$

The factor $2\pi \sin \theta$ is a solid angle weighting factor as explained in Fig. 9.4. Using the substitution $u = \cos \theta$ to evaluate the integrals gives

$$\bar{p}_{\parallel} = p \frac{\int_{-1}^{+1} u \, e^{ux} \, du}{\int_{-1}^{+1} e^{ux} \, du}, \quad (9.22)$$

where $x = pE_L/k_B T$ is a dimensionless measure of the electric field strength. A neat way of proceeding is to note that Eq. (9.22) can be written

$$\bar{p}_{\parallel} = p \frac{1}{I} \frac{dI}{dx} = p \frac{d \ln I}{dx}$$

where

$$I = \int_{-1}^{+1} e^{ux} \, du = \left[\frac{1}{x} e^{ux}\right]_{-1}^{+1} = \frac{2 \sinh x}{x}.$$

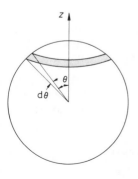

Fig. 9.4   The shaded ring on the surface of the sphere of *unit* radius is the solid angle for all the possible directions between θ and $\theta + d\theta$ from the z axis. The circumference of the ring is $2\pi \sin \theta$ so that the area is $2\pi \sin \theta d\theta$

Thus

$$\bar{p}_{\parallel} = p(\coth x - 1/x) = pL(x). \tag{9.23}$$

The **Langevin function** $L(x)$, defined by this equation, is the Brillouin function of Eq. (7.16) in the limit of infinite J; this is appropriate since a large angular momentum can point in almost any direction and therefore behaves analogously to a freely rotating electric dipole moment.

Using typical values of 10^{-29} C m for a permanent dipole moment (equivalent to the displacement of one electronic charge through 0.6 Å) and 10^7 V m^{-1} for the largest electric field that can be applied to a solid without breakdown gives $x \approx 0.02$ at room temperature. As molecular rotation in solids is usually absent at significantly lower temperatures, the small x limit of Eq. (9.23) is appropriate; using $\coth x \approx 1/x + x/3 + \cdots$ gives

$$\bar{p}_{\parallel} \approx \frac{p^2 E_{\mathrm{L}}}{3 k_{\mathrm{B}} T}.$$

The equivalent atomic polarizability defined by Eq. (9.3) is

$$\alpha = \frac{p^2}{3 k_{\mathrm{B}} T}. \tag{9.24}$$

The $1/T$ dependence of α is the electrical equivalent of Curie's law in magnetism (Eq. (7.18)). Using $p \approx 10^{-29}$ C m we find

$$\alpha/4\pi\varepsilon_0 \approx 10^{-28} \ \mathrm{m}^{-3} \tag{9.25}$$

at $T = 300$ K. We have divided by $4\pi\varepsilon_0$ in order to make comparison with Eq. (9.15) easier; we thus see that orientational polarizabilities are normally much larger than those associated with relative displacement of electrons and nuclei in atoms.

For such large values of α the Clausius–Mossotti relation (Eq. (9.6)) often gives erroneous results. To see this we use the example of water.† Inverting Eq. (9.6) to obtain the dielectric constant,

$$\varepsilon = \frac{2 \sum \alpha_i/\varepsilon_0 + 3}{3 - \sum \alpha_i/\varepsilon_0}, \tag{9.26}$$

predicts the divergence of ε when $\sum \alpha_i/\varepsilon_0 \to 3$ and negative values of ε for $\sum \alpha_i/\varepsilon_0 > 3$. The permanent dipole moment of the water molecule is 0.62×10^{-29} C m and the molecular density in water is 4×10^{28} m^{-3}, giving $\sum \alpha_i/\varepsilon_0 = 12$ at $T = 300$ K. However, the measured static dielectric constant is positive ($\varepsilon \approx 80$). The failure of Eq. (9.26) is an indication that the Lorentz local field relation (Eq. (9.5)) is inadequate to describe the large space- and time-dependent electric field in the neighbourhood of a polar molecule.

---

† The Clausius–Mossotti result cannot be applied to ice since the molecules are not in positions of cubic symmetry.

The slowness of molecular rotation in solids means that the orientational contribution to the dielectric constant disappears at a low frequency compared to the contribution from the relative displacement of electrons and nuclei discussed in the previous section. Debye suggested the use of the **relaxation equation**

$$\frac{dP}{dt} = \frac{P_E - P}{\tau} \tag{9.27}$$

to calculate the frequency-dependent susceptibility $\chi(\omega)$ of liquids containing polar molecules. P_E is the equilibrium value of the polarization for a static field with a value equal to the instantaneous applied field; thus $P_E = \chi(0)\varepsilon_0 E$ where $\chi(0)$ is the static susceptibility. The physical meaning of Eq. (9.27) is that the polarization is always relaxing towards its instantaneous equilibrium value at a rate determined by the **relaxation time** τ; we expect τ to represent the time scale for molecular rotation. We will assume that Eq. (9.27) can also be used to describe polar molecules in solids.

Taking $E = E_0 e^{i\omega t}$ we find that Eq. (9.27) has the solution

$$P = \frac{\varepsilon_0 \chi(0)}{1 + i\omega\tau} E_0 e^{i\omega t}$$

from which $\chi(\omega)$ and hence $\varepsilon(\omega)$ can be deduced using Eqs. (9.1) and (9.2). Thus

$$\varepsilon(\omega) = 1 + \frac{\chi(0)}{1 + i\omega\tau}; \tag{9.28}$$

the frequency dependence of the real and imaginary parts of ε is shown in Fig. 9.5. Note the qualitative difference to the curves in Fig. 9.3; since there is no

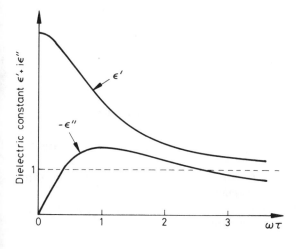

Fig. 9.5 Variation of the real and imaginary parts of the dielectric constant with frequency as predicted by Eq. (9.28)

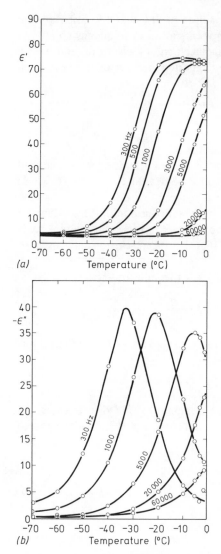

Fig. 9.6 Real and imaginary parts, ε' and ε'', of the dielectric constant of ice as a function of temperature at different frequencies (in Hz). (Reprinted with permission from C. P. Smyth and C. S. Hitchcock, *J. Am. Chem. Soc.* **54**, 4631 (1932). Copyright (1932) American Chemical Society)

equilibrium position for the rotating dipoles with associated restoring forces there is no resonant-like behaviour in Fig. 9.5.

Fig. 9.6 shows the dielectric constant of ice as a function of temperature at various frequencies. At each frequency the decrease in ε' with decreasing temperature and the maximum in $-\varepsilon''$ indicate the region in which $\omega\tau \approx 1$; the results can therefore be used to investigate the temperature dependence of the relaxation time (problem 9.4). Note that the relevant frequencies are low (10^2–10^4 Hz), indicating that molecular rotation in solids is very slow.

★9.1.4 Dielectric constant and lattice vibrations of ionic crystals

An electric field applied to an ionic crystal causes a displacement of the positive and negative ions in opposite directions; the resulting polarization contributes to the dielectric constant. Since the optical lattice vibration modes of long wavelength in ionic crystals have positive and negative ions moving in antiphase (section 2.3.2) an electric field of appropriate frequency can couple strongly to these modes. To investigate this coupling and its effect on the dielectric properties of ionic crystals we must insert the electric field in the equations of motion of the ions. We will find that there is a contribution to the electric field from the ionic displacements, which enables us to take into account the long range of the Coulomb force between the ions that we ignored in section 2.3.2; the existence of this contribution modifies strongly the frequency of the *longitudinal* optical modes.

Coupling of the electric field to *transverse* optical lattice modes also affects the propagation of electromagnetic waves in an ionic crystal. The dispersion relations for light and lattice vibrations in the absence of coupling are shown in Fig. 9.7. Since the velocity of light exceeds that of sound by a factor of order 10^5, the intersection of the dispersion relation for light and that for the optical modes of the lattice, indicated by point A on Fig. 9.7, occurs at a very small wavenumber (wavelength long compared to the atomic spacing). The coincidence of both frequency and wavelength at this point provides essentially a resonance condition, which produces profound changes to the dispersion curves of the electromagnetic wave and the *transverse* optical modes.

Consider first the one-dimensional ionic crystal of Fig. 2.6. We now modify the equations of motion, Eqs. (2.15) and (2.16), of the ions to allow for the

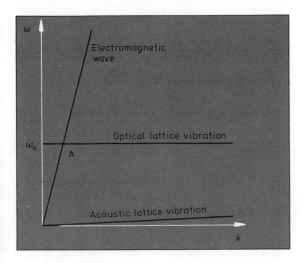

Fig. 9.7 Dispersion relations near $k = 0$ for uncoupled atomic vibrations and light in a diatomic crystal

presence of an electric field **E**. Since we are interested in *optical* modes in the *long-wavelength* limit we will approximate the terms representing the interaction of an ion with its neighbours by assuming that both nearest neighbours have the same displacement. The main effect of this approximation is to eliminate the k dependence of the optical mode frequency in Fig. 9.7. Thus, taking u_+ and u_- to represent the local displacements of positive and negative ions, Eqs. (2.15) and (2.16) become

$$M\ddot{u}_+ = 2K(u_- - u_+) + eE \qquad (9.29)$$

$$m\ddot{u}_- = 2K(u_+ - u_-) - eE.$$

We have taken the positive and negative ions to have masses M and m respectively and the final terms on the right-hand sides are the forces exerted by the electric field on ions of charge $\pm e$. If we subtract $M/(M + m)$ times the second of Eqs. (9.29) from $m/(M + m)$ times the first we obtain a single equation for the relative displacement $w = u_+ - u_-$:

$$\ddot{\mathbf{w}} = -\omega_0^2 \mathbf{w} + \frac{e}{M^*} \mathbf{E}. \qquad (9.30)$$

Here $M^* = mM/(M + m)$ is the reduced mass and $\omega_0 = (2K/M^*)^{1/2}$ is the optical mode frequency at $k = 0$ as calculated in section 2.3.2 using the assumption of short-range forces only. We have written **w** and **E** as vectors because we wish to apply Eq. (9.30) to a three-dimensional crystal in order to discuss both longitudinal and transverse modes. In our generalization to three dimensions we assume that the interaction of an ion with its nearest neighbours can be described by an isotropic spring constant K; this means that the transverse and longitudinal optical modes are degenerate in the absence of the field **E**.

By setting $\ddot{\mathbf{w}} = 0$, Eq. (9.30) gives the relative displacement of the ions due to a static electric field as

$$\mathbf{w} = \frac{e}{M^*\omega_0^2} \mathbf{E} = \frac{e}{2K} \mathbf{E}. \qquad (9.31)$$

Displacement of the ions causes an electric polarization

$$\mathbf{P} = Ne\mathbf{u}_+ - Ne\mathbf{u}_- + [\varepsilon(\infty) - 1]\varepsilon_0 \mathbf{E} = Ne\mathbf{w} + [\varepsilon(\infty) - 1]\varepsilon_0 \mathbf{E}, \qquad (9.32)$$

which on inserting Eq. (9.31) becomes

$$\mathbf{P} = \left(\frac{Ne^2}{2K} + [\varepsilon(\infty) - 1]\varepsilon_0\right) \mathbf{E}. \qquad (9.33)$$

N is the density of the positive ions in the crystal and the term $[\varepsilon(\infty) - 1]\varepsilon_0 \mathbf{E}$ is the contribution to **P** from the polarization of the electron clouds on each ion by the electric field as discussed in section 9.1.2; the notation $\varepsilon(\infty)$ indicates that this contribution determines the dielectric constant $\varepsilon(\omega)$ at frequencies large compared to those of the lattice vibrations. Eq. (9.33) allows us to identify the contribution of ionic displacement to the static dielectric constant $\varepsilon(0)$,

$$\varepsilon(0) = \frac{Ne^2}{2K\varepsilon_0} + \varepsilon(\infty) = \frac{Ne^2}{M^*\omega_0^2\varepsilon_0} + \varepsilon(\infty), \qquad (9.34)$$

and thus to write Eq. (9.30) as

$$\ddot{\mathbf{w}} = -\omega_0^2\mathbf{w} + \omega_0^2\frac{\varepsilon_0[\varepsilon(0)-\varepsilon(\infty)]}{Ne}\mathbf{E}. \qquad (9.35)$$

Note that in this section we are ignoring the distinction between the local electrical field $\mathbf{E}_L$ at the positions of the ions and the macroscopic electric field $\mathbf{E}$. Equations of the form of (9.29) are still valid when $\mathbf{E}$ is the macroscopic field rather than the local field, but the effective values of the spring constant K and the charge e on the ions are modified slightly. The important results of our calculation, Eqs. (9.40), (9.42), (9.43) and (9.44), are unaffected by this approximation.

The electromagnetic fields inside the crystal are related by Maxwell's equations:

$$\text{curl } \mathbf{E} = -\dot{\mathbf{B}}, \qquad \text{curl}\left(\frac{\mathbf{B}}{\mu_0}\right) = \dot{\mathbf{D}} = \varepsilon_0\dot{\mathbf{E}} + \dot{\mathbf{P}}. \qquad (9.36)$$

Taking the curl of the first of Eqs. (9.36) and the time derivative of the second enables us to eliminate $\mathbf{B}$ between them. Then, using Eq. (9.32) to eliminate $\mathbf{P}$, we obtain

$$\text{curl curl } \mathbf{E} = -\frac{\varepsilon(\infty)}{c^2}\ddot{\mathbf{E}} - \frac{Ne}{c^2\varepsilon_0}\ddot{\mathbf{w}}, \qquad (9.37)$$

where $c = (\varepsilon_0\mu_0)^{-1/2}$ is the velocity of light in free space. Eqs. (9.35) and (9.37) are coupled equations for the electric field $\mathbf{E}$ and the relative ionic displacement $\mathbf{w}$; the coupling is through the second terms on the right-hand side of each equation, and in the absence of these terms the equations lead to the dispersion relations shown on Fig. 9.7 for the optical mode and light wave respectively. The nature of the coupling is such that transverse and longitudinal lattice vibrations couple only to transverse and longitudinal electric field waves respectively.

We consider first the case of longitudinal lattice vibrations and the associated longitudinal electric field wave. Since curl $\mathbf{E} = 0$ for a longitudinal electric field wave, there is no magnetic field associated with such a wave and no corresponding electromagnetic wave in free space. From the second of Eqs. (9.36) we deduce $\mathbf{D} = \varepsilon_0\mathbf{E} + \mathbf{P} = 0$ for a longitudinal wave and hence Eq. (9.32) becomes

$$\mathbf{E} = -\frac{Ne}{\varepsilon_0\varepsilon(\infty)}\mathbf{w}. \qquad (9.38)$$

Using this to eliminate $\mathbf{E}$ from Eq. (9.35) gives the following equation for $\mathbf{w}$:

$$\ddot{\mathbf{w}} = -\omega_0^2\left(1 + \frac{\varepsilon(0) - \varepsilon(\infty)}{\varepsilon(0)}\right)\mathbf{w} = -\frac{\varepsilon(0)\omega_0^2\mathbf{w}}{\varepsilon(\infty)}. \tag{9.39}$$

This is a simple harmonic oscillator equation predicting an angular frequency

$$\omega_L = \left(\frac{\varepsilon(0)}{\varepsilon(\infty)}\right)^{1/2}\omega_0 \tag{9.40}$$

for the long-wavelength longitudinal optical modes. The effect of the long range of the Coulomb field is therefore to change the frequency by a factor $[\varepsilon(0)/\varepsilon(\infty)]^{1/2}$. The second term in the brackets in Eq. (9.39) provides the modification of the longitudinal optic mode frequency due to the electric field of distant ions. To see this we identify the electric field associated with this term from Eq. (9.38); taking the divergence of Eq. (9.38) gives

$$\mathrm{div}\,[\varepsilon_0\varepsilon(\infty)\mathbf{E}] = -\mathrm{div}\,(N e \mathbf{w}),$$

which is Gauss' law for the electric field due to a charge density $-\mathrm{div}\,(N e \mathbf{w})$. Recalling that $-\mathrm{div}\,\mathbf{P}$ is the charge density equivalent to a polarization $\mathbf{P}$, we deduce from Eq. (9.32) that $-\mathrm{div}\,(N e \mathbf{w})$ is the charge density associated with the ionic displacements.

We now consider *transverse* waves, for which $\mathrm{div}\,\mathbf{E} = 0$, so that curl curl $\mathbf{E} = -\nabla^2\mathbf{E}$. For a transverse wave in the x direction Eq. (9.37) becomes

$$\frac{\partial^2\mathbf{E}}{\partial x^2} = \frac{1}{c^2}\left(\varepsilon(\infty)\ddot{\mathbf{E}} + \frac{Ne}{\varepsilon_0}\ddot{\mathbf{w}}\right). \tag{9.41}$$

If we now try a solution

$$\mathbf{E} = \mathbf{E}_0\,\mathrm{e}^{\mathrm{i}(kx-\omega t)}, \qquad \mathbf{w} = \mathbf{w}_0\,\mathrm{e}^{\mathrm{i}(kx-\omega t)},$$

Eq. (9.41) gives

$$k^2 c^2\mathbf{E}_0 = \omega^2\left[\varepsilon(\infty)\mathbf{E}_0 + \frac{Ne}{\varepsilon_0}\mathbf{w}_0\right]$$

and Eq. (9.35) becomes

$$(\omega_0^2 - \omega^2)\mathbf{w}_0 = \varepsilon_0[\varepsilon(0) - \varepsilon(\infty)]\omega_0^2\mathbf{E}_0/Ne.$$

Solving these equations we find the dispersion relation

$$k^2 = \frac{\omega^2}{c^2}\left(\frac{\varepsilon(0)\omega_0^2 - \varepsilon(\infty)\omega^2}{\omega_0^2 - \omega^2}\right) \tag{9.42}$$

for transverse waves, which is sketched in Fig. 9.8, together with that of the longitudinal mode we have already discussed. The general form of the dispersion

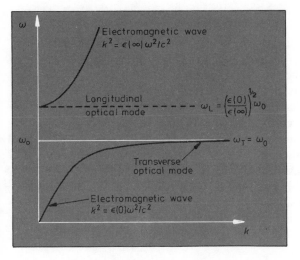

Fig. 9.8   The effect of the coupling of lattice vibrations to an electric field on the dispersion relations in an ionic crystal (----) longitudinal mode; (——) transverse modes

curve is easily obtained by noting the following limits:

$$\omega \ll \omega_0, \qquad\qquad k^2 = \varepsilon(0)\omega^2/c^2;$$
$$\omega \gg \omega_0, \qquad\qquad k^2 = \varepsilon(\infty)\omega^2/c^2;$$
$$\omega \to \omega_{\mathrm{L}} = [\varepsilon(0)/\varepsilon(\infty)]^{1/2}\omega_0, \qquad k^2 \to 0;$$
$$\omega \to \omega_0, \qquad\qquad k^2 \to \infty.$$

The coupling of the transverse optical modes to the electric field has produced a large change in the dispersion relations at long wavelength; along each of the two transverse branches of the dispersion relation the motions change smoothly from being purely lattice vibrations at one end to purely electromagnetic waves at the other. In the changeover region the quantum associated with the motions must be seen as a linear combination of a photon and a transverse optical phonon; this entity is called a **polariton**. Polaritons have been detected by **Raman effect** experiments in which the momentum and energy change of photons inelastically scattered by the crystal are measured and used to deduce the energy–momentum relation of the excitations within it.

The dispersion relation can also be observed through its effect on the refractive index and dielectric constant, the predicted values of which, from Eqs. (9.10), (9.11) and (9.42), are

$$\varepsilon(\omega) = n^2(\omega) = \frac{c^2}{(\omega/k)^2} = \left(\frac{\varepsilon(0)\omega_0^2 - \varepsilon(\infty)\omega^2}{\omega_0^2 - \omega^2}\right). \tag{9.43}$$

Note that Eq. (9.43) has the appropriate values, $\varepsilon(0)$ and $\varepsilon(\infty)$, as $\omega \to 0$ and $\omega \to \infty$ respectively. Note also that $\varepsilon(\omega) \to \infty$ as $\omega \to \omega_0$ and $\varepsilon(\omega) \to 0$ as ω tends to the longitudinal optic mode frequency ω_{L} of Eq. (9.40). Fig. 9.9(a) shows the measured refractive index of LiF as a function of wavelength. The predicted divergence of $n(\omega)$ at ω_0 is clearly seen; the rapidly varying refractive index near this frequency is made use of in constructing prisms for infrared spectroscopy. The variation of $n(\omega)$ is the typical anomalous dispersion associated with resonance absorption at ω_0 (compare Fig. 9.3). Our model contains no damping and thus does not predict the absorption, which is a quantum process in which a photon is absorbed and its energy transferred to the lattice vibrations. It is apparent from Fig. 9.9(a) that the dielectric constant for $\omega \ll \omega_0$ exceeds that for $\omega \gg \omega_0$; the difference according to the theory we have outlined is the contribution of the ionic motion.

The further rise in refractive index at the left of Fig. 9.9(a) heralds the approach of absorption associated with the electronic energy levels of an individual ion as discussed in section 9.1.2. Comparison of Figs. 9.9(a) and 9.3 shows a strong qualitative similarity between the frequency dependence for the contributions to the dielectric constant from relative ionic motion (an *inter*-ionic effect) and relative motion of electrons and nuclei inside each ion (an *intra*-ionic

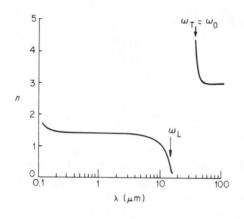

(a) Experimental refractive index of LiF as a function of wavelength (plotted logarithmically) (Reproduced with permission from H. W. Hohls, *Ann. Physik* **29**, 433 (1937))

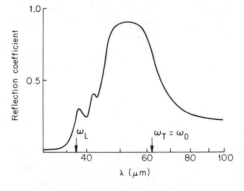

(b) Reflection coefficient of a thick NaCl crystal (Reproduced with permission from A. Mitsuishi *et al.*, *J. Opt. Soc. Am.*, **52**, 14 (1962))

Fig. 9.9

effect). The frequency ranges are however very different; the former contribution to ε' normally disappears at frequencies in the infrared region of the spectrum whereas the latter does so in the ultraviolet or long-wavelength x-ray region.

No values of the refractive index are shown in Fig. 9.9(a) for the frequency range $\omega_0 < \omega < \omega_L$, where ω_L is the frequency of the *longitudinal* optical mode given by Eq. (9.40). In this region we have $k^2 < 0$ so that k and hence $n(\omega)$ are imaginary; this means that no wave can propagate. Instead there is an **evanescent wave** that decays exponentially as it goes into the crystal and there is *total external reflection* of radiation incident on the crystal from outside. Fig. 9.9(b) shows that measurements of reflection coefficient for NaCl bear this out qualitatively, but the reflection is never total. This is another consequence of our neglect of damping; there is no transmission through a crystal more than a few wavelengths thick in this frequency range, but some of the energy is absorbed rather than reflected. Reflection from ionic crystals is a useful way of selecting a band of wavelengths of infrared radiation; the selected radiation is known as Reststrahlen (residual waves).

Provided the wavenumber is not so small as to take us into the polariton region of the dispersion curve of Fig. 9.8 we see that the frequencies of long-wavelength longitudinal and transverse optical vibrations are related by

$$\frac{\omega_L}{\omega_T} = \frac{\omega_L}{\omega_0} = \left(\frac{\varepsilon(0)}{\varepsilon(\infty)}\right)^{1/2} \tag{9.44}$$

This is the **Lyddane–Sachs–Teller** (LST) relationship, which is rather more general than our derivation would suggest. It shows that the relationship between ω_L and ω_T is determined only by the macroscopic dielectric constant.

In section 2.7.1 we explained that anharmonic effects cause the frequencies of lattice vibration modes to vary with temperature and that a displacive phase transition can occur when the frequency of a transverse optic mode of infinite wavelength goes to zero. The Lyddane–Sachs–Teller relation predicts that the disappearance of the frequency of the soft mode will be accompanied by a divergence of the static dielectric constant, $\varepsilon(0) \to \infty$. Essentially the anharmonicity causes the *effective* spring constant K in Eq. (9.29) to vanish at the critical temperature; the results $\varepsilon(0) \to \infty$ and $\omega_T \to 0$ follow directly from this.

9.2   PYROELECTRIC MATERIALS

Pyroelectric materials possess an electric polarization in the absence of an applied electric field. Each primitive unit cell has a dipole moment associated with it. The prefix *pyro-* comes from the Greek word for fire and is used in this context because a macroscopic dipole moment is only seen when the material is heated; the dipole moment is normally neutralized by ions and electrons that collect on the surface of the sample. Heating removes some of the surface charges and also causes the volume polarization to change so that the masking is no longer complete. The macroscopic dipole moment can also be small because of the existence of domains in the material in each of which the polarization is in a different direction.

Ferroelectric materials are pyroelectric only below a certain temperature, which is characteristic of the material. They are so called because they are the electrical analogues of ferromagnets, not because they are associated with iron. Ferroelectric transitions can arise in a number of different ways. One possibility occurs in molecular solids containing polar molecules that are able to rotate at high temperatures (see section 9.1.3); if rotation ceases on cooling the dipole moments can align to give a spontaneous electric polarization. Ferroelectricity can also arise as a result of a displacive phase transition (section 2.7.1); the ferroelectricity of barium titanate ($BaTiO_3$), which occurs through this mechanism, is illustrated in Fig. 9.10. Above 120°C $BaTiO_3$ has the cubic perovskite structure and is paraelectric; below this temperature the structure distorts and becomes tetragonal ($a = b \neq c$, $\alpha = \beta = \gamma = 90°$) due to a relative displacement

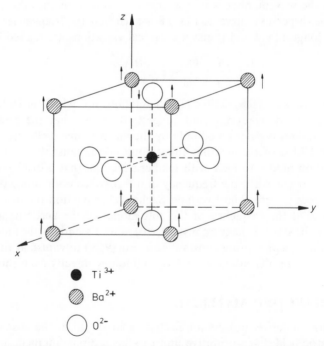

$\bullet$ Ti^{3+}

$\oslash$ Ba^{2+}

$\bigcirc$ O^{2-}

Fig. 9.10 The cubic perovskite structure of $BaTiO_3$ above 120°C. Below 120°C a displacive phase transition occurs to a structure of tetragonal symmetry ($a = b \neq c$, $\alpha = \beta = \gamma = 90°$). The arrows indicate the *relative* sizes and directions of the displacement of the ions along the z direction; the oxygen ions at $z = c/2$ are taken as fixed. The crystal also shrinks slightly in the xy plane. The displacement of positive and negative ions in opposite directions is responsible for the spontaneous polarization

of the positive (Ba^{2+} and Ti^{4+}) and negative (O^{2-}) ions along the $[1\,0\,0]$ direction† as indicated in the figure. As the displacement grows so does the associated spontaneous electric polarization along the $[0\,0\,1]$ direction. At lower temperatures, $BaTiO_3$ undergoes further displacive phase transitions to structures of even lower symmetry; below $-5°C$ it becomes orthorhombic with polarization along the $[0\,1\,1]$ direction of the original cube and below $-90°C$ it becomes rhombohedral with polarization along $[1\,1\,1]$. In the following section we describe a simple model that reproduces many of the observed features of ferroelectric phase transitions.

★9.2.1 The Landau model

Close to the transition the spontaneous polarization P is often small and the basic assumption of the Landau model is that under these circumstances it is possible to expand the energy density of the material as a power series in P‡

$$F = F_0 + \alpha P^2 + \beta P^4 + \cdots \qquad (9.45)$$

where F_0, α and β are temperature-dependent coefficients. Only even powers of P need be included if the crystal has a centre of symmetry in the unpolarized state ($P = 0$); the energy density must then be invariant under the change $P \to -P$. We will assume that terms of order higher than P^4 in the expansion can be neglected (but see problem 9.6). This will only be the case if $\beta > 0$, otherwise the minimum of F will correspond to $P \to \pm \infty$, which is unphysical.

The form of F as a function of P for positive and negative α is shown in Fig. 9.11. For positive α, F takes its minimum value for $P = 0$, whereas for negative α, the minimum occurs at a finite value of P. Thus, if we assume that the value of P in the equilibrium state is that which minimizes F, then *a transition to a ferroelectric state will occur at the temperature T_C at which α decreases through zero*. We are interested in the behaviour close to this temperature and will therefore assume that the temperature dependences of α and β can be adequately represented by the lowest-order terms in their Taylor series expansion about $T = T_C$:

$$\alpha = a(T - T_C)$$
$$\beta = b \qquad (9.46)$$

where a and b are *constants*. The term independent of $T - T_C$ in α is absent because α vanishes at $T = T_C$. According to our discussion above b is positive

---

† Because of the cubic symmetry the spontaneous displacement is equally likely to occur along the $[1\,0\,0]$ or $[0\,1\,0]$ directions. In practice a single crystal often becomes divided into a number of domains, each containing one of the possible polarization directions. The presence of an electric field on cooling through the transition can produce a single-domain ferroelectric.

‡ See Landau and Lifshitz, *Statistical Physics*, Chapter 14 for a discussion of this expansion, which can also be applied to other types of phase transition.

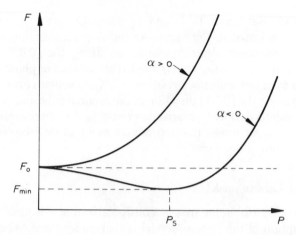

Fig. 9.11   Effect of the change of sign of α on the Landau energy density

and, if the ferroelectric state is to be the stable one below T_C, then a is positive also.

The spontaneous polarization P_S is obtained by explicit minimization of F. From Eq. (9.45),

$$\partial F/\partial P = 2\alpha P + 4\beta P^3 = 0,$$
$$P_S = (-\alpha/2\beta)^{1/2} = [a(T_C - T)/2b]^{1/2}, \tag{9.47}$$

where we have used Eqs. (9.46). Our calculation thus predicts a spontaneous polarization growing as $(T_C - T)^{1/2}$ as T falls below T_C. Since P_S increases continuously from zero a second-order phase transition is predicted. Note however that the ferroelectric phase transition at 120°C in $BaTiO_3$ is first order (see problem 9.6).

The Landau model can be used to predict the behaviour of the susceptibility above T_C if we add to Eq. (9.45) the interaction energy density $-EP$ of the polarization with an applied field E. Above the transition, in weak fields, P is small and the quartic term in Eq. (9.45) can be ignored. Minimization of the energy density then gives

$$\partial F/\partial P = 2\alpha P - E = 0$$
$$P = \frac{E}{2\alpha} = \frac{E}{2a(T - T_C)},$$

from which the susceptibility can be identified by comparison with Eq. (9.1) as

$$\chi = \frac{1}{2a\varepsilon_0(T - T_C)}. \tag{9.48}$$

This is the electrical equivalent of the Curie–Weiss law in magnetism (Eq. (8.9)).

From Eq. (9.48) we see that the static susceptibility and hence dielectric constant diverge as $T \to T_c$. The divergence of the dielectric constant at a displacive phase transition has already been noted in section 9.1.4. We saw there that the transition was associated with the vanishing of the frequency of a transverse optic mode in the lattice vibration spectrum, and this implied a divergent dielectric constant through the Lyddane–Sachs–Teller equation, Eq. (9.44). The mode associated with the displacive phase transformation in the perovskite structure can be directly observed by inelastic neutron scattering measurements of the lattice vibration frequencies. Fig. 9.12 shows such measurements for $SrTiO_3$ as a function of temperature.

The above model of ferroelectricity, like the Weiss model of ferromagnetism, is a mean field theory. The polarization is assumed to be constant and equal to the value that minimizes the energy density. Fluctuations of the free energy about this value are ignored. However, such fluctuations are large near a second-order phase transition and modify the temperature dependences in Eqs. (9.47) and (9.48) just as in the corresponding equations in ferromagnetism (see section 8.3.2).

9.3 PIEZOELECTRICITY

In **piezoelectric** materials the polarization is changed by applying a stress as well as by changing the electric field; for small stresses the polarization depends linearly on the stress. Conversely, in these materials the application of an electric field causes a strain. All ferroelectric materials are piezoelectric but the converse

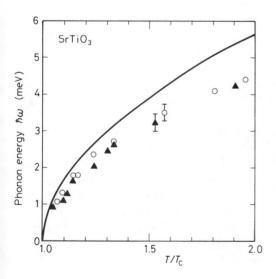

Fig. 9.12 The circles and triangles show inelastic neutron scattering measurements of the temperature dependence of the frequency of the soft mode associated with the displacive phase transition in $SrTiO_3$. T_C is the temperature at which the transition occurs. The full curve is theoretical. (Reproduced with permission from J. Feder and E. Pytte, *Phys. Rev.* B **1**, 4803 (1970))

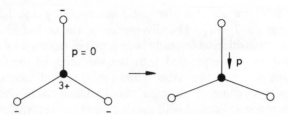

Fig. 9.13   Illustration of the way in which a molecule can obtain a dipole moment as a result of uniaxial strain

is not true; quartz is the best known example of a material that is piezoelectric but not ferroelectric. Fig. 9.13 illustrates the way in which a uniaxial stress can produce a dipole moment in a molecule with a symmetry that implies zero dipole moment in the absence of stress. It is essential that the molecule should not possess a centre of symmetry and this is also a restriction on the crystal structures that can exhibit piezoelectricity.

Piezoelectric materials are widely used as transducers for the interconversion of electrical and mechanical energy. Important applications include the generation and detection of ultrasonic waves, the electrical control of small displacements (in optical systems and in the scanning tunnelling microscope, for example) and surface acoustic wave (SAW) devices.

The behaviour of piezoelectrics is described by equations giving the strain e and polarization P produced by simultaneously applied stress Z and electric field E. These are of the form

$$P = -dZ + \varepsilon_0 \chi E$$

$$e = -sZ + dE \qquad (9.49)$$

where d, χ and s are the piezoelectric constant, susceptibility and elastic compliance of the material. Note that the polarization produced by a stress is described by the same coefficient d as the strain produced by an electric field. The strength and direction of the polarization induced by a stress depend on the type and direction of the stress; the coefficient d is therefore tensorial in nature. For example, three components, d_{13}, d_{33} and d_{15}, are required to specify completely the piezoelectric properties of ferroelectric tetragonal $BaTiO_3$; their values at 25°C are quoted in Fig. 9.14 where the polarization associated with some specific stress geometries is indicated. For an explanation of the notation see *Crystal Symmetry and Physical Properties* by S. Bhagavantam, Academic Press, London (1966). Values of d for non-ferroelectric piezoelectrics such as quartz are typically smaller by a factor of 100.

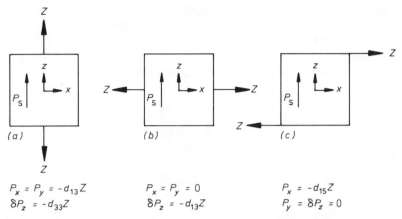

$$P_x = P_y = -d_{13}Z$$
$$\delta P_z = -d_{33}Z$$

$$P_x = P_y = 0$$
$$\delta P_z = -d_{13}Z$$

$$P_x = -d_{15}Z$$
$$P_y = \delta P_z = 0$$

Fig. 9.14 Changes in the electric polarization of tetragonal ferroelectric $BaTiO_3$ due to: (a) uniaxial stress along z; (b) uniaxial stress along x; (c) shearing stress in the xz plane. The stress Z is the applied force per unit area in each case. The values of the coefficients at 25°C in 10^{-10} m V^{-1} are $d_{13} = -0.35$, $d_{33} = 0.86$ and $d_{15} = 3.92$

PROBLEMS 9

9.1 Show that if $n(\omega)$ is defined by Eq. (9.10) and $\varepsilon(\omega)$ by Eqs. (9.1) and (9.2), then $n(\omega) = [\varepsilon(\omega)]^{1/2}$. Show that the energy of a wave travelling through a solid decays by a factor e in a distance

$$c/\{\omega[2(\varepsilon'^2 + \varepsilon''^2)^{1/2} - 2\varepsilon']^{1/2}\}.$$

9.2 Show that the group and phase velocities, v_g and v_p, for short-wavelength x-rays in solids satisfy $v_g v_p = c^2$.

9.3 Compare $\hbar\omega_0$, where ω_0 is given by Eq. (9.18), with the Bohr theory result for the binding energy of an electron to a nucleus of charge Ze.

9.4 Use the data of Fig. 9.6 to estimate the relaxation time, as a function of temperature, for the rotation of the molecules in ice. Show that the temperature dependence is consistent with $\tau = \tau_0 \exp(T_0/T)$ and find values for τ_0 and T_0. What is the physical basis for a temperature dependence of this kind?

9.5 NaCl has unit cell side 5.6 Å, static dielectric constant 5.89 and Young's modulus in the $[1\,0\,0]$ direction 5×10^{10} N m^{-2}. Estimate the frequency range over which electromagnetic radiation is strongly reflected by a NaCl crystal and compare your answers with the experimental data of Fig. 9.9(b).

9.6 The ferroelectric transition at 120°C in $BaTiO_3$ is first order; P_s jumps discontinuously at T_C to a finite value. Show that such behaviour can be predicted by a Landau model in which terms up to sixth order in P are included in the free energy density:

$$F = F_0 + \alpha P^2 + \beta P^4 + \gamma P^6.$$

By drawing graphs, as in Fig. 9.11, deduce appropriate signs for the coefficients. Evaluate P_s at $T = T_C$.

My own beliefs are that the road to a scientific discovery is seldom direct, and it does not necessarily require great expertise. In fact, I am convinced that often a newcomer to a field has a great advantage because he is ignorant and does not know all the reasons why a particular experiment should not be attempted.—*Ivan Giaever (discoverer of tunnelling between superconductors), Nobel prize address, 1973*

CHAPTER

Superconductivity

10.1 INTRODUCTION

Superconductivity was discovered by H. Kamerlingh Onnes in 1911, three years after his first liquefaction of helium. The availability of this liquid enabled him to investigate the electrical resistance of metals at low temperatues. He chose mercury for study since it could be readily purified by distillation and there was speculation at that time that the resistance of very pure metals might tend to zero at $T = 0$. As can be seen from Fig. 10.1. the observed behaviour was much more dramatic than this; an abrupt transition to a state of apparently zero resistance occurs at a temperature of about 4.2 K. Onnes described the new state as the superconducting state, and it was quickly established that there was no essential connection with high purity; adding substantial amounts of impurity often has little effect on the superconducting transition, although the resistance of the normal state (section 3.3.2) is increased considerably.

Subsequently many metals and alloys have been shown to become superconducting.† The superconducting transition can be very sharp, with a width of less than 10^{-3} K in well annealed single crystals of a metal such as tin. The element with the highest transition temperature, $T_c = 9.2$ K, is niobium (Nb). The search

---

† Among common metallic elements that *do not* become superconducting at temperatures currently accessible are copper, silver, gold, the alkali metals and magnetically ordered metals such as iron, nickel and cobalt.

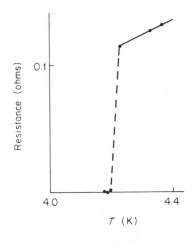

Fig. 10.1   Superconducting transition of mercury. (After H. Kamerlingh Onnes, *Leiden Commun.* **124c** (1911))

for materials with higher transition temperatures led to the investigation of alloys and compounds. In 1972 Nb_3Ge was found to have a T_c of 23 K. For the next 14 years this remained the record T_c and many researchers were misled, with some theoretical justification, into believing that it would not be possible to find materials with significantly higher transition temperatures. In 1986 there was a dramatic breakthrough when Bednorz and Muller found that $La_{2-x}Ba_xCuO_4$ had a T_c of about 35 K for $x \approx 0.15$. This discovery was followed by a frenetic search for other materials. In 1987 $YBa_2Cu_3O_{7-\delta}$ ($\delta \approx 0.1$) was found to have a T_c of 92 K and in 1988 $Bi_2Sr_{3-x}Ca_xCu_2O_{8+\delta}$ ($x \leqslant 1$) raised T_c to 110 K. At the time at which this book was written $Tl_2Ba_2Ca_2Cu_3O_{10}$, also discovered in 1988, has the highest known T_c of 125 K. These new **high-temperature superconductors** are discussed further in section 10.6.

No one has succeeded in measuring a finite resistance to small currents in the superconducting state. The most sensitive method for detecting a small resistance is to look for the decay of a current around a closed superconducting loop. If the resistance of the loop is R and the self-inductance L then the current should decay with time constant $\tau = L/R$. Failure to observe the decay of a **persistent current** has enabled an upper limit of about 10^{-26} Ω m to be put on the resistivity of superconductors as compared to a value of order 10^{-8} Ω m for copper at room temperature (problem 10.1).

10.2   MAGNETIC PROPERTIES OF SUPERCONDUCTORS

10.2.1   Type I superconductors

Superconductors divide into two classes according to their behaviour in a magnetic field. In this section we describe the simpler behaviour of **type I**

superconductors and in section 10.2.3 that of **type II superconductors**. All pure samples of superconducting elements, except Nb, exhibit type I behaviour and their superconductivity is destroyed by a modest applied magnetic field B_c, known as the **critical field**. B_c is shown as a function of temperature for mercury in Fig. 10.2. To a good approximation the temperature dependence of B_c is

$$B_c(T) = B_c(0)\left[1 - \left(\frac{T}{T_c}\right)^2\right].$$ (10.1)

It follows from the existence of a critical field that there will be a critical current for flow along a wire, which occurs when the field due to the current equals B_c; this is known as the **Silsbee hypothesis**.

In 1933, Meissner and Ochsenfeld investigated the variation in space of the magnetic field in the neighbourhood of a superconductor and discovered that the field distribution was consistent with the field inside the superconductor being zero. This exclusion of the magnetic flux from the superconductor is known as the **Meissner effect** and is due to electric currents, known as **screening** currents, flowing on the surface of the superconductor in such a way as to generate a field equal and opposite to the applied field. The expulsion of the flux when the field is reduced below B_c at constant temperature is illustrated in Fig. 10.3 for a sample in the form of a long cylinder; expulsion also occurs if the sample is cooled into the superconducting state in a steady applied field. For many purposes we can take account of the Meissner effect by regarding the superconductor as a magnetic material in which the screening currents are replaced by an equivalent magnetization; since we require $\mathbf{B} = \mu_0(\mathbf{H} + \mathbf{M}) = 0$ we must have

$$\mathbf{M} = -\mathbf{H}.$$ (10.2)

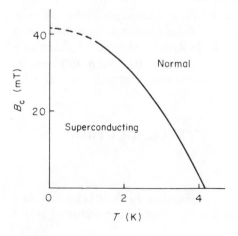

Fig. 10.2   Critical field curve of mercury

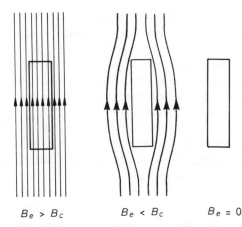

$B_e > B_c$ $B_e < B_c$ $B_e = 0$

Fig. 10.3   Expulsion of flux by a long superconducting cylinder when the field is reduced below B_c. In equilibrium there is no trapped flux

Comparison of Eq. (10.2) with Eq. (7.1) shows that a type I superconductor behaves as though it has a magnetic susceptibility $\chi = -1$ and is consequently often referred to as a **perfect diamagnet**. Fig. 10.4 illustrates how closely a well annealed long cylinder of lead conforms to the behaviour predicted by Eq. (10.2).

Non-annealed specimens often show an incomplete Meissner effect; magnetic flux is trapped within the material in metastable regions which remain the normal state when the field is reduced through B_c. Flux trapping offers a partial explanation of the 22 year delay between the first observation of superconductivity and the discovery of the Meissner effect. It was not realized that the trapped flux was only a manifestation of non-equilibrium behaviour; instead it was regarded as an inevitable consequence of the infinite conductivity of the

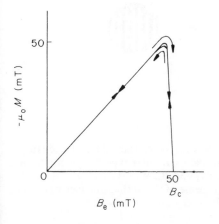

Fig. 10.4   Almost reversible magnetization curve of a well annealed rod of pure superconducting lead. (Reproduced with permission from J. P. Livingston, *Phys. Rev.* **129**, 1943 (1963))

superconducting state because of the following argument. Infinite conductivity implies vanishing of the electric field inside a superconductor and hence through Faraday's law, curl $\mathbf{E} = -\dot{\mathbf{B}}$, it indicates a time-independent magnetic field. This was erroneously interpreted as implying that any magnetic field within a sample would be trapped by a transition to the superconducting state. The discovery of the Meissner effect showed that the zero flux state was the true equilibrium state of a long cylindrical sample at all fields below B_c.

For other shapes of sample the complete exclusion of flux, even in well annealed specimens, does not occur at all fields less than B_c. To explain this, consider the spherical sample in Fig. 10.5. Because the flux is expelled from the interior of the sphere the field at the equator exceeds the applied field. Thus, when the applied field reaches the value $\frac{2}{3}B_c$, the field at the equator becomes B_c and the sphere can no longer remain in the Meissner state. It cannot make a transition to the normal state because this would reduce the field everywhere to $\frac{2}{3}B_c$, a value at which the normal state is not stable. For applied fields between $\frac{2}{3}B_c$ and B_c the sphere is in the **intermediate state** in which it consists of alternating *macroscopic* normal and superconducting regions, shown schematically in Fig. 10.5(b); the field is B_c in the normal regions and zero in the superconducting regions. The intermediate state of a type I superconductor should not be confused with the mixed state of a type II superconductor (sections 10.2.3 and 10.5.3).

The existence of the critical field B_c is a consequence of the Meissner effect. The energy stored in the field ($B^2/2\mu_0$ per unit volume) is greater for the Meissner state than for the normal state in which the field penetrates the

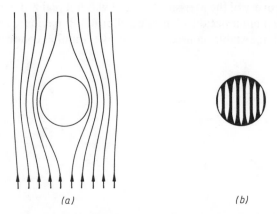

(a)   (b)

Fig. 10.5   (a) Superconducting sphere in the Meissner state. The field at the equator is 50% higher than the applied field. (b) Intermediate state of a type I superconducting sphere, appropriate to applied fields between $\frac{2}{3}B_c$ and B_c. With increasing field in this range the shaded normal regions grow at the expense of the unshaded superconducting regions. For simplicity the field lines are not shown

material uniformly† (we can usually ignore the weak magnetism of the normal state). Eventually, with increasing magnetic field, the increased magnetic field energy equals the energy difference between the normal and superconducting states and it becomes advantageous for the material to make a transition to the normal state. To quantify this argument we must do some simple thermodynamics

10.2.2.   Thermodynamics of the superconducting transition

The field B_c at which the normal (N) and superconducting (S) states are in equilibrium is indicated by the equality of their Gibbs free energies. We take the magnetic work term to be $-\mathbf{M}.\,d\mathbf{B}_e$, where $\mathbf{B}_e$ is the applied field, and the Gibbs free energy per unit volume is then‡

$$G = U - TS$$

where U and S are the internal energy and entropy per unit volume. To see this we calculate

$$dG = dU - T\,dS - S\,dT = T\,dS - \mathbf{M}.\,d\mathbf{B}_e - T\,dS - S\,dT$$
$$= -\mathbf{M}.\,d\mathbf{B}_e - S\,dT. \tag{10.3}$$

Thus G is the thermodynamic function that is minimized in thermal equilibrium at fixed temperature and applied field.

Consider a long cylinder of superconductor parallel to the applied field. Eq. (10.3) can be integrated at constant temperature to deduce the effect of an applied field on the free energy G_S of a superconductor,

$$G_S(B_e, T) = G_S(0, T) - \int_0^{B_e} \mathbf{M}.\,d\mathbf{B}_e. \tag{10.4}$$

For a long cylinder we show in appendix B that $\mathbf{B}_e = \mu_0 \mathbf{H}$, where $\mathbf{H}$ is the field inside the superconductor. Inserting $\mathbf{M} = -\mathbf{H}$ (Eq. (10.2)) for a superconductor in its Meissner state, we obtain

$$G_S(B_e, T) = G_S(0, T) + \int_0^{B_e} \frac{B_e}{\mu_0}\,dB_e = G_S(0, T) + \frac{B_e^2}{2\mu_0}, \tag{10.5}$$

---

† The magnetic energy *inside* the material is smaller for the Meissner state because $B = 0$ there, but the increased energy outside more than compensates.

‡ See Mandl[2] for a discussion of magnetic work. It is more common in superconductivity, although not in magnetism, to take the work term to be $+\mathbf{B}_e.\,d\mathbf{M}$, appropriate to an internal energy, $U' = U + \mathbf{M}.\mathbf{B}_e$, which includes the energy of interaction, $+\mathbf{M}.\mathbf{B}_e$, of the specimen with the sources of the external field. In this approach G is written $G = U' - TS - \mathbf{M}.\mathbf{B}_e$. By not including a term PV in G we are ignoring the effect of changes in pressure and volume on the superconducting transition.

where the final term represents the additional magnetic energy associated with the exclusion of the magnetic field, as discussed at the end of the previous section. If we ignore the weak magnetism of the normal state then the Gibbs free energy G_N of this state is field-independent,

$$G_N(B_e, T) = G_N(0, T).$$

Equating the Gibbs free energies at the critical field then gives

$$G_N(0, T) - G_S(0, T) = \frac{B_c^2}{2\mu_0} \tag{10.6}$$

so that the critical field is directly related to the difference in free energies between the normal and superconducting states in zero field; for this reason B_c is often referred to as the thermodynamic critical field. The positive value of $G_N - G_S$ explains why the superconducting state is more stable than the normal state in zero field; this quantity is the **condensation energy** of the superconducting state.

Experimentally it is found approximately that $B_c \propto T_c$ with a constant of proportionality of order 0.01 T K^{-1}; thus, from Eq. (10.6), the condensation energy is of order $40T_c^2$ J m^{-3}. This energy difference corresponds to a fraction $k_B T_c/\varepsilon_F$ of the conduction electrons having their energy reduced by an amount $k_B T_c$ as result of the transition to superconductivity, and is therefore smaller by a factor $(k_B T_c/\varepsilon_F)^2 \sim 10^{-7}$ than the total kinetic energy of the electrons.

Two important exact results for type I superconductors can be obtained from Eq. (10.6). Using $S = -(\partial G/\partial T)$ (from Eq. (10.3)) we find that the difference in entropy density between the two states in zero field is

$$\Delta S = S_S - S_N = \frac{1}{2\mu_0} \frac{dB_c^2}{dT} = \frac{B_c}{\mu_0} \frac{dB_c}{dT}, \tag{10.7}$$

and using $C = T\partial S/\partial T$, the difference in heat capacity per unit volume in zero field is

$$\Delta C = C_S - C_N = \frac{T}{2\mu_0} \frac{d^2B_c^2}{dT^2} = \frac{T}{\mu_0} \left[B_c \frac{d^2B_c}{dT^2} + \left(\frac{dB_c}{dT} \right)^2 \right]. \tag{10.8}$$

Using Eqs. (10.7) and (10.8) in conjunction with a critical field curve of the form shown in Fig. 10.2 enables us to make some important qualitative deductions:

(1) The entropy difference ΔS vanishes at T_c since $B_c = 0$ there, but the heat capacity difference ΔC is finite since $dB_c/dT > 0$; the discontinuity of the specific heat at T_c is clearly seen in Fig. 10.6, which shows the measured heat capacity of aluminium. The superconducting transition in zero applied field is therefore a second-order phase transition.

(2) ΔS and ΔC vanish at $T = 0$ in accordance with the third law of thermodynamics.

(3) For $0 < T < T_c$, dB_c/dT is negative so that $\Delta S < 0$; the superconducting

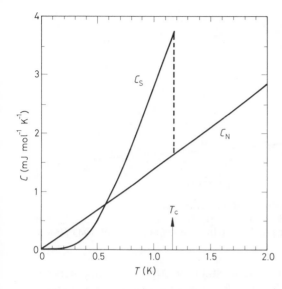

Fig. 10.6   Heat capacity of normal and superconducting aluminium. The normal state measurements were made by applying a field greater than B_c. The high Debye temperature of aluminium means that the lattice contribution to the heat capacity is small in this temperature range and the electronic contribution is dominant. Note the discontinuity in heat capacity at T_c and the exponential fall-off in C_S at low temperature. (After N. E. Phillips, *Phys. Rev.* **114**, 676 (1959))

state is therefore more ordered than the normal state. We discuss the nature of the ordering in section 10.4.

(4) Because ΔS is finite for $0 < T < T_c$ there is a latent heat at the superconducting transition in a finite field given by $T\Delta S$; strictly Eq. (10.7) gives the entropy difference in zero field but ΔS is field-independent. S_N is field-independent because normal state magnetism is very weak and S_S is field-independent because the Meissner effect means that the interior of the superconductor remains in zero field up to B_c.

Deductions (1), (2) and (3) remain valid for a type II superconductor but they must be proved by a different method, as type II superconductors do not exhibit a sharp transition from the Meissner state to the normal state at a field B_c.

10.2.3   Type II superconductors

Although Nb is the only element that is type II in its pure state, other elements generally become type II when the electron mean free path is reduced sufficiently by alloying. Fig. 10.7 compares the magnetization curves of thin cylinders of pure Pb and a Pb–In alloy; with increasing field the alloy shows a complete Meissner effect only up to a field B_{c1} that is less than the thermodynamic critical field of pure Pb. Above B_{c1} there is partial flux penetration into the alloy although it retains the ability, characteristic of the superconducting state, to support dissipationless current flow.† The transition to the normal state and

† The critical current I_c is *not* related to a critical field by the Silsbee hypothesis but depends on the metallurgical state: the more inhomogeneous the material, the higher I_c (see section 10.5.3 for an explanation).

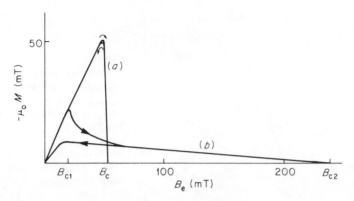

Fig. 10.7   Almost reversible magnetization curves for well annealed long rods of: (a) pure lead (as in Fig. 10.4); (b) lead made type II by alloying with 8.23% indium. (Reproduced with permission from J. P Livingston, *Phys. Rev.* **129**, 1943 (1963))

complete flux penetration occur at the substantially higher field B_{c2}. Between its **lower** and **upper critical fields**, B_{c1} and B_{c2}, the alloy is in the **mixed state**, the nature of which will be explained in section 10.5.3.

According to Eq. (10.4) the increase in Gibbs free energy associated with the exclusion of magnetic flux by a superconductor is equal to the area, $\int(-\mathbf{M}).d\mathbf{B}_e$, under the magnetization curve. This equation is strictly applicable only to equilibrium states, characterized by reversible magnetization curves. We can however apply it approximately to the almost reversible curve for the Pb–In alloy in Fig. 10.7. The area under this curve is almost equal to that under the curve for pure Pb; we deduce that alloying produces no substantial change in the condensation energy. The partial flux penetration in the mixed state allows the superconductivity to persist to significantly higher fields in the alloy. With increasing indium concentration B_{c1} decreases and B_{c2} increases.

In extreme type II superconductors B_{c1} is so small and the flux penetration in the mixed state so nearly complete that very large values of B_{c2} are reached before the area under the magnetization curve becomes equal to the condensation energy. For large B_{c2} our thermodynamic approach must be generalized to allow for the decrease in Gibbs free energy associated with the weak paramagnetism of the normal state. This decrease puts a fundamental upper limit on B_{c2}, the **Clogston limit**, of about $1.8T_c$ tesla (see problem 10.4). Fig. 10.8 shows values of B_{c2} as a function of temperature for some extreme type II superconductors.

10.3   THE LONDON EQUATION

We saw in section 7.3.2 that the rigidity of an electron wavefunction against perturbation by a magnetic field led directly to diamagnetism with the field being excluded from the region occupied by the electron except for a surface

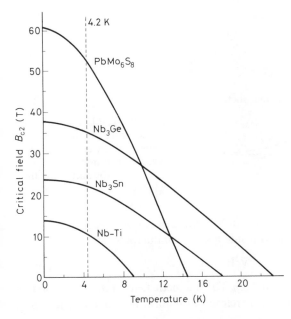

Fig. 10.8   Upper critical field, B_{c2}, as a function of temperature for some extreme type II superconductors. (Reproduced with permission from R. Chevrel, *Superconductor Materials Science: Fabrication and Applications*, ed. S. Foner and B. Schwartz, Plenum, New York (1980))

layer about 100 Å thick. The weakness of diamagnetic effects in most materials is then explained because ordinary *atomic* wavefunctions are small in extent compared to this screening distance. The perfect diamagnetism of superconductors implies that there are wavefunctions extending *throughout* the material that are not readily perturbed by a magnetic field.

This possibility was first suggested by Fritz London, who proposed that the currents responsible for the screening should be described by

$$\operatorname{curl} \mathbf{j} = -\frac{n_S e^2}{m} \mathbf{B}. \tag{10.9}$$

Eq. (10.9) is known as the **London equation** and it is the curl of

$$\mathbf{j} = -\frac{n_S e^2}{m} \mathbf{A}, \tag{10.10}$$

which is Eq. (7.32) with the replacement $n \to n_S$ to allow for the possibility that only a fraction n_S/n of the electrons (the **superconducting fraction**) have a rigid wavefunction. Eq. (10.9) (or its equivalent, Eq. (10.10)) can be regarded as a replacement for Ohm's law, $\mathbf{j} = \sigma \mathbf{E}$, as a description of the behaviour of the superconducting electrons.

To see that Eq. (10.9) explains the Meissner effect we apply it to a plane boundary ($x = 0$) separating a superconductor ($x > 0$) from a vacuum ($x < 0$), when there is a magnetic field $\mathbf{B} = B_e \hat{\mathbf{z}}$ parallel to the boundary in the vacuum

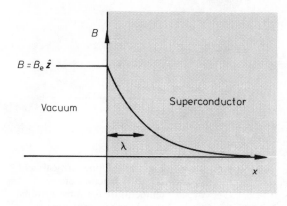

Fig. 10.9   The London
equation predicts the
exponential decay of a
magnetic field into a
superconductor occupying the
region $x > 0$

(Fig. 10.9). By combining Eq. (10.9) with Maxwell's equations† curl $\mathbf{B} = \mu_0\mathbf{j}$ and div $\mathbf{B} = 0$, we find that the field $\mathbf{B}$ inside a superconductor satisfies

$$\lambda^2 \nabla^2 \mathbf{B} = \mathbf{B} \tag{10.11}$$

where $\lambda^2 = m/\mu_0 n_s e^2$ as in section 7.3.2. The magnetic field in the superconductor in the geometry of Fig. 10.9 is therefore of the form $\mathbf{B} = B(x)\hat{\mathbf{z}}$, where $B(x)$ satisfies

$$\lambda^2 \frac{\mathrm{d}^2 B}{\mathrm{d}x^2} = B.$$

The solution of this equation is

$$B(x) = a\,\mathrm{e}^{-x/\lambda} + b\,\mathrm{e}^{+x/\lambda}, \tag{10.12}$$

where a and b are constants of integration. The second term, which has B increasing exponentially with x at large distances from the boundary, is unphysical and we reject it. To satisfy $B = B_e$ at $x = 0$ then requires $a = B_e$ so that

$$B(x) = B_e\,\mathrm{e}^{-x/\lambda}. \tag{10.13}$$

The magnetic field thus decays exponentially with distance into the superconductor with a characteristic length scale λ, kown as the **penetration depth**, as shown in Fig. 10.9. To estimate λ at $T = 0$ we suppose that all the electrons are superconducting at this temperature and set $n_s = n = 10^{29}$ m^{-3}, a typical conduction electron concentration in a superconducting metal, to obtain

$$\lambda = \lambda_L(0) = \left(\frac{m}{\mu_0 n e^2}\right)^{1/2} \approx 170\ \text{Å} \tag{10.14}$$

---

† It is important to note that in this section we take the screening currents explicitly into account rather than replacing them by their equivalent magnetization. In this approach, which is more appropriate when investigating the behaviour of superconductors at a microscopic level, we put $\mathbf{M} = 0$ and $\mathbf{B} = \mu_0\mathbf{H}$.

where the notation $\lambda_L(0)$ indicates that this is the penetration depth as predicted by the London equation at $T = 0$. The small size of λ means that the magnetic flux is effectively excluded from the interior of macroscopic samples of superconductors and the Meissner effect is explained. Note that in the geometry of Fig., 10.9 the screening currents flow in the y direction and also decay exponentially with characteristic depth λ from the surface of the superconductor.

At higher temperatures we expect n_S to decrease and λ to increase. This is seen to be the case in Fig. 10.10, which shows the measured temperature dependence of λ for tin. The temperature dependence is often well described by

$$\lambda = \frac{\lambda(0)}{[1 - (T/T_c)^4]^{1/2}}$$

where $\lambda(0)$ is the value of λ at $T = 0$; λ thus diverges as $T \to T_c$ and $n_S \to 0$.

The measured value of $\lambda(0)$ is often greater than $\lambda_L(0)$. This does not signify a fundamental defect of the London theory; the discrepancy can be explained by modifying Eq. (10.10) slightly so that the current density $\mathbf{j}$ at a point $\mathbf{r}$ does not depend only on the vector potential $\mathbf{A}$ at $\mathbf{r}$ but on the average of $\mathbf{A}$ taken over all points in the neighbourhood of $\mathbf{r}$. This modification converts the **local** current–field relation of Eq. (10.10) into a **non-local** relation. A similar change has to be made to Ohm's law in normal metals when the electric field varies rapidly on the length scale of the electron mean free path l. Such a situation occurs in pure normal metals at high frequencies and low temperature where the electromagnetic skin depth (which gives the length scale for variation of $\mathbf{E}$) is normally shorter than l, and the current density at a point $\mathbf{r}$ then depends on the average of the electric field over a region of size $\sim l$ surrounding $\mathbf{r}$; the necessary generalization of Ohm's law is a non-local relation between $\mathbf{j}$ and $\mathbf{E}$. Pippard exploited the analogy with the normal state to propose that the penetration depth of pure superconductors could be explained if there was a non-local

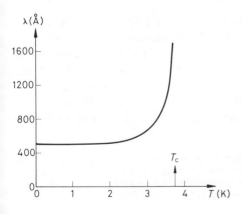

Fig. 10.10 Superconducting penetration depth λ in tin. The value at $T = 0$ is 510 Å, which has to be compared with the London prediction $\lambda_L(0) = 340$ Å

relation between **j** and **A**† in which the vector potential was averaged over a distance ξ, where

$$\xi \approx \frac{\hbar v_F}{k_B T_c} \qquad (\xi \approx 10^4 \text{ Å for } v_F = 10^6 \text{ m s}^{-1} \text{ and } T_c = 10 \text{ K}). \quad (10.15)$$

We show later (section 10.4) that a characteristic distance of this form arises naturally in the theory of superconductivity. In impure superconducting metals where l is less than ξ, the mean free path takes over from ξ in determining the range of the non-locality and λ then depends on l. Pippard's proposals were later confirmed in essence by the microscopic theory of superconductivity.

It is interesting to investigate the extent to which the London equation can be deduced from an assumption of infinite conductivity; to do so we allow the electron scattering time τ in Eq. (3.23) to become infinite. The resulting acceleration equation

$$m_e \, dv/dt = -eE$$

together with $\mathbf{j} = -n_s e\mathbf{v}$ and Faraday's law, curl $\mathbf{E} = -\dot{\mathbf{B}}$, lead to the time derivative of Eq. (10.9). To obtain the London equation by integration of this equation involves making an assumption about the integration constant, which is equivalent to assuming the Meissner effect. This again demonstrates that superconductivity is more than just infinite conductivity.

10.4   THE THEORY OF SUPERCONDUCTIVITY

We will give only a brief qualitative description of the very successful microscopic theory of superconductivity that was proposed by Bardeen, Cooper and Schrieffer (BCS) in 1957; the quantitative details of the BCS theory involve techniques that are too advanced for this book.‡

10.4.1   The energy gap and electron pairing

We saw in the previous section that the temperature dependence of the penetration depth suggests a density n_s of superconducting electrons that increased from zero at T_c to the full electron density at $T = 0$. The behaviour is consistent with the existence of an energy gap Δ separating the states of the superconducting electrons from those of the 'normal' electrons. There is a considerable amount of evidence for such a gap; both experiment and theory indicate that Δ is temperature-dependent, vanishing at T_c and attaining its

---

† Note that high frequencies are not required to cause **A** to vary rapidly in space in a superconductor; even for a dc field, the Meissner effect ensures that **A** varies on a length scale λ.

‡ For an excellent series of review articles on superconductivity the reader is recommended to consult *Superconductivity*, ed. R. D. Parks, Marcel Dekker, New York (1969).

maximum value $\Delta(0)$ at $T = 0$. At low temperatures ($T \ll T_c$) one would expect that the number of excited (normal) electrons would fall off as $\exp\left[-\Delta(0)/k_B T\right]$ and that this temperature dependence would be reflected in the electronic contribution to the heat capacity; this is indeed found to be the case (see Fig. 10.6) and $\Delta(0)$ turns out to be of order $k_B T_c$.

Direct evidence for an energy gap is provided by measurements of the absorption of electromagnetic waves. At low temperature ($T \ll T_c$) the absorption is vanishingly small at low frequencies but increases sharply when the photon energy is sufficient to excite electrons across the energy gap. The frequency for the onset of absorption is given by

$$h\nu = 2\Delta(0). \tag{10.16}$$

The factor 2 arises because absorption of a photon creates two excited electrons. A natural explanation for this is provided by the BCS theory of superconductivity, according to which the superconducting electrons are bound together in pairs, known as **Cooper pairs**. Thus 2Δ is the binding energy of a Cooper pair so that Eq. (10.16) describes the breaking of a pair by absorption of a photon. The attractive interaction that binds the pairs is due to the lattice vibrations (section 10.4.3).

The wavefunction of all the pairs has to be identical to maximize the energy reduction due to the attractive interaction; the binding energy of a Cooper pair is largest when all the pairs are in the same state. Superconductivity is therefore said to be a **cooperative phenomenon**; ferromagnetism is another example of a cooperative phenomenon since the better the alignment of the spins, the greater the molecular field that is responsible for the alignment (see section 8.3.1). The existence of a common wavefunction for the Cooper pairs provides the rigidity of the wavefunction that leads to the Meissner effect and it is also responsible for the infinite conductivity (section 10.4.5).

At $T = 0$ all the electrons are paired but at $T > 0$ some pairs are broken by thermal excitation. Because of the cooperative nature of superconductivity, the binding energy of the remaining pairs falls. The resulting decrease in the measured energy gap can be seen in Fig. 10.11; $\Delta(T)$ falls to zero with infinite slope at $T = T_c$. The sharing of a common wavefunction by the pairs is present at all temperatures below T_c and the resulting order is responsible for the lower entropy of the superconducting state.

The average distance between the electrons for the Cooper pair wavefunction in a pure metal at $T = 0$ is of order

$$\xi_0 = \hbar v_F / \pi \Delta(0); \tag{10.17}$$

ξ_0 is known as the **BCS coherence length** and it plays an important role in the theory of superconductivity. Since $\Delta(0) \approx k_B T_c$ (the BCS theory predicts $\Delta(0) = 1.76 k_B T_c$), it is essentially ξ_0 that determines the range of non-locality (Eq. (10.15)) in the current-field relation of the superconducting electrons in a pure

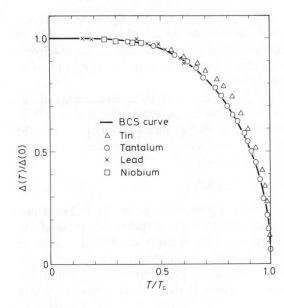

Fig. 10.11   Temperature dependence of the superconducting energy gap. The full curve is the BCS theory prediction. (Reproduced with permission from P. Townsend and J. Sutton, *Phys. Rev.* **128**, 591 (1962))

metal; the current is a flow of Cooper pairs and each Cooper pair responds to the vector potential averaged over its wavefunction.

★10.4.2   The Cooper problem

By solving a simple problem in 1956, Cooper provided the inspiration for the BCS theory. Cooper solved the Schrödinger equation for two interacting electrons in the presence of a Fermi sphere of non-interacting electrons, as shown in Fig. 10.12. This calculation cannot be applied directly to a real metal since it is impossible to turn off the interaction between all but two of the conducting electrons, but it serves to indicate the kind of effect that the interaction might produce. The wavefunction of the two electrons can be

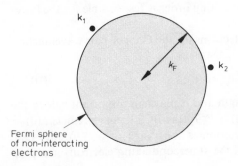

Fig. 10.12   The Cooper problem. Two interacting electrons are restricted to states, k_1 and k_2, outside the Fermi surface by the Fermi sphere of non-interacting electrons

expanded as a linear combination of plane waves (see Eq. (3.3))

$$\psi(\mathbf{r}_1, \mathbf{r}_2) = \sum_{\mathbf{k}_1} \sum_{\mathbf{k}_2} f(\mathbf{k}_1, \mathbf{k}_2)\, e^{i\mathbf{k}_1 \cdot \mathbf{r}_1}\, e^{i\mathbf{k}_2 \cdot \mathbf{r}_2} \qquad (10.18)$$

where the role of the non-interacting electrons is to restrict the summation to plane wave states outside the Fermi sphere ($|\mathbf{k}_1|, |\mathbf{k}_2| > k_F$). Cooper looked for states of this form with an energy less than $2\varepsilon_F$, the energy of two 'normal' electrons at the Fermi surface. Such states would correspond to bound states of the two electrons and their existence would indicate that the normal state, as represented by the Fermi sphere, was unstable against the formation of bound pairs of electrons.

For the lowest energy, the centre of mass of the two electrons is at rest and this is achieved by including only states with equal and opposite momentum, $\mathbf{k}_1 = -\mathbf{k}_2 = \mathbf{k}$, in the expansion of Eq. (10.18), which then simplifies to

$$\psi(\mathbf{r}_1, \mathbf{r}_2) = \sum_{\mathbf{k}} g(\mathbf{k})\, e^{i\mathbf{k} \cdot (\mathbf{r}_1\, \mathbf{r}_2)} \qquad (10.19)$$

where the summation is again restricted to states $\mathbf{k}$ outside the Fermi surface. Cooper found that bound states existed if the interaction between the two electrons was attractive, no matter how weak the attraction; this was surprising in that bound states exist for two particles in a vacuum only if the attractive potential exceeds a threshold value. BCS made the bold extrapolation from Cooper's result that bound Cooper pairs would still result when all the electrons interacted with each other.

★10.4.3   Origin of the attractive interaction

An attractive interaction between electrons seems an unlikely possibility in view of the large repulsive force between two *isolated* electrons. We shall see in Chapter 13 however that the effective Coulomb interaction between two electrons in a metal is much reduced by the presence of the other electrons and the positive ions. Each electron repels other electrons from its neighbourhood and thereby creates a hole in the electron 'fluid' which is of order one atom in size and on average contains a positive charge from the ions equal and opposite to the electronic charge (Fig. 10.13(a)). The net charge in the neighbourhood of the electron is therefore approximately zero and the effective interaction of the electron with another electron outside the screening hole is weak.

The attractive force arises because an electron attracts the positive ions so that, as it moves through the metal, it leaves a wake of enhanced positive charge density behind it (Fig. 10.13(a)). Because ions move more slowly than electrons, the wake persists after the electron moves away and can attract another electron. The attraction is of very short range since the wake is only of the order of an atomic spacing in width, but it is retarded because the electron causing the wake

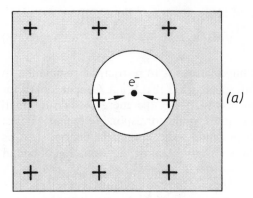

(a) Because of Coulomb repulsion an electron is surrounded by a screening hole in the electron fluid, which on average contains a positive charge equal and opposite to the electronic charge. The electron attracts the positively charged ion cores in its neighbourhood. The enhanced positive charge density persists after the electron moves away and can attract another electron

(b) The attractive interaction between two electrons can be pictured as the exchange of a virtual phonon. The phonon is virtual since an electron cannot undergo a sufficient energy change ($\sim \hbar\omega_D$) at low temperatures ($T \ll \theta_D$) to create a real phonon of short wavelength. A virtual phonon is emitted by one electron and absorbed by another within such a short time ($< 1/\omega_D$) that its 'existence' is allowed by the energy–time uncertainty relation. Momentum is conserved in the individual emission and absorption processes but energy is only conserved overall

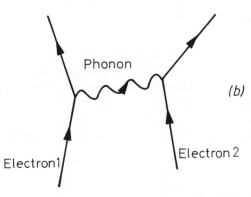

Fig. 10.13

has already moved away. Since ionic motion communicates the interaction between the two electrons, the attraction is said to result from the exchange of virtual phonons (Fig. 10.13(b)). The detailed nature of the interaction is important in determining the transition temperature of the superconductor but the qualitative behaviour of the superconductor below T_c is determined almost entirely by the existence of the Cooper pairs. Indeed, BCS calculated successfully most of the properties of superconductors by replacing the real short-range retarded interaction by a fictitious but simpler instantaneous interaction spread out to a range $\sim v_F/\omega_D$ to allow for the distance moved by an electron during the characteristic time ($\sim 1/\omega_D$) for ionic motion.

★10.4.4   Nature of the superconducting ground state

According to the BCS theory all the electrons are paired at $T = 0$. Since the wavefunctions of all the pairs are identical, superconductivity is often described as arising because of a Bose condensation of Cooper pairs† (see Mandl,[2] p. 292). It is instructive to see how it is possible to write down a wavefunction that corresponds to such a ground state. The common wavefunction of all the pairs can be expanded in plane waves as in Eq. (10.19) except that the restriction to states with $|\mathbf{k}| > k_F$ is removed as there is now no non-interacting Fermi sphere. In most (perhaps all) known superconductors the pair wavefunction is essentially spherically symmetric, $\psi(\mathbf{r}_1, \mathbf{r}_2) = \psi(|\mathbf{r}_1 - \mathbf{r}_2|)$, so that the Cooper pair possess no orbital angular momentum; the spherical symmetry is distorted slightly by the anisotropy of the crystal structure but we will ignore this. The spherical symmetry corresponds to $g(\mathbf{k})$ depending only on the magnitude of $\mathbf{k}$ and the wavefunction is therefore symmetric under interchange of $\mathbf{r}_1$ and $\mathbf{r}_2$. An antisymmetric pair wavefunction $\phi(1, 2)$ can be obtained by combining this space wavefunction with the antisymmetric spin singlet wavefunction. Thus

$$\phi(1, 2) = \psi(|\mathbf{r}_1 - \mathbf{r}_2|)\frac{1}{\sqrt{2}}(\uparrow \downarrow - \downarrow \uparrow). \tag{10.20}$$

The electrons in the Cooper pair therefore have opposite spins. A wavefunction for N electrons which has $N/2$ pairs all in the same state can be written

$$\Psi(1, 2, 3, 4, \ldots, N) = P\{\phi(1, 2)\phi(3, 4) \ldots \phi(N - 1, N)\} \tag{10.21}$$

where P is an operator that makes the product wavefunction in the curly brackets antisymmetric under interchange of any two electrons. We will not discuss how this is done in general but will demonstrate how it works for two pairs by writing down tne wavefunction explicitly for this case:

$$\Psi(1, 2, 3, 4) = P\{\phi(1, 2)\phi(3, 4)\}$$

$$= \frac{1}{\sqrt{3}}[\phi(1, 2)\phi(3, 4) - \phi(1, 3)\phi(2, 4) - \phi(1, 4)\phi(3, 2)].$$

Eq. (10.21) is essentially the ground state wavefunction of the BCS theory.

In all superconductors where it has been possible to elucidate unambiguously the nature of the pairing, the Cooper pairs have been found to have zero orbital angular momentum. However, the nature of the pairing in high-temperature superconductors has not yet been established. Some heavy fermion superconductors‡ may also have finite angular momentum pairing. Liquid ^{3}He undergoes a superfluid transition due to Cooper pairing into a state with $L = 1$ and

---

† Although a tightly bound pair of fermions behaves like a boson, there are dangers in pushing this simple idea too far in the case of Cooper pairs; these are weakly bound and there is a strong overlap of the wavefunctions of neighbouring pairs.

‡ Heavy fermion materials such as UPt_3 and UBe_{13} are so called because they have a very large electronic heat capacity at low temperatures, equivalent to a large heat capacity effective mass for the electrons (section 3.2.3). This seems to arise because of a contribution to the density of states at the Fermi surface from the 5f electrons of the U atoms.

the neutrons in neutron stars are also believed to be in a Cooper paired state of finite angular momentum; these *neutral* Fermi systems cannot however be described as superconductors!

10.4.5 Explanation of infinite conductivity

To give a qualitative explanation of infinite conductivity we must first describe how it is possible to obtain a current-carrying state by giving all the pairs a finite centre-of-mass momentum. A uniform current density corresponds to a pair wavefunction of the form†

$$\phi = e^{i\mathbf{q}\cdot\mathbf{r}}\phi_0 \tag{10.22}$$

where $\mathbf{r} = (\mathbf{r}_1 + \mathbf{r}_2)/2$ is the centre-of-mass position of the two electrons and ϕ_0 is a wavefunction for a pair at rest. Eq. (10.22) corresponds to a centre-of-mass momentum $\hbar\mathbf{q}$ and hence to a velocity $\mathbf{v}$, where

$$\hbar\mathbf{q} = 2m\mathbf{v}.$$

As the charge on a Cooper pair is $-2e$ the resulting current density is

$$\mathbf{j} = -\frac{n_S}{2}2e\frac{\hbar\mathbf{q}}{2m} \tag{10.23}$$

for n_S superconducting electrons per unit volume ($n_S/2$ pairs).

Consider a wire carrying a Cooper pair current of this kind. We must explain why the scattering of electrons by phonons and impurities is ineffective in producing electrical resistance. The process in which a Cooper pair absorbs a phonon of energy of order $2\Delta(T)$ and two normal electrons are created (Fig. 10.14) undoubtedly occurs, as does the inverse process in which two normal electrons combine with the emission of a phonon to form a Cooper pair. Indeed these processes occur with equal rates in order to preserve a dynamic equilibrium between the concentrations of Cooper pairs and normal electrons.

Because the energy is lower when all the Cooper pairs are in the same state, the pairs created by phonon emission always have the wavefunction of Eq. (10.22); *unless their centre-of-mass motion is the same as that of the existing pairs, their binding energy vanishes.* The current is thus unaffected by phonon scattering. Since impurity scattering is elastic, impurities cannot scatter Cooper pairs at all; a change of momentum for a single Cooper pair involves the loss of its binding energy and is therefore an inelastic process. The pair current can only be changed by an influence that affects all the pairs equally such as an electric field.

---

† We assume a uniform current density for simplicity. Note that a spatially uniform current density can only be obtained in practice in a conductor (such as a thin film or fine wire) with one or more dimensions small compared to the penetration depth λ.

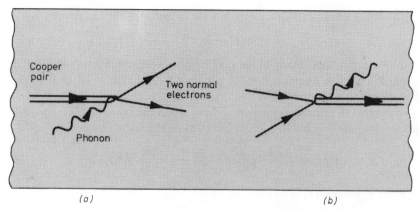

Fig. 10.14   Phonon scattering processes in a wire carrying a supercurrent: (*a*) absorption of a phonon by a Cooper pair of momentum **q** creates two normal electrons; (*b*) two normal electrons combine with the emission of a phonon to form a Cooper pair of momentum **q**

10.5   MACROSCOPIC QUANTUM PHENOMENA

10.5.1   The superconducting order parameter

Since the Cooper pairs share a common wavefunction, the behaviour of the superconducting electrons is *completely* specified by this function; that a function of only two position variables is needed to describe $\sim 10^{29}$ electrons/m^3 is in complete contrast to the situation in a normal metal where the behaviour is only determined by specifying all of the single-particle states occupied. The coherence in the wavefunction associated with macroscopic occupation of the same quantum state by Cooper pairs causes superconductors to exhibit quantum mechanical effects on a macroscopic scale. A similar situation occurs for photons; the macroscopic occupation of a single quantum state leads to a macroscopically observable electric field.

For many purposes the relative motion of the two electrons in the pair can be ignored and the pair regarded as a point particle. Only the dependence of the wavefunction on the centre-of-mass coordinate needs to be considered and this is given by the **order parameter** $\psi(\mathbf{r})$;† thus, for example, we see from Eq. (10.22)

---

† Like the Weiss theory of ferromagnetism (section 8.3) the BCS theory is a mean field theory; the order parameter $\psi(\mathbf{r})$ is the mean field of the theory and is thus analogous to the magnetization of the ferromagnet. The mean field theory of superconductivity is more successful than that of ferromagnetism because fluctuation effects in macroscopic samples of superconductor occur so close to T_c that they are difficult to observe.

that the order parameter describing a state of uniform current density is

$$\psi(\mathbf{r}) = \psi_0 \, e^{i\mathbf{q} \cdot \mathbf{r}} \tag{10.24}$$

where A is a constant. Many of the properties of superconductors follow if $\psi(\mathbf{r})$ is regarded as the wavefunction of a particle of charge $-2e$ and mass $2m$ (appropriate to a Cooper pair).

The current density associated with such a wavefunction is given by making the substitutions $e \to 2e$, $m \to 2m$ in Eq. (C8) of appendix C:

$$\mathbf{j}(\mathbf{r}) = + \frac{i\hbar e}{2m}(\psi^*\nabla\psi - \psi\nabla\psi^*) - \frac{2e^2}{m}\psi^*\psi\mathbf{A}. \tag{10.25}$$

The most general form of $\psi(\mathbf{r})$ is

$$\psi(\mathbf{r}) = |\psi(\mathbf{r})| \, e^{i\theta(\mathbf{r})} \tag{10.26}$$

and inserting this in Eq. (10.25) we find

$$\mathbf{j}(\mathbf{r}) = -(e/m)|\psi(\mathbf{r})|^2(\hbar\nabla\theta + 2e\mathbf{A}). \tag{10.27}$$

This equation will be the starting point for our discussion of macroscopic quantum phenomena, but first we will use it to rederive two of our previous results:

(1) Inserting $\theta(\mathbf{r}) = \mathbf{q} \cdot \mathbf{r}$ (Eq. (10.24)) and $\mathbf{A} = 0$ (see problem 10.8) into Eq. (10.27) gives Eq. (10.23) if the order parameter is normalized so that

$$|\psi(\mathbf{r})|^2 = n_s/2 = \text{Cooper pair density.}$$

(2) Taking the curl of Eq. (10.27) and assuming that the Cooper pair density $|\psi(\mathbf{r})|^2$ is independent of position (i.e. that the wavefunction is rigid) gives the London equation (10.9).†

10.5.2 Flux quantization

Far from the surface of a superconductor in its Meissner state we have $\mathbf{j} = 0$. Eq. (10.27) then becomes

$$\hbar\nabla\theta = -2e\mathbf{A}. \tag{10.28}$$

We integrate this equation around a closed curve C inside the superconductor,

$$\hbar \oint_C \nabla\theta \cdot d\mathbf{l} = \hbar\Delta\theta = -2e \oint_C \mathbf{A} \cdot d\mathbf{l}. \tag{10.29}$$

Since the order parameter $\psi(\mathbf{r})$ behaves like a wavefunction, it must be single-valued and the phase change $\Delta\theta$ around a closed loop must be $\pm 2\pi n$

---

† The price we pay for ignoring the internal structure of the Cooper pair wavefunction is to obtain the local London current–field relation rather than the true non-local relation. For an explanation of the difference between Eqs. (10.10) and (10.27) see problem 10.9.

where n is a positive (or zero) integer. The integral $\oint_C \mathbf{A} \cdot d\mathbf{l}$ may be transformed by Stokes' theorem as in Eq. (7.28) to show that it is equal to the magnetic flux Φ through the curve C. We thus obtain

$$\Phi = \pm \frac{2\pi n h}{2e} = \pm \frac{nh}{2e} = \pm n\Phi_0, \qquad (10.30)$$

which shows that the flux through any closed curve on which $\mathbf{j} = 0$ within a superconductor is quantized in units of the flux quantum $\Phi_0 = h/2e = 2.07 \times 10^{-15} \text{ T m}^2$.

Applying this result to the flux associated with the persistent current flowing around a superconducting ring (Fig. 10.15) we see that the current is also quantized and this sheds new light on its stability. A change in current corresponding to a change in flux through the ring of one quantum involves a change in $\Delta\theta$ of 2π. Such a change can only be achieved if the coherence of the superconducting wavefunction is temporarily destroyed in some way, with the consequent loss of condensation energy of the Cooper pairs. There is thus a large energy barrier against such a change. Because of the energy associated with the current and the trapped flux, a state with a finite persistent current is strictly only metastable, but with an effectively infinite lifetime.

Fig. 10.16 illustrates schematically an experiment that used a superconducting ring to measure the flux quantum. The specimen was in the form of a thin film of tin electroplated onto a fine copper wire a few millimetres long and about 10 μm diameter (remember that copper is an insulator in comparison with superconductors!); the small diameter was used so that one flux quantum corresponded to a reasonable field ($\sim 10 \ \mu$T) within the ring. The sample was placed in a magnetic field of this order and cooled through the transition temperature; the field was then removed and the trapped flux measured by vibrating the sample between two search coils connected in series opposition. The experiment was repeated a number of times and the trapped flux as a function of the initial

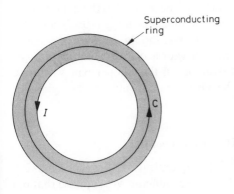

Superconducting ring

I

C

Fig. 10.15   Integration of Eq. (10.28) round the curve C proves that the magnetic flux through the superconducting ring is quantized. The persistent current I that gives rise to the flux flows on the inner surface of the ring

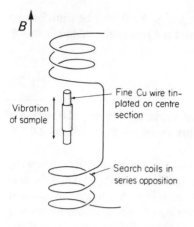

(a) Experimental arrangement for measuring the flux quantum

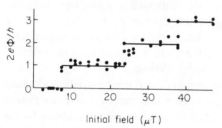

Initial field (μT)

(b) Flux trapped in the ring after it had been cooled through the superconducting transition in a magnetic field, which was then removed. (Reproduced with permission from B. S. Deaver and W. M. Fairbank, *Phys. Rev. Lett.* **7**, 43 (1961))

Fig. 10.16

applied field is shown in Fig. 10.16(b). Quantization in units of $h/2e$ is apparent; the number of quanta is such as to make the trapped field as close as possible to the initial applied field. The higher quanta in Fig. 10.16(b) become less well defined probably because of a flaw in the tin film part way along its length through which one or more flux quanta could pass.

The magnitude of the flux quantum provides very strong evidence of the presence of Cooper pairing in superconductors. The factor 2 in the denominator of $h/2e$ comes from the 2 in the second term in brackets in Eq. (10.27), and thus directly from the charge on a Cooper pair. We should reassure the reader worried about the lack of rigour in our derivation of flux quantization (for example in our neglect of the internal structure of the Cooper pair wavefunction) that a rigorous derivation can be given, based only on the symmetry properties of the order parameter.

10.5.3 Quantized flux lines and type II superconductivity

We next consider the implications of flux quantization through a curve C surrounding a region completely filled by superconductor. We suppose that one

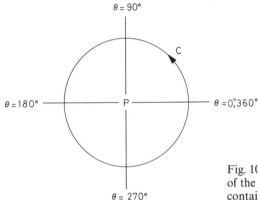

Fig. 10.17   Contours of constant phase of the order parameter for a curve C containing one quantum of magnetic flux

flux quantum passes through C so that the phase θ of the superconducting order parameter changes by 2π in one complete circuit of C. Contours of constant phase could then appear as in Fig. 10.17 and this creates a problem at a point P within C where θ must take on all values between 0 and 2π simultaneously. As this is inconsistent with the requirement of a single-valued order parameter it would appear to rule out the passage of quantized magnetic flux through the interior of a superconductor, thereby implying that the superconductor is in the Meissner state.

There is an alternative possibility. If we allow $|\psi|$ to go to zero at point P, then the order parameter is again single-valued there (the single value is zero); the phase of the order parameter is undefined at a point where $|\psi| = 0$. If we repeat this argument for other sections through the superconductor then we find that $|\psi|$ must vanish along a continuous line and we are thus led to the concept of a **quantized flux line**. The structure of such a line is shown in Fig. 10.18. The density of Cooper pairs $|\psi|^2$ falls to zero on the line (Fig. 10.18(*a*)), which can therefore be pictured as a filament of non-superconducting material. There is a circulating current around the line (Fig. 10.18(*b*)), which generates the magnetic field (Fig. 10.18(*c*)) associated with the quantized flux.

An array of quantized flux lines provides the mechanism for the flux penetration in the mixed state of type II superconductors (section 10.2.3); electron microscopy studies† indicate that the flux lines tend to form a regular triangular lattice. In principle it is possible to have lines containing more than one quantum of flux but they would have a higher energy and only singly quantized lines are found in practice. We see from Fig. 10.18 that there are two length scales associated with a flux line. From section 10.3 we expect the length scale for the current and field variation (Figs. 10.18(*b*) and (*c*)) to be the penetration depth λ. We might expect the length scale ξ for the variation of $|\psi|^2$

---

† See, for example, U. Essman and H. Traüble, *Scientific American*, **224** (March), 74 (1971)).

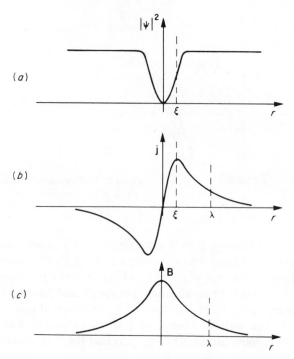

Fig. 10.18   Variation of $|\psi|^2, j$ and B through a quantized flux line

(Fig. 10.18(a)) to be associated with the size of the Cooper pair wavefunction; this indeed turns out to be the case and in a pure superconductor

$$\xi \approx \frac{\xi_0}{(1 - T/T_c)^{1/2}} \tag{10.31}$$

where ξ_0 is the BCS coherence length of Eq. (10.17).

We can now answer qualitatively the question of why some superconductors are type I and others type II by estimating the energy cost of forming a plane boundary between a superconducting and normal region in a type I superconductor as shown in Fig. 10.19; since the superconducting and normal phases are in equilibrium at the applied field B_c the free energies per unit volume of the bulk uniform regions on either side of the boundary are equal. In the boundary region itself however there is a loss of condensation energy over a distance ξ at the boundary, resulting in an *increase* in free energy

$$\Delta G_C \approx (G_N - G_S)\xi \tag{10.32}$$

per unit area of boundary, where $G_N - G_S$ is the condensation energy per unit volume. The presence of the boundary allows the field B_c to penetrate the

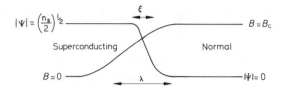

Fig. 10.19   Boundary between a normal and superconducting region in a type I superconductor. The boundary can only be in equilibrium at an applied field B_c. The order parameter decays in a distance of order ζ and the field penetrates a distance of order λ into the superconducting region

superconducting region a distance of order λ, resulting in a *decrease* in free energy

$$\Delta G_B \approx -\frac{1}{2\mu_0} B_c^2 \lambda \tag{10.33}$$

per unit area of boundary. We expect type I behaviour only if the energy associated with the formation of the boundary, $\Delta G_C + \Delta G_B$, is positive. From Eq. (10.6), $(G_N - G_S)$ and $B_c^2/2\mu_0$ are equal, so that the condition for type I behaviour is approximately $\xi > \lambda$. When $\xi < \lambda$ it is energetically favourable for the superconductor in an applied field of order B_c to break up into a mixture of normal and superconducting regions; the energy decrease associated with the penetration of the field into the superconducting regions more than compensates for the loss of condensation energy. The arrangement of normal and superconducting regions with the lowest energy is the lattice of quantized flux lines; if $\xi < \lambda$ type II behaviour is therefore expected. When the mean free path of the electrons is decreased, λ increases and ξ decreases, and this explains the change in behaviour from type I to type II that is produced by alloying in many metals.

The existence of superconductivity up to fields of order 40 T in some type II alloys and compounds (Fig. 10.8) explains the use of these materials in the construction of solenoids for the generation of large magnetic fields. The major problem is to find materials that will carry a large dissipationless current in high fields. To explain the problem we consider a solenoid with its ends connected together to form a continuous superconducting circuit; the field is generated by a persistent current flowing in this circuit. In type II superconductors such a field can unfortunately decay by the passage of quantized flux lines across the windings and out of the coil and this is equivalent to the coil having a finite electrical resistance. Some mechanism is required to prevent the free migration of flux lines. This is usually done by making the material inhomogeneous, either by precipitation or work hardening; regions where the flux line energy is low are thereby produced and these act as **pinning centres** for the flux lines. Such materials are characterized by highly irreversible magnetization curves. High-temperature superconductors have even larger values of B_{c2} than the materials shown in Fig. 10.8 but the problem of flux pinning at liquid nitrogen temperatures has yet to be solved in these materials.

Another important problem in superconducting solenoids is the possibility that a small region may revert to the normal state, which has a high resistivity. The consequent heating rapidly causes the whole magnet to become normal; the energy stored in the magnetic field is dumped in the liquid helium bath with disastrous consequences. In practice the superconducting wire is a composite of superconductor and copper, such that, if a small region does become normal, the copper carries the current with little dissipation, thus preventing rapid growth of the normal region.

10.5.4   Josephson effects

Josephson effects are probably the most striking manifestation of macroscopic quantum phenomena. They occur when two macroscopic superconducting regions are **weakly coupled**. To explain what this means we consider first two isolated samples of a superconductor with spatially constant order parameters $|\psi_1| \exp(\mathrm{i}\theta_1)$ and $|\psi_2| \exp(\mathrm{i}\theta_2)$ as shown in Fig. 10.20(a). If the temperature of both samples is the same then

$$|\psi_1|^2 = |\psi_2|^2 = n_S/2.$$

In the absence of interaction between the two samples however the phases θ_1 and θ_2 will in general be different; all that is required is that the phase should be spatially constant within each region corresponding to the Cooper pairs being at rest. Strongly coupling the two samples by bringing them into contact over a large area causes the phase to equalize, $\theta_1 = \theta_2$, so that all the Cooper pairs can be in the same state; this equality is then very difficult to disturb. If there is weak coupling, the lowest energy state is still one with $\theta_1 = \theta_2$, but it is possible to generate a phase difference between the two regions by passing a small current though the coupling or applying a small voltage across it. Two superconductors, weakly coupled in this sense, are said to form a **Josephson junction**; the coupling between them is decribed as a **weak link**.

There is more than one way of achieving weak coupling but we will restrict our discussion to two superconductors separated by an oxide barrier of a few atoms thickness as shown in Fig. 10.20(b); the coupling arises because electrons can cross the barrier by a quantum mechanical tunnelling process. When the metal is in its normal state the tunnelling current through the barrier is proportional to the voltage across it; such behaviour is described as **ohmic** and a typical junction resistance is 1 Ω.

Below T_c it is possible for Cooper pairs to tunnel through the oxide barrier; a net flow can take place in the absence of an applied potential difference and this

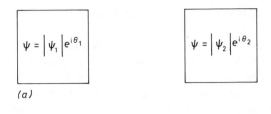

(a)

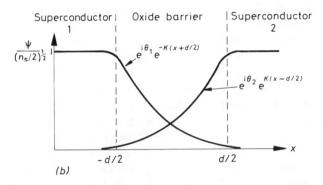

(b)

Fig. 10.20 (a) Two isolated samples of a superconductor. (b) The contributions to the superconducting order parameter within the oxide barrier associated with the tunnelling of Cooper pairs through the barrier

corresponds to a dissipationless supercurrent, which we now calculate. Because of the tunnelling of the pairs, the superconducting order parameter extends throughout the barrier; inside the barrier we regard it as being the sum of the contributions shown in Fig. 10.20(b): one contribution originates in region 1 and decays exponentially within the barrier, and the other originates in region 2 and decays within the barrier. We assume that the contribution from region 1 is very small by the time it reaches region 2 and vice versa so that we can regard the order parameter within the superconducting region as retaining its 'bulk' value up to the edge of the barrier. We therefore write the order parameter within the barrier as

$$\psi = (n_{\mathrm{s}}/2)^{1/2}(e^{i\theta_1 - K(x+d/2)} + e^{i\theta_2 + K(x-d/2)}) \qquad (10.34)$$

where the barrier extends from $x = -d/2$ to $x = d/2$ and K^{-1} is the characteristic length for decay of the order parameter within the barrier. θ_1 and θ_2 are the phases of the order parameter on the two sides of the junction. To calculate the pair current density through the barrier we use Eq. (10.25) with $\mathbf{A} = 0$ and the order parameter of Eq. (10.34) to find

$$j = \frac{ie\hbar n_{\mathrm{s}}}{2m} K \, e^{-Kd}(-e^{i(\theta_1 - \theta_2)} + e^{i(\theta_2 - \theta_1)}) = j_0 \sin \delta, \qquad (10.35)$$

where $\delta = \theta_1 - \theta_2$ is the phase difference between the two sides of the junction and $j_0 = e\hbar n_s K \exp(-Kd)/m$.

If a current is caused to flow through the junction the phase difference adjusts itself so that the **Josephson equation** (10.35) is satisfied. The existence of dissipationless flow of Cooper pairs through a weak link is called the **dc Josephson effect** and experimental confirmation of this effect is seen in Fig. 10.21. The maximum current density in the oxide barrier is j_0, corresponding to a phase difference δ of $\pi/2$. What happens when this current is exceeded depends on the load line of the circuit used to provide the current; the behaviour for the circuit used to obtain the results of Fig. 10.21 is indicated in the figure.

The current observed at finite voltages in Fig. 10.21 coresponds to tunnelling of normal electrons through the oxide barrier. At low temperatures where all the electrons on both sides of the barrier are paired, the tunnelling of a normal electron requires the breaking of a pair. This can only occur if the electron tunnelling through the barrier gains an energy 2Δ from the voltage difference across the barrier. The current is therefore small until the voltage reaches a value $2\Delta(T)/e$. The increase in current when this condition is satisfied is apparent on Fig. 10.21 and normal electron tunnelling provides an accurate and direct method for measuring $\Delta(T)$; the measurements of Fig. 10.11 were obtained by this method. At voltages above $2\Delta(T)/e$ the current–voltage relation reverts to the ohmic behaviour characteristic of the normal state.

What happens to the Cooper pair tunnelling at finite voltages? To answer this question we must consider the time dependence of the superconducting order parameter. Since the order parameter acts as the wavefunction of the Cooper

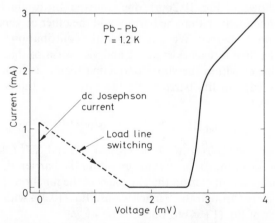

Fig. 10.21   Current–voltage characteristic of a Pb–PbO–Pb tunnel junction at 1.2 K. The current spike at $V = 0$ is the dc Josephson effect. (Reproduced with permission from D. N. Langenburg *et al.*, *Proc. IEEE* **54**, 560 (1966). © 1966 IEEE)

pairs we might expect a dependence of the form

$$\psi \propto e^{-i\mu t/\hbar},$$

where μ is the energy of a pair; the relevant energy turns out to be the chemical potential of the pair. More generally if μ depends on time we have

$$\psi \propto e^{i\theta(t)}$$

where

$$\hbar\, \partial\theta/\partial t = -\mu. \tag{10.36}$$

Ordinarily, because a superconductor cannot sustain a potential difference, μ is uniform and Eq. (10.36) has no observable consequences. It is however possible to maintain a potential difference V between two weakly coupled superconductors, in which case we deduce from Eq. (10.36) that

$$\hbar\frac{\partial\theta_1}{\partial t} - \hbar\frac{\partial\theta_2}{\partial t} = -\mu_1 + \mu_2 = 2eV$$

or

$$\hbar\frac{\partial\delta}{\partial t} = 2eV \tag{10.37}$$

where δ is the phase difference across the junction as in Eq. (10.35).

If V is a constant we can integrate Eq. (10.37) to obtain

$$\delta = \frac{2eV}{\hbar} t + \delta_0 \tag{10.38}$$

where δ_0 is the value of δ at $t = 0$. The phase difference thus increases linearly with time and inserting this in Eq. (10.35) for the current gives

$$j = j_0 \sin\left(\frac{2eV}{\hbar} t + \delta_0\right). \tag{10.39}$$

At finite potential difference therefore there is an ac supercurrent of Cooper pairs at a frequency $v = \omega/2\pi = 2eV/h$ and this is known as the **ac Josephson effect**; because the current is alternating it is not seen in the dc current–voltage characteristic of Fig. 10.21. The ratio of the voltage to the frequency is $h/2e$ = the flux quantum = 2.07×10^{-15} V Hz^{-1}, and the ac Josephson effect provides a very accurate method of measuring this ratio of fundamental constants.

One way of observing the ac Josephson effect is to irradiate the junction with microwaves of frequency ω in addition to applying a dc potential V_0. The total potential difference is then $V_0 + v \cos(\omega t)$ and integrating Eq. (10.37) gives

$$\delta = \frac{2e}{\hbar}\left(V_0 t + \frac{v}{\omega}\sin(\omega t)\right) + \delta_0.$$

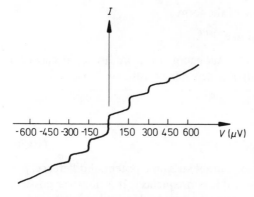

Fig. 10.22 Steps induced on the current–voltage relation of a point-contact Josephson junction by microwave radiation of frequency 72 GHz. The junction is formed by contact between a sharply pointed piece of niobium and a flat niobium surface. (Reproduced with permission from C. C. Grimes and S. Shapiro, *Phys. Rev.* **169**, 397 (1968))

The resulting pair current through the junction, from Eq. (10.35), is

$$j = j_0 \sin \left[\frac{2e}{\hbar} \left(V_0 t + \frac{v}{\omega} \sin (\omega t) \right) + \delta_0 \right]$$

which is a frequency-modulated current containing components at frequencies $(2e/\hbar)V_0 \pm n\omega$, where n is any integer. Thus there is a dc current (zero frequency) if

$$V_0 = \frac{n\hbar\omega}{2e}. \tag{10.40}$$

Fig. 10.22 shows the current–voltage characteristic of a microwave-irradiated Josephson junction, which shows well defined steps at the voltages predicted by Eq. (10.40). It is the steepness of the steps that enables $h/2e$ to be determined with precision† (see W. H. Parker, *et al.*, *Phys. Rev.* **177**, 639 (1969)).

★10.5.5 Quantum interference

Consider a superconductor ring containing two identical Josephson junctions, labelled a and b, as shown in Fig. 10.23(*a*). From Eq. (10.35) the current I flowing through the junctions in parallel is

$$I = Aj_0 \sin \delta_a + Aj_0 \sin \delta_b = 2Aj_0 \cos \left(\frac{\delta_a - \delta_b}{2} \right) \sin \left(\frac{\delta_a + \delta_b}{2} \right) \tag{10.41}$$

where δ_a and δ_b are the phase differences across junctions a and b respectively and A is the area of each junction. We now show that $\delta_a - \delta_b$ is determined by

---

† The position of the steps can be determined with such great precision that the accuracy of the $h/2e$ measurement is limited by the accuracy with which standard voltage sources can be calibrated. This has led to the use of the Josephson junction as a means of establishing a voltage standard by *defining* a value of $h/2e$; the defined value is of course consistent with the best known value of this ratio.

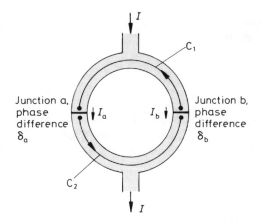

(a) Current flow through two Josephson junctions, a and b, in parallel

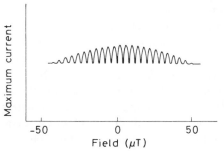

(b) Maximum current passed by the junctions as a function of applied magnetic field. The amplitude of the variation falls off at higher fields because the field causes the phase difference within each junction to vary with position. (Reproduced with permission from R. C. Jaklevic et al., Phys. Rev. **140**, A1628 (1965))

Fig. 10.23

the magnetic flux through the ring. We use an approach similar to that used to prove flux quantization in section 10.5.2. Because the current density vanishes along the curves C_1 and C_2 in the bulk superconducting regions, Eq. (10.28) is valid and integrating this along these curves we find

$$\theta_{a1} - \theta_{b1} = \frac{2e}{\hbar} \int_{C_1} \mathbf{A} \cdot d\mathbf{l} \qquad \text{and} \qquad \theta_{b2} - \theta_{a2} = \frac{2e}{\hbar} \int_{C_2} \mathbf{A} \cdot d\mathbf{l},$$

where θ_{a1}, θ_{b1}, θ_{a2} and θ_{b2} are the phases are at the ends of curves C_1 and C_2 close to the junctions indicated by the subscripts. Adding these equations gives

$$\delta_a - \delta_b = \frac{2e}{\hbar} \oint_C \mathbf{A} \cdot d\mathbf{l} = \frac{2e\Phi}{\hbar}, \tag{10.42}$$

where Φ is the flux through the ring and $\delta_a = \theta_{a1} - \theta_{a2}$ and $\delta_b = \theta_{b1} - \theta_{b2}$ are the phase differences across the two junctions. In order to obtain the integral of $\mathbf{A}$ around a closed curve we have had to include the small contributions from the junctions themselves; this introduces negligible error since **A** varies smoothly through the very narrow junction region.

Inserting Eq. (10.42) into Eq. (10.41) gives

$$I = 2Aj_0 \cos\left(\frac{e}{\hbar}\Phi\right) \sin\left(\frac{\delta_a + \delta_b}{2}\right). \tag{10.43}$$

This resembles the supercurrent (Eq. (10.35)) through a single junction; for the double junction it is $(\delta_a + \delta_b)/2$ that varies to match the current I fed into the ring. The maximum supercurrent that the junction can carry is now

$$I_{max} = 2Aj_0 \left| \cos\left(\frac{e}{\hbar}\Phi\right) \right| \tag{10.44}$$

and thus varies periodically with Φ; the period is just the flux quantum $h/2e$. The measured variation of maximum supercurrent for a double junction can be seen in Fig. 10.23(b). If the two junctions are not identical then the maximum current varies periodically with Φ but does not fall to zero as predicted by Eq. (10.44).

We designate this effect **quantum interference** because of the analogy with the Young's slits interference experiment in optics (Smith and Thomson,[5] p. 127). The difference $\delta_a - \delta_b$ is analogous to the phase difference between the rays of light from the slits to the screen on which the interference pattern is observed; Eq. (10.44) thus corresponds to the cosine dependence of the light amplitude with position on the screen. Experiments with superconducting interferometers have been performed with junctions separated by distances of order 1 cm, impressive evidence that the superconducting order parameter is phase coherent over truly macroscopic distances.

Because of the smallness of the flux quantum, a pair of junctions as in Fig. 10.23(a) embracing an area of 1 cm^2 would change from maximum to minimum critical current for a change of field of only 10^{-11} T. The dc SQUID (superconducting quantum interference device) is an instrument that exploits this geometry to measure very small magnetic fields with great precision.

10.6 HIGH-TEMPERATURE SUPERCONDUCTORS

High-T_c superconductors are all oxides and have many other features in common. We use the widely studied $YBa_2Cu_3O_{7-\delta}$ to illustrate their behaviour; this material has $T_c = 92$ K and is referred to as a 1–2–3 superconductor because of the relative numbers of metal atoms in its chemical formula. The yttrium can be replaced by various other trivalent atoms (e.g. holmium and neodymium) without any significant effect on the superconducting properties. The crystal structure of $YBa_2Cu_3O_{7-\delta}$ is shown in Fig. 10.24(a). It contains planes of Cu and O atoms with the chemical formula CuO_2 as indicated; all superconductors with a T_c greater than 50 K discovered up to 1990 possess CuO_2 (or NiO_2)

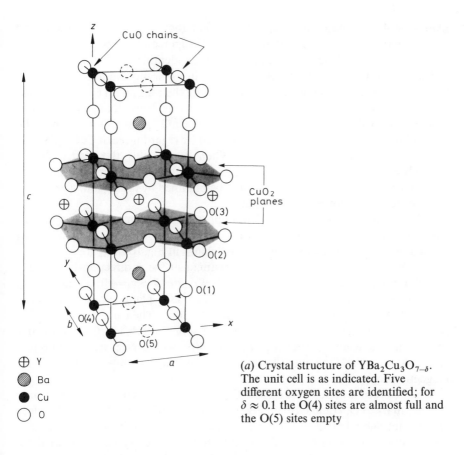

⊕ Y
▨ Ba
● Cu
○ O

(a) Crystal structure of $YBa_2Cu_3O_{7-\delta}$. The unit cell is as indicated. Five different oxygen sites are identified; for $\delta \approx 0.1$ the O(4) sites are almost full and the O(5) sites empty

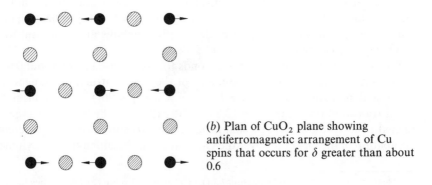

(b) Plan of CuO_2 plane showing antiferromagnetic arrangement of Cu spins that occurs for δ greater than about 0.6

Fig. 10.24

planes similar to these and it is believed that they play a crucial role in the conductivity and superconductivity of high-T_c superconductors. $YBa_2Cu_3O_{7-\delta}$ also has chains of alternate Cu and O atoms as identified in Fig. 10.24(a).

The electrical resistivity of $YBa_2Cu_3O_{7-\delta}$ in its normal state is very aniso-tropic, being much higher for current flow along the z axis than for current flow in the xy plane. This is normally regarded as evidence that conduction is predominantly due to motion of carriers in the CuO_2 planes. Discussion of the behaviour of $YBa_2Cu_3O_{7-\delta}$ is often simplified by regarding each CuO_2 plane as an isolated two-dimensional system. When we do this the reader should remember that this is a gross oversimplification; a complete understanding of the properties of $YBa_2Cu_3O_{7-\delta}$ can only be obtained by taking into account its complicated three-dimensional structure.

It is instructive to consider what happens as the amount of oxygen in $YBa_2Cu_3O_{7-\delta}$ is varied. We start with $YBa_2Cu_3O_6$, corresponding to $\delta = 1$. In this material the oxygen atoms in the CuO chains in Fig. 10.24(a) are completely absent. Since there is then nothing to distinguish the x direction from the y direction the structure is tetragonal ($a = b \neq c$, $\alpha = \beta = \gamma = 90°$). $YBa_2Cu_3O_6$ is an electrical insulator; in this material the CuO_2 plane can be considered approximately as being made of Cu^{2+} and O^{2-} ions. The Cu^{2+} ions have nine 3d electrons in their outer shell with a total spin $S = \frac{1}{2}$.† The Cu spins order antiferromagnetically, as shown in Fig. 10.24(b), with a Néel temperature just above 400 K. The O^{2-} ions have a filled 2p outer shell and therefore no magnetic properties.

When oxygen is added to $YBa_2Cu_3O_6$, the additional atoms initially occupy the sites marked O(4) and O(5) on Fig. 10.24(a) randomly; the structure therefore remains tetragonal. The added oxygen atoms act like acceptor impurities in a semiconductor (section 5.3) and thus add holes to the crystal. Some of these holes are located on the CuO_2 planes but for small concentrations there is no conduction; $YBa_2Cu_3O_{7-\delta}$ remains an antiferromagnetic insulator until δ decreases to about 0.6. This can be understood by assuming that the holes are localized on oxygen atoms in the CuO_2 planes. An oxygen atom with a hole has an outer shell with five 2p electrons and thus spin $S = \frac{1}{2}$. The localization of the holes is an indication that electron–electron interactions are important in the CuO_2 layers (see sections 4.3.2 and 13.5.6).

When the additional oxygen corresponds to a reduction in δ to about 0.6 two important changes occur: the symmetry of the crystal structure changes from tetragonal to orthorhombic ($a \neq b \neq c$, $\alpha = \beta = \gamma = 90°$) and an insulator–metal transition occurs (section 13.5.6). The extent to which these changes are related is not yet known. The change in crystal structure is due to preferential occupation of the O(4) sites over the O(5) sites, thus breaking the x-y symmetry

---

† This follows form Hund's rules (section 7.2.1). Presumably the orbital angular momentum of the ion is quenched by the crystal field.

and leading to the formation of the CuO chains in Fig. 10.24(a). The onset of conduction is due to delocalization of the holes; it is not clear if it is better to view the conduction as arising because of the hopping of a hole from one oxygen atom to another or as being linked with the formation of a two-dimensional energy band associated with the hybridization (section 4.3.4) of 3d states on the Cu atoms with 2p states on the oxygen atoms.

For δ just less than 0.6 the metallic $YBa_2Cu_3O_{7-\delta}$ undergoes a superconducting transition at about 40 K, but as δ decreases further T_c increases and reaches 92 K at $\delta \sim 0.1$. It has proved impossible to prepare $YBa_2Cu_3O_{7-\delta}$ with the structure shown in Fig. 10.24(a) with values of δ any smaller than about 0.1. The superconductivity is interpreted as arising because of Cooper pairing of the holes; flux quantum measurements indicate that pairing of particles with a charge of magnitude e is involved. The interaction responsible for pair formation has not yet been identified; the binding energy of the pairs is rather too high to be explained only by the mechanism involving the lattice vibrations that is responsible for Cooper pairing in 'conventional' superconductors. The antiferromagnetic order of the Cu atoms disappears at the insulator–metal transition but it is possible that the antiferromagnetic interactions between the Cu spins may play a role in the superconducting transition.

Our discussion would suggest that the superconductivity of $YBa_2Cu_3O_{7-\delta}$ is essentially two-dimensional. In practice this means that the properties of $YBa_2Cu_3O_{7-\delta}$ are very anisotropic. The critical current, for example, is much larger for flow of current in the xy plane than for flow along z. The high T_c and small Fermi velocity of $YBa_2Cu_3O_{7-\delta}$ mean that the coherence length (Eq. (10.17)), which measures the size of the Cooper pair wavefunction, is small, comparable to the size of the unit cell. In contrast the low carrier density implies through Eq. (10.14) that the penetration depth is large. The high-T_c superconductors are therefore extreme type II with very large values of B_{c2}.

Because of this and the fact that they are superconducting at the temperature of liquid nitrogen (77 K) there are many potential applications for these materials.† Difficult problems must however be overcome before the materials come into widespread use. Paramount among the problems for $YBa_2Cu_3O_{7-\delta}$ is that it is most easily prepared as a ceramic, that is as many small crystallites bonded together. Although the critical current parallel to the xy plane within each crystallite is high, the performance of the ceramic is degraded by poor contact between crystallites; it is possible to improve this by aligning the crystallites so that the xy planes in neighbouring crystallites are parallel. If the materials are to carry large currents in high magnetic fields some means of pinning the quantized flux lines must be devised. This problem is more acute in high-T_c superconductors operating at liquid nitrogen temperature because more

---

† See 'The new superconductors: prospects for applications' by A. M. Wolsky, R. F. Geise and E. J. Daniels, in *Scientific American*, February 1989.

thermal energy is available to allow the flux line to escape from its pinning centre.

PROBLEMS 10

10.1   A current is induced to flow around the walls of the thin lead tube shown at 4.2 K (not to scale, all dimensions in cm):

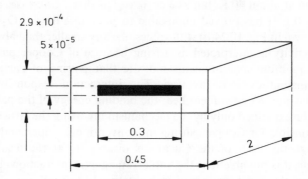

The current decays by less than 2% (the experimental sensitivity) in a time of 7 h. Deduce an upper limit for the electrical resistivity of superconducting lead. Assume a value 5×10^{-8} m for the penetration depth of lead. (This problem is based on the experiment of Quinn and Ittner, *J. Appl. Phys.* **33**, 748 (1962).)

10.2   The superconductor tin has $T_c = 3.7$ K and $B_c = 30.6$ mT at $T = 0$. Calculate the critical current for a tin wire of diameter 1 mm at $T = 2$ K. What diameter of wire would be required to carry a current of 100 A?

10.3   Use the approximate form of Eq. (10.1) for B_c to deduce approximate temperature dependences for the differences of the free energy, entropy and heat capacity between the normal and superconducting states. What is the discontinuity in the heat capacity at the superconducting transition in zero applied field?

10.4   Show that the Clogston limiting value of B_{c2} for a type II superconductor is given by $\mu_B B_{c2} \approx k_B T_c$.

10.5   Use the London equation to show that the penetration of a parallel magnetic field into a superconducting film of thickness d in the xy plane is described by

$$B = B_e \cosh(z/\lambda)/\cosh(d/2\lambda)$$

where B_e is the applied field and the centre of the film is at $z = 0$. Calculate the field at which the Gibbs free energies of the normal and superconducting states are equal for the film.

10.6   The effect of the non-locality of the current–field relation on the zero-temperature penetration depth of a pure type I superconductor in the limit $\lambda \ll \xi$ may be estimated by saying that, as the field decays on a length scale λ but the current depends on the average of $\mathbf{A}$ over a length scale ξ, the effective value of $\mathbf{A}$ to insert in Eq. (10.10) is $\lambda \mathbf{A}/\xi$. Show that this approach predicts

$$\lambda^3 = \lambda_L^2(0)\xi.$$

(The exact result from the BCS theory is $\lambda^3 = 0.62\lambda_L^2(0)\xi_0$.)

10.7 Suggest reasons for the following:

(a) At $T = 1$ K tin strongly absorbs electromagnetic radiation of wavelength 0.9 mm but only weakly absorbs radiation of wavelength 1.1 mm.

(b) Superconductors are poor conductors of heat for $T \ll T_c$.

(c) The critical field at $T = 0$ of different superconductors is *approximately* proportional to T_c.

(d) For different isotopes of the same element T_c depends on the isotopic mass.

10.8 A supercurrent, corresponding to the order parameter $(n_S/2)^{1/2} \exp(iqx)$, flows in a thin film in the xy plane of thickness $d \ll \lambda$. Calculate the vector potential within the film in a gauge for which $A = 0$ in the centre of the film and div $A = 0$. Show that in this gauge the second term in Eq. (10.27) is smaller by a factor $\sim d^2/\lambda_L^2(T)$ than the first term.

10.9 Eq. (10.10) cannot be generally valid since the left-hand side must be invariant under a gauge change $A \to A + \nabla\chi$ of the vector potential whereas the right-hand side obviously is not (both A's give the same field B). The correct gauge-invariant equation is Eq. (10.27). Explain why the gauge in which Eq. (10.10) is valid satisfies div $A = 0$. Use Eq. (10.27) to deduce the change in the order parameter due to the gauge transformation $A \to A + \nabla\chi$.

10.10 Deduce:

(a) the condensation energy, $G_N - G_S$, of lead from Fig. 10.4;

(b) dB_c/dT at T_c for aluminium (molar volume 10^{-5} m^3) from Fig. 10.6;

(c) the cross-sectional area of the tin cylinder from Fig. 10.16(b);

(d) the energy gap of lead from Fig. 10.21;

(e) the flux quantum from Fig. 10.22;

(f) the area of the loop containing the double junction from Fig. 10.23(b).

X-rays will turn out to be a hoax.—*Lord Kelvin (1893)*

CHAPTER

Waves in crystals

11.1 INTRODUCTION

The lattice dynamics of a chain of atoms (section 2.3.1) has several features in common with the electron states of the chain (section 4.3.3) and also with the dynamics of a chain of magnetic moments (section 8.5). In each case it is necessary to solve N coupled equations (Eqs. (2.7), (4.9) and (8.26)), one for each atom in the chain, and the wavelike solution is of the form $\exp\left[\mathrm{i}(kna - \omega t)\right]$, where the atomic positions are $x = na$; the frequency ω is periodic in k with period $2\pi/a$.

In this chapter we discuss the generalization of these ideas to arbitrary crystal structures in three dimensions. We begin by considering, in more detail than in section 1.4, the diffraction of waves incident on the crystal from an external source. This enables us to introduce the important and useful concept of the reciprocal lattice. We then explain why the internal motions of the crystal are expected to be wavelike and how the regularity of the dispersion relations of the waves in **k**-space is determined by the reciprocal lattice.

11.2 ELASTIC SCATTERING OF WAVES BY A CRYSTAL

11.2.1 Amplitude of the scattered wave

The Bragg law, Eq. (1.3), identifies the angles of the incident radiation relative to the lattice planes for which diffraction peaks occur, but it gives no informa-

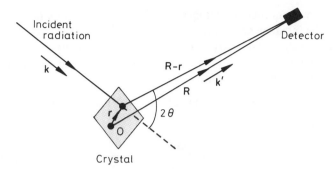

Fig. 11.1   Geometry of an experiment to look at radiation scattered from a crystal. For elastic scattering $|\mathbf{k}'| = |\mathbf{k}|$, and $\hbar\mathbf{K} = \hbar(\mathbf{k}' - \mathbf{k})$ is the momentum change of the incident particles. The notation 2θ for the angle of deflection is consistent with that used in section 1.4

tion on the intensities of the diffracted beams. To calculate the intensities a more detailed approach is required. Suppose that a plane wave $A_0 \exp[i(\mathbf{k}.\mathbf{r} - \omega t)]$ is incident on the atom at position $\mathbf{r}$ within the crystal as shown in Fig. 11.1. To make our treatment as general as possible we do not specify the nature of the radiation; x-rays, electrons and neutrons are the probes most often used in diffraction studies. Some of the incident radiation is elastically scattered by the atom.† If the scattering is weak then the contribution of the atom at $\mathbf{r}$ to the scattered wave at a detector at a large distance $|\mathbf{R} - \mathbf{r}|$ from the atom (Fig. 11.1) can be written as the product of three factors:

$$A_r = A_0\, e^{i(\mathbf{k}.\mathbf{r} - \omega t)} \times f \times \frac{e^{ik|\mathbf{R}-\mathbf{r}|}}{|\mathbf{R} - \mathbf{r}|}. \tag{11.1}$$

The first factor is the incident wave; for weak scattering the amplitude A_0 of this is effectively the same for all atoms in the crystal. The second factor f is the **atomic form factor** or **atomic scattering factor**, and it depends on the details of the interaction of the radiation with the atom; the magnitude of f reflects the strength of the interaction and in general f depends on the scattering angle (2θ in Fig. 11.1). The final factor in Eq. (11.1) represents the amplitude decrease and phase change associated with a point source at the position of the atom. For a distant detector the amplitude decrease is effectively the same for all atoms in the crystal and we can replace $|\mathbf{R} - \mathbf{r}|$ in this denominator by R, the distance of the detector from the origin of the crystallographic axes; we must not make the same replacement in the phase factor since the variation of this from one atom to another causes large changes in the signal at the detector. Note that for elastic

---

† Because the crystal is a massive object it can absorb the momentum change of the incident particle without taking any energy from it (see problem 11.1). Inelastic scattering is considered in Chapter 12.

scattering the wavenumber and frequency of the scattered radiation are identical to those of the incident radiation.

The wave at a distant detector approximates to a plane wave of wavevector $\mathbf{k}'$ ($|\mathbf{k}'| = |\mathbf{k}|$) which is approximately parallel to both $\mathbf{R}$ and $\mathbf{R} - \mathbf{r}$. To a good approximation, therefore, we can write

$$k|\mathbf{R} - \mathbf{r}| \approx \mathbf{k}' . (\mathbf{R} - \mathbf{r}) = \mathbf{k}' . \mathbf{R} - \mathbf{k}' . \mathbf{r} \approx kR - \mathbf{k}' . \mathbf{r}.$$

Eq. (11.1) can then be written

$$A_r \approx A_0 \frac{e^{i(kR - \omega t)}}{R} f\, e^{-i\mathbf{K} . \mathbf{r}}$$

where $\mathbf{K} = \mathbf{k}' - \mathbf{k}$ is known as the **scattering vector**. The factor $A_0 exp\,[i(kR - \omega t)]/R$ is the same for all atoms in the crystal and factorizes out when the total scattered wave at the detector is evaluated by summing the contributions of all the atoms; we can therefore ignore this factor. The important term is $f\, exp\,(-i\mathbf{K} . \mathbf{r})$, which contains the phase differences between the contributions of the different atoms. The amplitude of the scattered wave is then proportional to

$$A = \sum_n f_n\, e^{-i\mathbf{K} . \mathbf{r}_n} \tag{11.2}$$

where the sum is over all the atoms in the crystal and f_n is the atomic scattering factor of the nth atom.†

The regularity of the atomic structure in a crystal means that, for certain special directions of the incident radiation, the scattering from the atoms adds up in phase to give a large scattered amplitude in a particular direction. We will use Eq. (11.2) to calculate the directions and intensities of the diffracted beams. Since the structure can be made up by associating a basis of atoms with each lattice point (section 1.2), the position of atom n may be written

$$\mathbf{r}_n = \mathbf{r}_l + \mathbf{r}_p \tag{11.3}$$

where $\mathbf{r}_l$ is the position of the lattice point with which atom n is associated and $\mathbf{r}_p$ is the position of the atom relative to the lattice point as shown in Fig. 11.2. Using Eq. (11.3) enables Eq. (11.2) to be factorized as

$$A = \left(\sum_l e^{-i\mathbf{K} . \mathbf{r}_l} \right) \times \left(\sum_p f_p\, e^{-\mathbf{K} . \mathbf{r}_p} \right) \tag{11.4}$$

$$= \left(\begin{array}{c} \text{sum over} \\ \text{lattice points} \end{array} \right) \times \left(\begin{array}{c} \text{sum over atoms} \\ \text{in basis} \end{array} \right).$$

---

† Eq. (11.2) is valid only for weak scattering since multiple scattering and attenuation of the incident wave by the crystal are ignored; this is normally a good approximation for neutrons and x-rays but not for electrons.

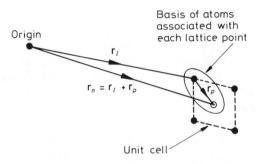

Fig. 11.2   The position $\mathbf{r}_n$ of any atom in the crystal can be written as $\mathbf{r}_l + \mathbf{r}_p$, where $\mathbf{r}_l$ is the position of the lattice point with which the atom is associated and $\mathbf{r}_p$ is the position of the atom relative to the lattice point. The boundaries of the primitive unit cell containing the atom are also shown

The first term, which can also be regarded as a sum over the primitive unit cells in the crystal, contains the information on the crystal lattice, and consequently it is this term which determines the directions for which diffraction occurs. The second term, which is a sum over the contents of a primitive unit cell, is thus a sum over a relatively small number of atoms (for many structures only one atom) and is the same for all lattice points; this term is known as the **structure factor** and it determines the relative intensities of the diffracted beams. We consider these two terms in turn.

11.2.2   Laue conditions for diffraction and the reciprocal lattice

Using Eq. (1.2)

$$\mathbf{r}_l = u\mathbf{a} + v\mathbf{b} + w\mathbf{c}$$

for the positions of the lattice points enbles us to write the first term in Eq. (11.4) as

$$\sum_l e^{-i\mathbf{K}\cdot\mathbf{r}_l} = \sum_u e^{-i\mathbf{K}\cdot\mathbf{a}u} \sum_v e^{-i\mathbf{K}\cdot\mathbf{b}v} \sum_w e^{-i\mathbf{K}\cdot\mathbf{c}w} \tag{11.5}$$

A large scattering amplitude is obtained when the contributions from all the lattice points are in phase and this is the case if

$$\mathbf{K}\cdot\mathbf{a} = 2\pi h \tag{11.6a}$$

$$\mathbf{K}\cdot\mathbf{b} = 2\pi k \tag{11.6b}$$

$$\mathbf{K}\cdot\mathbf{c} = 2\pi l \tag{11.6c}$$

where h, k and l are integers. Eqs. (11.6) are the **Laue conditions** for diffraction. When they are satisfied, each term in Eq. (11.5) is unity and the sum is then equal to $N_1 N_2 N_3$ for a crystal of extent N_1, N_2 and N_3 lattice spacings in the x, y and z directions respectively: $N_1 N_2 N_3$ is of course just the number of primitive unit cells within the crystal. Eqs. (11.6) are the conditions for diffraction off a three-dimensional diffraction grating; comparison with the results for a one-dimensional optical grating (Smith and Thomson,[5] chapter 11) shows that the scattered amplitude falls off very rapidly as K varies from a value satisfying these equations.

The directions of the diffracted beams are given by the set of vectors K that satisfy Eqs. (11.6). These can be represented in an elegant way using the **reciprocal lattice** concept. To explain this important idea we first note that k-space (sometimes called **reciprocal space**) is the appropriate space for plotting wavevectors. The values of the scattering vector K that satisfy the Laue conditions lie on a regular lattice in this space; this is the reciprocal lattice. All the points of the reciprocal lattice can be generated from three primitive reciprocal lattice vectors a^*, b^* and c^* by using the equation

$$G_{hkl} = ha^* + kb^* + lc^*, \qquad (11.7)$$

where h, k and l are integers; this is analogous to the use of the three primitive translation vectors a, b and c to define the lattice points r_l of a crystal in real three-dimensional space using Eq. (1.2). We will now prove that, if

$$K = G_{hkl}, \qquad (11.8)$$

where G_{hkl} is *any* reciprocal lattice vector, and we choose the following values for a^*, b^* and c^*

$$a^* = \frac{2\pi(b \times c)}{a.(b \times c)}, \qquad b^* = \frac{2\pi(c \times a)}{a.(b \times c)}, \qquad c^* = \frac{2\pi(a \times b)}{a.(b \times c)}, \qquad (11.9)$$

then K satisfies the Laue conditions. To do this we evaluate $K.a$ to obtain

$$K.a = G_{hkl}.a = (ha^* + kb^* + lc^*).a = 2\pi h,$$

where we have used the relations

$$a^*.a = 2\pi, \qquad b^*.a = c^*.a = 0, \qquad (11.10)$$

which follow from the definitions (11.9).† If K is equal to G_{hkl} it therefore satisfies Eq. (11.6a); evaluating $G_{hkl}.b$ and $G_{hkl}.c$ shows that K also satisfies Eqs. (11.6b) and (11.6c).

The relations (11.10) mean that b^* and c^* are perpendicular to a. It does not follow that a and a^* are parallel; this is only the case if crystal axes are orthogonal. It is easier to remember the definitions (11.9) if it is noted that, once

---

† Relations (11.10) are true because $(b \times c).a = a.(b \times c)$ and $(c \times a).a$ and $(a \times b).a$ vanish; the scalar product of two perpendicular vectors is zero.

a* has been defined, b* and c* can be written down by cyclic permutation of **a**, **b** and **c**. Note that the denominator **a**.(**b** × **c**) in the definitions is unchanged by a cyclic permutation; this quantity is in fact the volume of a unit cell. It follows from the symmetry of Eqs. (11.9) that Eqs. (11.10) can be supplemented by **b***.**b** = 2π, **c***.**c** = 2π and relations which indicate that **a*** and **c*** are perpendicular to **b** and that **a*** and **b*** are perpendicular to **c**.

Since the scattering vector of each diffracted beam corresponds to a point in the reciprocal lattice of the form of Eq. (11.7) we can use the integers (*hkl*) to label that beam. We now show that this labelling is identical to that introduced in section 1.4, where the diffracted beams were labelled by the Miller indices of the lattice planes with which they were associated. In doing this we will also demonstrate that Eqs. (11.6) or Eq. (11.8) are equivalent to the Bragg law, Eq. (1.3). We first establish the relationship between the reciprocal lattice vector $\mathbf{G}_{hkl}$ and the set of lattice planes with Miller indices (*hkl*). The plane of the (*hkl*) set nearest the origin is shown in Fig. 11.3 and we see that a vector **d** perpendicular to the planes with length equal to the plane spacing satisfies

$$\mathbf{d}.\mathbf{a}/h = \mathbf{d}.\mathbf{b}/k = \mathbf{d}.\mathbf{c}/l = d^2. \tag{11.11}$$

These equations for **d** are similar in form to Eqs. (11.6), which determine the scattering vectors of the diffracted beams. Exploiting this similarity we find that, by analogy with Eq. (11.8), **d** can be written

$$\mathbf{d} = \frac{d^2}{2\pi}(h\mathbf{a}^* + k\mathbf{b}^* + l\mathbf{c}^*) = \frac{d^2}{2\pi}\mathbf{G}_{hkl}. \tag{11.12}$$

This may be checked by direct substitution in Eqs. (11.11) and using Eqs. (11.9). Eq. (11.12) shows that the reciprocal lattice vector $\mathbf{G}_{hkl}$ is perpendicular to the lattice planes with Miller indices (*hkl*) and has a length $2\pi/d$, where d is the spacing of the planes.

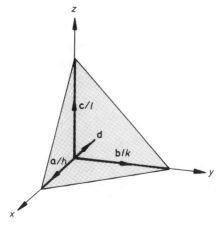

Fig. 11.3   The first (*hkl*) plane out from the origin has intercepts a/h, b/k and c/l on the x, y and z axes respectively. The vector **d** is perpendicular to the plane and its length is equal to the plane spacing. The relation $\mathbf{a}.\mathbf{d}/h = d^2$ (Eq. (11.11)) follows because the component of $\mathbf{a}/h$ along the **d** direction is $|\mathbf{d}|$

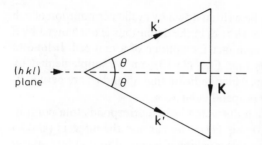

Fig. 11.4   The Bragg formulation of diffraction has the incident and diffracted beams making equal angles θ to the lattice planes. Hence, since $|\mathbf{k}| = |\mathbf{k}'|$, $\mathbf{K}$ is perpendicular to the planes

The interpretation of the Bragg law as 'reflection' of waves off lattice planes is illustrated in Fig. 11.4. The use of some simple geometry (recall $|\mathbf{k}'| = |\mathbf{k}|$) establishes that the scattering vector $\mathbf{K}$ is perpendicular to the planes and thus parallel to the vector $\mathbf{G}_{hkl}$ as is required by Eq. (11.8). The magnitude of $\mathbf{K}$ ($|\mathbf{K}| = 2|\mathbf{k}| \sin \theta$ from Fig. 11.4) is equal to that of $\mathbf{G}_{hkl}$ if $2|\mathbf{k}| \sin \theta = 2\pi/d$, that is if $2d \sin \theta = \lambda$, which is just the Bragg law for first-order diffraction. The Bragg and Laue formulations of the conditions for diffraction are thus completely equivalent, as is the use of reciprocal lattice vectors and Miller indices to label the diffracted beams. As already indicated in section 1.4, when a primitive unit cell is used to describe the lattice, higher orders of diffraction are indicated by the appearance of a common factor in (hkl). Thus (1 0 0), (2 0 0) and (3 0 0) correspond to first-, second- and third-order diffraction respectively.

The reciprocal lattice is such an important and widely used concept for discussing diffraction of waves by a crystal that it is worth while repeating the main conclusion of this section. For incident radiation of wavevector $\mathbf{k}$, the directions of the diffracted beams are given by wavevectors $\mathbf{k}' = \mathbf{k} + \mathbf{K}$, where the scattering vector $\mathbf{K}$ is any one of the points of the reciprocal lattice in $\mathbf{k}$-space. The reciprocal lattice points are given by Eq. (11.7) and the primitive lattice vectors of the reciprocal lattice are determined from the real space primitive lattice vectors by using Eqs. (11.9).

11.2.3   Examples of reciprocal lattices

In discussing the following examples we will find it useful to refer to a set of Cartesian axes defined by mutually perpendicular unit vectors $\mathbf{i}$, $\mathbf{j}$ and $\mathbf{k}$.

Simple cubic real space lattice

In terms of the Cartesian axes, we have

$$\mathbf{a} = a\mathbf{i}, \qquad \mathbf{b} = a\mathbf{j}, \qquad \mathbf{c} = a\mathbf{k}. \qquad (11.13)$$

Thus $\mathbf{a} \cdot (\mathbf{b} \times \mathbf{c}) = a^3$ and, using Eqs. (11.9),†

$$\mathbf{a}^* = \frac{2\pi}{a}\mathbf{i}, \qquad \mathbf{b}^* = \frac{2\pi}{a}\mathbf{j}, \qquad \mathbf{c}^* = \frac{2\pi}{a}\mathbf{k}. \qquad (11.14)$$

† Recall that $\mathbf{i} \times \mathbf{j} = \mathbf{k}$, $\mathbf{j} \times \mathbf{k} = \mathbf{i}$ and $\mathbf{k} \times \mathbf{i} = \mathbf{j}$.

The reciprocal lattice is therefore also simple cubic with side $2\pi/a$ and is in the same orientation as the real space lattice.

Face-centred cubic real space lattice

We take the Cartesian axes to be along the sides of the conventional cubic unit cell. The primitive translational vectors, shown in the right-hand cube in Fig. 11.5, are then

$$\mathbf{a} = \frac{a}{2}(\mathbf{j} + \mathbf{k}), \qquad \mathbf{b} = \frac{a}{2}(\mathbf{k} + \mathbf{i}), \qquad \mathbf{c} = \frac{a}{2}(\mathbf{i} + \mathbf{j}), \qquad (11.15)$$

where a is the side of the conventional cubic unit cell. Thus $\mathbf{a} \cdot (\mathbf{b} \times \mathbf{c}) = a^3/4$ and, using Eqs. (11.9), we find

$$\mathbf{a}^* = \frac{2\pi}{a}(\mathbf{j} + \mathbf{k} - \mathbf{i}), \qquad \mathbf{b}^* = \frac{2\pi}{a}(\mathbf{k} + \mathbf{i} - \mathbf{j}), \qquad \mathbf{c}^* = \frac{2\pi}{a}(\mathbf{i} + \mathbf{j} - \mathbf{k}).$$
$$(11.16)$$

These are the primitive translational vectors for a *body*-centred cubic reciprocal lattice with a cubic unit cell of side $4\pi/a$. That the reciprocal lattice is bcc is demonstrated in Fig. 11.5. The vector $\mathbf{a} \times \mathbf{b}$ which is parallel to $\mathbf{c}^*$ is clearly directed towards the body-centered position of the left-hand cube. It follows that the conventional cubic unit cells of the real space lattice and the reciprocal lattice have the same orientation. Note that labelling of diffracted beams based on the above primitive translation vectors of the reciprocal lattice corresponds to labelling based on Miller indices for the primitive unit cell and therefore differs from labelling based on Miller indices for the conventional cubic unit cell (see problem 11.5).

Body-centred cubic real space lattice

It is a general result that taking the reciprocal lattice of the reciprocal lattice gives the real space lattice back again (problem 11.2); the reciprocal lattice of a bcc lattice is therefore an fcc lattice. If the conventional unit cell has side a, the

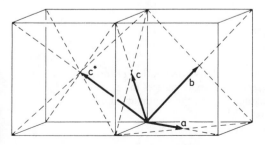

Fig. 11.5   *Pairs of primitive lattice vectors of the fcc lattice lie in* {1 1 1} planes. The reciprocal lattice vectors are perpendicular and are therefore in directions parallel to the body diagonals of the cube (for example the [1 1 1] direction). These are the directions of the primitive lattice vectors of a bcc lattice

primitive translation vectors of the real space lattice are

$$\mathbf{a} = \frac{a}{2}(\mathbf{j} + \mathbf{k} - \mathbf{i}), \qquad \mathbf{b} = \frac{a}{2}(\mathbf{k} + \mathbf{i} - \mathbf{j}), \qquad \mathbf{c} = \frac{a}{2}(\mathbf{i} + \mathbf{j} - \mathbf{k}) \quad (11.17)$$

so that $\mathbf{a} \cdot (\mathbf{b} \times \mathbf{c}) = a^3/2$ and

$$\mathbf{a}^* = \frac{2\pi}{a}(\mathbf{j} + \mathbf{k}), \qquad \mathbf{b}^* = \frac{2\pi}{a}(\mathbf{k} + \mathbf{i}), \qquad \mathbf{c}^* = \frac{2\pi}{a}(\mathbf{i} + \mathbf{j}). \quad (11.18)$$

These are the primitive translation vectors for an fcc reciprocal lattice with a cubic unit cell of side $4\pi/a$.

Hexagonal real space lattice

This is, for example, the lattice of the hexagonal close-packed structure of Fig. 1.11 and the primitive lattice vectors, in terms of the Cartesian axes, are

$$\mathbf{a} = a\mathbf{i}, \qquad \mathbf{b} = a\left(-\frac{\mathbf{i}}{2} + \frac{\sqrt{3}\mathbf{j}}{2}\right), \qquad \mathbf{c} = c\mathbf{k}. \quad (11.19)$$

Thus $\mathbf{a} \cdot (\mathbf{b} \times \mathbf{c}) = (\sqrt{3}/2)a^2 c$ and

$$\mathbf{a}^* = \frac{4\pi}{\sqrt{3}a}\left(\frac{\sqrt{3}\mathbf{i}}{2} + \frac{\mathbf{j}}{2}\right), \qquad \mathbf{b}^* = \frac{4\pi\mathbf{j}}{\sqrt{3}a}, \qquad \mathbf{c}^* = \frac{2\pi\mathbf{k}}{c}. \quad (11.20)$$

These are the primitive translation vectors of a hexagonal reciprocal lattice. The relative orientation of the real space and reciprocal space lattices is shown in Fig. 11.6.

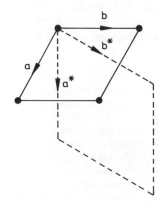

Fig. 11.6   Relative orientation of the real and reciprocal space lattices for a hexagonal lattice; the vectors $\mathbf{c}^*$ and $\mathbf{c}$ are parallel

11.2.4   The structure factor

The second factor in Eq. (11.4), the structure factor

$$S = \sum_p f_p\, e^{-i\mathbf{K}\cdot\mathbf{r}_p},\qquad(11.21)$$

is a sum over the atoms in the basis, where $\mathbf{r}_p$ is the position of the atom relative to the lattice point. The simplest situation is for a basis consisting of one atom on each lattice point. S then contains one term, for which $\mathbf{r}_p = 0$. Thus

$$S = f$$

and the only variation in the intensity of the diffracted peaks is due to the angular variation of the atomic form factor f. This is usually smooth and monotonic so that neighbouring diffraction peaks have similar intensities. Fig. 11.7 shows the angular dependence of f for x-rays for cubic close-packed aluminium as deduced from the diffraction peaks indicated; f is shown as a function of $(\sin\theta)/\lambda$ where θ is the Bragg angle.

The hexagonal close-packed structure of Fig. 1.11 has a basis of two identical atoms at $\mathbf{r}_1 = 0$ and $\mathbf{r}_2 = \tfrac{1}{3}\mathbf{a} + \tfrac{2}{3}\mathbf{b} + \tfrac{1}{2}\mathbf{c}$. Thus for the (hkl) diffracted beam

$$S = f\{\exp(i0) + \exp[-i(h\mathbf{a}^* + k\mathbf{b}^* + l\mathbf{c}^*)\cdot(\tfrac{1}{3}\mathbf{a} + \tfrac{2}{3}\mathbf{b} + \tfrac{1}{2}\mathbf{c})]\}$$

$$= f\{1 + \exp[-i\pi(\tfrac{2}{3}h + \tfrac{4}{3}k + l)]\}$$

$$= f\{1 + \exp[-i\pi((h + k + l) + \tfrac{1}{3}(k - h))]\}\qquad(11.22)$$

where we have used Eqs. (11.7) and (11.8) and the properties of the primitive reciprocal lattice vectors (see Eqs. (11.10)). The intensity of the diffracted beam is proportional to $|S|^2$ and this can take four possible values depending on h, k and l (problem 11.3). For some diffracted beams, in particular the (0 0 1) beam,

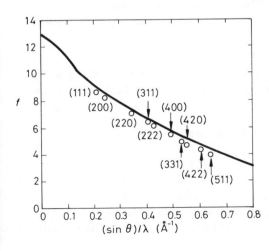

Fig. 11.7   Atomic form factor for aluminium determined from the intensities of the diffraction peaks indicated. The labelling is based on the conventional cubic unit cell. The full curve is a theoretical prediction. (Reproduced with permission W. Batterman, D. R. Chipman and J. J. de Marco, *Phys. Rev.* **122**, 68 (1961))

the intensity vanishes. Thus some of the possible diffracted beams predicted by Eq. (11.8) are absent because of destructive interference of the scattering from the two atoms in the basis.

This is worthy of further comment. The vanishing of S for the hcp structure depends crucially on the assumption of identical form factors for both atoms in the basis. Although the atoms are chemically identical, their environments within the crystal are different; electron states within the atoms are distorted slightly by the neighbouring atoms and this distortion is reflected in the angular dependence of f. Thus the form factors of the two atoms differ slightly and very weak diffracted beams do occur in the 'forbidden' directions. For the approximately spherically symmetric atoms and ions that form hexagonal close-packed structures, these diffracted beams are normally too weak to be seen, but a beam of this kind is observed for the covalently bonded diamond stucture, as discussed below. The important lesson to be learned is that it is necessary to distinguish between results that are derived rigorously by symmetry arguments (such as prediction of the direction of diffracted beams using the reciprocal lattice) and results that follow from an approximation of some kind (such as the near-vanishing of predicted diffraction beams due to approximate equality of the form factors of the atoms in the basis).

Before calculating the structure factors of the diamond and sodium chloride structures, both of which have an fcc lattice, it will be helpful for us to discuss diffraction from a simple fcc structure when the conventional cubic unit cell rather than the primitive unit cell is used to determine the reciprocal lattice. To do this we regard the fcc structure as being built up from a simple cubic lattice of side a with the face-centred atoms forming part of the basis of atoms associated with each lattice point. The reciprocal lattice of the simple cubic lattice is simple cubic with side $2\pi/a$ (Eq. (11.14)). The intensities of the diffracted beams are given by a structure factor obtained by summing over basis atoms at: $\mathbf{r}_1 = 0$, $\mathbf{r}_1 = \frac{1}{2}(\mathbf{a} + \mathbf{b})$, $\mathbf{r}_3 = \frac{1}{2}(\mathbf{b} + \mathbf{c})$, $\mathbf{r}_4 = \frac{1}{2}(\mathbf{c} + \mathbf{a})$, where $\mathbf{a}$, $\mathbf{b}$ and $\mathbf{c}$ are the lattice vectors for the conventional unit cell. Thus, proceeding as for the hcp structure,

$$S = f(1 + e^{i\pi(h+k)} + e^{i\pi(k+l)} + e^{i\pi(l+h)}) \tag{11.23}$$

where f is the atomic form factor of the atoms in the basis; note that f is strictly equal for the four atoms since their environments are identical. A few moments thought suffices to show that S can take only two values:

$$S = \begin{cases} 4f & \text{when } h, k \text{ and } l \text{ are all odd or all even,} \\ 0 & \text{otherwise.} \end{cases} \tag{11.24}$$

We see from Fig. 11.8 that the diffracted beams eliminated in this way are precisely those required to convert the simple cubic reciprocal lattice of side $2\pi/a$ into the body-centred cubic reciprocal lattice of cube side $4\pi/a$ of Eq. (11.16), obtained using the *primitive* unit cell of the fcc lattice. Both approaches therefore

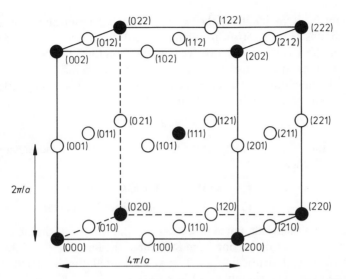

Fig. 11.8   Selecting points of a simple cubic lattice of side $2\pi/a$ for which the coordinates (hkl) are all odd or all even generates a body-centred cubic lattice (full circles) of side $4\pi/a$

lead to the same diffracted beams although the labelling is different (see problem 11.5).

For the diamond and sodium chloride structures we will use labelling appropriate to the conventional cubic unit cell. As the diffracted beams must correspond to the reciprocal lattice of the fcc lattice, we know from our analysis above that diffracted beams will only occur if h, k and l are all even or all odd. For the diamond structure (Fig. 1.15), the intensities of the beams are determined by the structure factor of the basis, $\mathbf{r}_1 = 0$, $\mathbf{r}_2 = \frac{1}{4}(\mathbf{a} + \mathbf{b} + \mathbf{c})$, of two atoms associated with each lattice point. Thus, proceeding as for the hcp structure,

$$S = f\{1 + \exp[-i\tfrac{1}{2}\pi(h + k + l)]\}$$

where f is the form factor for the atoms. Hence:
(1) for h, k and l all odd,

$$|S|^2 = 2f^2;$$

(2) for h, k and l all even,

$$|S|^2 = \begin{cases} 4f^2 & \text{if } h + k + l \text{ is a multiple of 4,} \\ 0 & \text{if } h + k + l \text{ is not a multiple of 4.} \end{cases}$$

The observed diffraction is in general agreement with these predictions but a very weak (2 2 2) x-ray beam is observed for diamond in contradiction with the rules. This indicates failure of the assumption that the form factors of the two atoms in the basis are identical; the two atoms have their tetrahedrally arranged covalent bonds (Fig. 1.15(b)) in different orientations and, since a significant fraction of the electrons are involved in the bonding, the form factors are sufficiently different to cause an observable diffracted beam.

In the sodium chloride structure, the basis of atoms is Na^+ at $r_1 = 0$ and Cl^- at $r_2 = \frac{1}{2}a$. The structure factor is thus

$$S = f_+ + f_- e^{-i\pi h}$$

where f_+ and f_- are the form factors of the positive and negative ions respectively. Hence:

$$|S|^2 = \begin{cases} (f_+ - f_-)^2 & \text{for } h, k \text{ and } l \text{ all odd,} \\ (f_+ + f_-)^2 & \text{for } h, k \text{ and } l \text{ all even.} \end{cases}$$

This generalizes the result, noted in section 1.4, that the (1 1 1) diffracted beam is not observed in x-ray diffraction from KCl. Since the form factors of the K^+ and Cl^- ions are almost identical ($f_+ \approx f_-$); we see that all the diffracted beams for odd h, k and l are vanishingly small in KCl.

11.3   WAVELIKE NORMAL MODES—BLOCH'S THEOREM

We now turn our attention to the electron states and lattice vibrations of a crystal. We are interested in motions where the time dependence is of the form $e^{-i\omega t}$ throughout the crystal, that is in the normal modes. The motions are therefore specified by a function of the form

$$\Psi(\mathbf{r}, t) = \psi(\mathbf{r}) e^{-i\omega t}. \tag{11.25}$$

We will show that the motions are wavelike in character. We consider first the situation where the function $\Psi(\mathbf{r}, t)$ only needs to be defined at the lattice sites. This is the case, for example, for the classical (Newtonian) dynamics of the atoms in a crystal in which the lattice and stucture are identical (only one atom in each primitive unit cell); in this case the function $\Psi(\mathbf{r}_l, t)$ gives the displacement of the atom that occupies the lattice position $\mathbf{r} = \mathbf{r}_l$ when the crystal is at rest.

The lattice looks identical from each lattice point and the N equations of motion needed to determine $\Psi(\mathbf{r}_l, t)$ (one for each atom in the crystal) must therefore be identical in form. This implies that if the function $\psi(\mathbf{r}_l)$ changes by a factor A in going from the nth to the $(n + 1)$th lattice point in any direction it must change by the same factor in going from the $(n + 1)$th to the $(n + 2)$th lattice point in that direction. This is illustrated for a two-dimensional example in Fig. 11.9, where we suppose that $\psi(\mathbf{r}_l)$ is equal to unity at the origin and changes by a factor A for displacement by the primitive lattice vector $\mathbf{a}$ and a factor B for displacement by the primitive lattice vector $\mathbf{b}$. It can be seen that for a general

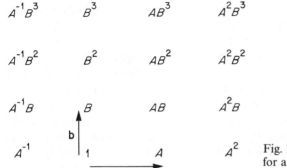

Fig. 11.9   Normal mode amplitude for a two-dimensional rectangular lattice

displacement $\mathbf{r}_l = u\mathbf{a} + v\mathbf{b}$ we have

$$\psi(\mathbf{r}_l) = A^u B^v.$$

Unless we choose A and B such that $|A| = |B| = 1$, we will have an unphysical situation in which $\psi(\mathbf{r}_l)$ increases indefinitely in some directions. We can therefore set

$$A = \mathrm{e}^{i\theta} \qquad B = \mathrm{e}^{i\phi}.$$

More general solutions are acceptable near a boundary, but not in an infinite crystal. We thus have

$$\psi(\mathbf{r}_l) = \mathrm{e}^{i(u\theta + v\phi)}.$$

If we *define* a vector $\mathbf{k}$ such that $\mathbf{k}\cdot\mathbf{a} = \theta$ and $\mathbf{k}\cdot\mathbf{b} = \phi$, then

$$\Psi(\mathbf{r}_l, t) = \psi(\mathbf{r}_l)\,\mathrm{e}^{-i\omega t} = \mathrm{e}^{i\mathbf{k}\cdot(u\mathbf{a}+v\mathbf{b})}\,\mathrm{e}^{-i\omega t} = \mathrm{e}^{i(\mathbf{k}\cdot\mathbf{r}_l - \omega t)}. \qquad (11.26)$$

The wavelike nature of the solution is thus a consequence of the translational symmetry of the lattice.

   We must now generalize this result to the case where there is more than one atom in each primitive unit cell. A more general situation has already been considered in section 2.3.2, namely the vibrations of a linear chain consisting of two types of atom. There we had two types of equation of motion, one for each type of atom, and the solution was specified by the ratio $\alpha(\mathbf{k})$, giving the relative displacements of the two atoms in each cell, as well as by a wavelike factor equivalent to that of Eq. (11.26). For each value of $\mathbf{k}$ there were two possible values of $\alpha(\mathbf{k})$ and consequently two different normal mode frequencies $\omega(\mathbf{k})$. This result is easily generalized to the case of s atoms per primitive unit cell: there will be s distinct types of equation of motion; $s - 1$ ratios $\alpha_2(\mathbf{k}), \ldots, \alpha_s(\mathbf{k})$ will be required to specify, for *each* normal mode, the relative displacement of the atoms in the unit cell; there will be s normal modes for each value of $\mathbf{k}$ and thus s branches of the dispersion relation $\omega(\mathbf{k})$. The argument is not specific to lattice vibrations; it applies to any situation in which an amplitude has to be defined at s points in the primitive unit cell.

Now consider the case in which $\Psi(\mathbf{r}, t)$ is an electron wavefunction, as in Chapter 4, which has to be defined at all points in space. The appropriate generalization of Eq. (11.26) for this case may be obtained by taking the limit $s \to \infty$ in the discussion of the previous paragraph. This gives an infinite number of branches of $\omega(\mathbf{k})$ corresponding to an infinite number of energy bands for the electrons; fortunately only a small number of these will be important. In the limit $s \to \infty$, the set of numbers $\alpha_2, \ldots, \alpha_s$ tends to a continuous function $u(\mathbf{r})$ defined at all points within a unit cell; the values of $\mathbf{u}(\mathbf{r})$ like those of $\alpha_2, \ldots, \alpha_s$ repeat within the next cell, so that $u(\mathbf{r})$ is a periodic function with the period of the lattice. The appropriate generalization of Eq. (11.26) is therefore

$$\Psi(\mathbf{r}, t) = \frac{1}{\sqrt{V}} u_{\mathbf{k}}(\mathbf{r})\, e^{i(\mathbf{k} \cdot \mathbf{r} - \omega t)} \tag{11.27}$$

where we have added a suffix $\mathbf{k}$ to indicate that the form of the function $u(\mathbf{r})$ depends on $\mathbf{k}$; it is also different for each branch of the frequency spectrum. For convenience we have written a normalization factor $1/\sqrt{V}$ in Eq. (11.27), where V is the volume of the crystal (cf. Eq. (3.3)). The wavefunction (11.27) is known as a **Bloch wavefunction** and the fact that wavefunctions in a crystal can be expressed in this form is **Bloch's theorem**.

11.4   NORMAL MODES AND THE RECIPROCAL LATTICE

11.4.1   Periodicity of the dispersion relation

In this section we derive the important result that *a normal mode of the crystal described by a function of the form of Eq. (11.26) or (11.27) for some wavevector* $\mathbf{k}$ *can also be described by a function of the same form but with a different wavevector* $\mathbf{k}'$ *related to* $\mathbf{k}$ *by*

$$\mathbf{k}' = \mathbf{k} + \mathbf{G}_0 \tag{11.28}$$

where $\mathbf{G}_0$ *is any vector of the reciprocal lattice of the crystal as given by Eq.* (11.7). This result is easy to establish for Eq. (11.26) where the function is defined only at the lattice points. From Eq. (1.2), $\mathbf{r}_l = u\mathbf{a} + v\mathbf{b} + w\mathbf{c}$, so that, by using Eqs. (11.7) and (11.9), we find $\mathbf{G}_0 \cdot \mathbf{r}_l = 2\pi(uh + vk + wl)$. Hence

$$e^{i\mathbf{k}' \cdot \mathbf{r}_l} = e^{i\mathbf{G}_0 \cdot \mathbf{r}_l}\, e^{i\mathbf{k} \cdot \mathbf{r}_l}$$
$$= e^{i2\pi(uh + vk + wl)}\, e^{i\mathbf{k} \cdot \mathbf{r}_l}$$
$$= 1 \times e^{i\mathbf{k} \cdot \mathbf{r}_l}$$

and Eq. (11.26) is unaffected by the substitution $\mathbf{k} \to \mathbf{k}'$.

This result can be extended to the functions of Eq. (11.27) that are defined at all positions. Since this extension involves more complicated algebra and the introduction of new ideas, we will first indicate the importance of the result by pointing out some of its consequences:

(1) Since the same motions are described by any $\mathbf{k}'$ that satisfies Eq. (11.28) it follows that the mode frequency is the same for all such wavevectors and therefore that *for any branch of the dispersion relation the frequency is periodic in* $\mathbf{k}$-*space with the same periodicity as the reciprocal lattice*. We have already encountered this result for the lattice vibrations and electron states of a one-dimensional chain (sections 2.3.1, 2.3.2 and 4.3.3), where we found the dispersion relations to be periodic in k with period $2\pi/a$, which is the reciprocal lattice spacing for a one-dimensional real space lattice of spacing a.

(2) The flexibility of being allowed to add any reciprocal lattice vector to the wavevector used to represent a normal mode means that *any mode can be represented by a wavevector inside a single primitive unit cell of the reciprocal lattice*; Eq. (11.28) relates any wavevector $\mathbf{k}$ outside this cell to one $\mathbf{k}'$ inside for a suitably chosen reciprocal lattice vector $\mathbf{G}_0$. In our one-dimensional calculations in sections 2.3 and 4.3.3 we saw that all physically distinct lattice vibrations and electron states could be represented by wavevectors within a range $2\pi/a$, the size of the primitive cell of the reciprocal lattice. We use results (1) and (2) in the following section to generalize to more than one dimension the repeated, reduced and extended zone plotting schemes for dispersion relations of Fig. 4.9.

(3) *The number of normal modes associated with any branch of the dispersion relation is equal to the number* N_c *of the primitive unit cells in the crystal*. For the periodic boundary conditions that are consistent with running waves of the form of Eqs. (11.26) and (11.27) we have already shown (Eq. (2.41)) that possible $\mathbf{k}$ vectors are distributed uniformly in $\mathbf{k}$-space with a density $V/(2\pi)^3$, where V is the volume of the crystal. All the distinct modes can be represented by a $\mathbf{k}$ vector inside one primitive cell of the reciprocal lattice which has a volume $\mathbf{a}^* . (\mathbf{b}^* \times \mathbf{c}^*)$; using Eq. (11.9), this can be written $(2\pi)^3/[\mathbf{a} . (\mathbf{b} \times \mathbf{c})] = (2\pi)^3/v_c$ where v_c is the volume of the primitive unit cell of the real space lattice. Thus there are

$$\frac{V}{(2\pi)^3} \frac{(2\pi)^3}{v_c} = \frac{V}{v_c} = N_c \qquad (11.29)$$

$\mathbf{k}$ values for each branch of the dispersion relation.

To extend our proof of the statement at the beginning of this section to cover functions of the form of Eq. (11.27) we must first generalize to more than one dimension the Fourier series expansion of a periodic function of position. We begin by looking at the one-dimensional expansion in a different way to that normally used. If a function of x has period a then we can associate a lattice of spacing a with it; the function looks identical when viewed from points this distance apart. The basis functions of a Fourier series expansion are all the sine and cosine functions with periodicity a; these have the general form $\cos(G_n x)$ and $\sin(G_n x)$ where

$$G_n = 2n\pi/a \qquad (11.30)$$

and n is an integer. We use the notation G_n because Eq. (11.30) just gives the reciprocal lattice vectors of the one-dimensional lattice of spacing a. Instead of $\cos(G_n x)$ and $\sin(G_n x)$ the complex exponentials $\exp(iG_n x)$ and $\exp(-iG_n x)$ can be used, and thus any function can be expanded

$$u(x) = \sum_G a_G\, e^{iG_n x} \qquad (11.31)$$

where the sum is over all the reciprocal lattice vectors and a_G are the Fourier coefficients; note that we do not need to include terms in $\exp(-iG_n x)$ explicitly since the sum is over all positive and negative values of n.

The function $u_k(\mathbf{r})$ in Eq. (11.27) can be expanded using the three-dimensional equivalent of Eq. (11.31),

$$u_k(\mathbf{r}) = \sum_G a_G(\mathbf{k})\, e^{i\mathbf{G}\cdot\mathbf{r}} \qquad (11.32)$$

where the sum is over all reciprocal lattice vectors, and the $a_G(\mathbf{k})$ are the Fourier coefficients. If we insert Eq. (11.32) into Eq. (11.27) and use Eq. (11.28) to substitute for $\mathbf{k}$, we obtain

$$\Psi(\mathbf{r}, t) = \frac{1}{\sqrt{V}} \sum_G a_G(\mathbf{k}' - \mathbf{G}_0)\, e^{i\mathbf{G}\cdot\mathbf{r}}\, e^{i[(\mathbf{k}' - \mathbf{G}_0)\cdot\mathbf{r} - \omega t]}$$

$$= \frac{1}{\sqrt{V}} \sum_G a_G(\mathbf{k}' - \mathbf{G}_0)\, e^{i(\mathbf{G} - \mathbf{G}_0)\cdot\mathbf{r}}\, e^{i(\mathbf{k}'\cdot\mathbf{r} - \omega t)}.$$

The difference $\mathbf{G}' = \mathbf{G} - \mathbf{G}_0$ is also a reciprocal lattice vector and since the sum is over *all* reciprocal lattice vectors it can be written as a sum over $\mathbf{G}'$, i.e.

$$\Psi(\mathbf{r}, t) = \frac{1}{\sqrt{V}} \sum_{G'} a_G(\mathbf{k}' - \mathbf{G}_0)\, e^{i\mathbf{G}'\cdot\mathbf{r}}\, e^{i(\mathbf{k}'\cdot\mathbf{r} - \omega t)}$$

$$= \frac{1}{\sqrt{V}} \sum_{G'} b_{G'}(\mathbf{k}')\, e^{i\mathbf{G}'\cdot\mathbf{r}}\, e^{i(\mathbf{k}'\cdot\mathbf{r} - \omega t)},$$

where the replacement of $a_G(\mathbf{k}' - \mathbf{G}_0)$ by $b_{G'}(\mathbf{k}')$ represents a relabelling of the Fourier coefficients. Thus $\Psi(\mathbf{r}, t)$ is again in the form of Eq. (11.27), namely

$$\Psi(\mathbf{r}, t) = \frac{1}{\sqrt{V}} \times \begin{pmatrix} \text{function of position} \\ \text{with the periodicity} \\ \text{of the real space} \\ \text{lattice} \end{pmatrix} \times e^{i(\mathbf{k}'\cdot\mathbf{r} - \omega t)},$$

except that it is represented by the wavevector $\mathbf{k}'$ rather than $\mathbf{k}$.

It is possible to use Eq. (11.28) to change the **k** vector used to represent the free electron wavefunction of Eq. (3.3) by rewriting this equation as

$$\Psi(\mathbf{r}, t) = \frac{1}{\sqrt{V}} e^{i(\mathbf{k}\cdot\mathbf{r}-\omega t)} = \frac{1}{\sqrt{V}} e^{-i\mathbf{G}_0\cdot\mathbf{r}} e^{i(\mathbf{k}'\cdot\mathbf{r}-\omega t)}.$$

Since $\exp(-i\mathbf{G}_0\cdot\mathbf{r})$ has the periodicity of the crystal lattice the rewritten wavefunction is in the form of Eq. (11.27). This is not a sensible thing to do because there is an obvious choice of **k** to be used for the free electron wavefunction. When the effect of the periodic lattice potential is introduced there is in general no longer an obvious choice of **k** and the freedom to be able to choose which primitive unit cell of the reciprocal lattice is to be used becomes very useful. We discuss different possible choices in the following section.

11.4.2   Brillouin zones and the plotting of dispersion relations

In plotting one-dimensional electron dispersion relations in Chapter 4 we found it useful to split **k**-space into Brillouin zones with boundaries at the k values for which Bragg diffraction of the electron wave occurs. The diffraction produces standing waves rather than running waves at the zone boundaries, with the consequent vanishing of the group velocity of wavelike normal modes. The energy gaps, produced by a periodic lattice potential, in the free electron dispersion curve also appear at the Brillouin zone boundaries. In a one-dimensional crystal the boundaries of the nth zone are given in terms of the lattice spacing a by

$$(n-1)\pi/a < |k| < n\pi/a,$$

and each zone consists of two regions each of width π/a symmetrically disposed about the origin (see Fig. 4.2). The first zone plays a special role in that it is a primitive unit cell of the reciprocal lattice† and is the unit cell normally used when dispersion curves from different branches of the dispersion curve are plotted in the same cell (as in Fig. 4.9(b)).

We now extend the Brillouin zone concept to two and three dimensions and find analogues for most of the properties mentioned in the previous paragraph. Although our discussion is phrased in terms of the electron states of the crystal it can be applied equally well to the other wavelike normal modes. We take the Brillouin zone boundaries to be given by the wavevectors **k** that satisfy the diffraction condition, Eq. (11.8); to identify these we write this equation as $\mathbf{k}' = \mathbf{k} + \mathbf{G}$, which on squaring becomes $k'^2 = k^2 + 2\mathbf{k}\cdot\mathbf{G} + G^2$. Recalling that $k'^2 = k^2$ then gives $2\mathbf{k}\cdot\mathbf{G} = -G^2$; if **G** is a reciprocal lattice vector then so is $-\mathbf{G}$ and this allows us change the sign in this equation and to write the

---

† Note that the lattice point is in the middle of the cell, not at a corner.

condition that $\mathbf{k}$ must satisfy for diffraction to occur as

$$2\mathbf{k} \cdot \mathbf{G} = G^2 \tag{11.33}$$

where $\mathbf{G}$ is any vector of the reciprocal lattice.

Values of $\mathbf{k}$ that satisfy this equation have the simple geometrical interpretation shown in Fig. 11.10; they lie on the plane that perpendicularly bisects the reciprocal lattice vector $\mathbf{G}$. The Brillouin zone boundaries for a two-dimensional square lattice are therefore as shown in Fig. 11.11. Successive zones are indicated by numbering; as in one dimension we regard the nth zone as comprising those regions of $\mathbf{k}$-space that can be reached from the origin by crossing a minimum of $n - 1$ boundaries. In contrast to the one-dimensional case the higher zones become increasingly fragmented. The first Brillouin zone is always continuous however; by its construction it is the locus of points nearer to the origin than to any other lattice point. It is thus the coordination polyhedron of a reciprocal lattice point, that is the Wigner–Seitz unit cell of the reciprocal lattic. Fig. 11.12 shows how the Wigner–Seitz cells of a general two-dimensional reciprocal lattice stack together to fill space. As in one dimension it is the first Brillouin zone that is identified as the 'special' zone for the plotting of dispersion relations.

To illustrate the different plotting schemes for dispersion relations we consider the curves predicted by the nearly free electron theory for the two-dimensional reciprocal lattice of Fig. 11.11. Fig. 11.13 shows contours of constant energy superimposed on the Brillouin zone structure; these curves have been obtained by distorting the free electron curves slightly to ensure that the contours intersect the zone boundaries at right angles, thus ensuring that the component of group velocity perpendicular to the boundary vanishes. By comparison with Fig. 4.9 we see that Fig. 11.13 is the generalization of the extended zone scheme to two dimensions; successively higher zones contain successively higher energy bands. Two difficulties arise in connection with this plotting scheme. First, the fragmentation of the higher zones makes it difficult to

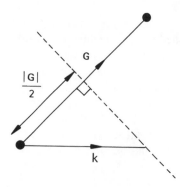

Fig. 11.10   Geometrical interpretation of Eq. (11.33). If $\mathbf{k}$ lies on the plane that perpendicularly bisects $\mathbf{G}$, then the component of $\mathbf{k}$ along $\mathbf{G}$ is $\mathbf{k}.\mathbf{G}/|\mathbf{G}| = |\mathbf{G}|/2$; hence $2\mathbf{k}.\mathbf{G} = G^2$

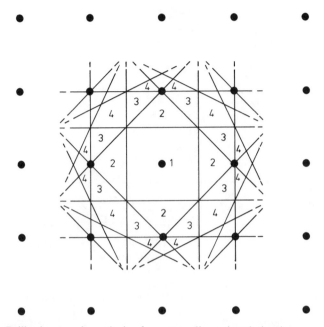

Fig. 11.11 Brillouin zone boundaries for a two-dimensional simple square reciprocal lattice. The boundaries are the perpendicular bisectors of the reciprocal lattice vectors for the lattice points shown. The numbering of successive Brillouin zones is achieved by counting the number of boundaries that must be crossed to reach them from the origin

see what is going on in the higher energy bands. Secondly, it is only for the nearly free electron model that it is possible to identify particular zones with particular energy bands in this way; in this model the electron states are obtained by perturbation theory from free electron states and the free electron **k** vector can be used to identify the Brillouin zone appropriate to the state.

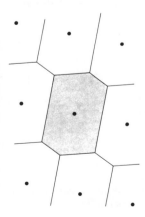

Fig. 11.12 The first Brillouin zone (shaded) is the Wigner–Seitz unit cell of the reciprocal lattice

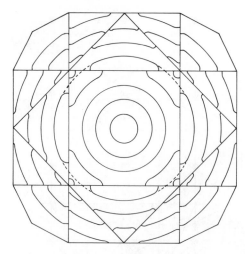

Fig. 11.13   Contours of equal electron energy superimposed on the Brillouin zone boundaries of Fig. 11.11. The first, second, third and part of the fourth zone are shown. The contours are obtained by distorting free electron circles slightly to obtain perpendicular intersections with zone boundaries. The broken curve shows an undistorted circle

To overcome these difficulties the reduced zone scheme can be used in which the dispersion relation of all energy bands is plotted in the first Brillouin zone. The lowest energy band in Fig. 11.13 is already plotted in this zone. Fig. 11.14 shows how the electron contours from the second zone can be remapped into the first zone by translation through the reciprocal lattice vectors indicated; the

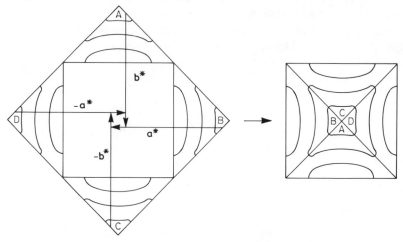

Fig. 11.14   Remapping of the energy contours from the second Brillouin zone into the first by translation through the reciprocal lattice vectors indicated

justification for this procedure is given in the previous section. The continuity of the resulting contours is guaranteed because the states that become adjacent are separated by a reciprocal lattice vector in the original extended zone scheme and are therefore the same state. By comparison with the original free electron circles on Fig. 11.13 we see that there is a maximum in the energy at the centre of the zone for this band.

Because the dispersion relations are periodic with periodicity of the reciprocal lattice we can continue the contours for each band throughout **k**-space to obtain the repeated zone scheme for the first two bands as shown in Fig. 11.15. Points labelled M in Fig. 11.15(a) are energy maxima, and since they differ by reciprocal lattice vectors they all correspond to the same state; this is also true for the maxima labelled M in Fig. 11.15(b). Points labelled m_1 in Fig. 11.15(b) are energy minima corresponding to the same state; the points labelled m_2 correspond to a different state which is degenerate with that labelled m_1, the degeneracy arising because of the symmetry of the lattice. We see that a great deal of information about the likely form of the dispersion curves can be obtained from the symmetry of the reciprocal lattice. Our approach in this section is largely two-dimensional but in Chapter 13 we show that it can be extended to predict the likely forms of the Fermi surface of real three-dimensional metals.

Thus far we have considered only the effects of the reciprocal lattice on the dispersion relations. For crystal structures with a basis of more than one atom associated with each real space lattice point, we saw in section 11.2.4 that the structure factor had a big influence on the intensities of the diffracted beams; in particular if the basis consists of two chemically identical atoms, the intensity of the beams associated with some reciprocal lattice vectors can become immeasurably small. The weakness of Bragg diffraction of electron waves associated with certain reciprocal lattice vectors would result in the appearance of a very small energy gap on the corresponding Brillouin zone boundary. Thus, for

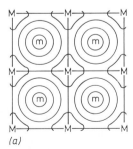

(a)

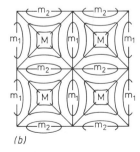

(b)

Fig. 11.15   Repeated zone scheme representation of (a) the lowest and (b) the next lowest energy band of the free electron model of Fig. 11.13. Labels M and m represent energy maxima and minima respectively

example, the energy gap at the Brillouin zone boundary corresponding to the (0 0 1) reciprocal lattice vector would be very small in an hcp metal.

PROBLEMS 11

11.1 Apply Newtonian mechanics to the elastic scattering of a light object off a stationary heavy object to show that the momentum change of the light object can be large whilst its energy change is negligible.

11.2 Show that taking the reciprocal lattice of the reciprocal lattice gives the real space lattice back again.

11.3 Calculate the four possible values of $|S|^2$, where S is the structure factor of Eq. (11.22), for the hcp structure.

11.4 Derive the conditions that must be satisfied by h, k and l for diffraction from a bcc structure when the conventional cubic unit cell is used to label the beams. Show that these conditions convert the simple cubic reciprocal lattice of the conventional cell into the fcc reciprocal lattice of the primitive cell.

11.5 Diffracted beams from a cubic close-packed structure are labelled as (1 0 0), ($\bar{1}$ 0 0), (1 1 1), ($\bar{1}$ 1 1), (1 1 0) and ($\bar{1}$ 1 0) using the primitive translation vectors of the reciprocal lattice given by Eq. (11.16). Deduce the labels that are attributed to these beams if the conventional cubic unit cell is used. Calculate values of $a(\sin \theta)/\lambda$ for the beams where a is the side of the conventional cell. Discuss the relative merits of the two labelling systems.

11.6 Calculate the structure factor for the CsCl structure (Fig. 1.14).

11.7 The Mo Kα radiation used to obtain the data in Fig. 11.7 has a wavelength 0.713 Å. Calculate the length of the side of the conventional cubic unit cell of aluminium from the figure.

11.8 Show how the third and fourth Brillouin zones on Fig. 11.11 can be mapped into the first zone by translation through appropriate reciprocal lattice vectors. Sketch the nearly free electron energy contours from the third zone when this has been done. Identify maxima and minima.

11.9 Calculate the ratio k_F/k_m for metals of valency 1 for both the bcc and fcc structures, where k_F is the free electron Fermi wavenumber and k_m is the minimum distance in $\mathbf{k}$-space from the origin to the boundary of the first Brillouin zone. What is the relevance of your results to the Fermi surfaces of sodium (bcc) and copper (fcc)?

We became very excited about this experimental challenge and the opening up of new possibilities. Astonishingly it took us a couple of weeks to realise that, not only would we have a local spectroscopic probe, but that scanning would deliver spectroscopic and topographic images, i.e. a new microscope.—*Gerd Binnig and Heinrich Rohrer—on the discovery of the scanning tunnelling microscope—Nobel prize address, 1986*

CHAPTER

Scattering of neutrons and electrons from solids

12.1 INTRODUCTION

In section 11.2 we presented a general approach to the problem of elastic scattering of waves by crystals and showed how measurements of the directions and intensities of diffracted beams enable the structure to be determined.† Experimental arrangements used in x-ray diffraction have already been described in Chapter 1. In this chapter we describe methods used to study the scattering of neutrons and electrons. We explain how inelastic scattering of neutrons can be used to provide information on the energy–momentum relation for the excitations in a solid, such as phonons and magnons. Finally we give a brief survey in section 12.6 of a number of important techniques, in most of which electrons are used, for studying the surfaces of solids.

12.2 COMPARISON OF X-RAYS, NEUTRONS AND ELECTRONS

12.2.1 Interaction of x-rays, neutrons and electrons with atoms

There are important differences in the way x-rays, neutrons and electrons interact with the atoms in a solid. X-rays are scattered primarily by the atomic

---

† We use the term *scattering* to encompass both elastic and inelastic processes. We reserve the term *diffraction* to refer to elastic scattering from an ordered structure.

electrons, and the atomic scattering factor f (see Eq. (11.1)), which describes the strength of the scattering, therefore increases steadily with increasing atomic number (Fig. 12.1); this makes it difficult for x-rays to detect light atoms, particularly in solids in which there are also much heavier atoms. Because of destructive interference between radiation scattered by different parts of the electron cloud, f also falls with increasing scattering angle (problem 12.1).

Neutrons interact with atoms in two ways. Because of the strong nuclear force they are scattered by the nucleus and the strength of this scattering is conventionally represented by the **scattering length** b such that the total cross section for scattering is $4\pi b^2$; the scattered wave is of the form of Eq. (11.1) with f replaced by b. As can be seen from Fig. 12.1 the scattering length does not increase monotonically with atomic number; neutrons are thus more useful than x-rays for determining the structure of solids containing light elements (particularly hydrogen) and for distinguishing between elements of similar atomic number. Because the nucleus is much smaller than the neutron wavelength it acts as a point scatterer and the scattering length does not fall off with increasing scattering angle. The second type of interaction of neutrons with atoms is the magnetic force between the magnetic moment of the neutron and that (if any) of

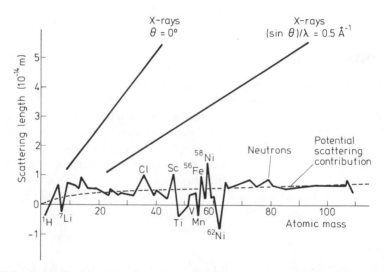

Fig. 12.1   Scattering length for the nuclear scattering of neutrons as a function of atomic number. The broken curve corresponds to scattering by a hard sphere of radius equal to that of the nucleus. The irregular variation about this curve is associated with resonant scattering; this occurs when the energy is such that, in the scattering process, a neutron and nucleus can combine to form an intermediate compound nucleus. The scattering length for x-rays is also shown; this is a function of $(\sin \theta)/\lambda$ where θ is the angle of scattering and λ the x-ray wavelength (problem 12.1). (Reproduced by permission of the Oxford University Press from G. E. Bacon, *Neutron Diffraction*, Third Edition, 1975)

the atom.† The force depends on the direction of the atomic moment and this allows neutron diffraction to be used to determine the nature of magnetic ordering in solids (section 12.5).

Like x-rays, electrons are scattered by the atomic electrons. The scattering of electrons is however very much stronger, and this has two important consequences. First, a low-energy electron beam, suitable for diffraction studies, is rapidly attenuated on entering a solid and such electrons are therefore used mainly for studying the chemical and physical nature of the surfaces of solids. Secondly, the simple scattering theory of section 11.2 is not valid because it assumes that the wave scattered by an atom is so weak that the amplitude of rescattering by another atom can be ignored. The analysis of electron scattering experiments is considerably complicated by the need to take account of multiple scattering. Because electrons are charged, electron beams can be focused by electric and magnetic fields; electrons can therefore be used to study a small localized region of the sample surface.

12.2.2   Inelastic scattering

To use inelastic scattering of x-rays, neutrons and electrons to investigate dispersion curves of excitations in solids it is necessary to measure the energy (frequency) change of the radiation. The relative difficulty of achieving this can be investigated by calculating the energies of photons, neutrons and electrons of wavelength 2 Å ($k = 10^{10}\pi$ m^{-1}). This wavelength is comparable to the atomic spacing and therefore ideal for structural studies; it is also well suited to investigating excitations of comparable wavelength (why?). These energies are:

x-rays	neutrons	electrons
$E = h\nu = hc/\lambda$	$E = \dfrac{\hbar^2 k^2}{2M_n}$	$E = \dfrac{\hbar^2 k^2}{2m_e}$
$\simeq 5 \times 10^4$ eV	$\simeq 0.02$ eV	$\simeq 40$ eV

Phonons are typical examples of excitations in a solid and have energies up to about the Debye energy (section 2.6.4), which is usually in the range 0.01–0.1 eV; the energy change of an inelastically scattered particle will be of this order. Comparing this with the incident energy we see that much higher energy resolution is required to measure the energy change for x-ray photons than for neutrons. If photons of lower energy are used their wavelength is no longer comparable to the interatomic spacing.

From what we have said here and in the previous section it might be thought that neutrons are a generally superior tool to x-rays. However, small and cheap

---

† Magnetic scattering by the nuclear magnetic moment can be ignored since nuclear magnetic moments are typically smaller by a factor 2000 than electronic moments.

x-ray sources are widely available. Neutrons are obtained either from a fission reactor or by bombardment of a suitable target with high-energy electrons or protons from a particle accelerator; in either case the sources are large and very expensive. Also the neutron intensity from such sources is relatively low (typically 10^{15} cm^{-2} s^{-1} from a reactor) and the range of neutron energies large, so that the intensity has often to be reduced further by selecting a narrow band of energy. In practice in a neutron scattering experiment there is always some sacrifice of collimation and energy resolution for the sake of intensity. In contrast, for x-rays an intense monochromatic incident beam is easily obtained since the output of an x-ray tube is a mixture of characteristic lines and continuous background; with a suitable choice of operating voltage a large proportion of line emission can be obtained. X-rays are therefore normally used, wherever possible, for precision structure determination.

12.3   NEUTRON SCATTERING TECHNIQUES

A neutron scattering experiment in general requires a neutron source and detector, collimation to define the directions of the incident and scattered beams, and some means of determining the energies of the incident and scattered particles. An apparatus for studying neutron scattering is known as a **neutron spectrometer**.

12.3.1   Neutron sources

A continuous neutron flux can be obtained from a fission reactor. Collisions of the neutrons within the moderator (typically graphite) of the reactor give the neutrons a Maxwell–Boltzmann distribution of speeds appropriate to a temperature of order room temperature; such neutrons are described as **thermal neutrons**. Although the corresponding mean energy of 0.025 eV is very suitable for many scattering experiments (see previous section), the broadness of the speed distribution means that some method of energy selection must be incorporated into the experimental set-up.

In some experiments a short pulse of neutrons is used. Pulses can be obtained from a reactor source (see section 12.3.3) but it is often better to create them by collision of a pulse of high-energy protons or electrons with a suitable target. In a **proton spallation source** pulses of protons with an energy of order 800 MeV from a synchrotron are used. If uranium is the target material then each proton produces as many as 30 neutrons. The neutrons are produced as 'chips' knocked off the uranium nuclei by the protons;† additional neutrons are produced by fission of the nuclei.

---

† The word 'spallation' comes from the verb *to spall*, meaning to splinter or chip.

The energy of the emitted neutrons is typically 1 MeV—too large for scattering studies of solids. A slab of polyethylene can be used as a moderator; the thickness of the slab is chosen to obtain a neutron energy distribution appropriate to the experiment being performed. Spallation sources are particularly useful if neutrons of higher energy than the thermal neutrons from a reactor are required; neutrons do not remain within the moderator long enough to become thermalized to the moderator temperature and are said to be **epithermal**. The neutron pulse is typically of duration 10 μs and contains a broad spectrum of neutron energies.

12.3.2   Neutron detectors

Neutrons are uncharged and therefore difficult to detect directly. Low-energy neutrons can however induce nuclear reactions in which high-energy ions are produced; these cause strong ionization, which can be detected. One such reaction that has a high cross section (2100×10^{-28} m^2 for neutrons of 1 Å wavelength) is with the boron isotope of mass 10,

$$^{10}\text{B} + \text{n} = {}^{7}\text{Li} + {}^{4}\text{He} + 2.3 \text{ MeV}.$$

This reaction is exploited in the BF$_3$ proportional neutron counter in which the ionization in BF$_3$ gas enriched with the ^{10}B isotope is detected.

12.3.3   Time-of-flight methods

Two methods are used for defining and measuring neutron energies: in this section we describe time-of-flight techniques and in the following section we discuss crystal monochromators. The velocity of a neutron of wavelength 2 Å is

$$v = \frac{\hbar k}{M_\text{n}} \approx \frac{10^{-34} \times 10^{10}\pi}{1.6 \times 10^{-27}} \text{ m s}^{-1} \approx 2000 \text{ m s}^{-1}. \tag{12.1}$$

A velocity of this order can readily be determined by measuring the time of flight of a neutron over a distance of a few metres. The time at which the flight ends can be determined by noting the time of arrival at the detector. The time of the beginning of the flight is known if the neutrons are introduced into the spectrometer in pulses. Time-of-flight methods are therefore particularly suited for use with pulsed neutron sources; neutrons of different energy within each pulse arrive at the detector at different times, and energy discrimination in pulsed experiments is thereby achieved.

A simple spectrometer for neutron diffraction studies based on time-of-flight methods is shown in Fig. 12.2. The count rate at the detector is measured as a function of the time of arrival. For elastically scattered neutrons the time of flight of neutrons of wavelength λ (momentum p) is given by

$$t = \frac{L}{v} = \frac{M_n L}{p} = \frac{\lambda M_n L}{h},$$ (12.2)

where L is the total length of the neutron path from the source to the detector via the specimen. Diffraction occurs when the Bragg law (Eq. (1.3))

$$n\lambda = 2d \sin \theta$$

is satisfied. For a fixed detector position, θ is determined by the experimental geometry as indicated in Fig. 12.2.

To see diffraction from a given set of lattice planes two conditions must be satisfied: the planes must be at the correct angle to the incident beam and the beam must include neutrons of the wavelength required to satisfy the Bragg law. The first condition can be satisfied by using a powder specimen (see section 1.4) so that planes at all possible incident angles occur. The second condition is fulfilled if the incident pulses contain neutrons with a wide energy range. By combining Eqs. (1.3) and (12.2) we find that the times of flight for neutrons undergoing Bragg diffraction are given by

$$t = \frac{M_n L \lambda}{h} = \frac{M_n L}{h} \frac{2d}{n} \sin \theta.$$ (12.3)

Peaks therefore occur in the neutron count rate at times satisfying this equation. Fig. 12.3 shows a plot of count rate against flight time for such an experiment. Eq. (12.3) has been used to relabel the horizontal axis with the wavelength λ of the neutrons corresponding to a particular flight time.

Neutron pulses can be obtained using a spallation source as described in section 12.3.1. They can also be obtained from the continuous beam from a

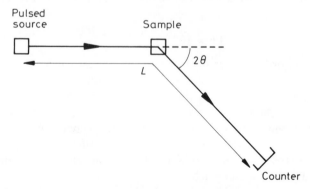

Fig. 12.2   Time-of-flight spectrometer for studying elastic scattering of neutrons. To investigate inelastic scattering it is necessary to add a monochromator to define the energy of either the incident or scattered neutrons

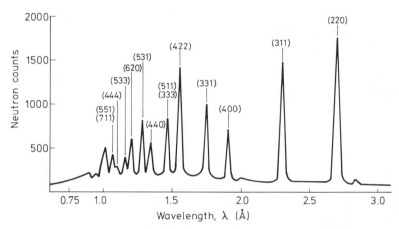

Fig. 12.3 Diffraction pattern of silicon powder, measured at a scattering angle 2θ of $90°$ with a time-of-flight spectrometer. (From Risø Report R164 (1967). Reproduced by permission of Bente Lebech, Risø National Laboratory, Denmark)

reactor by using a **chopper**. Fig. 12.4 shows two types of chopper. In the disc type, a disc opaque to neutrons rotates about an axis parallel to the beam, allowing bursts of neutrons to pass through one or more slots near the periphery. The Fermi-type chopper consists of a multilayer sandwich of a material with high neutron absorption (e.g. cadmium for low-energy neutrons, boron for neutrons of higher energy) and a material with low neutron absorption (e.g. aluminium) rotated about an axis perpendicular to the beam; neutrons are transmitted only when the sandwich is parallel to the incident beam.

To perform inelastic scattering measurements using time-of-flight methods requires the addition to the spectrometer of Fig. 12.2 of a monochromator to define the energy of either the incident or the scattered neutrons; together with the total time of flight this enables the energy of both incident and scattered neutrons to be determined. Monochromation of the incident beam can be achieved by placing a chopper in the incident beam at some distance from the source of the neutron pulse. The chopper only transmits those neutrons which

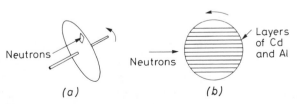

Fig. 12.4 (a) Disc-type chopper. (b) Fermi-type chopper

arrive at a time when the chopper is transparent and in this way velocity selection is achieved. Monochromation can also be obtained by placing a crystal monochromator in either the incident or scattered beam. Monochromation reduces the number of neutrons in the pulse and makes data collection slow. To compensate for this problem several detectors can be used so that many scattering angles are studied simultaneously.

12.3.4 Crystal monochromators

An alternative to time-of-flight methods for neutron energy selection is to use Bragg reflection from a suitably oriented single crystal. This technique is used to make the monochromator and analyser in the **triple-axis spectrometer** shown in Fig. 12.5. The monochromator diffracts neutrons of the desired wavelength (varied by changing θ_M) towards the sample, allowing the undiffracted remainder of the neutron beam to be absorbed by shielding, not shown in Fig. 12.5. Commonly a (1 1 1) reflection from a Ge crystal is used because there is effectively no second-order (2 2 2) reflection from the diamond structure (section 11.2.4); reflection of neutrons of half the desired wavelength is thereby avoided. Pyrolitic graphite and Be single crystals are also used for neutron monochromation. Single crystals are often too perfect and need to be strained a little to introduce some mosaic structure; this increases the range of wavelengths reflected by the crystal and thus increases the intensity (Bacon[33]). If polarized neutrons are required, a magnetized ferromagnetic crystal can be used for the monochromator (section 12.5).

The analyser consists of another crystal adjusted for Bragg reflection and is used to define the energy of the scattered neutrons reaching the detector; the energy is varied by varying θ_A. For elastic scattering studies the analyser is not required. For inelastic scattering the triple-axis spectrometer suffers the disadvantage with respect to the time-of-flight spectrometer that only one incident and outgoing energy and one scattering angle are studied at any particular time. On the other hand the neutron flux is continuous, not pulsed, and as we shall now explain the triple-axis spectrometer allows the user the convenience of being able to study the energy changes of neutrons for a fixed value of scattering vector $\mathbf{K} = \mathbf{k}' - \mathbf{k}$. In the time-of-flight spectrometer, for neutrons of given incident wavevector $\mathbf{k}$, the scattered wavevector depends on the energy of the scattered neutrons and hence so does $\mathbf{K}$. In the triple-axis spectrometer the incident energy can be fixed by keeping θ_M constant. The outgoing energy is adjusted by changing θ_A, but the consequent change in $|\mathbf{k}'|$ can be compensated by simultaneously altering the scattering angle ϕ to keep $|\mathbf{K}| = |\mathbf{k}' - \mathbf{k}|$ constant; also the angle ψ of the sample can be varied so that $\mathbf{K}$ maintains its orientation with respect to the crystal axes (see Fig. 12.5(b)). Thus by simultaneously changing the three angles, θ_A, ϕ and ψ, it is possible to investigate the

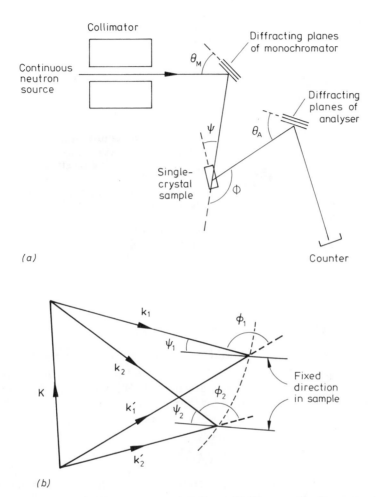

Fig. 12.5   (a) Triple-axis neutron spectrometer, so called because simultaneous rotations about three axes are used in its operation (see text). (b) Diagram to illustrate the constant **K** mode of operation of the triple-axis spectrometer. For a fixed incident energy (θ_M = constant, $|\mathbf{k}_1| = |\mathbf{k}_2|$) but varying scattered energy ($\theta_A \neq$ constant, $|\mathbf{k}_1'| \neq |\mathbf{k}_2'|$) it is possible to keep $|\mathbf{K}|$ constant in magnitude and direction relative to the sample by suitable changes in ψ and ϕ

variation of neutron count rate with the energy E' ($= \hbar^2 k'^2/2M_n$) of the scattered neutrons at fixed values of the incident neutron energy E ($= \hbar^2 k^2/2M_n$) and scattering vector **K**. Fig. 12.6 shows data obtained in this way for scattering from a single crystal of magnesium; the explanation of the data is given in section 12.4.

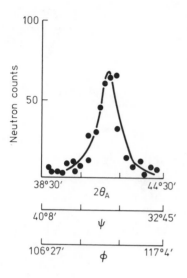

Fig. 12.6 Inelastic neutron scattering from a magnesium crystal observed using a triple-axis neutron spectrometer in the constant **K** mode of operation. The values of θ_A, ψ and ϕ required to keep **K** constant are also indicated. (Reproduced with permission from P. K. Iyengar in *Thermal Neutron Scattering*, ed. P. A. Egelstaff, Academic Press, London (1965))

12.4 DETERMINATION OF PHONON SPECTRA

In section 11.2 we considered diffraction of a wave by a perfectly periodic crystal. At finite temperature a real crystal is not perfectly periodic because thermally excited lattice vibrations are present. We now generalize our approach to allow for this possibility; for simplicity we suppose that a single lattice wave of wavevector **q** and frequency ω is present so that the position of the nth atom within the crystal is†

$$\mathbf{r}_n = \mathbf{r}_n^0 + \mathbf{u}_0 \cos{(\mathbf{q}\cdot\mathbf{r}_n^0 - \omega t)} \qquad (12.4)$$

where $\mathbf{r}_n^0$ is the equilibrium position, and the magnitude and direction of $\mathbf{u}_0$ give the amplitude and polarization direction of the lattice wave. We shall find that this periodic perturbation of the perfect crystal gives rise to additional scattering maxima analogous to the 'ghosts' that occur with optical diffraction gratings with a periodic ruling error (Born and Wolf, *Principles of Optics*, Pergamon, London (1959), p. 407).

The amplitude of the scattered wave is given by Eq. (11.2) as

$$A = \sum_n f_n \exp{[-i(\mathbf{K}\cdot\mathbf{r}_n + \Omega t)]}, \qquad (12.5)$$

where, since we wish to consider inelastic scattering, we have introduced the time dependence associated with the energy $\hbar\Omega$ of the incident neutron.

---

† In this chapter we follow the usual practice of calling the phonon wavevector **q** to distinguish it from the electron or neutron wavevector **k**.

Substituting the atomic displacements of Eq. (12.4) gives

$$A = \sum_n f_n \exp\{-i[\mathbf{K}.(\mathbf{r}_n^0 + \mathbf{u}_0 \cos(\mathbf{q}.\mathbf{r}_n^0 - \omega t)) + \Omega t]\}$$

$$\approx \sum_n f_n \exp[-i(\mathbf{K}.\mathbf{r}_n^0 + \Omega t)]\{1 - i\mathbf{K}.\mathbf{u}_0 \cos(\mathbf{q}.\mathbf{r}_n^0 - \omega t) - \cdots\} \quad (12.6)$$

where the term in curly brackets on the second line has been obtained by expanding $\exp[-i\mathbf{K}.\mathbf{u}_0 \cos(\mathbf{q}.\mathbf{r}_n^0 - \omega t)]$; this is justified if the amplitude of the vibrations is small. By writing the cosine in terms of complex exponentials, Eq. (12.6) becomes

$$A = \sum_n f_n \exp[-i(\mathbf{K}.\mathbf{r}_n^0 + \Omega t)] - \tfrac{1}{2}i\mathbf{K}.\mathbf{u}_0 \sum_n f_n \exp\{-i[(\mathbf{K}-\mathbf{q}).\mathbf{r}_n^0 + (\Omega + \omega)t]\}$$

$$- \tfrac{1}{2}i\mathbf{K}.\mathbf{u}_0 \sum_n f_n \exp\{-i[(\mathbf{K}+\mathbf{q}).\mathbf{r}_n^0 + (\Omega - \omega)t]\}$$

$$+ \text{(terms of order } u_0^2 \text{ and higher).} \quad (12.7)$$

We showed in section 11.2 that the first term gives a sharp Bragg diffraction peak when $\mathbf{K} = \mathbf{G}$, where $\mathbf{G}$ is any reciprocal lattice vector; this term oscillates at the same frequency Ω as the incident radiation and thus corresponds to elastic neutron scattering from a perfectly periodic crystal.

Using a similar argument to that used in section 11.2 we see that the second term gives a sharp maximum when $\mathbf{K} - \mathbf{q} = \mathbf{G}$, or

$$\mathbf{k}' = \mathbf{k} + \mathbf{q} + \mathbf{G}. \quad (12.8)$$

The amplitude of the maximum is proportional to $\mathbf{K}.\mathbf{u}_0$ and oscillates at a frequency Ω' given by

$$\Omega' = \Omega + \omega. \quad (12.9)$$

Eqs. (12.8) and (12.9), when multiplied by $\hbar$, have the appearance of the laws of conservation of momentum and energy for a process in which a neutron of wavevector $\mathbf{k}$ absorbs a phonon of wavevector $\mathbf{q}$ and is scattered into a state of wavevector $\mathbf{k}'$. This is the quantum interpretation of the above classical calculation. We should, however, be cautious in interpreting Eq. (12.8) as a momentum conservation equation. It is certainly true that in the scattering process the crystal receives an impulse $\hbar(\mathbf{k} - \mathbf{k}') = -\hbar\mathbf{K} = \hbar(-\mathbf{q} - \mathbf{G})$, and that this impulse is transmitted to the crystal mounting. It is pure convention to divide this into a part $-\hbar\mathbf{G}$ given to the whole lattice and a part $-\hbar\mathbf{q}$ associated with absorption of a phonon. The division is arbitrary in that the phonon of wavevector $\mathbf{q}$ is equally well represented by a wavevector $\mathbf{q} + \mathbf{G}'$ where $\mathbf{G}'$ is any reciprocal lattice vector (see section 11.4). This possibility can be used, by setting $\mathbf{G}' = \mathbf{G}$, to associate the whole of the momentum change with the phonon and none with the lattice as a whole! As we have already mentioned the quantity $\hbar\mathbf{q}$ is called the crystal momentum because it behaves as a momentum in equations such as Eq. (12.8).

The third term in Eq. (12.7) represents neutron scattering in which a phonon of wavevector $\mathbf{q}$ is *emitted*. The terms of order u_0^2 and higher correspond to quantum processes in which two or more phonons are emitted or absorbed.[†]

We see from Eq. (12.7) that the scattered amplitude due to single-phonon emission and absorption processes is proportional to u_0, and hence the intensity is proportional to u_0^2, i.e. to the phonon intensity or phonon number. Our classical calculation is not quite right here; a quantum mechanical calculation gives the result that, if n is the number of phonons present initially in a lattice mode, the emission probability is proportional to $(n + 1)$ and the absorption probability to n.[‡] At low temperatures, when very few phonons are present ($n \ll 1$) it follows that only the phonon emission process can occur; there are no phonons to be absorbed.

Because the triple-axis spectrometer can be used to investigate neutron scattering at fixed values of the scattering vector $\mathbf{K} = \mathbf{k'} - \mathbf{k}$, it is an excellent tool for studying phonon dispersion relations. The procedure is to set the fixed value of $\mathbf{K}$ equal to the wavenumber $\mathbf{q}$ for which the energy is to be determined (or to $\mathbf{q} + \mathbf{G}$ where $\mathbf{G}$ is any reciprocal lattice vector); this ensures that Eq. (12.8) is always satisfied. The neutron count rate as a function of the energy of the scattered neutrons will peak whenever Eq. (12.9) is also satisfied. Such a peak is seen in Fig. 12.6 and from its position the energy of the phonons of wavenumber $\mathbf{q}$ can be determined. The appearance of more than one peak is an indication that phonons of more than one energy occur for this wavenumber. Once the energies of the phonons of a particular wavenumber have been determined, the experiment can be repeated at other values of $\mathbf{K}$ and hence of the phonon wavenumber. It is thus possible to obtain the complete phonon spectrum. Fig. 12.7 shows phonon dispersion relations for potassium that have been obtained in this way.

12.5 MAGNETIC SCATTERING

12.5.1 Determination of magnetic structure

We have already mentioned that, in addition to the nuclear scattering, a neutron may be scattered magnetically by the electrons in a solid. This scattering has been calculated by treating the potential energy of the neutron in the magnetic field of the electrons as a perturbation. The calculation is lengthy and the detailed interpretation of experimental results is complicated by the

---

[†] A measurable consequence of the second-order terms is the reduction in the intensity of the Bragg diffraction peaks with increasing temperature by the **Debye–Waller factor**; this factor can be estimated as $\exp\left(-k_B T |\mathbf{K}|^2/M\omega_D^2\right)$ where M is the atomic mass and ω_D is the Debye frequency.

[‡] This is shown for any particle obeying Bose–Einstein statistics in chapter 4 of Feynman.[6] In the particular case of photons (black-body radiation, see Mandl[2]) the 1 in $(n + 1)$ corresponds to spontaneous emission.

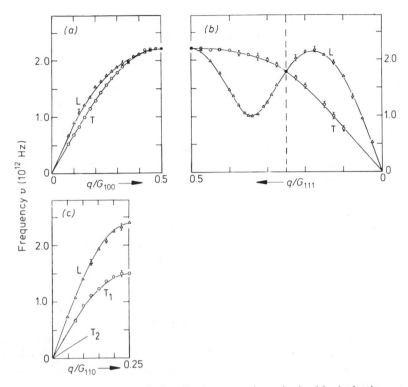

Fig. 12.7   Phonon dispersion relations for bcc potassium obtained by inelastic neutron scattering using a triple-axis spectrometer. Curves are shown for the phonon wavevector **q** in the (a) [1 0 0], (b) [1 1 1] and (c) [1 1 0] directions. The triangles and circles denote longitudinal and transverse modes respectively. In the [1 0 0] and [1 1 1] directions the two transverse modes are degenerate. In the [1 1 0] direction only the initial slope of one of the transverse modes is shown. In (a) and (c) the curves are plotted up to the first Brillouin zone boundary. In (b) the position of the first Brillouin zone boundary is indicated by the vertical broken line; the parts of the curves to the left of this line can be translated into the first Brillouin zone by subtracting a suitable reciprocal lattice vector. The conventional cubic unit cell of the fcc reciprocal lattice has been used to define the reciprocal lattice vectors $\mathbf{G}_{100}$, $\mathbf{G}_{111}$ and $\mathbf{G}_{110}$. To help understanding of this figure the reader is recommended to attempt problem 12.4. (Reproduced with permission from R. A. Cowley, A. D. B. Woods and G. Dolling, *Phys. Rev.* **150**, 487 (1966))

interplay between magnetic and nuclear scattering (see Bacon[33]). For the determination of magnetic structure the important result is that the magnetic scattering of a neutron by an atom with a magnetic moment is described by a contribution

$$b_{\mathrm{M}} = \pm p \sin \alpha \qquad (12.10)$$

to the scattering length, where α is the angle between the atomic magnetic moment and the scattering vector **K**. The + or − sign depends on the orientation of the neutron spin.† The coefficient p is proportional to the atomic magnetic moment and the neutron moment, and it falls off with increasing scattering angle in a similar manner to the atomic scattering factor f for x-rays (see problem 12.1).

In a paramagnetic crystal the atomic magnetic moments are randomly orientated with respect to $\mathbf{K}$; α thus varies randomly from atom to atom so that the total scattering length for a neutron varies randomly between $b + p$ and $b - p$, where b is the contribution of nuclear scattering to the scattering length. We saw in section 11.2 that sharp Bragg diffraction peaks occur because a crystal looks like a lattice of identical repeat units to the incident waves; because the magnetic contribution to the scattering changes randomly from atom to atom, magnetic scattering does not contribute to the amplitude of the Bragg peaks in a paramagnetic solid. The magnetic scattering produces a finite neutron intensity between the Bragg peaks which varies smoothly with scattering angle. Magnetic scattering by paramagnetic solids is said to be **incoherent**.‡

In a ferromagnetically ordered crystal, however, α does not vary randomly from atom to atom and the magnetic scattering adds **coherently** to the nuclear scattering and does contribute to the Bragg peaks. The essential features of the Bragg peaks are however unmodified in this case.§ In antiferromagnetic crystals more drastic effects are seen, because the value of α is equal and opposite for the magnetic moments on the A and B sublattices (section 8.4); these behave towards a neutron as *two different types of atom* with coherent scattering lengths $b + p \sin \alpha$ and $b - p \sin \alpha$. This results in a magnetic unit cell that is in general larger than the chemical unit cell.

To illustrate this we consider MnO, which has the NaCl structure (Fig. 1.13). MnO is antiferromagnetic with a Néel temperature of 120 K. The results of neutron scattering experiments on a powdered sample of MnO at 80 K and at room temperature are shown in Fig. 12.8. Below the Néel temperature extra diffraction peaks are visible, in particular a strong peak at about 12°. This is approximately half the scattering angle of the (1 1 1) peak that appears both above and below the Néel temperature, and thus indicates scattering from lattice planes with twice the spacing of the (1 1 1) planes.

---

† The neutron spin can be either parallel or antiparallel to the direction perpendicular to both $\mathbf{K}$ and the atomic moment.

‡ A fraction of the nuclear scattering may also be incoherent, i.e. at angles between the Bragg peaks. Incoherent nuclear scattering arises because some elements have more than one isotope, with quite different nuclear scattering lengths, and also because, for nuclei with a finite spin, the scattering length depends on the relative orientation of the nuclear and neutron spins. The distribution of isotopes and nuclear spin orientations normally varies randomly within a crystal.

§ Ferromagnetic crystals can be used to produce a monochromatic beam of *polarized* neutrons. A Bragg peak is chosen for which the total scattering length $b - p \sin \alpha$ for neutrons of one polarization is close to zero; the diffracted beam then consists almost entirely of neutrons of the opposite polarization.

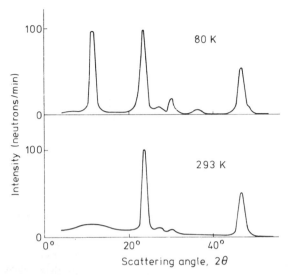

Fig. 12.8   Neutron diffraction from powdered MnO above and below the antiferromagnetic Néel temperature of 120 K. Note particularly the extra reflection at 12°, which is the (1 1 1) reflection of the doubled magnetic unit cell. (Reproduced with permission from C. G. Shull and J. S. Smart, *Phys. Rev.* **76**, 1256 (1949))

These results can be explained by the structure shown in Fig. 12.9, in which alternate (1 1 1) planes of Mn have oppositely aligned magnetic moments; the moments within a (1 1 1) plane are aligned ferromagnetically with respect to each other but antiferromagnetically with respect to the moments on the neighbouring planes. Because of the opposite orientation, the Mn atoms on successive planes have different scattering lengths for neutrons, and a diffraction maximum can occur when the phase difference between successive planes is only π; hence the peak at the scattering angle of 12° in Fig. 12.8. The magnetic unit cell below the Néel temperature has double the linear dimension of the chemical unit cell shown in Fig. 12.9. Precision x-ray measurements, of greater resolution than the neutron measurements, show that the unit cell is no longer strictly cubic below the Néel temperature; it is distorted by extension along a [1 1 1] direction. This is consistent with the symmetry of the magnetic structure deduced from the neutron measurements.

   Although the determination of magnetic structures has been done predominantly by neutron diffraction it is also possible to use x-rays for this purpose. The scattering of circularly polarized x-rays by an atom does depend weakly on the spin and orbital angular momentum of the atom. The high-intensity x-rays from synchrotron sources (section 1.4.2) can be used to exploit this possibility. Although the radiation in the direction tangential to the particle motion is strongly plane polarized in the plane of the synchrotron orbit, circularly

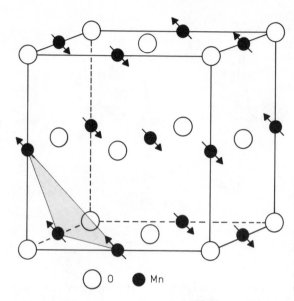

Fig. 12.9 The antiferromagnetic structure of MnO. A chemical unit cell of the structure is shown; the magnetic unit cell has twice the linear dimensions. The Mn atoms have magnetization oppositely directed in alternate (1 1 1) planes

○ O   ● Mn

polarized x-rays of much lower intensity are emitted at a small angle to this direction. The intensity is still sufficiently high to enable the investigation of magnetic structures.

12.5.2 Determination of magnon spectra

Because the magnetic scattering of neutrons depends on the orientation of the atomic magnetic moment, the presence of a spin wave leads to a modulation of the scattering properties of the lattice. In view of our discussion of scattering of neutrons by lattice vibrations in section 12.4, the reader will not be surprised to learn that the scattering of a neutron can be pictured as occurring by emission or absorption of a magnon with conservation of momentum and energy:

$$\mathbf{k'} = \mathbf{k} \pm \mathbf{q} + \mathbf{G}, \tag{12.11}$$

$$\Omega' = \Omega \pm \omega, \tag{12.12}$$

where the upper sign refers to absorption, the lower to emission. The inelastic scattering of neutrons thus makes it possible to determine experimentally the relation between the magnon frequency ω and wavevector $\mathbf{q}$. The appearance of the reciprocal lattice vector $\mathbf{G}$ in Eq. (12.11) is, as in Eq. (12.8), a consequence of the ambiguity of the wavevector of the excitations in a periodic lattice. A magnon of wavenumber $\mathbf{q}$ can be equally well represented by a wavenumber $\mathbf{q} + \mathbf{G}$; $\hbar\mathbf{q}$ is the crystal momentum of the magnon.

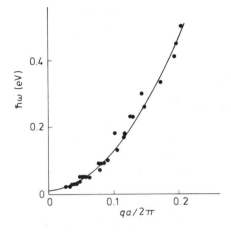

Fig. 12.10   Spin wave spectrum of ferromagnetic $Co_{0.92}Fe_{0.08}$ obtained using inelastic neutron scattering. Note the finite energy gap at $q = 0$ due to crystalline anisotropy. (Reproduced with permission from R. N. Sinclair and B. N. Brockhouse, *Phys. Rev.* **120**, 1638 (1960))

As for phonons, it is convenient to use the triple-axis spectrometer in its 'constant **K**' mode to determine magnon dispersion curves. From Eq. (12.11) we see that this is equivalent to looking for absorption of magnons of wavenumber $\mathbf{q} = \mathbf{K} - \mathbf{G}$ (or emission of magnons of wavenumber $\mathbf{G} - \mathbf{K}$). The frequency of magnons of known wavenumber can thus be determined from the value of $\Omega' - \Omega$ at which a peak in scattering intensity is observed. Inelastic neutron scattering was used to determine the spin wave spectrum of Fig. 12.10 for a ferromagnetic cobalt–iron alloy. In contrast to the prediction of Eq. (8.31) the magnon energy does not vanish at $\mathbf{q} = 0$. This is a consequence of crystalline anisotropy, which we neglected in section 8.5.2. There is a preferred direction for the spontaneous magnetization within the lattice and a finite amount of energy is required to cause a spatially uniform ($\mathbf{q} = 0$) rotation of the magnetization from this direction.

12.6   ELECTRON SCATTERING

As we have already mentioned, electron beams with an energy (≈ 100 eV) suitable for diffraction studies penetrate only a short distance (≈ 5 Å) into a crystal. The distribution of the scattered electrons therefore gives information only about the region of the crystal that lies within about two atomic diameters of the surface. Electron scattering can thus be used to study the surfaces of solids.

Electrons with a well defined energy of order 100 eV are readily produced by accelerating the electrons from a hot filament. They can be electrostatically or magnetically focused onto the surface being studied. The choice of detector is determined by the requirements of the experiment. For electron *diffraction* experiments the detector must eliminate the inelastically scattered electrons (usually the majority) and allow measurement of the angular distribution of the elastically scattered electrons. In *inelastic scattering* experiments the distribution over energy of the scattered electrons must be determined. The widespread use of electrons for surface studies has led to the development of a number of standard experimental configurations, some of which we mention below. Surface studies are very important for technological purposes; the nature of the surface plays a crucial role for example in semiconductor physics, in corrosion and in catalysis.

For studies of surface physics it is desirable to be able to produce a 'clean' surface. Such a surface is normally very reactive and impinging atoms and molecules tend to stick to it, a process known as **adsorption**. A 'dirty' layer one atom thick can form on the surface with great rapidity (problem 12.7) even if the solid is in an 'evacuated' container. To maintain a 'clean' surface for a sufficient time to study it requires a pressure in the range of $10^{-12} - 10^{-13}$ atmospheres; this region of pressure is known as **ultrahigh vacuum** (UHV). At such low pressures scattering of the electrons by gas molecules is also avoided.

Various methods have been used to produce a 'clean' surface. Some crystals may be cleaved inside the UHV system. Alternatively bombardment of the surface by gas ions of energy of order 10^2 to 10^3 eV may be used to erode the impure surface layers. More recently the molecular beam epitaxy technique (section 6.6) has allowed chemically pure surfaces to be prepared under UHV conditions. Other techniques are also available.

The use of electron scattering techniques for the study of surfaces is analogous to the use of neutrons for studies of the interior of crystals described earlier in this chapter. Thus the structure of the two-dimensional crystal formed by the surface atoms can be investigated using **low-energy electron diffraction** (LEED).† The directions of the diffracted beams are related to the lattice by the two-dimensional version of the theory presented in section 11.2; calculation of the intensities of the beams is more difficult for electrons because of the need to allow for multiple scattering.

Inelastic electron scattering studies give information on the energy spectrum of the excitations localized to the surface; the technique for doing this is known as **high-resolution electron energy loss spectroscopy** (HREELS or sometimes just EELS). An example of a surface excitation is a vibrational mode of an adsorbed molecule, so HREELS can detect chemical contamination of the surface. Typically HREELS investigates energy losses up to about 1 eV in 5 eV electrons.

The chemical composition of the surface can also be investigated by **Auger electron spectroscopy** (AES). In this technique an incident electron of energy typically 3 keV excites an atom on the surface by knocking an electron out of an inner shell state. The hole in the inner shell is then filled by an electron falling from a higher level; the energy lost by this electron can be released either as a

---

† Diffraction of ~ 100 eV electrons is designated low-energy to distinguish it from **high-energy electron diffraction** (HEED) which uses ~ 10 keV electrons.

photon† or by emission of a secondary electron. This latter possibility is the Auger process and in AES the energy distribution of the emitted electrons is measured. An example of the Auger process would be the filling of a K shell vacancy by an electron from the L shell (this electron loses energy $\Delta E = |E_L - E_K|$) with emission of an M shell electron of kinetic energy $\Delta E - |E_M|$. Since the energies of the inner shell states vary from one element to another, it is possible to identify the elements on the surface and their approximate concentrations from the Auger emission spectrum.

One of the most widely studied crystal surfaces is the (1 1 1) surface of silicon. Continuation of the bulk structure of silicon up to this surface would suggest the structure shown in Fig. 12.11 (compare with Fig. 1.15(b), which shows a (1 1 1) slice of the crystal). Each of the surface atoms in Fig. 12.11 has a 'dangling bond' and such a surface will seek to reduce its energy by reducing the number of these. This process is known as **reconstruction** and involves a shift in position of the atoms near to and on the surface. The reduction in energy associated with the decrease in the number of unsatisfied bonds is counterbalanced by the extra strain energy associated with distortion of the structure below the surface.

More than one ordered reconstruction of the Si (1 1 1) surface has been observed using LEED. The different reconstructions are characterized by the dimensions of the two-dimensional unit cell of the surface structure. These are conveniently specified in terms of the number of repeat units of the unreconstructed cell. The most stable structure of a clean Si (1 1 1) surface at room temperature is believed to be a 7×7 reconstruction; the repeat distance of the reconstructed structure is seven times the unreconstructed unit cell side along both crystallographic axes.

Although the structure of the 7×7 reconstruction of the Si (1 1 1) surface has been determined by LEED, the most impressive confirmation of this structure is

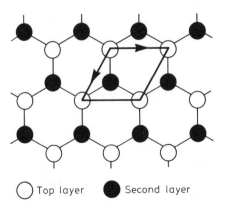

○ Top layer   ● Second layer

Fig. 12.11   Plan view of the unreconstructed (1 1 1) surface of silicon. Each atom in the top layer forms only three covalent bonds; the fourth bond in the direction perpendicular to the paper is broken in forming the surface. The unit cell of the two-dimensional surface structure is indicated

† The spectrum of the emitted photons contains the same information as that of the Auger electrons. Studying the photon spectrum is known as **appearance potential spectroscopy** (APS).

provided by the **scanning tunnelling microscope** (STM). This is another device that uses electrons to investigate surfaces, although it is not strictly scattering that is involved. In the STM (Fig. 12.12) electrons tunnel quantum mechanically across the small gap between a sharp tungsten tip and the surface under investigation. Piezoelectric transducers enable the position of the tip to be adjusted. The effective radius of the tip is typically 10 Å and the gap between the surface and the tip is also of this order. If the tip is moved parallel to the surface the tunnelling current varies depending on whether the tip is over an atom on the surface or a space between atoms; more precisely the current varies with the electron density on the surface. The variations in current thus measure the profile of the surface with atomic resolution.

The normal mode of operation is in fact to use feedback to vary the spacing between the tip and the surface in order to keep the tunnelling current constant as the probe is moved across the surface; the feedback signal then contains the information on the surface. Fig. 12.13(b) shows an image of the 7 × 7 Si (1 1 1) surface obtained using an STM. Fig. 12.13(a) shows the complex atomic structure of this surface; the reader should not worry too much about the details but should note that the reconstructed surface differs significantly from the unreconstructed surface of Fig. 12.11.

It is interesting to point out the difference between the LEED and STM methods for determining surface structures. The directions of the diffracted beams in LEED essentially allow the determination of the two-dimensional *reciprocal* lattice of the surface structure through the use of the two-dimensional equivalent of Eq. (11.8). The STM allows the direct determination of the *real space* structure.

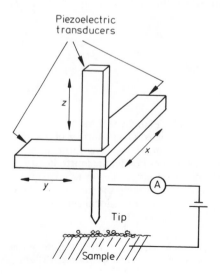

Fig. 12.12   Schematic diagram of the scanning tunnelling microscope. The peizoelectric transducers allow independent motion in the x, y and z directions by the application of electric fields

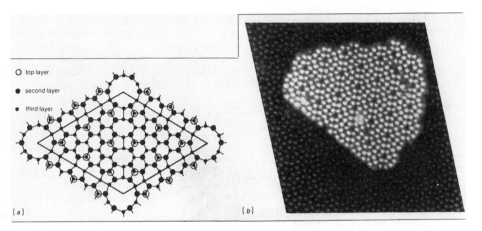

Fig. 12.13   The 7 × 7 reconstruction of the (1 1 1) surface of silicon. (*a*) The positions of the atoms in the first three layers. The reader should check that the atoms in the top layer are in the positions indicated by the STM image. (*b*) STM image of the surface. The lighter region is an island on the surface one monolayer thick. The white spots are the positions of the top layer of atoms. (Reproduced with permission from D. King, *Physics World*, March 1989, p. 45)

The STM can only be used to study the surfaces of conducting solids. A related device that can be used to investigate the surfaces of insulators is the **atomic force microscope** (AFM). In the AFM a probe with a sharp tip is mounted on a cantilevered beam and this enables the variations in the force between the surface and the tip to be monitored as the latter is scanned across the sample surface; atomic resolution is also possible using this technique.†

In our brief survey of methods that use electrons to study surfaces we have mentioned only a small fraction of the techniques that are available for this purpose. As readers will have gathered, the field of surface studies is riddled with acronyms. To allow the reader to learn the jargon we have listed in table 12.1 the acronyms for and brief details of some of the more important techniques. One problem with surface studies is that only a small fraction of the atoms in a bulk specimen reside on or near the surface. Particles used as probes for looking at the surface must therefore either interact very strongly with the surface atoms (as is the case with electrons) or be available in beams of very high intensity so that the small amount of scattering associated with the surface can be detected. The high intensities available from synchrotron sources has led to a revival in the use of x-rays for probing surfaces.

† For more details of the STM, AFM and other devices for probing surfaces with atomic resolution see 'Scanned probe microscopes' by H. K. Wickramasinghe in *Scientific American*, October 1989.

TABLE 12.1   *Some* of the techniques used to study the surfaces of solids

Technique	Acronym	Description	Examples of use
Low-energy electron diffraction	LEED	Elastic scattering of ~ 100 eV electrons	Surface structure determination
Reflection high-energy electron diffraction	RHEED	Scattering of $\sim$ 30 keV electrons at glancing incidence	Surface topography crystal growth
High-resolution electron energy loss spectroscopy	HREELS	Inelastic scattering of ~ 5 eV electrons	Surface vibrations particularly of adsorbed molecules
Auger electron spectroscopy	AES	Secondary electron emission after excitation of inner shell electron by $\sim$ 3 keV electron	Chemical composition of surface
Appearance potential spectroscopy	APS	Emission of x-ray photons after excitation of inner shell electron by $\sim$ 3 keV electron	Chemical composition of surface
Ultraviolet (x-ray) photoemission spectroscopy	UPS (XPS)	Emission of electrons caused by ultraviolet (x-ray) photons	Energy distribution of surface electrons
Inverse photoemission spectroscopy	IPS	Photon emission caused by incident 10–100 eV electrons	Energies of *unoccupied* surface electron states
Surface extended x-ray absorption fine structure	SEXAFS	Absorption of x-rays as a function of energy	Determines the local environment of surface atoms
Scanning tunnelling microscopy	STM	Tunnelling of electrons from tungsten tip to sample surface	Measures variation of local electron density of states
Atomic force microscopy	AFM	Measures force between fine tip and surface	Surface studies of insulators

PROBLEMS 12

12.1 (a) Calculate the dependence on scattering angle of the atomic form factor f for scattering of x-rays of wavelength λ from an atom. Assume that the atom consists of Z electrons uniformly distributed within a sphere of radius R. Assume that the contribution to the scattered wave from a point within the atom is proportional to the electron density at that point.

(b) Calculate the angular dependence of magnetic scattering of neutrons from an atomic magnetic moment μ. Assume that the contribution from a point within the atom is proportional to the local magnetic moment density and that this is uniformly distributed over a *spherical surface* of radius R and zero elsewhere. This is often a reasonable approximation since the atomic moment is associated with a particular shell of electrons (e.g. 3d or 4f).

12.3 Neutrons of energy 0.02 eV are scattered at an angle of 10° from solid helium (speed of sound $= 300$ m s^{-1}) with emission of a phonon. Estimate the energy loss of the neutrons. What is the time of flight over a 10 m path of unscattered and scattered neutrons?

Why will a small-angle scattering experiment of this type not work for a crystal such as sapphire, for which the speed of sound (10^4 m s^{-1}) is greater than the speed of the neutrons? How would you use neutrons to investigate the phonon spectrum of sapphire?

12.3 Show that the data of Fig. 12.3 are consistent with the structure of silicon and calculate the unit cell dimensions.

12.4 Calculate G_{100}, G_{110} and G_{111} and explain the positions of the first Brillouin zone boundaries on Fig. 12.7. Given that the bcc unit cell of potassium has $a = 5.23$ Å, use the data in the figure to calculate the velocities of transverse and longitudinal sound waves in the [1 0 0], [1 1 1] and [1 1 0] directions.

Explain why:
(i) the T and L modes are degenerate when $\mathbf{q} = \frac{1}{2}\mathbf{G}_{100}$,
(ii) the right-hand end of (a) matches the left-hand end of (b),
(iii) the slope of the dispersion curves does not vanish at the first Brillouin zone boundary in (b).

12.5 Metallic dysprosium has a hexagonal structure. It is alleged that the atomic moments are aligned ferromagnetically in the basal plane, but that the direction of alignment rotates about the c axis through an angle of order 40° from one layer to the next. What neutron scattering experiments would you make to confirm this, and what results would you expect?

12.6 Identify, with respect to (a) the magnetic unit cell and (b) the chemical unit cell, the Miller indices of the magnetic reflections at 11.9°, 30.2° and 36.4° on Fig. 12.8.

12.7 Estimate the time it takes for a monolayer of oxygen molecules to adsorb on a surface at room temperature if the surface is exposed to oxygen gas at a pressure of (a) 1 bar, (b) 10^{-6} bar and (c) 10^{-12} bar (1 bar $= 10^5$ N m^{-2}).

Life gets harder the smarter you get, the more you know.—
Katherine Hepburn (1987)

CHAPTER

Real metals

13.1 INTRODUCTION

In Chapter 4 we discussed the effect of a periodic lattice potential on the electron states in one- and two-dimensional metals, and in Chapter 11 we derived the general form for an electron wavefunction in a crystal and discussed the implications of this for the electron dispersion relation. In this chapter we combine the knowledge gained in these two previous chapters to discuss the properties of electrons in real metals. In addition, in section 13.5, we give a simple explanation of why this independent electron approach gives reasonable answers despite the strong Coulomb repulsion between the electrons.

13.2 FERMI SURFACES

13.2.1 Fermi surface of a nearly free electron two-dimensional metal

The free electron model predicts that many properties of metals are determined by electrons close to the Fermi surface. This is still the case when the effect of the periodic lattice potential is taken into account. A knowledge of Fermi surface geometry is therefore essential in any calculation of the properties of a metal. The electron energies can be close to the free electron values, even though the wavefunctions do not look like those of free electrons (section 4.3.3); this suggests the use of the nearly free electron approach (section 4.1) to predict

Fermi surface geometry. It turns out that this method often gives the topology of the surface correctly, even when it does not give a good approximation to the energy.

We have already used this approach in section 4.1 to generate the Fermi surface of a two-dimensional divalent metal with a simple square crystal structure; the Fermi surface is shown in the extended zone scheme in Fig. 4.5(b) and in the repeated zone scheme in Fig. 4.10. We also used nearly free electron theory to generate the constant energy contours of Figs. 11.13, 11.14 and 11.15 for a two-dimensional metal with a simple square reciprocal lattice. To predict the Fermi surface it is necessary to identify the constant energy contour that contains just enough electron states to accommodate all the conduction electrons. For a two-dimensional metal the constant energy contours in the extended zone scheme of Fig. 11.13 are close to the free electron circles and the Fermi surface is thus close to the free electron Fermi surface. To obtain an approximate Fermi surface it is therefore only necessary to draw a circle of the appropriate radius centred on the origin of reciprocal space.

The appropriate radius is conveniently obtained by expressing the area of the circle as multiples of the area of the first Brillouin zone. From section 11.4.1 we know that, for a crystal of N_c primitive unit cells, the first Brillouin zone contains $2N_c$ electron states (recall that the electron has two spin states). Thus, for example, for a two-dimensional metal with four conduction electrons in each primitive unit cell, the free electron Fermi surface is a circle with an area equal to twice that of the first Brillouin zone. This circle is shown in Fig. 13.1(a) together with the Brillouin zone boundaries for a simple square reciprocal lattice (as in Fig. 11.11). The different shadings indicate the zones to which the electrons inside the Fermi surface are allocated.

In section 11.4.2 we showed that, by translation through suitably chosen reciprocal lattice vectors, the energy contours from any Brillouin zone in Fig. 11.13 could be replotted to form a periodic structure in k-space as in Fig. 11.15. Figs. 13.1(b)–(d) show the sections of the free electron Fermi surface from the second, third and fourth Brillouin zones in Fig. 13.1(a) replotted into the periodic zone scheme in this way.† In Figs. 13.1(c) and (d) we have pockets of occupied states and we describe these sections of Fermi surface as electron-like; in Fig. 13.1(b) there are pockets of unoccupied states and these sections of Fermi surface are hole-like. The reason for this notation will become clear in section 13.4. Fig. 13.2 illustrates the way in which the sections of Fermi surface in Figs. 13.1(b)–(d) can be obtained by drawing circles centred on all reciprocal lattice points with radii equal to that of the free electron Fermi circle; this is the basis of the **Harrison construction** for producing the free electron Fermi surface.

The Harrison construction is the most straightforward way of generating the free electron Fermi surface of a three-dimensional metal and we shall use it for

---

† In the example we have chosen the Fermi surface does not intersect the first Brillouin zone.

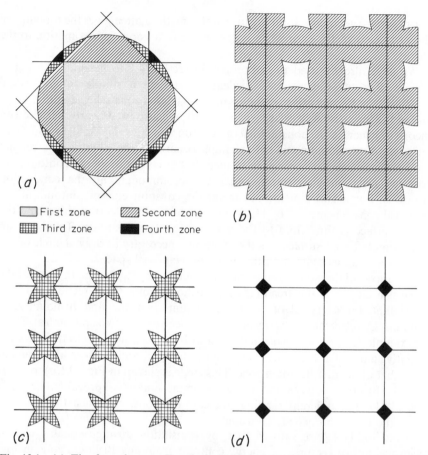

First zone ☐   Second zone ▨   (b)
Third zone ▦   Fourth zone ■

Fig. 13.1 (a) The free electron Fermi circle for a two-dimensional metal with four conduction electrons per primitive unit cell superimposed on the Brillouin zone boundaries for a simple square reciprocal lattice. The sections of the circle are associated with the first, second, third and fourth zones as indicated. By translation through appropriate reciprocal lattice vectors the Fermi circle can be remapped into the repeated zone scheme to give: (b) second zone hole-like Fermi surface, (c) third zone electron-like surface and (d) fourth zone electron-like surface

this purpose in the following section. To understand the construction we will explain its use in the two-dimensional case. Note first that the circles in Fig. 13.2 divide k-space up into regions covered by one, two, three and four circles. The free electron Fermi surface is obtained by applying the following rules:

(1) The Fermi surface in the nth Brillouin zone is the boundary dividing regions covered by n circles from regions covered by $n - 1$ circles.

(2) If the region covered by the larger number of circles is inside this boundary we have an electron Fermi surface and, if it is outside, a hole surface.

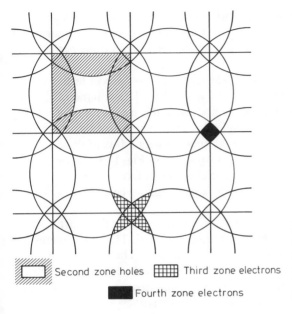

Second zone holes   Third zone electrons

Fourth zone electrons

Fig. 13.2   Harrison construction for obtaining the repeated zone representation of the Fermi surface of the two-dimensional free electron metal of Fig. 13.1

The reader should check that the use of these rules does indeed identify the Fermi surface from each zone as indicated in Fig. 13.2.

The identification of different parts of the free electron Fermi surface with different Brillouin zones only becomes significant when the energy gaps at the Brillouin zone boundaries produced by the periodic lattice potential are taken into account. The sections of Fermi surface in different zones then correspond to different energy bands. According to the nearly free electron theory the Fermi surface is distorted slightly from its free electron form so that it intersects the Brillouin zone boundaries at right angles (Fig. 11.13). Comparison of Fig. 13.1(a) with Fig. 11.13 shows that this causes rounding of the cusps in the Fermi surface of Figs. 13.1(b)–(d). Even with this rounding the Fermi surface from each zone is very non-circular in character. Increasing the perturbation due to the lattice potential causes both hole and electron regions of the Fermi surface to shrink. A sufficiently strong perturbation removes the Fermi surface from a particular zone completely. When there is an even number of electrons per primitive unit cell, the ultimate result of this process is an insulator with no free Fermi surface; thus for four electrons per primitive unit cell the first two Brillouin zones become full and all the others empty.

13.2.2   Fermi surface of three-dimensional metals

Extension of the Harrison approach to three dimensions is straightforward. A sphere of volume equal to that of the free electron Fermi sphere is centred on each reciprocal lattice point and the two rules from the previous section are applied to the overlapping regions. To illustrate this we consider the particular case of a cubic close-packed metal with four conduction electrons per unit cell (lead, for example).†

The first Brillouin zone of the body-centred cubic *reciprocal* lattice of this structure is the truncated octahedron of Fig. 1.12(b). The volume of the free electron Fermi sphere is twice that of the first zone; this means that it is just big enough to enclose the first zone completely. Two spheres overlap near zone faces, three near zone edges and four near zone corners. Most of these features can be seen on a (1 1 0) cross section of k-space as in Fig. 13.3; the origin of a second zone hole surface and a third zone electron surface can be seen in Fig. 13.3(b). The (1 1 0) cross section misses the zone corners where small pockets of electrons in the fourth zone occur.

Fig. 13.4 shows schematic views of the Fermi surface after the sharp corners have been rounded off by the periodic lattice potential. The most interesting feature is the third zone surface known as the 'monster'. The three-dimensional view makes it clear that this cannot be unambiguously called an 'electron' surface or a 'hole' surface because it is multiply connected; cross sections can be drawn that look electron-like (as in Fig. 13.3(b)) or hole-like. We consider the experimental consequences of this type of Fermi surface in section 13.4. Note that monsters are essentially three-dimensional creatures; they cannot exist in a smaller number of dimensions.

13.2.3 Density of states at the Fermi surface

A number of properties of the metal depend only on the density of states at the Fermi surface. Thus Eq. (3.19),

$$C_V = \tfrac{1}{3}\pi^2 g(\varepsilon_F)k_B^2 T,$$

gives quite generally the electronic heat capacity at low temperatures. To calculate $g(\varepsilon_F)$ in general we consider the constant energy surface, $\varepsilon_F + d\varepsilon$. This surface is displaced from the Fermi surface by a perpendicular distance (see Fig. 13.5)

$$dk_\perp = \frac{d\varepsilon}{(\partial\varepsilon/\partial k_\perp)} = \frac{\partial\varepsilon}{\hbar v_F},$$

where the Fermi velocity v_F is defined as the group velocity, $\hbar^{-1}\partial\varepsilon/\partial k$, of the electron waves at the Fermi surface and will in general vary with position on the surface. For the element of area dS_F of Fermi surface shown in Fig. 13.5 the volume of k-space between the two surfaces is $dk_\perp \, dS_F$. The total volume of

---

† For pictures of Fermi surfaces obtained using the Harrison construction, see *Pseudopotentials in the Theory of Metals*, by W. A. Harrison, Benjamin, New York (1966).

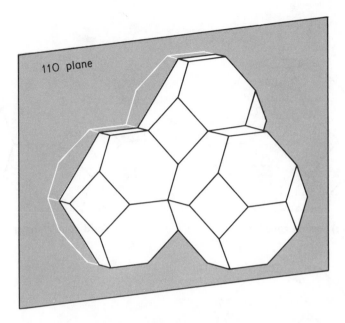

(a) Wigner–Seitz unit cells of the body-centred reciprocal lattice of a cubic close-packed structure, sectioned by a (1 1 0) plane

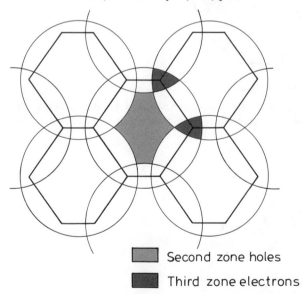

Second zone holes

Third zone electrons

(b) Harrison construction in a (1 1 0) plane for the reciprocal lattice of (a), The radius of the spheres is appropriate to a cubic close-packed metal with four conduction electrons per atom

Fig. 13.3

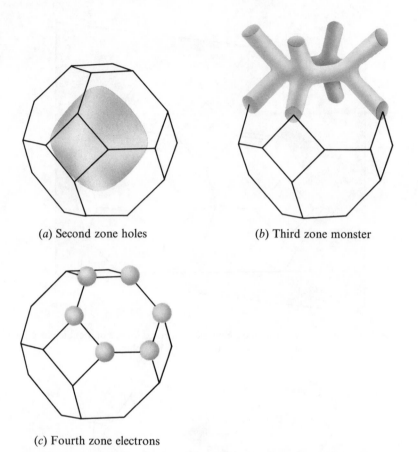

(a) Second zone holes

(b) Third zone monster

(c) Fourth zone electrons

Fig. 13.4   Fermi surface for a cubic close-packed metal with four conduction electrons per atom, as predicted using the nearly free electron approach

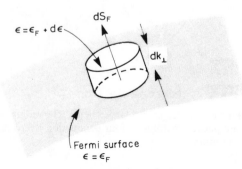

Fig. 13.5   Volume element of **k**-space between the Fermi surface and the surface of energy $\varepsilon_F + d\varepsilon$

k-space between the two contours is therefore

$$\int dk_\perp \, dS_F = \int \frac{d\varepsilon}{\hbar v_F} \, dS_F,$$

where the integral is over the Fermi surface.† From Eq. (2.41) this volume of k-space contains

$$\frac{V}{(2\pi)^3} \int \frac{d\varepsilon}{\hbar v_F} \, dS_F$$

k states where V is the volume of the metal. The number of electron states for electrons of *both* spins per unit energy range is therefore

$$g(\varepsilon_F) = \frac{V}{4\pi^3} \int \frac{1}{\hbar v_F} \, dS_F. \tag{13.1}$$

The density of states thus increases as the Fermi surface area increases, with the largest contributions coming from the regions of the surface with the smallest Fermi velocity. This explains the large electronic heat capacity coefficients γ for some transition metals; where the Fermi energy falls within the narrow 3d bands the small Fermi velocity associated with these bands gives a large contribution to the density of states.

13.3   ELECTRON DYNAMICS IN A THREE-DIMENSIONAL METAL

13.3.1   Equation of motion and effective mass

We showed in section 4.4 that the equation of motion of an electron wavepacket in a one-dimensional crystal in a electric field was given by Eq. (4.24) as

$$\hbar \, dk/dt = -eE$$

in the absence of collisions. This is generalized to three dimensions by replacing k and E by vectors (we have already used this generalization in Chapter 5)

$$\hbar \frac{d\mathbf{k}}{dt} = -e\mathbf{E}. \tag{13.2}$$

We introduce the effect of collisions in section 13.3.2.

Eq. (13.2) equates the change of momentum of the crystal to the force on the electron. It is tempting to identify $\hbar\mathbf{k}$ as the momentum of the electron, but this is

---

† To avoid overcounting of states the area can be taken over the Fermi surface in the extended zone scheme or in the reduced zone scheme by summing over energy bands.

misleading since the momentum change due to the field is shared between the electron and the lattice as a whole in a way that in general cannot be unambiguously unravelled.

To see this we evaluate the true momentum of an electron with a Bloch wavefunction, which, from Eqs. (11.27) and (11.32), can be written

$$\Psi(\mathbf{r}, t) = \frac{1}{\sqrt{V}} \sum_{\mathbf{G}} a_{\mathbf{G}}(\mathbf{k})\, e^{i[(\mathbf{k}+\mathbf{G})\cdot\mathbf{r} - \omega t]}. \tag{13.3}$$

This is not a momentum eigenfunction but we can calculate the expectation value of the momentum as

$$\langle \mathbf{p}_{el} \rangle = \int dV\, \Psi^*(\mathbf{r}, t)(-i\hbar\nabla)\Psi(\mathbf{r}, t)$$

$$= \frac{1}{V} \int dV \sum_{\mathbf{G}} a_{\mathbf{G}}^*\, e^{-i(\mathbf{k}+\mathbf{G})\cdot\mathbf{r}} \sum_{\mathbf{G}'} a_{\mathbf{G}'}\,\hbar(\mathbf{k} + \mathbf{G}')\, e^{i(\mathbf{k}+\mathbf{G}')\cdot\mathbf{r}}$$

$$= \frac{1}{V} \sum_{\mathbf{G}} \sum_{\mathbf{G}'} a_{\mathbf{G}}^* a_{\mathbf{G}'}\,\hbar(\mathbf{k} + \mathbf{G}') \int dV\, e^{i(\mathbf{G}'-\mathbf{G})\cdot\mathbf{r}}.$$

From the othogonality of plane waves, the integral is equal to V if $\mathbf{G}' = \mathbf{G}$ and is zero otherwise, so that

$$\langle \mathbf{p}_{el} \rangle = \sum_{\mathbf{G}} |a_{\mathbf{G}}|^2 \hbar(\mathbf{k} + \mathbf{G}) = \hbar\left(\mathbf{k} + \sum_{\mathbf{G}} \mathbf{G}|a_{\mathbf{G}}|^2\right) \tag{13.4}$$

where the last step follows from the normalization condition, $\sum_{\mathbf{G}} |a_{\mathbf{G}}|^2 = 1$, for the wavefunction of Eq. (13.3).

The result (13.4) is just what we would expect from the fact that the plane wave $e^{i(\mathbf{k}+\mathbf{G})\cdot\mathbf{r}}$ is a momentum eigenstate of eigenvalue $\hbar(\mathbf{k} + \mathbf{G})$, and $|a_{\mathbf{G}}|^2$ is the probability of finding the electron in this state for the wavefunction of Eq. (13.3). If the electron is now accelerated so that its wavevector changes to $\mathbf{k} + \delta\mathbf{k}$, the expectation value of its momentum changes by

$$\delta\mathbf{p}_{el} = \hbar\left(\delta\mathbf{k} + \sum_{\mathbf{G}} \mathbf{G}\delta\mathbf{k} \cdot \frac{\partial}{\partial\mathbf{k}} |a_{\mathbf{G}}|^2\right). \tag{13.5}$$

A physical interpretation of the second term in Eq. (13.5) may be made by recalling what happens when a free electron, incident on the crystal from outside, undergoes Bragg diffraction. According to Eq. (11.8) its wavefunction changes from $e^{i\mathbf{k}\cdot\mathbf{r}}$ to $e^{i(\mathbf{k}+\mathbf{G})\cdot\mathbf{r}}$ so that the lattice receives a recoil momentum

$$\delta\mathbf{p}_{latt} = -\hbar[(\mathbf{k} + \mathbf{G}) - \mathbf{k}] = -\hbar\mathbf{G}.$$

The more general process in which the coefficients $a_{\mathbf{G}}(\mathbf{k})$ in Eq. (13.3) change by amounts other than 1 can be regarded as a series of partial Bragg reflections.

The corresponding recoil momentum is then

$$\delta \mathbf{p}_{\text{latt}} = -\hbar \sum_{\mathbf{G}} \mathbf{G}\delta(|a_{\mathbf{G}}|^2),$$ (13.6)

which is just the negative of the second term in Eq. (13.5), so that the total change in momentum of the crystal is given by

$$\delta \mathbf{p}_{\text{tot}} = \delta \mathbf{p}_{\text{el}} + \delta \mathbf{p}_{\text{latt}} = \hbar \delta \mathbf{k}.$$ (13.7)

This can be used to justify Eq. (13.2); because $\hbar \delta \mathbf{k}$ is the change in the total momentum of the system, it is correct to equate the applied force to $\hbar\, d\mathbf{k}/dt$.

The ambiguity in the allocation of momentum between the electron and the lattice as a whole arises because an electron of wavevector $\mathbf{k}$ can equally well be represented by a wavevector $\mathbf{k} + \mathbf{G}$ and does not therefore have a well defined momentum. As we have already indicated $\hbar \mathbf{k}$ is known as the crystal momentum of the electron.

Just as in the one-dimensional case, it is possible to rewrite Eq. (13.2) in the form of Newton's law for a particle of effective mass m_e,

$$m_e \frac{d\mathbf{v}}{dt} = -e\mathbf{E}.$$ (13.8)

Thus, proceeding as in section 4.4, we find that, for an isotropic dispersion relation, m_e is given by

$$m_e = \hbar^2 \left(\frac{\partial^2 \varepsilon}{\partial k^2} \right)^{-1}$$ (13.9)

As in one dimension, m_e is negative near the top of an energy band and the dynamics of such regions of $\mathbf{k}$-space are best treated by focusing on the empty states, which behave like positive charged particles with a positive effective mass as described in section 5.2.

Eqs. (13.8) and (13.9) must be modified if the dispersion relation is anisotropic, as is clearly the case for the energy surfaces of Fig. 13.4. In general $\partial \mathbf{v}/\partial t$ and $\partial \mathbf{k}/\partial t$ are not parallel; more precisely

$$\frac{\partial v_i}{\partial t} = \frac{1}{\hbar} \frac{\partial}{\partial t}\left(\frac{\partial \varepsilon}{\partial k_i} \right) = \frac{1}{\hbar} \sum_j \frac{\partial^2 \varepsilon}{\partial k_i \partial k_j} \frac{\partial k_j}{\partial t} \qquad \text{(subscripts } i, j \equiv x, y \text{ or } z\text{)}$$

$$= -\frac{1}{\hbar} \sum_j \frac{\partial^2 \varepsilon}{\partial k_i \partial k_j} eE_j,$$ (13.10)

where the final step follows from Eq. (13.2). Comparison with Eq. (13.8) allows the definition of an effective-mass tensor

$$\left(\frac{1}{m_e} \right)_{ij} = \frac{1}{\hbar^2} \frac{\partial^2 \varepsilon}{\partial k_i \partial k_j}$$ (13.11)

such that

$$\frac{\partial v_i}{\partial t} = -\sum_j \left(\frac{1}{m_e}\right)_{ij} eE_j. \tag{13.12}$$

The meaning of a tensorial effective mass is that the electron has more inertia for acceleration in some directionsthan others. If an electric field is applied in the direction of greatest or least inertia, acceleration is parallel to the field, but for a general field direction acceleration is preferentially along the direction of least inertia, as can be seen from Eq. (13.12). Eq. (5.58) for the frequency of cyclotron resonance of conduction band electrons in silicon can equally well be derived using the tensorial effective-mass concept (problem 13.2(b)).

★13.3.2   Relation of the electrical conductivity to the Fermi surface

In section 3.3 we explained how the equation of motion could be modified to allow for collisions and that the effect of a dc electric field was then to cause a small displacement of the free electron Fermi sphere in **k**-space (Eq. (3.34)). The modification of Eq. (13.2) to allow for a relaxation time τ due to collisions is

$$\hbar\left(\frac{d\mathbf{k}}{dt} + \frac{\delta\mathbf{k}}{\tau}\right) = -e\mathbf{E}. \tag{13.13}$$

A dc electric field thus causes a static displacement of a point on the Fermi surface

$$\delta\mathbf{k} = -\frac{e\mathbf{E}\tau}{\hbar} \tag{13.14}$$

as shown in Fig. 13.6; $\delta\mathbf{k}$ may vary over the Fermi surface if τ is not constant. We now calculate the electrical conductivity from this Fermi surface perturbation directly without using the effective-mass concept.

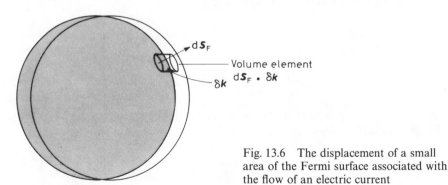

Fig. 13.6   The displacement of a small area of the Fermi surface associated with the flow of an electric current

Consider the element dS_F of Fermi surface displaced by δk on Fig. 13.6; the electrons in the volume $dS_F . \delta k$ of k-space move with the local Fermi velocity v_F and thus carry a current density

$$dj = \frac{2}{(2\pi)^3} (-e) v_F (dS_F . \delta k) \tag{13.15}$$

where we have used Eq. (2.41) to obtain the value $2/(2\pi)^3$ for the number of electron states per unit volume of metal in unit volume of k-space (recall that there are two spin states). Since the equilibrium electron distribution in the absence of a field carries no current, the current is obtained by integrating dj over the Fermi surface. As v_F is $\hbar^{-1} \partial\varepsilon/\partial k$, it is perpendicular to the Fermi surface, and thus parallel to dS_F. The vectors v_F and dS_F in Eq. (13.15) can therefore be interchanged and Eq. (13.14) used to give

$$j = -\frac{1}{4\pi^3} \int e \, dS_F (v_F . \delta k) = \frac{e^2}{4\pi^3\hbar} \int \tau \, dS_F (v_F . E). \tag{13.16}$$

This equation demonstrates that the conductivity in an anisotropic crystal, like the effective mass, is a tensor quantity, with the current and field being parallel only along directions of high symmetry. The quantity $\sigma_{eff} = |j|/|E|$ is an effective scalar conductivity that depends on the direction of E. It is given by

$$\sigma_{eff} = \frac{e^2}{4\pi^3\hbar} \left| \int \tau \, dS_F (v_F . \hat{E}) \right| \tag{13.17}$$

where $\hat{E}$ is a unit vector in the direction of E. The integral in Eq. (13.17) can be regarded as defining a mean free path l_{el} averaged over the Fermi surface such that

$$\sigma_{eff} = \frac{e^2 l_{el} S_F}{4\pi^3\hbar} \tag{13.18}$$

where S_F is the total free area of Fermi surface. This formula is more informative than Eq. (3.27); when the Fermi surface is restricted by contact with zone boundaries (e.g. Fig. 13.4) it is the total free area that matters rather than the total number of electrons. The free area of the Fermi surfaces of the pentavalent elements, arsenic, antimony and bismuth, are much reduced from their free electron values by the periodic lattice potential. Consequently these materials have small electrical conductivities and low electronic heat capacities; they are referred to as **semimetals**.

13.4 EXPERIMENTAL DETERMINATION OF THE FERMI SURFACE

Because the force on an electron in a magnetic field is perpendicular to its velocity, the field does not change the energy of the electron. An electron, which is initially in a state on the Fermi surface, moves through a sequence of states of constant energy under the influence of the field and therefore remains on the Fermi surface; measurements that provide information on the electron trajectory (referred to as its cyclotron orbit) thus give information on Fermi surface geometry. In this section we consider two examples of such measurements:

(1) cyclotron resonance, which we have already discussed in section 5.5.3 for semiconductors, where the carriers are non-degenerate and have a parabolic dispersion relation;

(2) the de Haas–van Alphen effect, which is an oscillatory variation of the conduction electron diamagnetism with magnetic field associated with the quantization of the cyclotron orbits.

13.4.1 Cyclotron orbits

The collisionless equation of motion of an electron wavepacket in **k**-space in a magnetic field is

$$\hbar \frac{d\mathbf{k}}{dt} = -e\mathbf{v} \times \mathbf{B} = -e\frac{d\mathbf{r}}{dt} \times \mathbf{B}. \tag{13.19}$$

For an electron on the Fermi surface, $\mathbf{v} = \mathbf{v}_F = \hbar(\partial\varepsilon/\partial\mathbf{k})_F$ is normal to the surface, so that $d\mathbf{k}/dt$ must lie in the surface. Since $d\mathbf{k}/dt$ is also perpendicular to **B**, it follows that the electron trajectory in **k**-space is along the Fermi surface contour in a plane perpendicular to **B** as shown in Fig. 13.7(a). From the final part of Eq. (13.19) we see that the motion in real space is closely related to that in **k**-space; more precisely the projection of the real space motion onto a plane perpendicular to **B** is an orbit of the same shape and rotation direction as the **k**-space orbit but rotated through 90° as shown in Fig. 13.7(b). The orbit in real space differs in size from that in **k**-space by a scale factor $\hbar/eB$.

Using Eq. (13.19), the period T of the orbit is given by

$$T = \oint dt = \oint \frac{dt}{dk}\, dk = \oint \frac{\hbar\, dk}{e|\mathbf{v} \times \mathbf{B}|} = \oint \frac{\hbar\, dk}{ev_\perp B} \tag{13.20}$$

where $v_\perp$ is the component of the electron group velocity in the plane perpendicular to **B**; this can be written as

$$v_\perp = \hbar^{-1} \frac{\delta\varepsilon}{\delta k_\perp} \tag{13.21}$$

where $\delta k_\perp$ is the perpendicular distance between the energy contours ε_F and $\varepsilon_F + \delta\varepsilon$ in this plane as shown in Fig. 13.7(a). Thus, inserting Eq. (13.21) in Eq. (13.20),

$$T = \frac{\hbar^2}{eB} \frac{\oint \delta k_\perp\, dk}{\delta\varepsilon};$$

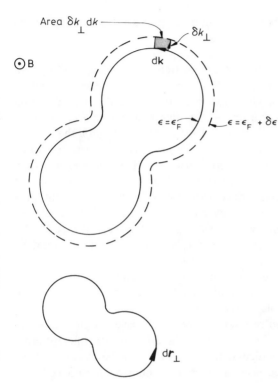

Area $\delta k_\perp \, dk$ ── $\delta k_\perp$

$\odot B$

dk

$\epsilon = \epsilon_F^-$ $\epsilon = \epsilon_F + \delta\epsilon$

(a) A cyclotron orbit in **k**-space is around the Fermi surface in a plane perpendicular to **B**

$dr_\perp$

(b) The projection into the plane perpendicular to the field of the real space orbit corresponding to the **k**-space orbit of (a) has the same shape and rotation sense but is rotated by 90° about the field direction

Fig. 13.7

$\delta k_\perp \, dk$ is the shaded area in Fig. 13.7(a) so that $\oint \delta k_\perp \, dk$ is the total area δA_k in **k**-space between the two contours in the plane perpendicular to **B**. In the limit $\delta\epsilon \to 0$, $\delta A_k/\delta\epsilon$ can be written $dA_k/d\epsilon$ and A_k can be interpreted as the area of the cyclotron orbit in **k**-space. The cyclotron frequency can then be written

$$\omega_C = \frac{2\pi}{T} = \frac{2\pi eB}{\hbar^2} \frac{d\epsilon}{dA_k}. \qquad (13.22)$$

Comparison with the value $\omega_C = eB/m$ for free electrons enables a cyclotron effective mass

$$m_C = \frac{\hbar^2}{2\pi} \frac{dA_k}{d\epsilon} \qquad (13.23)$$

to be defined. The cyclotron effective mass differs from the effective mass m_e introduced in section 13.3.1: m_C is the property of an orbit and is proportional to the first derivative $(d\epsilon/dk_\perp)^{-1}$ averaged around the orbit; m_e depends on the second derivative $(\partial^2\epsilon/\partial k^2)$ at a point in **k**-space. The two masses are only identical for an isotropic parabolic dispersion relation $\varepsilon = \hbar^2 k^2 / 2m_e$ (problem 13.2(a)).

In the derivation of Eqs. (13.22) and (13.23) we paid no attention to signs, but we now demonstrate that these equations, as written, give the sense of rotation of the orbit as well as its frequency. We continue to use the convention introduced in section 5.5.3 that clockwise rotation about the field direction corresponds to a positive ω_C; according to this sign convention *free* electrons have a positive ω_C (see Fig. 5.10). From Eq. (13.19) we deduce that if the velocity **v** associated with the Fermi surface of Fig. 13.7(a) is outwards then ω_C will be positive. This is so if states outside the Fermi surface have a higher energy than states inside; $d\varepsilon/dA_k$ is then positive, and Eqs. (13.22) and (13.23) predict the sign of ω_C correctly. A section of Fermi surface with positive ω_C would appropriately be classified as electron-like by the rules given in section 13.2.1. Alternatively, if **v** is inwards, ω_C is negative; in this case $d\varepsilon/dA_k$ is also negative and Eqs. (13.22) and (13.23) again predict the correct sign for ω_C. Such a surface would be appropriately classified as hole-like.

We will assume that cyclotron resonance measurements, like many other electronic properties of metals, are dominated by electrons near the Fermi surface; in fact the Pauli principle prevents electrons in lower lying states from absorbing energy in a cyclotron resonance experiment. Even so, it is possible that the frequency predicted by Eq. (13.22) will be different for different cross sections of Fermi surface perpendicular to the field. It can be shown that the frequency is the same for all sections of a spherical or ellipsoidal piece of Fermi surface. More generally, however, the behaviour is dominated by those cross sections of the surface where the frequency goes through a maximum or a minimum as a function of position along the k_z axis; it is these maximum and minimum frequencies that are actually observed. This is an example of a general result of combining a continuous spectrum of frequencies, known as the **principle of stationary phase**.

We have assumed that sections through the Fermi surface perpendicular to the field produce a closed orbit as in Fig. 13.7(a). This is not always the case; for Fermi surfaces of complicated connectivity in a repeated zone scheme, such as that of Fig. 13.4(b), it is possible to have **open orbits** for which the electron never returns to its initial **k**-state. Cyclotron resonance cannot be seen for such orbits but their existence has important implications for some properties, for example, magnetoresistance, which is the dependence of the electrical resistivity on an applied magnetic field.

13.4.2 Cyclotron resonance in metals

Our treatment of cyclotron resonance in semiconductors (section 5.5.3) implicitly assumed that the microwave electric field was uniform over the electron orbit. For semiconductors this is a reasonable assumption since the

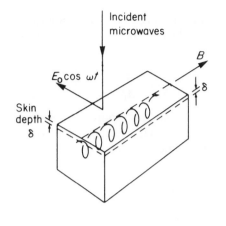

(a) Azbel–Kaner geometry for the observation of cyclotron resonance in metals. **B** is parallel to the sample surface and the electron is accelerated by the microwave field each time it enters the skin depth provided that the frequency ω is a harmonic of the orbit frequency ω_C

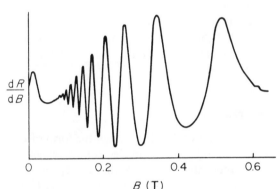

(b) Field derivative of the surface resistance of copper at 24 GHz. (Reproduced with permission from A. F. Kip, D. N. Langenberg and T. W. Moore, *Phys. Rev.* **124**, 359 (1962))

Fig. 13.8

small carrier density leads to a low conductivity and hence a large electromagnetic skin depth at microwave frequencies.† In metals the carrier density is much larger and the skin depth is usually smaller than the real space radius of the cyclotron orbit. This necessitates the use of a different experimental arrangement, the **Azbel–Kaner geometry** of Fig. 13.8(a), for observing cyclotron resonance in metals; this geometry resembles that of the cyclotrons used to accelerate fundamental particles.

The steady magnetic field **B** is applied parallel to plane surface of the specimen, so that in performing their cyclotron orbits in accordance with Eq. (13.19), some of the electrons enter the skin depth where they sense the microwave electric field $\mathbf{E}_0 \cos(\omega t)$ once in each revolution. Because the skin depth is much smaller than the orbit radius, the electrons 'see' the microwave field for a time short compared to the orbit period. The electrons absorb energy

---

† See section 13.6 for an explanation of the electromagnetic skin depth.

from the microwaves if the electric field is in the same direction each time they enter the skin depth; this is the case if the microwave frequency is equal to any harmonic of the cyclotron frequency

$$\omega = l\omega_C = l\frac{2\pi eB}{\hbar^2}\frac{d\varepsilon}{dA_k}, \tag{13.24}$$

where l is a positive integer. Since the normal procedure is to work at constant microwave frequency and to vary the dc magnetic field, we expect a power absorption that is periodic in $1/B$ with a period given by

$$\delta\left(\frac{1}{B}\right) = \frac{2\pi e}{\hbar^2\omega}\frac{d\varepsilon}{dA_k}. \tag{13.25}$$

Some measurements of the microwave surface resistance of copper (essentially a measure of the energy absorption within the skin depth) showing this periodicity are illustrated in Fig. 13.8(b). As in semiconductors it is essential for the observation of cyclotron resonance that the electronic mean free path is much longer than the real space radius of the orbit; the electron must undergo more than one complete orbit before being scattered.

13.4.3   Quantization of cyclotron orbits

The derivation of the equation of motion of the electrons in the form of Eqs. (4.24), (13.2) or (13.19) assumes that the trajectory of an electron wavepacket is that of a classical particle. Since the waves forming the wavepacket are essentially quantum mechanical in origin, this approach is often described as **quasi-classical**. The quasi-classical equations will certainly fail when B is large enough that $\hbar\omega_C \sim \varepsilon_F$ since the real space radius of the cyclotron orbit is then smaller than the size of the wavepacket (problem 13.4). Even in smaller fields we might expect that quasi-classical periodic motion at frequency ω_C would lead to quantization of electron energies in steps $\hbar\omega_C$. To investigate the nature of this quantization we must discuss the effect of a magnetic field on the electron energy levels in more detail.

We begin by considering *free* electrons. In the absence of a field the wavefunction is a plane wave (Eq. (3.3))

$$\psi(x, y, z) = \frac{1}{\sqrt{V}}e^{ik_x x}e^{ik_y y}e^{ik_z z}$$

with energy (Eq. (3.5))

$$\varepsilon = \frac{\hbar^2}{2m}(k_x^2 + k_y^2 + k_z^2).$$

In the presence of a field, Schrödinger's equation can be solved exactly (problem

13.5). For a field parallel to z the energy levels are of the form†

$$\varepsilon = (n + \tfrac{1}{2})\hbar\omega_C + \frac{\hbar^2 k_z^2}{2m} \tag{13.26}$$

where $\omega_C = eB/m$ is the cyclotron frequency of free electrons. As in the corresponding classical problem, the motion parallel to the field direction is unaffected by the field; the z dependence of the wavefunction remains as in Eq. (3.3) and the dependence of the energy on k_z is the same in Eqs. (3.5) and (13.26). Application of periodic boundary conditions in the z direction gives allowed k_z values, $2\pi r/L$, as before (Eq. (3.4)). The cyclotron orbits in the plane perpendicular to the field manifest themselves through a dramatic change in the x and y dependence of the wavefunctions; instead of being plane waves extending over the whole xy plane, the wavefunctions are localized in this plane on a length scale of order the cyclotron radius. The first term in the energy eigenvalue of Eq. (13.26) reflects the quantization associated with the periodic motion at frequency ω_C. The energy levels given by this equation are known as **Landau levels**.

It is possible (and very useful) to identify each Landau level with the set of quasi-classical cyclotron orbits in $\mathbf{k}$-space that have the same energy. The quasi-classical orbits are contours of constant (zero field) energy in a plane perpendicular to the field (i.e. constant k_z). Comparison of Eqs. (3.5) and (13.26) identifies the orbits corresponding to the nth Landau level as being given by

$$\frac{\hbar^2}{2m}(k_x^2 + k_y^2) = (n + \tfrac{1}{2})\hbar\omega_C. \tag{13.27}$$

This equation defines a series of concentric cylinders parallel to the field as shown in Fig. 13.9. The nth cylinder corresponds to the nth Landau level and, from Eq. (13.27), has an area

$$A_n = \pi(k_x^2 + k_y^2) = (n + \tfrac{1}{2})\frac{2\pi e B}{\hbar} \tag{13.28}$$

where we have used $\omega_C = eB/m$. From a quantum mechanical viewpoint, cyclotron resonance, as described by Eq. (13.24), is seen as the excitation of electrons from the nth to the $(n + l)$th Landau level cylinder by absorption of a photon of energy $\hbar\omega = l\hbar\omega_C$.

The degeneracy of the orbits‡ on each cylinder for a given k_z value can be obtained, as in the absence of a field, by applying suitable boundary conditions in the xy plane. It is found that the *average* density of states in $\mathbf{k}$-space is the same as in the absence of the field. This is illustrated in Fig. 13.10, which is a section through $\mathbf{k}$-space perpendicular to the z axis. Fig. 13.10(a) shows the

---

† For the moment we ignore the effect of electron spin (see section 13.4.4).

‡ The orbits can be degenerate without violating the Pauli principle because they are centred at different points in the xy plane in real space.

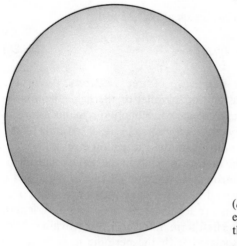

(a) In the absence of a magnetic field, *free* electrons are uniformly distributed throughout the Fermi sphere at $T = 0$

(b) A magnetic field causes quantization of the cyclotron orbits such that the electrons lie on a series of concentric cylinders in **k**-space. The occupied portions of the cylinders are essentially those parts that are inside the original Fermi surface. With increasing field the cylinders expand and move outwards through the Fermi surface. The occupied region of a cylinder shrinks most rapidly when it is just about to pass through the Fermi surface

Fig. 13.9

simple square lattice of electron states in this plane in the absence of a field, with the Landau cylinder structure superimposed. Fig. 13.10(b) shows the states on the Landau level cylinders in the presence of a field; each state in Fig. 13.10(a) has moved to the nearest Landau cylinder to leave the average density of states in **k**-space unchanged. According to Eq. (13.28) the area between successive cylinders in **k**-space is $2\pi eB/\hbar$, independent of n. The density of states per unit area in Fig. 13.10(a) is $(L/2\pi)^2$ so that the number of states g_n associated with the

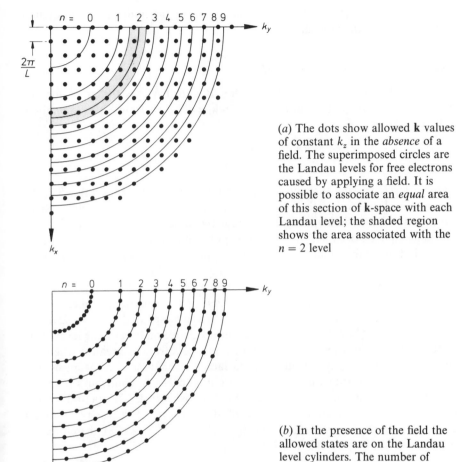

(a) The dots show allowed **k** values of constant k_z in the *absence* of a field. The superimposed circles are the Landau levels for free electrons caused by applying a field. It is possible to associate an *equal* area of this section of **k**-space with each Landau level; the shaded region shows the area associated with the $n = 2$ level

(b) In the presence of the field the allowed states are on the Landau level cylinders. The number of states associated with each level for each k_z value is the number from within the associated area as shown in (a)

Fig. 13.10   Effect of a magnetic field on the allowed free electron states

Landau level for each k_z value and each spin orientation is

$$g_n = \left(\frac{L}{2\pi}\right)^2 \left(\frac{2\pi e B}{h}\right) = \frac{eBL^2}{h},\tag{13.29}$$

which is also independent of n. From Eq. (10.30), $h/2e$ is the flux quantum associated with a Cooper *pair* of electrons in a superconductor; h/e can therefore be interpreted as a flux quantum for individual electrons. BL^2 is the magnetic flux passing through the crystal, so that g_n is just the number of flux quanta passing through the crystal.

13.4.4 The de Haas–van Alphen effect

If we take typical values of 10 T for B and $1\,\text{Å}^{-2} = 10^{20}\,\text{m}^{-2}$ for the cross-sectional area of the Fermi surface in $\mathbf{k}$-space then, from Eq. (13.28), there will be of order

$$n \approx \frac{\hbar A_n}{2\pi e B} \approx \frac{10^{-34} \times 10^{20}}{2\pi \times 10^{-19} \times 10} \approx 10^3$$

Landau levels passing through the free electron Fermi sphere in Fig. 13.9; the separation of the levels is thus small compared to the radius of the sphere. Since the density of states in $\mathbf{k}$-space is on average the same as in the absence of a field, the states that are occupied at $T = 0$ are approximately those on the portions of the cylindrical tubes that lie within the original Fermi sphere as shown in Fig. 13.9.

In an increasing magnetic field the radius of each Landau level cylinder increases and eventually becomes greater than that of the Fermi sphere; when this happens the number of electrons on it decreases and goes to zero as, according to Eq. (13.29), more states become available on lower lying levels. It is apparent from Fig. 13.9 that the rate of loss of electrons is greatest when the cylinder is about to pass through the surface. Under these circumstances it is perhaps not surprising that there is a small oscillatory contribution to the energy of the electrons as the magnetic field is varied. The period of the oscillation is determined by equality of the area A_n of successive Landau cylinders with the Fermi surface area $A_F = \pi k_F^2$; from Eq. (13.28), therefore, the oscillations will be periodic in $1/B$ with a period

$$\delta(1/B) = \frac{2\pi e}{\hbar A_F}. \tag{13.30}$$

The oscillatory dependence of energy on magnetic field implies that the magnetization is also oscillatory.† The resulting variation of the magnetic moment of a specimen can be detected and is known as the **de Haas–van Alphen effect**.

Fig. 13.11 shows de Haas–van Alphen oscillations for copper. To interpret these it is necessary to generalize the result (13.28) to electrons in a periodic lattice potential. This is difficult to do rigorously. The simplest way to quantize the quasi-classical cyclotron orbits is to use the correspondence principle that quantum mechanical results should go smoothly to classical results in the appropriate limit. Thus in the limit of large quantum numbers the difference in

---

† The magnetization $\mathbf{M}$ is $(\partial F/\partial \mathbf{B})_T$ where $F = U - TS$ is the Helmholtz free energy.

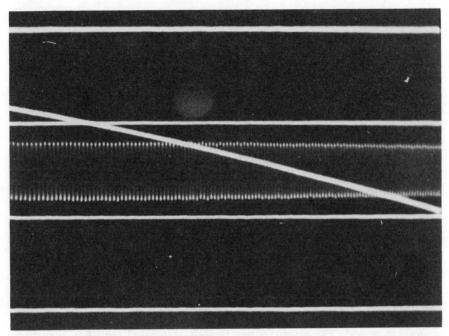

Fig. 13.11   De Haas–van Alphen oscillations in copper. The oblique line shows the magnetic field variation and the horizontal lines indicate fields of 10.47, 10.70, 10.93 and 11.16 T. The magnetic field was obtained by discharging a large capacitor through a liquid-nitrogen-cooled solenoid. The magnetic field decayed from a peak value of just over 11 T with a time constant of a few milliseconds; the trace on the oscilloscope is about 1 ms wide. The oscillations are the voltage in a pick-up coil containing the specimen. (Reproduced with permission from D. Shoenberg, *Phil. Trans. R. Soc.* A **255**, 85 (1962))

energy of successive quantized orbits should equal $\hbar$ times the quasi-classical orbit frequency of Eq. (13.22); that is

$$\varepsilon_n - \varepsilon_{n-1} = \hbar\omega_C = \frac{2\pi eB}{\hbar} \frac{d\varepsilon}{dA_k}. \qquad (13.31)$$

In the limit of large quantum numbers we can approximate the derivative as

$$\frac{d\varepsilon}{dA_k} = \frac{\varepsilon_n - \varepsilon_{n-1}}{A_n - A_{n-1}},$$

in which case Eq. (13.31) becomes

$$A_n - A_{n-1} = \frac{2\pi eB}{\hbar}.$$

This is satisfied if A_n is of the form

$$A_n = (n + \gamma)\frac{2\pi eB}{\hbar} \tag{13.32}$$

where γ is independent of n.

The similarity of this result to the free electron result, Eq. (13.28), means that the qualitative behaviour is little changed by the periodic lattice potential. The quantized cyclotron orbits lie on a series of tubes along the field direction; the cross section of each tube has an area given by Eq. (13.32)† and is a constant energy contour, but it is not circular in general and its shape can vary along the field direction. The periodicity of the de Haas–van Alphen oscillations is determined by the equality of the area of successive quantized orbits with the extremal cross-sectional areas of the Fermi surface perpendicular to the field as indicated in Fig. 13.12. Measurement of the period $\delta(1/B)$ of the oscillations enables the extremal areas to be deduced by the use of Eq. (13.30). Note that for hole-like regions of the Fermi surface the quantized orbit tubes will shrink rather than grow with increasing field.

The de Haas–van Alphen effect was first discovered in the semimetal bismuth, which has small pockets of Fermi surface containing only about 10^{-5} electrons per atom and consequently, from Eq. (13.30), gives a large and readily observable period. The much shorter periods associated with Fermi surfaces containing about one electron per atom mean that fields of order 10 T at 1 K are

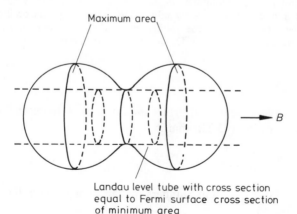

Maximum area

B

Landau level tube with cross section
equal to Fermi surface cross section
of minimum area

Fig. 13.12   De Haas–van Alphen oscillations are associated with the passage of the quantized cyclotron orbit tubes through extremal cross sections of the Fermi surface perpendicular to the field. The figure shows a portion of Fermi surface with both a maximum and minimum cross section

---

† This assumes that γ in Eq. (13.32) is independent of k_z.

needed to observe the effect satisfactorily; superconducting solenoids are normally used to provide the large fields required. A small ac field $B_1 \cos (\omega t)$ is added at a convenient frequency ω using auxiliary coils. Specimens are typically single-crystal wires a few millimetres long mounted in a small pick-up coil often with a similar dummy coil wound in series opposition to eliminate effects not associated with the specimen. The magnetization of the specimen in the modulated applied field can be expanded as a Taylor series about the average field B_0,

$$M(t) = M(B_0) + [B_1 \cos (\omega t)]\left(\frac{dM}{dB}\right)_{B_0} + \tfrac{1}{2}[B_1 \cos (\omega t)]^2 \left(\frac{d^2M}{dB^2}\right)_{B_0} + \cdots.$$

The emf in the pick-up coil is proportional to

$$-\frac{dM}{dt} = +\omega B_1 \left(\frac{dM}{dB}\right)_{B=B_0} \sin (\omega t) + \tfrac{1}{2}\omega B_1^2 \left(\frac{d^2M}{dB^2}\right)_{B=B_0} \sin (2\omega t) + \cdots. \quad (13.33)$$

The oscillations of the magnetization associated with the de Haas–van Alphen effect imply a non-zero value for (d^2M/dB^2), which oscillates as a function of B_0. The best way to observe the effect is therefore to employ phase sensitive detection at frequency 2ω; in this way the signal-to-noise ratio is improved and linear pick-up effects at frequency ω are eliminated. The modulation amplitude B_1 is best chosen to be comparable with the de Haas–van Alphen period.

The special value of the de Haas–van Alphen effect is that, for a complicated Fermi surface consisting of several pieces, it gives a separate periodic signal for each extremal area. The frequency of the signal shown in Fig. 13.11 is approximately that for a free electron sphere containing one electron per atom (problem 13.6), but other, quite different, frequencies are also seen for copper (and silver and gold, which are also cubic close-packed metals). This leads to the conclusion that in these metals the Fermi surface is stretched towards all the faces of the first Brillouin zone and actually makes contact with the hexagonal {1 1 1} faces as shown in Fig. 13.13(a). Thus as well as the main 'belly' orbit, there is a 'neck' orbit of minimum area at the contact with the Brillouin zone boundary. Fig. 13.13(b) shows a schematic model of the Fermi surface of copper in the repeated zone scheme; a hole orbit called the 'dog's bone' is indicated. The contact with the zone boundary makes the Fermi surface multiply connected in the repeated zone scheme, allowing various other hole orbits that cover bits of several spheres; open orbits also exist, which do not contribute to the de Haas–van Alphen effect.

The de Haas–van Alphen effect can also disentangle more complicated Fermi surfaces. Thus it has been used to show that the Fermi surface of lead is essentially as shown in Fig. 13.4. The observed variation of the measured periods with field orientation, from [1 0 0] to [1 1 0], is shown in Fig. 13.14. The α oscillations correspond to the second zone holes in Fig. 13.4; the β oscillations

Neck orbit

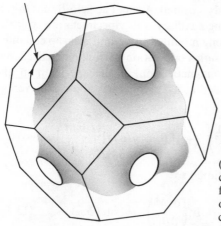

(a) Fermi surface of copper, showing contact with the hexagonal faces of the first Brillouin zone. De Haas–van Alphen oscillations are observed with a period corresponding to the neck orbit indicated

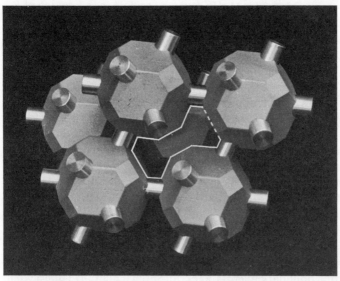

(b) Schematic model of the Fermi surface of copper in the repeated zone scheme showing the 'dog's bone' orbit. The model is constructed from rhombic dodecahedra. If a rhombic dodecahedron is inscribed inside the first Brillouin zone with its vertices just touching the centres of the faces then it exactly half-fills the zone. The Fermi surfaces of copper, silver and gold are closely related to this. They have large almost flat areas corresponding to the faces of the dodecahedron; the [1 0 0] vertices are rounded off within the zone, whereas the [1 1 1] vertices are rounded off across the zone boundaries to form the necks

Fig. 13.13

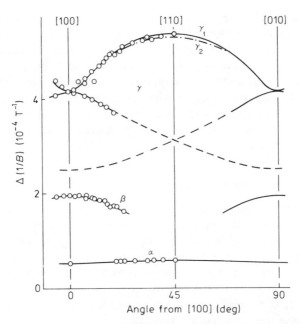

Fig. 13.14 Angular variation of the de Haas–Van Alphen periods in lead. (Reproduced with permission from A. V. Gold, *Phil. Trans. R. Soc.* A **251**, 85 (1958))

come from a hole orbit on one of the square 'faces' of the third zone monster; the γ oscillations come from the pockets of electrons in the fourth zone (in general these give three periods since pockets in three different orientations are seen for any field direction). Over limited ranges of angle, beats are seen on the γ oscillations, showing that two periods, γ_1 and γ_2, are present; the γ_2 period is due to orbits around limbs of the third zone monster.

Our discussion of the de Haas–van Alphen effect has ignored electron spin. Because of the splitting of the energies by $2\mu_B B$, there are two sets of Landau level tubes, one for each spin state. Each set moves through the Fermi surface at the frequency given by Eq. (13.30) so that the periodicity of the oscillations is unaffected. The magnitude of the oscillations is however significantly reduced if the spin up levels pass through the Fermi surface approximately half-way in between the spin down levels.

13.5 WHY DO ELECTRONS BEHAVE INDEPENDENTLY?

In our discussion of metals and semiconductors we have assumed that electrons behave independently. In particular the derivation of Bloch's theorem in Chapter 11 applies only to *single particle* wavefunctions;† the validity of the

† The consequences of lattice periodicity for a many particle wavefunction are less straightforward.

theory of metals developed earlier in this chapter therefore rests on the independent electron assumption. A comparison of the energies involved suggests that this is a bad approximation; the Coulomb repulsion energy $e^2/4\pi\varepsilon_0 r$ of two electrons 1 Å apart is about 10 eV and this is comparable to both the Fermi energy and the interaction energy of an electron with the nearest positive ion. Because of the Coulomb interaction the electrons behave more like a liquid than a non-interacting gas. We explain in this section why the behaviour of the liquid in its lowest lying excited states resembles that of a gas of almost independent particles.

13.5.1 Electrical neutrality in metals

We found in section 5.6.2 that a departure from electrical neutrality in a homogeneous semiconductor disappears on short distance and time scales through redistribution of the majority carriers. The higher electron density in metals makes the distance and time scales even smaller; they are so short that in deriving the equation of motion of the electron liquid it is possible to ignore collisions.† Thus, instead of using the electrical conductivity and diffusion constant to describe the response to an electric field and a concentration gradient, we use the acceleration equations

$$m\frac{d\mathbf{v}}{dt} = -e\mathbf{E} \qquad \text{and} \qquad mn\frac{d\mathbf{v}}{dt} = -\nabla p$$

respectively, where ∇p is the pressure gradient associated with the concentration gradient. Combining both terms gives

$$mn\frac{d\mathbf{v}}{dt} = -ne\mathbf{E} - \nabla p, \tag{13.34}$$

which is the equation of motion for an ideal (i.e. non-viscous) charged liquid.

By using the bare electron mass in Eq. (13.34) we are ignoring band structure effects and we therefore assume a spatially uniform charge density for the positive ions. Electron–electron repulsion enters the calculation through the use Gauss' law,

$$\text{div } \mathbf{E} = \frac{\rho}{\varepsilon_0} = -\frac{e(n - n_0)}{\varepsilon_0}, \tag{13.35}$$

for the electric field generated by departures of the electron density from its average value n_0. We assume that it is a reasonable approximation to take the

---

† This requires that the length scale should be short compared to the electron mean free path and the time scale short compared to the time between collisions. We explain in section 13.5.4 why the collisions are strong enough to cause liquid-like behaviour without introducing a very short relaxation time.

pressure in Eq. (13.34) as that for free electrons, which is calculated in problem 3.3.

$$p = \tfrac{2}{5} n \varepsilon_F. \tag{13.36}$$

Hence

$$- \nabla p = -\tfrac{2}{3} \varepsilon_F \nabla n, \tag{13.37}$$

where we have used Eq. (3.9) for ε_F as a function of n. By using Eqs. (13.35) and (13.37) together with the conservation law for particle number,

$$\partial n/\partial t = - \text{div}(n\mathbf{v}), \tag{13.38}$$

it is possible to eliminate $\mathbf{v}$, $\mathbf{E}$ and p from Eq. (13.34) to obtain the following equation for n:

$$\frac{\partial^2 n}{\partial t^2} = \frac{2}{3} \frac{\varepsilon_F}{m} \nabla^2 n - \frac{n_0 e^2}{m \varepsilon_0}(n - n_0). \tag{13.39}$$

We have linearized this equation by assuming that $n - n_0$, $\mathbf{v}$ and $\mathbf{E}$ are small.

Eq. (13.39) is the analogue for metals of Eq. (5.66) for semiconductors. It differs qualitatively in having a second- rather than first-order time derivative; this is because we have ignored collisions in its derivation. We now describe some of the implications of this equation.

13.5.2   Plasma oscillations

In the limit of slow spatial variation the $\nabla^2 n$ term can be ignored and Eq. (13.39) becomes

$$\frac{d^2 n}{dt^2} = -\omega_P^2 (n - n_0), \tag{13.40}$$

where

$$\omega_P^2 = \frac{n_0 e^2}{m \varepsilon_0}. \tag{13.41}$$

This is a simple harmonic oscillator equation with angular frequency ω_P. Any long-wavelength departure from electrical neutrality therefore oscillates at this frequency; these oscillations are known as **plasma oscillations** and ω_P is the **plasma frequency**. The finite frequency associated with the long-wavelength motions arises because the $1/r$ fall-off of the Coulomb interaction makes it a long-range interaction.†

---

† In contrast we have seen in Chapter 2 that, when there are short-range forces between particles, the long-wavelength motions are sound waves for which $\omega \to 0$ as $\lambda \to \infty$.

Inserting $n_0 \approx 10^{29}$ m^{-3} in Eq. (13.41) gives $\omega_P \approx 10^{16}$ s^{-1}. The quantum of energy $\hbar\omega_P$ associated with the plasma oscillations is therefore of order 10 eV and is sufficiently large that plasma oscillations are not normally thermally excited in metals and the electron liquid remains in its ground state as far as its long-wavelength motions are concerned. Quanta of plasma oscillations, known as **plasmons**, can however be excited by passing a beam of fast electrons through a thin metal foil. The electrons are found to emerge with discrete energy losses corresponding to the excitation of one or more plasmons (Fig. 13.15). Because plasma oscillations are long-range, many electrons are involved and hence plasmons are referred to as **collective excitations** of the electron liquid. Note that since a typical collision interval τ is of order 10^{-12} s, we have $\omega_P\tau \gg 1$ and our neglect of collisions in deriving Eq. (13.39) is justified.

13.5.3   Screening

Because the collective long-range motions of the electron liquid are not thermally excited, the motion of the electrons is highly correlated; the other electrons adjust in such a way as to cancel the long-range Coulomb field of any individual electron. At shorter distances the screening is incomplete and we can use Eq. (13.39) to investigate this. Suppose that there is a stationary *point* charge $-e$ at $\mathbf{r} = 0$. By putting $\partial^2 n/\partial t^2 = 0$ in Eq. (13.39) we find that the spherically symmetric *static* response of the electron liquid to this charge is given by

$$\nabla^2 n = \frac{1}{r^2}\frac{\mathrm{d}}{\mathrm{d}r}\left(r^2\frac{\mathrm{d}n}{\mathrm{d}r}\right) = \frac{(n-n_0)}{\lambda^2}, \tag{13.42}$$

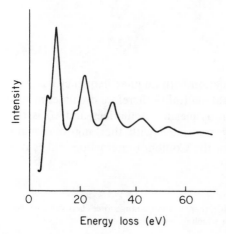

Fig. 13.15   Energy loss of 2020 eV electrons after scattering through 90° by a Mg film. The large peaks indicate the energy loss to be a multiple of $\hbar\omega_P$. The minor peaks are a surface effect (see Kittel[7]). (Reproduced with permission from C. J. Powell and J. B. Swann, *Phys. Rev.* **116**, 81 (1959))

where

$$\lambda = \left(\frac{2\varepsilon_F \varepsilon_0}{3n_0 e^2}\right)^{1/2} = \frac{1}{\omega_P}\left(\frac{2\varepsilon_F}{3m}\right)^{1/2} \approx \frac{v_F}{\omega_P} \approx 1 \text{ Å} \tag{13.43}$$

and we have used Eq. (13.41) for ω_P.

The solution of Eq. (13.42) is (see problem 13.9)

$$n - n_0 = -\frac{1}{4\pi\lambda^2}\frac{e^{-r/\lambda}}{r} \tag{13.44}$$

as plotted in Fig. 13.16. The electron liquid is excluded from a region of order λ in size around the charge at the origin. The total excluded charge

$$\int_0^\infty (n - n_0)(-e)4\pi r^2 \, dr = \lambda^2 e \int_0^\infty r \, e^{-r/\lambda} \, dr = e$$

just balances the charge at the origin and this explains the compensation of its long-range Coulomb field. From Eq. (13.43) the screening length λ is much less than the electron mean free path and this is further justification of our neglect of collisions in deriving Eq. (13.39).

Thus each electron in a metal is surrounded by a screening hole in the electron liquid. As the electron moves through the metal the screening hole moves with it; the large plasmon energy determines that the screening remains effective at finite velocity. It is the electron plus its associated screening hole which behaves as an almost independent particle; the combination forms an entity known as a **quasi-particle**. Because of the absence of collective excitations, the low lying excited states of metals correspond to a gas of weakly interacting quasi-particles. It is the wavefunction of a quasi-particle that obeys Bloch's theorem.

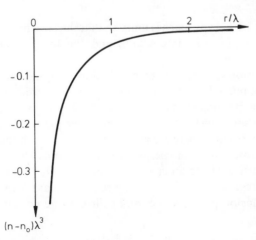

Fig. 13.16   The static response of the electron liquid to a point charge $-e$ at the origin

Eq. (13.44) is the **Thomas–Fermi equation** for the static screening of the long-range Coulomb field of an electron and λ is the **Thomas–Fermi screening length**. The result is normally derived in a slightly different way (Kittel,[7] p. 264) and is approximate in not allowing for the motion of the electron and in treating the electrons as a continuous liquid on an atomic length scale. It is also inconsistent in considering each electron both as an individual particle and as forming part of the background electron liquid. Improved theories of various degrees of sophistication for dealing with electron–electron interactions are available; the advantage of our simple-minded approach is that it provides a physical picture for what is happening.

13.5.4   The exclusion principle and scattering

After allowing for the screening of the Coulomb interaction discussed in the previous section, there is a residual interaction between two quasi-particles with a range of order 1 Å; essentially the interaction vanishes unless there is overlap of the screening holes associated with the quasi-particles. This residual interaction should result in a collision cross section A for two quasi-particles of order 1 Å^2. According to kinetic theory the corresponding mean free path l is $1/nA$ and the relaxation time is

$$\tau = \frac{l}{v_F} = \frac{1}{nAv_F}. \tag{13.45}$$

The resulting broadening of the single particle energy levels is given by the energy–time uncertainty relation as

$$\Delta\varepsilon = \frac{\hbar}{\tau} = \hbar nAv_F \approx \frac{nA}{k_F}\varepsilon_F. \tag{13.46}$$

With $A \approx 1$ Å^2, (nA/k_F) approaches unity so that the level broadening is of order ε_F, even with the screened interaction, and the independent quasi-particle picture appears to collapse.

The x-ray emission spectrum of metals (section 3.2.4) indicates that the energy of the particles is not subject to an uncertainty of this order and therefore that the lifetime must be longer than predicted by Eq. (13.45). The independent particle picture is saved by the Pauli exclusion principle. A collision between two particles can only occur if there are empty states into which the particles can be scattered with conservation of energy and momentum. This is the case only if *both* colliding particles have an energy within about $k_B T$ of the Fermi energy; this is the only region where both occupied and vacant levels may be found. Without the exclusion principle the total collision rate per unit volume is

$$w = \frac{n}{\tau} = n^2 Av_F.$$

When the exclusion principle is allowed for, each n in this expression is multiplied by a factor of order T/T_F by the above energy restriction, so that the actual collision rate is of order

$$w = n^2 A v_F (T/T_F)^2,$$

and the relaxation time of a single particle is

$$\tau = \frac{n}{w} = \frac{(T_F/T)^2}{n A v_F}.$$

From Eq. (13.46) the correct level broadening now becomes

$$\Delta\varepsilon = \frac{\hbar}{\tau} \approx \frac{nA}{k_F} \varepsilon_F \left(\frac{T}{T_F}\right)^2 \approx k_B T \left(\frac{T}{T_F}\right),$$

with $nA/k_F \approx 1$, as before. The level broadening is thus small compared to $k_B T$, the thermal broadening of the Fermi function, and the picture of independent quasi-particles is restored. At room temperature $T_F/T \sim 100$ and the level broadening is typically $k_B T/100 \approx 3 \times 10^{-4}$ eV with an associated relaxation time

$$\tau = \frac{\hbar}{\Delta\varepsilon} \approx 2 \times 10^{-12} \text{ s.}$$

This is a factor of ~ 100 longer than the relaxation time deduced from a typical electrical conductivity in section 3.3.2. Thus collisions between quasi-particles can normally be ignored.

★13.5.5   Fermi liquid effects

The properties of the nearly independent quasi-particles are modified slightly from those of electrons. For example, as a quasi-particle moves through the metal, the backflow of the screening electron liquid contributes to its effective mass (Fig. 13.17). More generally, whenever the occupied states are changed in

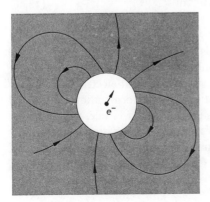

Fig. 13.17   Backflow of the screening electron liquid around an electron. The backflow contributes to the effective mass of the quasi-particle. In hydrodynamic theory the mass enhancement of an object due to backflow is called the hydrodynamic virtual mass. We have depicted a sharp cut-off in the screening liquid. Fig. 13.16 indicates that this is an approximation

some coherent way, by the application of an electric or magnetic field for example, an electron senses the perturbation not only directly but also through its interaction with the modified electron liquid. Landau introduced an approach, known as the **Landau Fermi liquid theory**, for taking into account the resulting change in quasi-particle energy; the change is usually small for the electronic quasi-particles in metals but is much larger for the atomic quasi-particles in liquid ^{3}He at low temperatures.

★13.5.6   The Mott transition

In section 4.3.2 we suggested that, because of electron–electron interactions, there might be a critical electron density required for the existence of a metallic state; below this density the electrons are localized on atoms. The transition from insulating to metallic behaviour is known as a **Mott transition** and we can now give a semi-quantitative explanation of it in terms of screening. Consider an array of hydrogen atoms;† in the insulating state each electron is bound into an atomic orbital by the Coulomb attraction of the nearest proton. In the metallic state the electrons are delocalized and the electron liquid screens the Coulomb field of the protons in the manner described in section 13.5.3. The resulting screened potential of a proton is (problem 13.10)

$$\frac{e}{4\pi\varepsilon_0 r}\,\mathrm{e}^{-r/\lambda},\tag{13.47}$$

which is just the bare Coulomb potential multiplied by a screening factor $\mathrm{e}^{-r/\lambda}$. Solving the Schrödinger equation with this screened potential shows that bound states exist only if the screening length λ is greater than the Bohr radius a_0. We therefore deduce the condition $\lambda > a_0$ for localized (insulating) behaviour. Inserting the value of λ from Eq. (13.43) and using Eq. (3.9) for ε_F gives the condition for insulating behaviour as

$$n_0 < \frac{\pi}{192 a_0^3} \approx 10^{29}\ \mathrm{m}^{-3},\tag{13.48}$$

where we have also used $a_0 = 4\pi\varepsilon_0\hbar^2/me^2$.

In fact the electron density in most metals is *less* than this, indicating that the bare ionic potential is not the simple Coulomb potential of a positive point charge. For a uniform positive charge density, the critical electron density for localization is smaller. Wigner showed that in this case localization is expected to occur by crystallization of the electrons into a regular lattice, which is now called a **Wigner lattice**. It is instructive to discuss why localization occurs at low densities since it might be expected to occur at high density where the effect of the interactions between electrons is larger. The contribution of the interactions

---

† We suppose that molecule formation can be suppressed by some means.

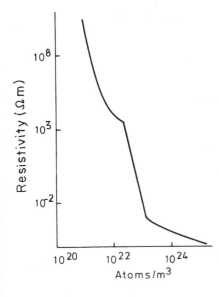

Fig. 13.18   Dependence of resistivity of germanium on impurity concentration at 2.5 K. For more than 10^{23} impurity atoms/m³ the resistivity is independent of temperature and metallic in character. (Reproduced with permission from N. F. Mott, *Phil. Mag. (8).* **6**, 287 (1961))

to the energy in fact increases as $n^{3/2}$ (this can be deduced by calculating the zero point energy of the plasma oscillations). However the kinetic energy of free electrons is proportional to $n\varepsilon_F$ and therefore varies as $n^{5/3}$. Since 5/3 > 3/2 the kinetic energy always dominates at high electron density and this favours extended states and metallic behaviour (see section 3.2.5).

A metal–insulator transition is most easily seen in semiconductors where the carrier concentration can be changed by varying the doping level. Thus in Fig. 5.9 the most heavily doped sample of germanium exhibits metallic behaviour at low temperatures, indicating that the electrons are no longer localized on the donor impurities. Fig. 13.18 shows an insulator–metal transition in the resistivity of germanium at 2.5 K as the impurity concentration is increased above 10^{23} m^{-3}.

13.6   ELECTROMAGNETIC WAVES IN METALS

Plasma oscillations are essentially longitudinal electromagnetic waves of long wavelength; such waves do not propagate in free space and for propagation in metals we deduce from Eq. (13.39) that the condition $\omega > \omega_P$ is required. In this section we consider transverse electromagnetic waves; for these div $\mathbf{E} = 0$ so that, from Eq. (13.35), there are no asssociated charge density fluctuations. If the relative electrical permittivity and magnetic permeability of the metal are unity, the propagation of transverse waves is described by the Maxwell equations:

$$\text{curl } \mathbf{H} = \mathbf{j} + \varepsilon_0 \dot{\mathbf{E}}, \tag{13.49}$$

$$\text{curl } \mathbf{E} = -\dot{\mathbf{B}} = -\mu_0 \dot{\mathbf{H}}. \tag{13.50}$$

Elimination of the magnetic field between these equations gives

$$-\text{curl curl } \mathbf{E} = \nabla^2 \mathbf{E} = \mu_0 \frac{\partial \mathbf{j}}{\partial t} + \frac{1}{c^2} \frac{\partial^2 \mathbf{E}}{\partial t^2}, \tag{13.51}$$

where $c = 1/(\varepsilon_0 \mu_0)^{1/2}$ is the velocity of light in free space.

At low frequencies $\mathbf{j}$ is given by Ohm's law, $\mathbf{j} = \sigma \mathbf{E}$, and the term containing $\partial \mathbf{j}/\partial t$ is much larger than the displacement current term on the right-hand side of Eq. (13.51). Eq. (13.51) is then a diffusion equation for $\mathbf{E}$ and has solutions with an oscillatory time dependence of the form

$$\mathbf{E} = \mathbf{E}_0 \, e^{-z/\delta} \cos(z/\delta - \omega t), \tag{13.52}$$

where

$$\delta = (2/\omega \mu_0 \sigma)^{1/2} \tag{13.53}$$

is the electromagnetic skin depth (Grant and Phillips,[3] p. 388). The amplitude of the wave thus decays exponentially with distance. For 1 MHz waves in copper at 300 K ($\sigma = 0.6 \times 10^8 \ \Omega^{-1} \ \text{m}^{-1}$) the decay length δ is 65 μm.

In pure metals at low temperatures the electronic mean free path l is large and can be comparable to or greater than the skin depth δ at microwave frequencies. The current density is then no longer given by Ohm's law but by a non-local relation in which $\mathbf{j}$ at point $\mathbf{r}$ depends on the value of $\mathbf{E}$ at all points in the neighbourhood of $\mathbf{r}$, within a distance of order l. Transverse electromagnetic waves are still attenuated but the skin depth tends to a finite value in the infinite conductivity ($\equiv$ infinite l) limit rather than tending to zero as predicted by Eq. (13.53). This change in behaviour is referred to as the **anomalous skin effect** and the limiting value of δ in single crystals has been used to provide information on the Fermi surface.

At infrared and optical frequencies the inertia of the electrons becomes more important than scattering and their behaviour is then described by the acceleration equation

$$\frac{d\mathbf{j}}{dt} = -ne\frac{d\mathbf{v}}{dt} = \frac{ne^2}{m_e}\mathbf{E}.$$

Inserting this into Eq. (13.51) gives

$$\nabla^2 \mathbf{E} = \frac{1}{c^2}\left(\omega_P^2 \mathbf{E} + \frac{\partial^2 \mathbf{E}}{\partial t^2}\right) \tag{13.54}$$

where ω_P is the plasma frequency of Eq. (13.41). If we look for wavelike solutions varying as $e^{i(\mathbf{k}\cdot\mathbf{r}-\omega t)}$ then, by substitution, we find the dispersion relation

$$k^2 = \frac{\omega^2 - \omega_P^2}{c^2}. \tag{13.55}$$

For $\omega < \omega_P$, k is imaginary, and the electric field decays exponentially with no phase change. Light incident on the crystal from outside is totally externally reflected as in Reststrahlen (section 9.1.4). For $\omega > \omega_P$, k is real and the wave can propagate through the metal. The plasma frequency is normally in the ultraviolet, and thin films of many metals do become transparent in this region of the spectrum. Eq. (13.55) predicts that a metal behaves as though it has a dielectric constant

$$\varepsilon = \frac{c^2}{(\omega/k)^2} = 1 - \frac{\omega_P^2}{\omega^2} \qquad (13.56)$$

for $\omega > \omega_P$, which is the same as Eq. (9.19). The effect of the electron scattering, which we have neglected, is to cause attenuation of the wave.

PROBLEMS 13

13.1 Show that Eqs. (13.1) and (13.17) reduce to the free electron results of Eqs. (3.8) and (3.27) when the dispersion relation is $\varepsilon = \hbar^2 k^2/2m$.

13.2 (a) Show that the cyclotron effective mass of Eq. (13.23) is equal to the dynamic effective mass of Eq. (13.9) for the isotropic dispersion relation $\varepsilon = \hbar^2 k^2/2m_e$.
(b) Use Eqs. (13.11) and (5.54) to calculate the effective-mass tensor for an electron near the conduction band minimum in silicon. Generalize Eq. (13.10) to the motion of electrons in a magnetic field and hence rederive Eq. (5.58) for the cyclotron resonance frequency.

13.3 In a cyclotron resonance experiment in potassium at 68 GHz, three consecutive resonances are observed at magnetic fields of 0.74, 0.59 and 0.49 T. Calculate the cyclotron effective mass of the orbit responsible for these resonances.

13.4 Show that $\hbar\omega_C \approx \varepsilon_F$ implies that the cyclotron orbit radius is smaller than an electron wavepacket.

13.5 A constant magnetic field $\mathbf{B}$ in the z direction can be represented by the vector potential $\mathbf{A} = (0, Bx, 0)$. Use the Hamiltonian for the kinetic energy of an electron in a magnetic field (Eq. (C2) of appendix C) to show that the Schrödinger equation for a free electron in this field is

$$-\frac{\hbar^2}{2m}\left[\frac{\partial^2\psi}{\partial x^2} + \left(\frac{\partial}{\partial y} - \frac{ieBx}{\hbar}\right)^2\psi + \frac{\partial^2\psi}{\partial z^2}\right] = E\psi.$$

Show that this equation has a solution of the form

$$\psi(x, y, z) = u(x)\exp\left[i(\beta y + k_z z)\right],$$

where $u(x)$ satisfies

$$-\frac{\hbar^2}{2m}\frac{\partial^2 u(x)}{\partial x^2} + \frac{m}{2}\left(\frac{eBx}{m} - \frac{\hbar\beta}{m}\right)^2 u(x) = E'u(x),$$

and $E' = E - \hbar^2 k_z^2/2m$. The equation for $u(x)$ is the Schrödinger equation for a simple harmonic oscillator centred on the point $x = \hbar\beta/eB$; deduce the frequency of the oscillator and hence the energy eigenvalues E' and E.

13.6 Estimate the cross-sectional area of Fermi surface responsible for the de Haas–van Alphen oscillations of Fig. 13.11. Compare your answer with the value for the free electron Fermi sphere. The atomic density in copper is $8.5 \times 10^{28}\,\mathrm{m}^{-3}$. Estimate the maximum temperature at which the de Haas–van Alphen effect will be observed in copper in a field of 10 T. If the impurity density n_i and the collision time τ are related by $n_i\tau = 10^{14}\,\mathrm{m}^{-3}\,\mathrm{s}$, up to what impurity density can the effect be observed?

13.7 Deduce the plasma frequency for magnesium from Fig. 13.15 and compare your answer with the value expected from Eq. (13.41) for the *atomic* density $4.3 \times 10^{28}\,\mathrm{m}^{-3}$.

13.8 Show that electron collisions cause damping of plasma oscillations. Show that critical damping occurs at a sufficiently small electron concentration. Estimate the critical concentration if the scattering time $\tau = 10^{-12}\,\mathrm{s}$. What happens when $\omega_p\tau \ll 1$?

13.9 Show, by substitution, that Eq. (13.44) satisfies Eq. (13.42) for $r \neq 0$. The point charge is at $r = 0$ and the solution of Eq. (13.44) diverges so that $\nabla^2 n$ cannot be readily calculated there. Demonstrate the correctness of the solution at this point by integrating Eq. (13.42) over a small sphere centred on $r = 0$. As the radius of the sphere goes to zero, the right-hand side is dominated by the point charge; the left-hand side can be converted to a surface integral. Equality of the two sides establishes the $1/4\pi\lambda^2$ factor in Eq. (13.44).

13.10 Show that Eq. (13.47) gives the screened potential of a proton.

13.11 Derive a modified form of Eq. (13.48) appropriate for the non-degenerate carriers in a semiconductor. Hence calculate the critical doping density for the formation of an impurity energy band in germanium. (Take $m_e = 0.22m$ and $\varepsilon =$ dielectric constant $= 16$). Are the experimental results of Fig. 13.18 consistent with this prediction?

Up to 1980 nobody expected that there exists an effect like the quantized Hall effect, which depends exclusively on fundamental constants and is not affected by irregularities in the semiconductor like impurities or lattice defects.—*Klaus von Klitzing, Nobel Prize address, 1986*

CHAPTER

Low-dimensional systems

14.1 INTRODUCTION

So far in this book our discussion has been concerned almost entirely with the *bulk* properties of crystalline solids. Such properties can normally be specified in terms of coefficients that are independent of the shape and size of the specimen. Thus, for example, the specific heat capacity (heat capacity per unit mass) is a coefficient that can be multiplied by the mass of the sample to give the heat capacity of any bulk specimen. When one or more of the dimensions of a solid are reduced sufficiently, the properties are no longer given by these bulk coefficients. The sample is then described as a **low-dimensional system** (LDS). Low-dimensional systems can be classified according to the number of dimensions that are small: thin films are two-dimensional since only the film thickness is reduced; fine wires are one-dimensional since only one dimension, the length, is large; dots or specks are zero-dimensional since all three dimensions are small in this case.

Departures from bulk behaviour occur when the size of the sample becomes comparable to the wavelength of the important excitations in the solid, a phenomenon sometimes described as a **quantum size effect**. The nature of the excitations then changes and as a result so does any property determined by those excitations. The properties of low-dimensional systems can be very different from those of bulk specimens, often in quite unexpected ways. In the following section we illustrate this by considering the specific example of electrons confined to a film; this system is called the **two-dimensional electron gas** (2DEG).

A less fundamental effect of reduced specimen size is observed in the transport properties of solids when the specimen size becomes comparable to the *mean free path* of the excitations. In this case the phenomenon is described as a **size effect**. An example of a size effect can be found in section 2.8.4. The phonon mean free path in pure single crystals at low temperatures becomes comparable to the dimensions of the crystal. The effective thermal conductivity then depends on the shape and size of the sample. Size effects can be calculated by taking into account the scattering of excitations by the sample surface. Provided the wavelength of the excitations remains short compared to the crystal size, the nature of the excitations themselves is unchanged.

14.2   THE TWO-DIMENSIONAL ELECTRON GAS

14.2.1   The electron states

For simplicity we assume initially that the electrons are confined to a film of thickness d by infinite potential barriers at $z = 0$ and $z = d$ as shown in Fig. 14.1(a). Motion of the electrons in the xy plane is assumed to be unconfined. We

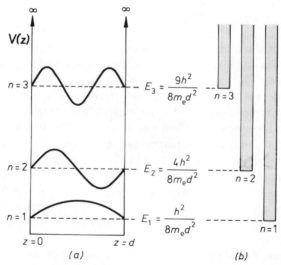

Fig. 14.1 (a) Wavefunctions and energies of the three lowest bound states of a one-dimensional infinite square potential well of width d. If electrons are confined to move in a film in the xy plane by such a well, the z dependence of their wavefunction corresponds to one of the bound states (Eq. (14.3)). (b) The shaded areas indicate the region of electron energies associated with each bound state for an electron in the film. Increasing energy within each shaded area corresponds to increasing motion in the xy plane (increasing $p^2 + q^2$ in Eq. (14.5))

use a free electron approach but take account of possible band structure and other effects by giving the electrons an effective mass m_e. The wavefunction of the electrons is determined, as in the three-dimensional case (section 3.2), by solving Schrödinger's equation but now with the boundary conditions

$$\Psi = 0 \quad \text{at} \quad z = 0 \quad \text{and} \quad z = d \quad (14.1)$$

at the edges of the film. For convenience we continue to use periodic boundary conditions in the xy plane (see Eq. (3.2)),

$$\Psi(x + L, y + L, z) = \Psi(x, y, z). \quad (14.2)$$

The resulting (unnormalized) wavefunction looks like a travelling wave for motion in the xy plane and a standing wave for motion along z,

$$\Psi(x, y, z) = e^{ik_x x} e^{ik_y y} \sin(k_z z). \quad (14.3)$$

To satisfy the boundary conditions of Eqs. (14.1) and (14.2)

$$k_x = 2\pi p/L, \qquad k_y = 2\pi q/L, \qquad k_z = n\pi/d, \quad (14.4)$$

where p, q and n are integers. The values of k_x and k_y are the same as for three-dimensional electrons. The z dependence of the wavefunction corresponds to the stationary states of a one-dimensional infinite square potential well as shown in Fig. 14.1(a) and the allowed values of k_z correspond to fitting an integral number of half-wavelengths into the well.

The energy associated with the wavefunction of Eq. (14.3) is

$$\varepsilon = \frac{\hbar^2}{2m_e}(k_x^2 + k_y^2 + k_z^2) = \frac{\hbar^2}{2m_e L^2}(p^2 + q^2) + \frac{n^2 h^2}{8m_e d^2}. \quad (14.5)$$

The final term corresponds to the energy levels

$$E_n = \frac{n^2 h^2}{8m_e d^2} \quad (14.6)$$

of the one-dimensional infinite square well potential as shown in Fig. 14.1(a); the term containing $p^2 + q^2$ is the additional energy associated with the motion in the xy plane. The implications of this energy level scheme can be seen from Fig. 14.1(b). The lowest state with $n = 2$ (that with $p = q = 0$) is higher in energy by $3h^2/8m_e d^2$ than the lowest $n = 1$ state. For d less than about 60 Å and $m_e = m$ this exceeds $k_B T$ at room temperature. We thus have the possibility that all the electrons can be frozen into the $n = 1$ state at low temperature—the z dependence of the wavefunction is then that of the ground state of the infinite square potential well. Motion of the electrons in the z direction is effectively frozen and their behaviour is that of free particles in a two-dimensional space consisting of the xy plane.

This discussion can be readily generalized to a more realistic potential variation in the z direction. An arbitrary one-dimensional potential well has one or more bound states and the wavefunctions and energies of these can be calculated. For a film with this potential variation in the z direction, the z dependence of the wavefunction corresponds to one of these bound states and the energy is of the form of Eq. (14.5) with the final term replaced by the corresponding one-dimensional eigenenergy. To determine the states of the 2DEG that are occupied at $T = 0$ we must take the Pauli principle into account, that only two electrons of opposite spin can have the same space wavefunction. To do this we must calculate the density of states associated with each bound state of the potential well.

14.2.2 Density of states of the two-dimensional electron gas

For states associated with the nth bound state of the potential well, the motion of the electrons in the xy plane is described by the k_x and k_y values of Eq. (14.4). Using the approach introduced in section 2.6.2, which was used in section 3.2 for 3D free electrons, we take k_x and k_y to be the components of a 2D vector **k** and plot the allowed values of this in a 2D **k**-space as in Fig. 14.2. The states lie on a simple square lattice of side $2\pi/L$ and the area of **k**-space per **k** state is therefore $(2\pi/L)^2$. The magnitude of the **k** vector is $k = (k_x^2 + k_y^2)^{1/2}$. In the area $2\pi k\,dk$ of k-space between circles of radius k and $k + dk$ there are

$$g(k)\,dk = \frac{2\pi k\,dk}{(2\pi/L)^2} = \frac{L^2 k\,dk}{2\pi} \tag{14.7}$$

allowed **k** states.

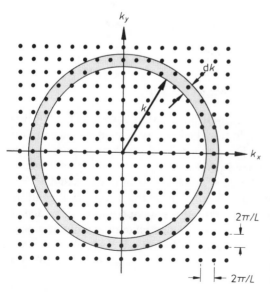

Fig. 14.2   The allowed wavevectors for 2D motion of electrons in the xy plane. The states form a simple square lattice of side $2\pi/L$ To calculate the density of states the number of wavevectors in the shaded circular ring between k and $k + dk$ must be determined

Eq. (14.5) for the energies of the states can be written

$$\varepsilon = \frac{\hbar^2 k^2}{2m_e} + E_n \tag{14.8}$$

where, for electrons confined by an infinite square well potential, E_n is given by Eq. (14.6). To obtain the density of states per unit energy range $g(\varepsilon)$ in *unit area of the film* ($L^2 = 1$) we use $g(\varepsilon)\, d\varepsilon = 2g(k)\, dk$ where the factor 2 arises because of the two possible spin states for the electron. Thus, for fixed n,

$$g(\varepsilon) = 2g(k) \frac{dk}{d\varepsilon} = \frac{m_e}{\pi \hbar^2}, \tag{14.9}$$

where we have used Eqs. (14.7) and (14.8). The density of states of the 2D electrons associated with each bound state of the potential well is thus independent of energy.

Adding together the densities associated with all the bound states gives the total density shown in Fig. 14.3. As the film thickness increases the bound state energies become more closely spaced so that the steps in the density of states are more difficult to observe. The density of states then approaches the smooth parabolic form appropriate to 3D free electrons as indicated by the broken curve in Fig. 14.3.

A potential well for confining electrons to a thin layer can be produced in semiconductor heterojunction structures prepared by the MBE technique (see section 6.6).† A potential well in the conduction band edge is obtained by sandwiching a layer of GaAs between macroscopic layers of $Ga_{1-x}Al_xAs$ as in Fig. 6.13. Fig. 14.4 shows the bound state energies for electrons in a one-

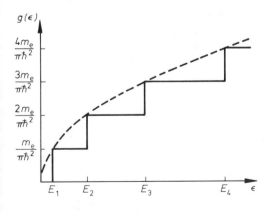

Fig. 14.3   Density of states per unit area for free electrons in a thin film. E_1, E_2,... are the bound state energies for the potential confining the electrons to the film (see Fig. 14.1(a)). The broken curve shows the energy dependence of the density of states for the three-dimensional electrons in a film of large thickness

---

† A two-dimensional electron gas can also be produced in the inversion layer of a MOSFET (section 6.5). This system has the advantage that the electron concentration can be easily varied by changing the gate potential.

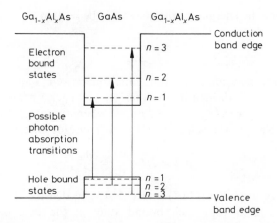

Fig. 14.4   Bound states associated with a thin layer of GaAs sandwiched between two thick $Ga_{1-x}Al_xAs$ layers (the energy levels are not drawn to scale). The allowed photon absorption transitions are shown and lead to onset of absorption at frequencies given by Eq. (14.10)

dimensional well of this shape. The step in the valence band edge acts as a potential well for holes (recall that the hole energy is the *negative* of the electron energy); the bound state energies for holes are also shown in Fig. 14.4.

The density of states in the GaAs layer can be investigated by measuring the absorption of electromagnetic radiation associated with the excitation of an electron from the valence band to the conduction band. Because the wavelength of the radiation is long compared to the width of the well, transitions only occur between states for which the spatial variation of the wavefunction is similar; this leads to the selection rule $\Delta n = 0$ for the absorption. The allowed transitions are therefore those indicated in Fig. 14.4. The frequency dependence of the absorption should reflect the step-like dependence on energy of the density of states; steps in the absorption should occur at frequencies ω_n given by

$$\hbar\omega_n = E_{Cn} - E_{Vn} \qquad (14.10)$$

where $E_{Cn} - E_{Vn}$ is the energy difference between the nth bound states in the conduction and valence bands as indicated in Fig. 14.4. The measured absorption spectra for GaAs layers of thicknesses 140, 210 and 4000 Å are shown in Fig. 14.5 and the expected step structure is clearly visible for the two thinner layers.

The arrows indicate the frequencies at which steps are expected (see problem 14.1). The onset of absorption is marked by peaks at energies just below these predicted values. This results from the existence of a bound state between the created electron and hole, and the reduction in energy of the peak relative to the expected value allows the binding energy of the electron–hole pair to be

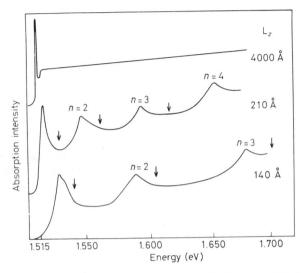

Fig. 14.5   Measured absorption of electromagnetic radiation as a function of photon energy for GaAs layers of thicknesses 4000, 210 and 140 Å. The arrows indicate the energies at which onset of absorption would be expected to occur for transitions involving the nth bound state as shown in Fig. 14.4. Above the excitonic peak the absorption curve for the 4000 Å layer is smooth, indicating 3D behaviour. (Reproduced with permission from R. Dingle, W. Wiegmann and C. H. Henry, *Phys. Rev. Lett.* **33**, 827 (1974))

determined. The bound electron–hole system is known as an **exciton** (see problem 14.2). An excitonic peak is clearly visible on the absorption curve for the 4000 Å GaAs layer but the step-like structure has disappeared, indicating that the density of states is the smooth curve appropriate to 3D behaviour.

14.3   THE QUANTUM HALL EFFECT

Perhaps the most remarkable property of 2D electron systems is the quantum Hall effect observed when a large magnetic field is applied perpendicular to the 2D layer at low temperatures. We consider the geometry of Fig. 14.6. A rectangular film of thickness d has dimensions L and W in the x and y directions respectively. A current I_x is applied in the x direction and voltmeters are used to measure the longitudinal and transverse voltages, V_x and V_y, when a magnetic field is applied in the z direction perpendicular to the film. From the measured voltages the longitudinal and transverse electric fields, E_x ($= V_x/L$) and E_y ($= V_y/W$), can be determined. It is convenient to present the results of the measurements by quoting values for longitudinal and transverse 2D resistivity

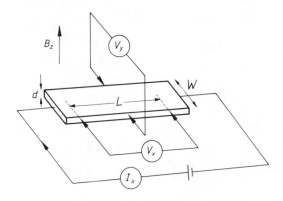

Fig. 14.6 Geometry for Hall effect measurement in a 2D electron system

coefficients, ρ_L and ρ_T, defined by

$$E_x = \rho_L J_x \qquad \text{and} \qquad E_y = \rho_T J_x, \qquad (14.11)$$

where J_x is a 2D current density defined as the current per unit width of film $(J_x = I_x/W)$. Note that the 2D resistivities defined in this way are measured in ohms not ohm metres as are 3D resistivities. The transverse resistivity ρ_T quantifies the Hall effect.

It will be helpful first to use our Hall effect calculation of section 3.3.5 to calculate the behaviour that would be expected for classical free particles. The electric fields in the x and y directions are given by Eqs. (3.37), which can be written

$$E_x = -\frac{m_e}{e\tau} v_x = \frac{m_e}{e\tau} \frac{j_x}{ne} = \frac{m_e}{ne^2\tau} \frac{I_x}{Wd} = \frac{m_e}{nde^2\tau} J_x,$$

$$E_y = v_x B = -\frac{j_x B}{ne} = -\frac{I_x B}{Wdne} = -\frac{B}{nde} J_x. \qquad (14.12)$$

Comparison of Eqs. (14.11) with Eqs. (14.12) enables us to identify the classical predictions for ρ_L nd ρ_T:

$$\rho_L = \frac{m_e}{n_A e^2 \tau}, \qquad \rho_T = -\frac{B}{n_A e}, \qquad (14.13)$$

where we have introduced the notation n_A for the 2D electron density; n_A is the number of electrons per unit area and is related to the volume density by $n_A = nd$. Eqs. (14.13) differ from their 3D equivalents, as given by Eqs. (3.27) and (3.38), only by the replacement $n \to n_A$. Eqs. (14.13) predict that ρ_L is independent of B and ρ_T increases linearly with B.

Fig. 14.7 shows measured values for ρ_L and ρ_T as functions of magnetic field for electrons in a semiconductor heterojunction structure like that of Figs. 6.13 and 14.4. The electrons are in the thin GaAs layer sandwiched between two

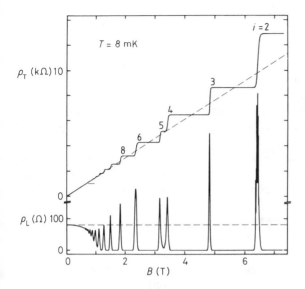

Fig. 14.7   Transverse and longitudinal resistivity components for a GaAs–Ga$_{0.71}$Al$_{0.29}$As heterostructure as a function of applied magnetic field at $T = 0.8$ mK. The dashed lines indicate the behaviour predicted by Eqs. (14.13). The integers i determine the values of ρ_T on the plateaux by Eq. (14.14). (Reproduced with permission from K. von Klitzing, *Physica* **126**, B + C, 242 (1984))

Ga$_{0.71}$Al$_{0.29}$As layers. The dashed lines indicate the classical behaviour predicted by Eqs. (14.13); th experimental results are very different. The transverse resistivity, which is a measure of the Hall effect, increases in steps at high fields. The value of ρ_T along the horizontal portions of the steps is very constant at a value related to the fundamental constants h and e by

$$\rho_T = -\frac{h}{ie^2} = -\frac{25812.8\ \Omega}{i},$$ (14.14)

where i is an integer. The values of the integer i for the steps on Fig. 14.7 are indicated. In the region where ρ_T is a constant, the longitudinal resistivity ρ_L is vanishingly small.

The reader might reasonably expect that the vanishing of ρ_L implies an infinite longitudinal conductivity. To see that this is not so we must generalize Eqs. (14.11) to a situation where the current flow is in an arbitrary direction in the xy plane:

$$E_x = \rho_L J_x - \rho_T J_y$$
$$E_y = \rho_T J_x + \rho_L J_y.$$ (14.15)

These equations can be inverted to obtain the current density produced by an electric field applied in an arbitrary direction. Thus we obtain

$$J_x = \sigma_L E_x - \sigma_T E_y$$
$$J_y = \sigma_T E_x + \sigma_L E_y,$$ (14.16)

where the 2D conductivity components are given by

$$\sigma_L = \frac{\rho_L}{\rho_L^2 + \rho_T^2} \quad \text{and} \quad \sigma_T = -\frac{\rho_T}{\rho_L^2 + \rho_T^2}. \tag{14.17}$$

Eqs. (14.17) have the intriguing property that if ρ_L vanishes but ρ_T remains finite then σ_L also vanishes. Thus the longitudinal conductivity and resistivity vanish simultaneously. What this means in practice is that an imposed current generates only a transverse electric field and an imposed electric field generates only a transverse current.

The strange behaviour of the 2D electron gas exhibited in Fig. 14.7 is known as the **quantum Hall effect**, and to understand it we must investigate the nature of the electron states in a two-dimensional electron gas in a magnetic field. We assume the electron density is sufficiently low that all the electrons are accommodated in states associated with the lowest bound state of the potential well. Motion of the electrons in the direction perpendicular to the 2D layer is then completely absent. In the presence of a perpendicular field the motion in the layer is no longer described by plane waves as in Eq. (14.3) for, as we have already described in section 13.4, the effect of a magnetic field on free electrons is to cause them to move in circular cyclotron orbits perpendicular to the field. The energies of these Landau levels are not given by Eq. (14.8) but by the 2D analogue of Eq. (13.26),

$$E = E_1 + (n + \tfrac{1}{2})\hbar\omega_C \pm \mu_B B, \tag{14.18}$$

where E_1 is the lowest bound state energy of the potential well and

$$\omega_C = \frac{eB}{m_e} \tag{14.19}$$

is the cyclotron frequency. The final term in Eq. (14.18) allows for the magnetic moment associated with the spin of the electron; μ_B is the Bohr magneton (Eq. (7.5)).

The density of states associated with the Landau levels is shown in Fig. 14.8 superimposed on top of the constant density (Eq. (14.9)) in the absence of the field. The number of states associated with each Landau level is finite but, because the width of the level is zero, the density of states is infinite. As in the three-dimensional case (section 13.4.3) the number of states associated with each Landau level can be determined by requiring that the average density of states is the same with and without the field. From Fig. 14.8 we see that there are two Landau levels (one for each spin state) in an energy range $\hbar\omega_C$; if each level contains N_L states then the averaged density of states in the presence of a field is $2N_L/\hbar\omega_C$. Equating this to the density of states in the absence of a field (Eq. (14.9)) gives

$$N_L = \frac{eB}{h} \tag{14.20}$$

for unit area of the 2D layer.

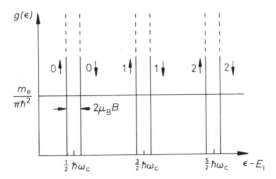

Fig. 14.8   Density of states in a 2D electron gas in a magnetic field. The vertical lines indicate the infinite density of states at the energies of the Landau levels given by Eq. (14.18). The arrows are the direction of the magnetic moment of the electron. The spin splitting and the cyclotron splitting are equal if $m_e = m$ (then $\hbar\omega_C = 2\mu_B B$). The horizontal line at $g(\varepsilon) = m_e/\pi\hbar^2$ shows the constant density of states in the absence of the field (cf. Fig. 14.3 for $E_1 < \varepsilon < E_2$). The Landau levels are identified by the value of n in Eq. (14.18) and the spin orientation; the level $1\uparrow$, for example, corresponds to $n = 1$ and the lower (negative) sign in Eq. (14.18)

With increasing magnetic field the number of states associated with each Landau level therefore increases and hence the number of levels required to accommodate the electrons decreases. If the temperature is sufficiently low (or the field sufficiently high) that thermal excitation of an electron from one Landau level to another does not occur, then we arrive at the situation illustrated in Fig. 14.9. In the field B_1 (Fig. 14.9(a)) the four lowest Landau levels are full and the fifth is partly filled. Since occupied and unoccupied states coexist at the energy of this level, the chemical potential of the electrons must coincide with this energy. The energy of the fifth level increases with increasing field and so therefore does the chemical potential. However, the number of electrons in this level decreases because more can be accommodated in the lower four levels. Fig. 14.9(b) illustrates the situation at the field B_2 at which the occupation of the fifth level goes to zero. At an even higher field B_3 (Fig. 14.9(c)) it is the fourth Landau level that is partly occupied and the chemical potential is then pegged to this level.

The resulting field dependence of the chemical potential is shown in Fig. 14.10 with the discontinuous glitches occurring whenever an integral number of Landau levels are exactly filled. If there are n_A electrons per unit area, this will occur when n_A/N_L is equal to an integer i; Eq. (14.20) then identifies the field B_i at which exactly i Landau levels are full as

$$B_i = \frac{N_L h}{e} = \frac{n_A h}{ie}. \qquad (14.21)$$

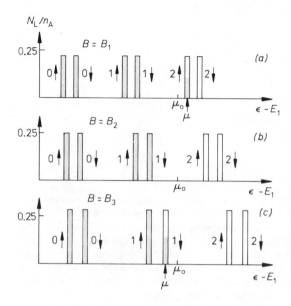

Fig. 14.9   Occupation of Landau levels in successively higher fields $B_1 < B_2 < B_3$. Here μ_0 is the chemical potential in the absence of the field. The labelling of the levels is explained in the caption of Fig. 14.8

This result allows us to demonstrate that the filling of Landau levels is linked to the quantum Hall effect. To do so we calculate, using our result for classical particles (Eq. (14.13)), the transverse resistivity at the fields given by Eq. (14.21),

$$\rho_T = -\frac{B_i}{n_A e} = -\frac{h}{ie^2}. \tag{14.22}$$

These are precisely the values of ρ_T along the horizontal portions of the steps on Fig. 14.7. That the longitudinal conductivity σ_L and hence the longitudinal

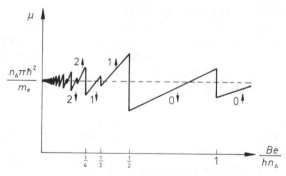

Fig. 14.10   Chemical potential as a function of magnetic field for a 2D electron gas. The vertical glitches occur when the chemical potential moves from one Landau level to another. The labels identify the Landau level in which the chemical potential lies; the caption to Fig. 14.8 explains the labelling

resistivity $\rho_{\rm L}$ should vanish when an integral number of Landau levels are exactly filled is not unexpected since, as we have already seen in section 4.2, electrical conduction requires the existence of unoccupied electron states at the chemical potential.

To understand fully the quantum Hall effect it is necessary to explain why the quantized value h/ie^2 of $\rho_{\rm T}$ and the vanishing of $\rho_{\rm L}$ occur not only at the field B_i predicted by Eq. (14.21) but over a range of fields centred on B_i. The full explanation is beyond the scope of this book but we will give an outline of it to the reader.

The ideally sharp Landau levels of Eq. (14.18) and Fig. 14.8 are expected only in a perfect 2D free electron gas. In practice the levels are broadened by imperfections of the material to produce a density-of-states curve like that in Fig. 14.11. The density of states still peaks at the energies given by Eq. (14.18) but neighbouring peaks overlap to a significant extent. States close to the peaks are believed to extend throughout the crystal and thus represent mobile electrons whereas states in the troughs are those of localized non-conducting electrons.†

With increasing magnetic field the spacing of the peaks and their size increases just as for the ideal case. The chemical potential is still oscillatory but no longer has the discontinuous glitches of Fig. 14.10. Instead it increases smoothly when it lies near the energy of one of the peaks in the density-of-states curve and decreases smoothly when it is near a trough. When the chemical potential lies within a region of localized states as in Fig. 14.11 the 2D electron

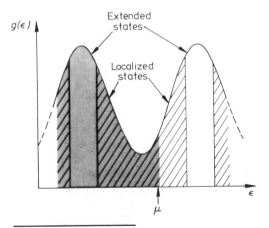

Fig. 14.11   Density of states associated with two adjacent Landau levels broadened by imperfections within the layer. Regions of localized and extended states are identified. Shading indicates which states are occupied

---

† An insight into why a low electron density leads to immobility is given in section 13.5.6. Why a low density of states should also have this effect can be understood by picturing all the states as essentially localized but with the possibility of conduction occurring by the hopping of an electron from one localized state to another. Such a process can only occur if there are sufficient states of approximately the same energy in the vicinity and thus only if the density of states at that energy exceeds a critical value.

system would be expected to behave as an insulator and it is thus possible to explain why the longitudinal conductivity σ_L vanishes over a range of magnetic field. To explain why the transverse resistivity ρ_T remains constant for the range of magnetic field for which the chemical potential lies in a region of localized states is more difficult. It is as though a Landau level behaves as if it is *exactly* filled whenever all the *mobile* states within it are completely occupied. It is remarkable that this not only appears to be the case but that the resulting value of ρ_T appears to be given rigorously in terms of fundamental constants by Eq. (14.22). The quantization of the Hall effect in a 2D electron system appears to depend on nothing less fundamental than the quantization of the elementary charge e!

Just as the Josephson effect in superconductivity provides an excellent method for defining a voltage standard (section 10.5.4), the quantum Hall effect of the 2D electron system can be used to define the unit of resistance. The value of ρ_T on one of the steps is constant to an accuracy greater than that with which h/e^2 is known, so it is necessary to define a value for h/e^2 in order to do this. The value that is chosen is of course consistent with the best known value of h/e^2.

14.4 RESONANT TUNNELLING DEVICES

In most of the semiconductor devices discussed in Chapter 6,† the dynamics of the carriers is described to a good approximation by classical equations of motion such as Eqs. (5.37); these predict correctly the trajectory of the wavepackets representing the particles and the wave nature of the particles plays no essential role in their behaviour. The continuing quest for faster and smaller semiconductor devices will inevitably result in the use of structures in which quantum effects are essential or unavoidable. Quantum effects become important if a carrier encounters structures in the direction of its motion which are comparable in size to its de Broglie wavelength. For an electron of energy $k_B T$ this wavelength at $T = 300$ K is

$$\lambda = \frac{h}{p} = \frac{h}{mv} = \frac{h}{m}\left(\frac{m}{k_B T}\right)^{1/2} = \frac{h}{(mk_B T)^{1/2}} \approx 10 \text{ nm}. \qquad (14.23)$$

For the flow of carriers *parallel* to a two-dimensional layer, such as in the high-electron-mobility transistor of Fig. 6.14, the geometry of the gate electrode may be constructed in such a way that the carriers are subjected to an electrostatic potential varying on the length scale of the de Broglie wavelength. Such a gate structure is shown in Fig. 14.12 and can be prepared by a process similar to that described for creating the electrode pattern in Fig. 6.12. The production of a mask that allows the definition of such small structures cannot however be achieved by optical lithography, since fundamental diffraction

† The tunnel diode is an exception.

effects prevent the focusing of light on a length scale of order nanometres. Very short-wavelength radiation is required to achieve the necessary resolution; focused electron beam or focused ion beam lithography have both been used successfully for preparing masks with a resolution of this order. Exposure of the photoresist through the mask as in Fig. 6.12. is normally done with x-rays when such small length scales are involved. That quantum effects are important in understanding the behaviour of the device of Fig. 14.12 is demonstrated by the observation that the grid acts as a diffraction grating for the electrons moving through the layer.

For carriers travelling *perpendicular* to the two-dimensional layers in hetero-junction structures, the thickness of the layer can be comparable to the wavelength of the carriers; **resonant tunnelling devices** exploit this possibility. Fig. 14.13(*a*) shows the conduction band in a resonant tunnelling diode. Two very thin layers of GaAlAs provide potential bariers, which lead to bound states, as indicated, within the central GaAs layer (cf. Fig. 14.4). The heavily doped GaAs regions on the outside of the sandwich are contacts through which carriers can be injected and removed. To flow from one contact region to the

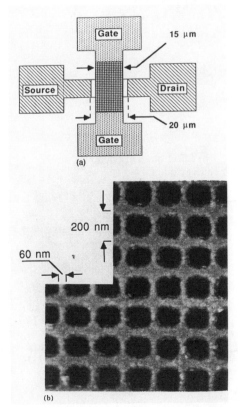

Fig. 14.12   (*a*) High-electron-mobility transistor in which the gate electrode is made from a Ti/Au alloy and has a grid structure. (*b*) Scanning tunnelling microscope image of the gate structure. (Reproduced with permission from Ismail *et al.*, *Appl. Phys. Lett.* **54**, 460 (1989))

other an electron must tunnel through the two potential barriers in series. The probability of this is normally low and the current through the device produced by an applied bias is small. The applied bias however alters the position of the bound state energy with respect to that of the tunnelling electrons as indicated in Fig. 14.13(*b*); the situation depicted is that in which the energy of the tunnelling electrons is equal to the energy of the lowest bound level. The wavelength of the electrons is then double the layer thickness and this leads to the phenomenon of

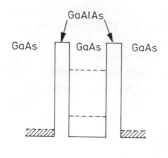

(*a*) Conduction band edge in resonant tunnelling diode structure. The energies of the bound states in the GaAs potential well are indicated

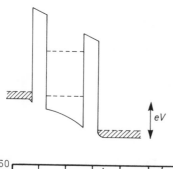

(*b*) Effect of an applied bias on the energy diagram

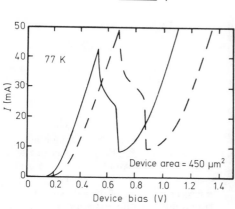

(*c*) Current–voltage characteristic of a resonant tunnelling diode. The broken curve indicates the behaviour for a bias applied in the opposite sense. (Reproduced with permission from Huang *et al.*, *Appl. Phys. Lett.* **51**, 121 (1987))

Fig. 14.13

resonant tunnelling. The probability of an electron tunnelling from one contact region to the other can approach unity at this bias and the resulting peak in the current is shown in Fig. 14.13(c).

The simplest way to understand resonant tunnelling is to exploit the analogy with the Fabry–Pérot etalon used in optical interferometry (Smith and Thomson,[5] chapter 13). The two potential barriers act as the low-transmissivity high-reflectivity mirrors of the etalon and the central GaAs layer is the gap between them. For normal incidence, the Fabry–Pérot etalon strongly transmits light only when the gap is an integral number of half-wavelengths. The amplitude of the light in the gap then becomes large through the constructive interference of multiply reflected beams. In the resonant tunnelling diode the corresponding build-up in electron charge in the bound level raises the energy of the level, with the result that, with increasing bias, it remains pinned to the incident electron energy over a wider range of bias than would be the case without resonant charge build-up. The negative resistance region of the current–voltage characteristic of Fig. 14.13(c) provides an application for the resonant tunnelling diode in oscillator circuits. Devices employing resonant tunnelling diodes have been constructed that will operate at frequencies of order 10^{12} Hz.

It is also possible to construct a resonant tunnelling transistor in which the position of the bound level in the potential well is influenced by a third (control) electrode. One possibility is to incorporate the resonant tunnelling structure of Fig. 14.13(a) into the base–emitter junction of a bipolar transistor so that the emitter–base bias determines the position of the level. Current peaks like those of Fig. 14.13(c) then appear in the collector current as a function of base–emitter voltage.

It is appropriate that this book should conclude with this glimpse into a potential growth area in solid state physics. The possibility of using solid state devices to do physical (as opposed to geometric) optics with electron waves is just one area in which solid state physics will continue to combine fundamental physics research with technological progress.

PROBLEMS 14

14.1 Calculate the photon energies at which you would expect onset of absorption to occur for the 140 and 210 Å GaAs layers in Fig. 14.5, and compare your answer with the observed values. The simplest approach is to assume that the bound state energies approximate to those of infinite square potential wells. Take the energy gap of GaAs to be 1.519 eV and the effective masses of electrons and holes to be $0.0665m$ and $0.45m$ respectively, where m is the bare electron mass.

To obtain more accurate values of the photon energies it is necessary to allow for the finite depth of the potential wells. If you wish to do this you will need to know that the depth of the wells in the conduction and valence band edges are 0.220 and 0.028 eV respectively.

14.2 Use the Bohr theory to calculate the binding energy of an exciton in bulk GaAs. Use the effective masses given in the previous question and take the value 13 for the dielectric constant.

14.3 Confirm that the values of ρ_T on the steps in Fig. 14.7 agree with Eq. (14.14). From the low-field data on this figure obtain values for the number of electrons per unit area and the scattering time of the electrons (assume $m_e = 0.07m$). Calculate the Fermi energy of the 2D electron gas. Estimate the field at which an electron completes one cyclotron orbit between collisions and comment on the answer. Estimate the maximum temperature at which you would expect to see the step-like structure in ρ_T.

APPENDIX A

Coupled probability amplitudes

In this section we derive the coupled probability amplitude equations (4.9) from the Schrödinger equation

$$i\hbar \, \partial\Psi/\partial t = H\Psi \tag{A1}$$

where H is the Hamiltonian operator. To do this we express the wavefunction $\Psi(\mathbf{r}, t)$ as a series in some set of functions $\psi_l(\mathbf{r})$ (which might, as in Eq. (4.7), be atomic eigenfunctions)

$$\Psi(\mathbf{r}, t) = \sum_l a_l(t)\psi_l(\mathbf{r}). \tag{A2}$$

Such an expansion is always possible provided the ψ_l form a **complete set** of functions. An example of a complete set is the sine and cosine functions; in this case the expansion (A2) is known as a Fourier series, or, in the limit in which Ψ cannot be regarded as a periodic function of position, as a Fourier transform. The expansion coefficients a_l in Eq. (A2) are time-dependent because Ψ is time-dependent. We will not worry here about the formal question of what constitutes a complete set, but will just assume for the derivation of Eqs. (4.9) that we have a set of functions ψ_l such that the expansion (A2) is both possible and unique. Later in this appendix we discuss how far the functions that we actually make use of in this book satisfy this condition.

If we assume that both $\Psi(\mathbf{r}, t)$ and the $\psi_l(\mathbf{r})$ are normalized, the probability of finding the system in the state ψ_n at time t is $|c_n(t)|^2$, where the probability amplitude $c_n(t)$ is given by

$$c_n(t) = \int \psi_n^*(\mathbf{r})\Psi(\mathbf{r}, t)\, d^3\mathbf{r} = \sum_l a_l(t) \int \psi_n^*(\mathbf{r})\psi_l(\mathbf{r})\, d^3\mathbf{r} \qquad (A3)$$

by use of Eq. (A2). If the functions ψ_l are orthogonal as well as normalized, the integral in Eq. (A3) is zero when $m \neq n$ and unity when $m = n$, so that $c_n(t) = a_n(t)$. We shall want to use atomic wavefunctions centred on adjacent lattice sites, which are not orthogonal to each other, and we shall therefore continue the argument without assuming orthogonality. A geometrical analogy may help to clarify Eq. (A3). Eq. (A2) is like writing a vector in terms of its components in a multidimensional space. In this analogy the a_n correspond to the components and the c_n to the projections on the coordinate axes; these are equal only if the coordinate axes are orthogonal (mutually perpendicular).†

We now substitute the expansion (A2) in the Schrödinger equation (A1) to obtain

$$i\hbar \sum_l \frac{da_l}{dt} \psi_l(\mathbf{r}) = \sum_l a_l H\psi_l(\mathbf{r}). \qquad (A4)$$

If Eq. (A4) is multiplied on the left by $\psi_n^*(\mathbf{r})$ and integrated over all space we obtain, by use of Eq. (A3), an expression for dc_n/dt,

$$i\hbar \frac{dc_n}{dt} = i\hbar \sum_l \frac{da_l}{dt} \int \psi_n^*(\mathbf{r})\psi_l(\mathbf{r})\, d^3\mathbf{r}$$

$$= \sum_l a_l \int \psi_n^*(\mathbf{r})H\psi_l(\mathbf{r})\, d^3\mathbf{r} = \sum_l a_l \int \psi_l(\mathbf{r})H\psi_n^*(\mathbf{r})\, d^3\mathbf{r}, \qquad (A5)$$

where the last step follows from the Hermitian property of the Hamiltonian operator. This property is obviously true for the potential energy term in H; to see that it is true for the kinetic energy term $-\hbar^2\nabla^2/2m$, note that successive integration by parts gives

$$\int \psi_1^* \nabla^2 \psi_2\, d^3\mathbf{r} = -\int (\nabla\psi_1^*) \cdot (\nabla\psi_2)\, d^3\mathbf{r} = \int (\nabla^2\psi_1^*)\psi_2\, d^3\mathbf{r}$$

provided only that ψ_1 and ψ_2 and their derivatives vanish at infinity, so that the surface integrals vanish.

Notice now that $H\psi_n^*(\mathbf{r})$ is just another function of position, so that it can also be expanded in a way analogous to Eq. (A2),

$$H\psi_n^*(\mathbf{r}) = \sum_m E_{nm}\psi_m^*(\mathbf{r}). \qquad (A6)$$

---

† Do not worry that the total probability $\sum_n |c_n|^2 \neq 1$ when the ψ_n are not orthogonal; this is because the possibilities are not mutually exclusive in this case.

The only differences are that the expansion coefficients E_{nm} are independent of time, and we have chosen to expand in terms of the complete set of functions $\psi_m^*(\mathbf{r})$, rather than the set $\psi_m(\mathbf{r})$. Substitution of the expansion (A6) in Eq. (A5) now gives

$$i\hbar \frac{dc_n}{dt} = \sum_l \sum_m E_{nm} a_l \int \psi_l(\mathbf{r})\psi_m^*(\mathbf{r}) \, d^3\mathbf{r} = \sum_m E_{nm} c_m, \qquad (A7)$$

by use of Eq. (A3); these are Eqs. (4.9). By comparing Eqs. (A5) and (A7) we see that the coefficients E_{nm} may be calculated by solving the set of simultaneous equations

$$\sum_m E_{nm} \int \psi_m^*(\mathbf{r})\psi_l(\mathbf{r}) \, d^3\mathbf{r} = \int \psi_n^*(\mathbf{r}) H \psi_l(\mathbf{r}) \, d^3\mathbf{r}. \qquad (A8)$$

In the special case where the $\psi_l(\mathbf{r})$ are all mutually orthogonal, Eq. (A8) reduces to

$$E_{nl} = \int \psi_n^*(\mathbf{r}) H \psi_l(\mathbf{r}) \, d^3\mathbf{r} = H_{nl} \qquad (A9)$$

so that in this case the E_{nl} are what are usually called the **matrix elements** of the operator H. In general if the non-orthogonality is small, the E_{nl} are quite close to these values. The formulation of quantum mechanics in terms of coupled probability amplitudes through Eqs. (A7) and (A9) is taken as basic by Feynman,[6] who uses it to make the Schrödinger equation plausible, the converse of our proof in this appendix.

We now consider how to obtain a complete set of wavefunctions $\psi_n(\mathbf{r})$ so that the expansions (A2) and (A6) can be made. All the energy eigenstates, bound and unbound, for an electron in the field of a single positive ion constitute a complete set, by a general theorem. But this is not a useful complete set, because very many terms would be required to describe an electron bound to another ion in a remote part of the crystal. We could try to meet this difficulty by considering all the wavefunctions centred on every ion in the crystal; but this enlarged set is clearly overcomplete, in that some of the functions can be expanded in terms of the others. This has the unfortunate consequence that the expansions (A2) and (A6) are not unique.

In this book we use a limited set of wavefunctions, consisting of the lowest unoccupied bound state on each ion. This set is incomplete so that the expansions (A2) and (A6) can only be made with limited accuracy. However, this simplification does enable us to obtain qualitatively correct results with the minimum of algebra. To understand the approximation involved, consider the application of Eq. (A6) to the problem of H_2^+, considered in section 4.3.2. The Hamiltonian for this problem is

$$H = -\frac{\hbar^2}{2m} \nabla^2 + V_1 + V_2, \qquad (A10)$$

where V_1 and V_2 are Coulomb potentials centred at the nuclear positions, $\mathbf{R}_1$ and $\mathbf{R}_2$ respectively. If $\psi_1(\mathbf{r})$ is the atomic ground state for an atom centred at $\mathbf{R}_1$ then

$$H\psi_1 = E_0\psi_1 + V_2\psi_1, \qquad (A11)$$

where E_0 is the ground state energy.

The coupled equations (4.13) for this problem can be derived by the above method provided that $H\psi_1$ can be expanded in the form of Eq. (A6),

$$H\psi_1 \approx B\psi_1 - A\psi_2 \qquad (A12)$$

where B and A are the coefficients that appear in Eqs. (4.13); note that ψ_1 and ψ_2 are real in this problem so that $\psi_1 = \psi_1^*$. Comparison of Eqs. (A11) and (A12) shows that the expansion is possible if $V_2\psi_1$ can be expanded in terms of ψ_1 and ψ_2. These functions are plotted in Fig. A.1, which shows that the expansion is roughly possible because $V_2\psi_1$ has peaks near $\mathbf{R}_1$ and $\mathbf{R}_2$ where ψ_1 and ψ_2 respectively have peaks. Detailed fitting of the peaks is clearly not possible since $V_2\psi_1$ diverges at $\mathbf{R}_2$ whereas ψ_2 remains finite there. Eqs. (A8) suggest that the criteria for choosing optimum values for A and B are that the values of

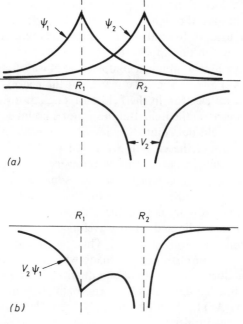

Fig. A.1 (a) Ground state wave functions ψ_1 and ψ_2 and potential V_2 for isolated hydrogen atoms at $\mathbf{R}_1$ and $\mathbf{R}_2$. (b) The function $V_2\psi_1$, which has to be represented as a linear combination of ψ_1 and ψ_2 (Eq. (A12))

$\int \psi_1 H \psi_1 \, d^3\mathbf{r}$ and $\int \psi_2 H \psi_1 \, d^3\mathbf{r}$ should be the same for both the exact result (A11) and the approximate result (A12).

A little thought suffices to show that when the protons are far apart

$$A = E_0 + V_2(\mathbf{R}_1) = E_0 - \frac{e^2}{4\pi\varepsilon_0 R}$$

where $R = |\mathbf{R}_1 - \mathbf{R}_2|$. The final term just cancels the internuclear repulsion so that the splitting of the bonding and antibonding orbitals at large R is associated with the coefficient B only. This explains why the splitting is symmetric in Fig. 4.7(b).

Truth is rarely pure and never simple.—*Oscar Wilde*

APPENDIX B

Electric and magnetic fields inside materials[†]

We consider magnetic materials first and will attempt to clarify the relation between:

(1) the local magnetic field $\mathbf{B}_L$, which determines the energy of an atomic dipole moment through Eq. (7.8);

(2) the macroscopically averaged field $\mathbf{B}_{mac}$ inside the material; and

(3) the applied magnetic field $\mathbf{B}_e$.

For simplicity we consider only sample shapes for which the magnetization $\mathbf{M}$ is uniform in space.

Because of the contribution of the atomic dipole moments the real microscopic magnetic field $\mathbf{B}_{mic}$ within a crystal varies rapidly on an atomic length scale. It satisfies the Maxwell equations

$$\text{div } \mathbf{B}_{mic} = 0 \qquad \text{and} \qquad \text{curl } \mathbf{B}_{mic} = \mu_0 \mathbf{j}_{mic}, \qquad \text{(B1)}$$

where $\mathbf{j}_{mic}$ is the *total* microscopic current density; the contribution of the atomic dipoles to $\mathbf{j}_{mic}$ is the source of the rapid variation of $\mathbf{B}_{mic}$. The field $\mathbf{B}_{mac}$ is the average of $\mathbf{B}_{mic}$ over a region containing many atoms. From Eqs. (B1), $\mathbf{B}_{mac}$ satisfies

$$\text{div } \mathbf{B}_{mac} = 0 \qquad \text{and} \qquad \text{curl } \mathbf{B}_{mac} = \mu_0 \mathbf{j}_{ave}, \qquad \text{(B2)}$$

[†] For a more extended discussion, see Grant and Phillips[3].

where j_{ave} is the average of j_{mic}. The macroscopic H field, used in the definition of the susceptibility χ, is defined by

$$\mathbf{B}_{mac} = \mu_0(\mathbf{H}_{mac} + \mathbf{M}).$$ (B3)

The atomic dipoles behave like little current loops (Fig. 7.1) and it follows that, if the magnetization is uniform, j_{ave} vanishes except at the surface of the sample where there is a current per unit length equal to the discontinuity in the component of M parallel to the surface (see Grant and Phillips,[3] chapter 5). The field $\mathbf{B}_{mac}$ can therefore be written as

$$\mathbf{B}_{mac} = \mathbf{B}_e + \mathbf{B}_1$$ (B4)

where $\mathbf{B}_1$ is the contribution from this surface current.

We now calculate $\mathbf{B}_{mac}$ and $\mathbf{H}_{mac}$ for the three important sample geometries in Fig. B.1.

Long cylinder parallel to applied field

$\mathbf{B}_1$ is the field inside a long solenoid carrying a current M per unit length in the direction shown in Fig. B.1(a). Thus

$$\mathbf{B}_1 = \mu_0 \mathbf{M}$$ (B5)

and

$$\mathbf{B}_{mac} = \mathbf{B}_e + \mu_0 \mathbf{M}.$$ (B6)

Comparing Eqs. (B3) and (B6), we see that, in this geometry,

$$\mu_0 \mathbf{H}_{mac} = \mathbf{B}_e.$$ (B7)

Thin disc perpendicular to applied field

The field at the centre of the current loop shown in Fig. B.1(c) is inversely proportional to the radius. The current flowing round the loop is $M \times$ (disc thickness). For a very thin disc we can therefore take

$$\mathbf{B}_1 = 0$$ (B8)

so that

$$\mathbf{B}_{mac} = \mathbf{B}_e$$ (B9)

and

$$\mu_0 \mathbf{H}_{mac} = \mathbf{B}_e - \mu_0 \mathbf{M}.$$ (B10)

Fig. B.1   Current distributions equivalent to uniform magnetization in various shapes of specimen for the field direction indicated. The currents on the outer surface determine the field $\mathbf{B}_1$ (Eq. (B4)). The small spherical surfaces are the regions within which the contribution $\mathbf{B}_3(0)$ (Eq. (B15)) of atomic dipoles to the field $\mathbf{B}_L$ is considered from a microscopic viewpoint

Sphere

We calculate the field $\mathbf{B}_1$ at the center of the sphere; the field is in fact uniform over the interior of the sphere although we will not prove this. We take $\mathbf{M}$ along the z axis in Fig. B.2 and consider the element of surface between θ and $\theta + d\theta$ shown. The discontinuity in the parallel component of $\mathbf{M}$ is $M \sin \theta$ so that the current flowing in the circular element is $M(\sin \theta)R\, d\theta$. Using the Biot–Savart

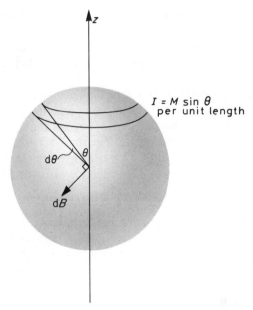

Fig. B.2   Calculation of the field at the centre of a spherical current sheet

law, the field due to a short length dl of the element is

$$dB = \frac{\mu_0}{4\pi} \frac{M(\sin\theta)r\, d\theta}{R^2} dl$$

in the direction shown in Fig. B.2. When the contributions from the whole of the circular element are added the components perpendicular to the z axis cancel giving a field along the z axis equal to

$$\frac{\mu_0}{4\pi} \frac{\mathbf{M}(\sin^2\theta)R\, d\theta}{R^2} 2\pi R \sin\theta = \tfrac{1}{2}\mu_0\mathbf{M}\sin^3\theta\, d\theta.$$

Integrating this over the sphere gives the total field

$$\mathbf{B}_1 = \int_0^\pi \tfrac{1}{2}\mu_0\mathbf{M}\sin^3\theta\, d\theta = \int_1^{-1} -\tfrac{1}{2}\mu_0\mathbf{M}(1-c^2)\, dc = \tfrac{1}{2}\mu_0\mathbf{M}\left[c - \frac{c^3}{3}\right]_{-1}^{1}$$

$$= \tfrac{2}{3}\mu_0\mathbf{M} \tag{B11}$$

where we have used the substitution $c = \cos\theta$. For a sphere therefore

$$\mathbf{B}_{\text{mac}} = \mathbf{B}_{\text{e}} + \tfrac{2}{3}\mu_0\mathbf{M} \tag{B12}$$

and

$$\mu_0\mathbf{H}_{\text{mac}} = \mathbf{B}_{\text{e}} - \tfrac{1}{3}\mu_0\mathbf{M}. \tag{B13}$$

To calculate $\mathbf{B_L}$, the value of $\mathbf{B_{mic}}$ at one of the atomic dipoles, we place our origin at the centre of the atom. The effect of distant atoms on $\mathbf{B_L}$ will depend only on the average magnetization and not on the detailed arrangement of those atoms. This suggests that it will be advantageous to divide the sample into two parts. The effect of the 'near' region, within a radius r, we calculate from the detailed microscopic distribution of atomic dipoles; the radius r is chosen large compared to atomic dimensions, but small compared to the size of the sample. The contribution to $\mathbf{B_L}$ from the 'far' region, beyond the radius r, we calculate from the magnetization $\mathbf{M}$ of that region. The small spherical surfaces separating the near and far regions are shown for the three sample shapes in Fig. B.1.

The field at position $\mathbf{r}$ relative to our origin may be written

$$\mathbf{B(r)} = \mathbf{B_e} + \mathbf{B_1} + \mathbf{B_2} + \mathbf{B_3(r)} + \mathbf{B_4(r)}, \tag{B14}$$

where $\mathbf{B_1}$ is the contribution from the magnetization current on the outer surface of the sample, as calculated above; $\mathbf{B_2}$ is the contribution from the magnetization current in the opposite sense on the inner boundary of the 'far' region; $\mathbf{B_3}$ is the field due to the atoms in the near region apart from the atom at the origin; and $\mathbf{B_4}$ is the contribution from the atom at the origin.

The effect of $\mathbf{B_4}$ on the atom at the origin has already been taken into account by including terms like the spin–orbit interaction in the energy of the atom; it is this interaction that is responsible for Hund's third rule. We must therefore omit $\mathbf{B_4}$ when calculating $\mathbf{B_L}$. It is usually assumed that the appropriate value of the field to insert in Eq. (7.8) is that at the nucleus of the atom† and we have therefore

$$\mathbf{B_L} = \mathbf{B_e} + \mathbf{B_1} + \mathbf{B_2} + \mathbf{B_3(0)}. \tag{B15}$$

We have already calculated $\mathbf{B_1}$ for the sample shapes of Fig. B.1. From our calculation of the field due to a spherical current sheet (Eq. (B12)) we see that

$$\mathbf{B_2} = -\tfrac{2}{3}\mu_0 \mathbf{M}. \tag{B16}$$

In general, $\mathbf{B_3(0)}$ depends on the arrangement of atomic dipoles within the sphere and must be calculated explicitly. A particular reason for using a spherical boundary to separate the near and far regions is that in this case $\mathbf{B_3(0)}$ vanishes for two important arrangements: (1) a random arrangement (as in a gas); (2) uncorrelated dipoles arranged with cubic symmetry about the point under consideration (as in many paramagnetic salts). We will not prove this result, but you may convince yourself of its validity for the random case by integrating the contributions from dipoles in spherical shells. For the case $\mathbf{B_3(0)} = 0$ therefore we have, from Eq. (B15):

---

† To do better than this we could expand the magnetic field as a Taylor series about the nuclear position. Eq. (7.8) would then correspond to the leading order in the expansion. The next term would represent a coupling between the magnetic field gradient and the *magnetic* quadrupole moment of the atom; parity conservation implies that the quadrupole moment vanishes so that this term is likely to be small. The next term would be coupling of second-order field gradients to the magnetic octopole moment of the atom.

Long cylinder parallel to applied field

$$\mathbf{B}_\mathrm{L} = \mathbf{B}_\mathrm{e} + \mu_0 \mathbf{M} - \tfrac{2}{3}\mu_0 \mathbf{M} + 0 = \mathbf{B}_\mathrm{e} + \tfrac{1}{3}\mu_0 \mathbf{M} \qquad\qquad\text{(B17)}$$

Thin disc perpendicular to applied field

$$\mathbf{B}_\mathrm{L} = \mathbf{B}_\mathrm{e} + 0 - \tfrac{2}{3}\mu_0 \mathbf{M} + 0 = \mathbf{B}_\mathrm{e} - \tfrac{2}{3}\mu_0 \mathbf{M} \qquad\qquad\text{(B18)}$$

Sphere

$$\mathbf{B}_\mathrm{L} = \mathbf{B}_\mathrm{e} + \tfrac{2}{3}\mu_0 \mathbf{M} - \tfrac{2}{3}\mu_0 \mathbf{M} + 0 = \mathbf{B}_\mathrm{e}. \qquad\qquad\text{(B19)}$$

Note that the relationships

$$\mathbf{B}_\mathrm{L} = \mathbf{B}_\mathrm{mac} - \tfrac{2}{3}\mu_0 \mathbf{M} = \mu_0 \mathbf{H}_\mathrm{mac} + \tfrac{1}{3}\mu_0 \mathbf{M} \qquad\qquad\text{(B20)}$$

are true for any sample shape provided $\mathbf{B}_3(0) = 0$. If $\mathbf{B}_3(0)$ is non-zero then it must be added to Eqs. (B17) to (B20). Note that Eq. (8.43) comes from using Eq. (B18) for the component of magnetization normal to a metal foil and Eq. (B17) for the component parallel to the foil.

We now turn our attention to the electric field inside materials. We will establish the Lorentz relation (Eq. (9.5)) between the electric field $\mathbf{E}_\mathrm{L}$ at an atom and the macroscopic electric field $\mathbf{E}_\mathrm{mac}$ inside the material. We use the analogy with the magnetic case to eliminate much of the detail. We consider only sample shapes for which the polarization $\mathbf{P}$ is uniform in space. The microscopic electric field $\mathbf{E}_\mathrm{mic}$ varies rapidly on an atomic scale and satisfies the Maxwell equations

$$\operatorname{div}\mathbf{E}_\mathrm{mic} = \rho_\mathrm{mic}/\varepsilon_0 \qquad \text{and} \qquad \operatorname{curl}\mathbf{E}_\mathrm{mic} = 0, \qquad\text{(B21)}$$

where ρ_mic is the microscopic charge density. $\mathbf{E}_\mathrm{mac}$ is the average of $\mathbf{E}_\mathrm{mic}$ over a region containing many atoms and therefore satisfies

$$\operatorname{div}\mathbf{E}_\mathrm{mac} = \rho_\mathrm{ave}/\varepsilon_0 \qquad \text{and} \qquad \operatorname{curl}\mathbf{E}_\mathrm{mac} = 0, \qquad\text{(B22)}$$

where ρ_ave is the average of ρ_mic.

For uniform polarization, the positive and negative regions of the dipoles contribute to ρ_ave with equal magnitude but opposite sign except at the surface of the sample, where there is a surface charge density equal to the discontinuity in the component of $\mathbf{P}$ perpendicular to the surface (see Grant and Phillips,[3] chapter 2). The field $\mathbf{E}_\mathrm{mac}$ can thus be written as

$$\mathbf{E}_\mathrm{mac} = \mathbf{E}_\mathrm{e} + \mathbf{E}_1, \qquad\qquad\text{(B23)}$$

where $\mathbf{E}_1$ is the contribution from this surface charge. The field $\mathbf{D}_\mathrm{mac}$ is defined by $\mathbf{D}_\mathrm{mac} = \varepsilon_0 \mathbf{E}_\mathrm{mac} + \mathbf{P}$. The major difference between the electric and magnetic case arises because it is the component of $\mathbf{P}$ *perpendicular* to the surface that determines the surface charge density whereas it is the component of $\mathbf{M}$ *parallel* to the surface that determines the surface current density. For the three important geometries discussed above we find the following.

Long cylinder parallel to applied field

ρ_{ave} is non-zero only on the distant ends of the cylinder. Hence

$$\mathbf{E}_1 = 0, \qquad \mathbf{E}_{mac} = \mathbf{E}_e, \qquad \mathbf{D}_{mac} = \varepsilon_0 \mathbf{E}_e + \mathbf{P}. \qquad (B24)$$

Thin disc perpendicular to applied field

$\mathbf{E}_1$ is the field inside a parallel-plate capacitor with a charge P per unit area on the plates (note that $\mathbf{E}_1$ is oppositely directed to $\mathbf{P}$):

$$\mathbf{E}_1 = -\mathbf{P}/\varepsilon_0, \qquad \mathbf{E}_{mac} = \mathbf{E}_e - \mathbf{P}/\varepsilon_0, \qquad \mathbf{D}_{mac} = \varepsilon_0 \mathbf{E}_e. \qquad (B25)$$

Sphere

Using Coulomb's law to evaluate the component of field parallel to $\mathbf{P}$ due to the charge $P\cos\theta$ per unit area on the element of surface between θ and $\theta + d\theta$ in Fig. B.2 gives

$$\mathbf{E}_1 = -\int_0^\pi \frac{\mathbf{P}\cos\theta}{4\pi\varepsilon_0 R^2}(\cos\theta)2\pi R^2 \sin\theta \, d\theta = -\tfrac{1}{3}\mathbf{P}/\varepsilon_0 \qquad (B26)$$

where we have used the substitution $c = \cos\theta$ as before. Hence

$$\mathbf{E}_{mac} = \mathbf{E}_e - \tfrac{1}{3}\mathbf{P}/\varepsilon_0, \qquad \mathbf{D}_{mac} = \varepsilon_0 \mathbf{E}_e + \tfrac{2}{3}\mathbf{P}. \qquad (B27)$$

$\mathbf{E}_1$ is known as the **depolarizing field**; in all three geometries it can be written in the form $-N\mathbf{P}/\varepsilon_0$, where N is the **depolarizing factor**. We deduce that N is 0 for the long cylinder, 1 for the thin disc and $\tfrac{1}{3}$ for the sphere.

To calculate $\mathbf{E}_L$, the value of $\mathbf{E}_{mic}$ at the centre of an atom, we proceed as in the magnetic case by dividing the material into 'far' and 'near' regions separated by a small spherical surface. By analogy with Eq. (B14) we can write

$$\mathbf{E}(\mathbf{r}) = \mathbf{E}_{mac} + \mathbf{E}_2 + \mathbf{E}_3(\mathbf{r}) + \mathbf{E}_4(\mathbf{r}), \qquad (B28)$$

where $\mathbf{E}_2$, $\mathbf{E}_3$ and $\mathbf{E}_4$ are the analogues of the corresponding magnetic contributions. We disregard $\mathbf{E}_4$ as in the magnetic case and assume that the appropriate value of the field is that at the centre of the atom,† $\mathbf{r} = 0$, so that

$$\mathbf{E}_L = \mathbf{E}_{mac} + \mathbf{E}_2 + \mathbf{E}_3(0). \qquad (B29)$$

---

† As in the magnetic case we can expand the electric field as a Taylor series about the nuclear position. The coupling between the electric field gradient and the electric quadrupole moment of the atom is important, for example, for f electrons since the quadrupole moment is non-zero for an f wavefunction. The orientation of such a wavefunction is therefore determined by this coupling as well as by the coupling of the magnetic dipole moment with the magnetic field.

From our above calculation of the field due to a charged spherical surface (Eq. (B26)) we deduce

$$\mathbf{E}_2 = +\tfrac{1}{3}\mathbf{P}/\varepsilon_0. \tag{B30}$$

$\mathbf{E}_3(0)$ like $\mathbf{B}_3(0)$ vanishes for a random arrangement of atomic dipoles and for uncorrelated atomic dipoles arranged with cubic symmetry about the point under consideration. If $\mathbf{E}_3(0) = 0$ the Lorentz local field relation

$$\mathbf{E}_L = \mathbf{E}_{mac} + \tfrac{1}{3}\mathbf{P}/\varepsilon_0 \tag{B31}$$

follows from Eqs. (B29) and (B30) and is the electrical analogue of Eq. (B20).

There is an even closer analogy between the electric and magnetic cases if a more old-fashioned approach to magnetic media is used in which the effects of the magnetization are allowed for by a density of fictitious magnetic poles on the surface of the sample equal to the discontinuity in the *perpendicular* component of $\mathbf{M}$. In this approach magnetic equivalents for all the equations we have derived for the electric case are obtained by the replacements: $\mathbf{E} \rightarrow \mathbf{H}$, $\mathbf{D} \rightarrow \mathbf{B}$, $\mathbf{P} \rightarrow \mu_0\mathbf{M}$, $\varepsilon_0 \rightarrow \mu_0$, $(\rho \rightarrow \mu_0\rho)$. We mention this because it explains why the reader may encounter the concept of **demagnetizing field** in connection with the magnetic field inside materials; this is the $\mathbf{H}_1$ field produced by the fictitious magnetic pole density on the surface of the sample. The demagnetizing factor is the same as the depolarizing factor for the same geometry. Readers should convince themselves that this alternative approach gives the same answers as that in which the effects of the magnetization are allowed for by surface currents.

APPENDIX C

Quantum mechanics of an electron in a magnetic field

According to classical mechanics the momentum conjugate to the velocity vector $\mathbf{v}$ of a particle is given by Eq. (7.25). Thus for an electron

$$\mathbf{p} = m\mathbf{v} - e\mathbf{A}. \tag{C1}$$

The transition from classical mechanics to quantum mechanics is made by replacing the momentum $\mathbf{p}$ by the operator $-i\hbar\nabla$. Thus the operator $\hat{K}$ for the kinetic energy is obtained as follows:

$$\tfrac{1}{2}mv^2 \rightarrow \frac{1}{2m}(\mathbf{p} + e\mathbf{A})^2 \rightarrow \frac{1}{2m}(-i\hbar\nabla + e\mathbf{A})^2 = \hat{K}. \tag{C2}$$

Expanding the bracket and recalling that div $\mathbf{A} = 0$ gives

$$\hat{K} = -\frac{\hbar^2}{2m}\nabla^2 + \frac{e}{m}\mathbf{A}.(-i\hbar\nabla) + \frac{e^2 A^2}{2m}. \tag{C3}$$

If the magnetic field is uniform we leave it as an exercise to the reader to check that the vector potential $\mathbf{A}$ that satisfies $\mathbf{B} = \text{curl }\mathbf{A}$ and div $\mathbf{A} = 0$ is

$$\mathbf{A} = -\tfrac{1}{2}\mathbf{r} \times \mathbf{B}. \tag{C4}$$

Using this enables the second term in $\hat{K}$ to be written

$$-\frac{e}{2m}(\mathbf{r} \times \mathbf{B}).(-i\hbar\nabla) = \frac{e}{2m}\mathbf{B}.[\mathbf{r} \times (-i\hbar\nabla)].$$

The operator $\mathbf{r} \times (-i\hbar\nabla)$ ($\equiv \mathbf{r} \times \mathbf{p}$) is the operator $\hbar\mathbf{l}$ for the orbital angular momentum of the electron, so that the kinetic energy (C3) in a uniform field is

$$\hat{K} = -\frac{\hbar^2}{2m} \nabla^2 + \mu_{\mathrm{B}} \mathbf{B}.\mathbf{l} + \frac{e^2 A^2}{2m}. \tag{C5}$$

where $\mu_{\mathrm{B}} = e\hbar/2m$ is the Bohr magneton.

The first term in $\hat{K}$ is the kinetic energy in the absence of a field. From Eq. (7.4), $-\mu_{\mathrm{B}}\mathbf{l}$ is the magnetic moment from the orbital angular momentum; the second term in $\hat{K}$ is thus the contribution of the orbital angular momentum of the electron to the part H_{P} (Eq. (7.8)) of the Hamiltonian that is responsible for paramagnetism. The third term is responsible for the induced diamagnetism discussed in section 7.3. The contribution of the electron to the diamagnetic susceptibility of Eq. (7.39) can be calculated by taking the expectation value of the third term and using Eq. (C4); thus

$$\langle H_{\mathrm{D}} \rangle = \left\langle \frac{e^2 A^2}{2m} \right\rangle = \frac{e^2 B^2}{8m} \langle \rho^2 \rangle \tag{C6}$$

where $\langle \rho^2 \rangle$ is the mean square distance of the electron from the z axis. The contribution $\delta\mu$ of the electron to the magnetization is then obtained from $\delta\mu = -\mathrm{d}\langle H_{\mathrm{D}} \rangle/\mathrm{d}B$, which gives the same answer for χ as our alternative approach in section 7.3.

The contribution of the electron to the electric current density is given by the expectation value of $-e\mathbf{v} = -e(\mathbf{p} + e\mathbf{A})/m$, that is by the integral

$$-\int \left(\frac{e}{2m} [\psi^*(-i\hbar\nabla + e\mathbf{A})\psi + \text{complex conjugate}] \right) \mathrm{d}V \tag{C7}$$

where we have taken the mean of the expectation value and its complex conjugate in order to ensure that the integrand is real. The integrand can be interpreted as the *local* current density $\mathbf{j}$ asociated with the wavefunction ψ. Hence

$$\mathbf{j} = \frac{i\hbar e}{2m} (\psi^*\nabla\psi - \psi\nabla\psi^*) - \frac{e^2}{m} A\psi^*\psi. \tag{C8}$$

This interpretation is confirmed in books on quantum mechanics where it is shown that this is the current density required to satisfy conservation of probability when ψ satisfies Schrödinger's time-dependent equation $\hat{K}\psi = i\hbar \, \partial\psi/\partial t$ with $\hat{K}$ given by Eq. (C2).

> Truth is the most valuable thing we have. Let us economize it.—*Mark Twain*

APPENDIX D

The exchange energy

We will use first-order perturbation theory to calculate the effect of the Coulomb interaction between two electrons. Our aim is to demonstrate that, although the Coulomb interaction does not depend explicitly on electron spin, the energy of the two electrons does depend on their relative spin. The unperturbed Hamiltonian is of the form

$$H = -(\hbar^2/2m)(\nabla_1^2 + \nabla_2^2) + V(\mathbf{r}_1) + V(\mathbf{r}_2) \tag{D1}$$

where ∇_1 and ∇_2 represent the operations of differentiation with respect to the positions $\mathbf{r}_1$ and $\mathbf{r}_2$ of the two electrons. The potential energy V can be that of a single atom, a group of atoms or the periodic potential of a crystal lattice.

We will take the unperturbed state to be one in which the electrons occupy states $\psi_a(\mathbf{r})$ and $\psi_b(\mathbf{r})$ of the unperturbed Hamiltonian. Thus $\psi_a(\mathbf{r})$ and $\psi_b(\mathbf{r})$ satisfy

$$\left(-\frac{\hbar^2}{2m} \nabla^2 + V(\mathbf{r}) \right) \psi_a(\mathbf{r}) = E_a \psi_a(\mathbf{r})$$

$$\left(-\frac{\hbar^2}{2m} \nabla^2 + V(\mathbf{r}) \right) \psi_b(\mathbf{r}) = E_b \psi_b(\mathbf{r}). \tag{D2}$$

For an atomic potential, the states ψ_a and ψ_b will be single particle atomic eigenstates; for a periodic potential they will be Bloch states (see Chapter 11). Taking into account the spin of the electrons, the unperturbed Hamiltonian has

four *degenerate* states of energy $E_a + E_b$. Four properly antisymmetrized†
wavefunctions corresponding to this energy, suitable for performing a first-order
perturbation calculation of the Coulomb energy, are:

$$\Psi_S = \tfrac{1}{2}[\psi_a(\mathbf{r}_1)\psi_b(\mathbf{r}_2) + \psi_a(\mathbf{r}_2)\psi_b(\mathbf{r}_1)][\alpha(s_1)\beta(s_2) - \alpha(s_2)\beta(s_1)]$$

$$\Psi_{T1} = (1/\sqrt{2})[\psi_a(\mathbf{r}_1)\psi_b(\mathbf{r}_2) - \psi_a(\mathbf{r}_2)\psi_b(\mathbf{r}_1)]\alpha(s_1)\alpha(s_2)$$

$$\Psi_{T2} = \tfrac{1}{2}[\psi_a(\mathbf{r}_1)\psi_b(\mathbf{r}_2) - \psi_a(\mathbf{r}_2)\psi_b(\mathbf{r}_1)][\alpha(s_1)\beta(s_2) + \alpha(s_2)\beta(s_1)] \qquad \text{(D3)}$$

$$\Psi_{T3} = (1/\sqrt{2})[\psi_a(\mathbf{r}_1)\psi_b(\mathbf{r}_2) - \psi_a(\mathbf{r}_2)\psi_b(\mathbf{r}_1)]\beta(s_1)\beta(s_2)$$

where s_1 and s_2 are the spin variables of the electrons and α and β are spin
eigenstates with $S_z = +\tfrac{1}{2}$ and $S_z = -\tfrac{1}{2}$ respectively. The unperturbed state Ψ_S is
the product of a symmetric space function with the antisymmetric spin singlet
wavefunction ($S = 0$) and thus corresponds to 'antiparallel' spins for the two
electrons. The other three states consist of products of an antisymmetric space
function with the three symmetric spin triplet wavefunctions ($S = 1$) with
$S_z = +1, 0$ and -1 respectively; these are states for which the electrons spins
are 'parallel'. Note that these three wavefunctions vanish if $\psi_a = \psi_b$ so that, in
accordance with the Pauli principle, two electrons can occupy the same state
only if they have opposite spins.

It is perhaps worth digressing slightly to point out why the above states are
suitable for the perturbation calculation. Because of the Coulomb interaction
the singlet state has a different energy to the three triplet states. When a
perturbation H' breaks the degeneracy of two unperturbed states, the splitting
can only be correctly calculated by using the two linear combinations, ψ_1 and
ψ_2, of those states that satisfy $\int \psi_1^* H' \psi_2 \, d\tau = 0$, where the volume element $d\tau$
contains in general summation over spin variables as well as integration over
space variables. Explicit calculation shows that

$$\sum_{s_1, s_2} \int d\mathbf{r}_1 \int d\mathbf{r}_2 \, \Psi_S^* \frac{e^2}{4\pi\varepsilon_0 |\mathbf{r}_1 - \mathbf{r}_2|} \Psi_{Ti} = 0$$

for $i = 1, 2$ or 3, so that the necessary condition is satisfied for each of the triplet
states. An example of two antisymmetric unperturbed states of energy $E_a + E_b$
which are *unsuitable* is

$$\Psi_1 = \frac{1}{\sqrt{2}}[\psi_a(\mathbf{r}_1)\alpha(s_1)\psi_b(\mathbf{r}_2)\beta(s_2) - \psi_a(\mathbf{r}_2)\alpha(s_2)\psi_b(\mathbf{r}_1)\beta(s_1)] = \frac{1}{\sqrt{2}}(\Psi_S + \Psi_{T2})$$

---

† Note that the wavefunctions must be antisymmetric under simultaneous interchange of both
space and spin coordinates: $\mathbf{r}_1, s_1 \leftrightarrow \mathbf{r}_2 \, s_2$.

and

$$\Psi_2 = \frac{1}{\sqrt{2}} [\psi_a(\mathbf{r}_1)\beta(s_1)\psi_b(\mathbf{r}_2)\alpha(s_2) - \psi_a(\mathbf{r}_2)\beta(s_2)\psi_b(\mathbf{r}_1)\alpha(s_1)]$$

$$= \frac{1}{\sqrt{2}}(-\Psi_S + \Psi_{T2}).$$

The expectation value of the Coulomb interaction depends only on the spatial part of the wavefunction. For each of the three triplet states we find an energy shift

$$\Delta E_T = \sum_{s_1,s_2} \int d\mathbf{r}_1 \int d\mathbf{r}_2 \, \Psi_{Ti}^* \frac{e^2}{4\pi\varepsilon_0|\mathbf{r}_1 - \mathbf{r}_2|} \Psi_{Ti}, \qquad i = 1, 2, 3$$

$$= E_1 - \mathscr{J} \tag{D4}$$

where

$$E_1 = \int d\mathbf{r}_1 \int d\mathbf{r}_2 \, |\psi_a(\mathbf{r}_1)|^2 \frac{e^2}{4\pi\varepsilon_0|\mathbf{r}_1 - \mathbf{r}_2|} |\psi_b(\mathbf{r}_2)|^2 \tag{D5}$$

is the 'obvious' contribution from a charge density $-e|\psi_a(\mathbf{r}_1)|^2$ interacting with a charge density $-e|\psi_b(\mathbf{r}_2)|^2$, and

$$\mathscr{J} = \int d\mathbf{r}_1 \int d\mathbf{r}_2 \, \psi_a^*(\mathbf{r}_1)\psi_b(\mathbf{r}_1) \frac{e^2}{4\pi\varepsilon_0|\mathbf{r}_1 - \mathbf{r}_2|} \psi_b^*(\mathbf{r}_2)\psi_a(\mathbf{r}_2) \tag{D6}$$

is the exchange contribution. The tendency for $\mathscr{J}$ to be positive can be seen by looking at the integrand in the region $\mathbf{r}_1 \approx \mathbf{r}_2$ where the Coulomb interaction is large. When $\mathbf{r}_1 = \mathbf{r}_2$ we see that the product of the four wavefunctions in Eq. (D6) becomes $|\psi_a(\mathbf{r}_1)|^2|\psi_b(\mathbf{r}_1)|^2$ so that the integrand tends to be positive in this limit. The expectation value of the Coulomb interaction for the singlet state is

$$\Delta E_S = \sum_{s_1,s_2} \int d\mathbf{r}_1 \int d\mathbf{r}_2 \, \Psi_S^* \frac{e^2}{4\pi\varepsilon_0|\mathbf{r}_1 - \mathbf{r}_2|} \Psi_S = E_1 + \mathscr{J} \tag{D7}$$

so that the difference in energy between the parallel spin state and the antiparallel spin state is

$$\Delta E_T - \Delta E_S = -2\mathscr{J}. \tag{D8}$$

Note that, for the exchange contribution (Eq. (D6)) to the energy to be important, it is necessary that there is some region of space in which $\psi_a^*(\mathbf{r})\psi_b(\mathbf{r})$ is appreciable. This requires that there should be a significant overlap between the wavefunctions $\psi_a(\mathbf{r})$ and $\psi_b(\mathbf{r})$. The highly localized nature of the 4f wavefunctions in rare-earth atoms explains why there is little direct exchange interaction energy between neighbouring rare-earth atoms. The overlap is large however for two 4f wavefunctions on the same atom where exchange energies are responsible for the first of Hund's rules.

Bibliography

BACKGROUND READING

1. B. H. Flowers and E. Mendoza, *Properties of Matter*, Wiley, London, 1970. (A good exposition of the basic atomistic view of matter including discussion of interatomic potentials and kinetic theory, presupposed in this book.)
2. F. Mandl, *Statistical Physics* (2nd edn), Wiley, London, 1988. (Contains the statistical mechanics and thermodynamics background.)
3. I. S. Grant and W. R. Phillips, *Electromagnetism* (2nd edn), Wiley, London, 1990. (Provides the background in electricity and magnetism.)
4. A. P. French and E. F. Taylor, *An Introduction to Quantum Physics*, Van Nostrand Reinhold, London, 1979. (Background in quantum mechanics.)
5. F. G. Smith and J. H. Thomson, *Optics* (2nd end), Wiley, London, 1988. (Background in optics.)
6. R. P. Feynman, *Lectures on Physics*, Addison-Wesley, Reading, Mass., 1965. (An interesting approach to physics as a whole in three volumes; volume 3 contains the approach to quantum mechanics that we adopt in Chapter 4.)

SELECTED BOOKS ON SOLID STATE PHYSICS

7. C. Kittel, *Introduction to Solid State Physics* (6th edn), Wiley, New York, 1986. (Standard student text, full of useful facts.)
8. H. P. Myers, *Introductory Solid State Physics*, Taylor and Francis, London, 1990. (Similar in level to this book.)

9. H. M. Rosenberg, *The Solid State* (3rd edn), Oxford University Press, Oxford, 1988. (By intention, a shorter and simpler book.)
10. G. Burns, *Solid State Physics*, Academic Press, Orlando, Florida, 1985. (More detailed than this book.)
11. N. W. Ashcroft and N. D. Mermin, *Solid State Physics*, Holt, Rinehart and Winston, New York, 1976. (Probably the best text to consult for a more advanced and detailed treatment.)
12. J. M. Ziman, *Principles of the Theory of Solids*, Cambridge University Press, Cambridge, 1964. (Good but now slightly dated introduction to more advanced theoretical methods.)
13. R. Dalven, *Introduction to Applied Solid State Physics* (2nd edn). Plenum, New York, 1990. (Applications involving semiconductors, superconductors, ferromagnets and non-linear optics.)
14. L. Solymar and D. Walsh, *Lectures on the Electrical Properties of Materials* (3rd edn), Clarendon Press, Oxford, 1984. (Covers material in Chapters 3, 4, 5, 6, 9 and 10 of this book.)

FURTHER READING ON SPECIFIC TOPICS

See also references given in the text.

Chapter 1

15. P. J. Brown and J. D. Forsyth, *The Crystal Structure of Solids*, Edward Arnold, London, 1973. (More details of crystallography at a simple level.)
16. R. C. Evans, *Crystal Chemistry* (2nd edn), Cambridge University Press, Cambridge, 1966. (An excellent account of the relation between chemical binding and crystal structure.)

Chapter 2

17. J. M. Ziman, *Electrons and Phonons*, Clarendon Press, Oxford, 1960. (Thorough, comprehensive, rigorous, mathematical treatment of electrons and phonons in solids and their interaction.)

Chapter 5

18. R. A. Smith, *Semiconductors* (2nd edn), Cambridge University Press, Cambridge, 1978. (Comprehensive treatment of properties of semiconductors.)
19. A. van der Ziel, *Solid State Physical Electronics* (3rd edn), Prentice-Hall, Englewood Cliffs, NJ, 1976. (Wide-ranging comprehensive textbook.)

Chapter 6

20. D. A. Fraser, *The Physics of Semiconductor Devices* (4th edn), Clarendon Press, Oxford, 1986. (Excellent simple account.)
21. S. M. Sze, *Semiconductor Devices: Physics and Technology*, Wiley, New York, 1985. (More advanced and detailed account.)
    See also reference 19 for an account of semiconductor device behaviour.
22. The October 1986 issue of *Physics Today* contains articles on the science of VLSI (very large-scale integration).
23. The February 1990 issue of *Physics Today* contains articles on nanoscale and ultrafast semiconductor devices.

Chapter 7

24. B. I. Bleaney and B. Bleaney, *Electricity and Magnetism* (3rd edn), Clarendon Press, Oxford, 1976. (Good elementary account of magnetic properties of materials.)

Chapter 8

25. J. Crangle, *The Magnetic Properties of Solids*, Edward Arnold, London, 1977. (More details of magnetic order in solids at an appropriate level.)

Chapter 9

26. H. Fröhlich, *Theory of Dielectrics* (2nd edn), Clarendon Press, Oxford, 1958. (Very careful and detailed approach.)
27. J. H. van Vleck, *The Theory of Electric and Magnetic Susceptibilities*, Oxford University Press, Oxford, 1932, but reprinted many times since. (Standard text for quantum calculation of susceptibilities.)

Chapter 10

28. D. R. Tilley and J. Tilley, *Superfluidity and Superconductivity* (2nd edn), Adam Hilger, Bristol, 1986. (For more details and an introduction to more advanced topics.)
29. P. G. de Gennes, *Superconductivity of Metals and Alloys*, Benjamin, New York, 1966. (Probably the best tutorial book for learning the theory of superconductivity.)
30. T. van Duzer and C. W. Turner, *Principles of Superconductive Devices and Circuits*, Elsevier, New York, 1981. (Good account of applications.)

Chapter 11

31. L. Brillouin, *Wave Propagation in Periodic Structures*, Dover, New York, 1953. (An excellent readable account by the inventor of the subject.)

Chapter 12

32. H. A. Hauptman, 'The phase problem of x-ray crystallography,' in *Physics Today*, November 1989, p. 24. (Discussion of the problem of deducing the structure from the diffraction pattern.)
33. G. E. Bacon, *Neutron Diffraction* (3rd edn), Clarendon Press, Oxford, 1975. (Comprehensive textbook.)

Chapter 13

34. A. B. Pippard, *The Dynamics of Conduction Electrons*, Blackie, London, 1965. (Beautiful exposition of the relation between conduction electron dynamics and Fermi surface geometry.)

Chapter 14

35. B. I. Halperin, 'The quantized Hall effect,' in *Scientific American*, April 1986, p. 40. (The quantum Hall effect without equations!)
   See ref. 23 for more discussion of resonant tunnelling and other quantum devices.

Being a philospher, I have a problem for every solution—
Robert Zend

Solutions to problems

CHAPTER 1

1.1 Three neighbouring atoms in one close-packed plane and their common neighbour in the adjacent plane form a regular tetrahedron of side a and height $c/2$. Simple geometry then gives $c/a = (8/3)^{1/2}$.

1.2 See Fig. 1.6.

1.3 The plane of the $(h\,k\,l)$ set nearest the origin has the equation $hx + ky + lz = a$ (this has the correct intercepts on the three axes). The plane parallel to this through the origin is $hx + ky + lz = 0$; an arbitrary vector in this plane is $(x, y, -(hx + ky)/l)$. The scalar product of this with $[h, k, l]$ is $hx + ky - (hx + ky) = 0$.

1.4 d is the normal distance from the origin of the plane $hx + ky + lz = a$ (see previous problem). The direction cosines of the normal to the plane are $d \div (a/h)$, etc. Therefore

$$\frac{h^2 d^2}{a^2} + \frac{k^2 d^2}{a^2} + \frac{l^2 d^2}{a^2} = 1 \qquad \text{or} \qquad d = \frac{a}{(h^2 + k^2 + l^2)^{1/2}}.$$

1.5

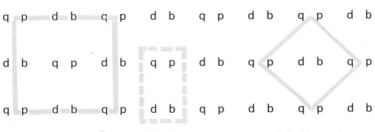

Rectangular cell   Basis   Primitive cell

1.6 (a) See Fig. 1.11(b) for help in doing this.

(b) fcc $(0, 0, 0)$; bcc $(0, 0, 0)$; hcp $(0, 0, 0)$, $(\frac{2}{3}, \frac{1}{3}, \frac{1}{2})$; diamond $(0, 0, 0)$, $(\frac{1}{4}, \frac{1}{4}, \frac{1}{4})$.

(c)

Structure	Sphere diameter	Number in cell	Fraction occupied
fcc	$a/\sqrt{2}$	4	0.740
bcc	$\sqrt{3}a/2$	2	0.680
hcp	a	2	as fcc
Diamond	$\sqrt{3}a/4$	8	0.340

1.7 Note that if $\mathbf{c}' = \mathfrak{z}\mathbf{\kappa}$ then $\mathbf{c} = \frac{1}{2}(\mathbf{a} + \mathbf{b} + \mathbf{c}')$, which is the body-centred position of a cubic unit cell defined by $\mathbf{a}$, $\mathbf{b}$ and $\mathbf{c}'$. The Bravais lattice is therefore bcc with a conventional cubic unit cell of volume 27 Å^3. The primitive cell is half this volume since the non-primitive cell contains two lattice points. The most densely packed planes are the $\{1\,1\,0\}$ planes of the cubic unit cell.

1.8 The primitive unit cells of fcc and bcc are shown in Figs. 1.8(b) and 1.12(b). For fcc the directions of the primitive $\mathbf{a}$, $\mathbf{b}$ and $\mathbf{c}$ are $[1\,1\,0]$, $[0\,1\,1]$ and $[1\,0\,1]$; thus $\cos\alpha = \mathbf{b}.\mathbf{c}/a.a = \frac{1}{2}$ and $\alpha = 60°$. For bcc the directions of the primitive $\mathbf{a}$, $\mathbf{b}$ and $\mathbf{c}$ are $[\bar{1}\,1\,1]$, $[1\,\bar{1}\,1]$ and $[1\,1\,\bar{1}]$, so $\cos\alpha = \mathbf{b}.\mathbf{c}/a.a = -\frac{1}{3}$ and $\alpha = 109°27'$.

1.9 Differentiating the Bragg law $\sin\theta = n\lambda/2d$ gives $\cos\theta\,\delta\theta = -n\lambda\,\delta d/2d^2$, which can be written $\delta d/d = -2d(\cos\theta)\,\delta\theta/n\lambda = -(\cot\theta)\,\delta\theta$; thus for the data given $\delta d/d = (\cot 47°)(1.15\pi/180) = 0.0187$ ($\delta\theta$ must be expressed in radians). The coefficient of linear thermal expansion is therefore $(1/d)(\delta d/\delta T) = 0.0187/(1273 - 293) = 1.91 \times 10^{-5}$ K^{-1}.

CHAPTER 2

2.1 The equation of motion of the atoms (2.7), as modified to include the forces due to the second nearest neighbours, is

$$M\ddot{u}_n = K(u_{n+1} - 2u_n + u_{n-1}) + K_2(u_{n+2} - 2u_n + u_{n-2}).$$

Inserting a wavelike solution of the form of Eq. (2.8) and proceeding as in section 2.3.1 leads to the dispersion relation

$$M\omega^2 = 2K[1 - \cos(ka)] + 2K_2[1 - \cos(2ka)].$$

(a) For $ka \ll 1$, this becomes $M\omega^2 \approx (K + 4K_2)k^2a^2$ corresponding to a sound velocity $v_S = \omega/k = a[(K + 4K_2)/M]^{1/2}$. The elastic modulus C can be determined from the increase in energy of the crystal $\Delta E = \frac{1}{2}NK(r - a)^2 + \frac{1}{2}NK_2(2r - 2a)^2$ on stretching the length L of the crystal from Na to Nr. (Note that there are equal numbers of each type of spring.) The force required is

$$F = \frac{\partial\Delta E}{\partial L} = \frac{1}{N}\frac{\partial\Delta E}{\partial r} = (K + 4K_2)(r - a) = C(r - a)/a,$$

from which $C = (K + 4K_2)a$. Thus the velocity of sound is $(C/\rho)^{1/2}$ (Eq. (2.14)) as required.

(b) $\partial\omega/\partial k$ is zero because both $\sin(ka)$ and $\sin(2ka)$ vanish at $k = \pm\pi/a$.

(c) $\cos(2ka)$ has period π/a.

Forces between an atom and its nth nearest neighbours introduce a term $K_n[1 - \cos(nka)]$ in the expression for $M\omega^2$.

2.2 The equations of motion are (cf. Eqs. (2.15) and (2.16))

$$M\ddot{u}_n = K_1 u_{n+1} - (K_1 + K_2)u_{n-1} + K_2 u_{n-1},$$
$$M\ddot{u}_{n-1} = K_2 u_n - (K_2 + K_1)u_n + K_1 u_{n-2}.$$

These are solved by taking wavelike solutions of the form:

$$u_n = A \exp\left[i(kna - \omega t)\right] \qquad \text{and} \qquad u_{n-1} = \alpha A \exp\left[i(kna - \omega t)\right].$$

Note that we can put kna in both exponents; the separation between the n and $n - 1$ atoms relative to the lattice spacing a does not enter the equations. The solution proceeds as in section 2.3.2 and the dispersion curves are as in Fig. 2.7, except that points A, B and C correspond to frequencies $(2K_1/M)^{1/2}$, $(2K_2/M)^{1/2}$ and $[2(K_1 + K_2)/M]^{1/2}$ respectively.

2.3 From Eqs. (2.29) and (2.33) the exact result is

$$C = \frac{2Nk_B}{\pi} \int_0^{2(K/M)^{1/2}} \left(\frac{4K}{M} - \omega^2\right)^{-1/2} \left(\frac{\hbar\omega}{k_B T}\right)^2 \frac{\exp(\hbar\omega/k_B T)}{[\exp(\hbar\omega/k_B T) - 1]^2} \, d\omega.$$

The Debye frequency determined from $\int_0^{\omega_D} g(\omega) \, d\omega = N$ (cf. Eq. (2.49)) is $\omega_D = \pi(K/M)^{1/2}$ so that $\Theta_D = \hbar\pi(K/M)^{1/2}/k_B$ (cf. Eq. (2.53)). In terms of Θ_D and $x = \hbar\omega/k_B T$ the exact heat capacity is

$$C = 2Nk_B \int_0^{2\Theta_D/\pi T} \left[\left(\frac{2\Theta_D}{T}\right)^2 - (\pi x)^2\right]^{-1/2} \frac{x^2 e^x \, dx}{(e^x - 1)^2}.$$

The constant density of states in the Debye approximation is $(2N/\pi)(M/4K)^{1/2}$ and in this approximation the heat capacity becomes

$$C = \frac{Nk_B T}{\Theta_D} \int_0^{\Theta_D/T} \frac{x^2 e^x \, dx}{(e^x - 1)^2}.$$

Because the average frequency is lower for the exact density of states the exact heat capacity is higher than the Debye approximation at all T. When $T \ll \Theta_D$ only low-frequency modes are excited for which the two densities of states are the same; the integrals may then be taken to infinity, so that

$$C = \frac{Nk_B T}{\Theta_D} \int_0^\infty \frac{x^2 e^x \, dx}{(e^x - 1)^2}.$$

2.4 In two dimensions the density of states $\rho_R(\mathbf{k})$ in $\mathbf{k}$-space for running waves in a system of size $L \times L$ is $(L/2\pi)^2$ (section 2.6.2) so that the number of states between circles of radii k and $k + dk$ is $(L/2\pi)^2 2\pi k \, dk = Ak \, dk/2\pi = g(k) \, dk$, where $A = L^2$ is the area of the system. For $\omega^2 = \sigma k^3/\rho$ the density of states per unit frequency range is therefore given by

$$g(\omega) = g(k) \frac{dk}{d\omega} = \frac{A}{3\pi}\left(\frac{\rho}{\sigma}\right)^{2/3} \omega^{1/3}.$$

With Eq. (2.26) for the average energy of a harmonic oscillator this gives

$$E = E_0 + \frac{1}{3\pi}\left(\frac{\rho}{\sigma}\right)^{2/3} \hbar\left(\frac{k_B T}{\hbar}\right)^{7/3} \int_0^{\omega = \omega_D} \frac{x^{4/3} \, dx}{e^x - 1}$$

for the energy per unit area of surface; E_0 is the zero point energy, $x = \hbar\omega/k_B T$ and ω_D is a suitable cut-off frequency. This cut-off is hard to determine since we do not know how many degrees of freedom to associate with the surface, though we expect the number to be of the same order as the number of atoms per unit area of surface. However in the limit of low temperature this does not matter since we can take the integral to infinity to obtain $E = E_0 + aT^{7/3}$, where a is independent of temperature. The surface heat capacity at low temperatures is then $C = dE/dT = \frac{7}{3}aT^{4/3}$. The entropy S is $\int_0^T C \, dT/T = \frac{7}{4}aT^{4/3}$ so that σ, the surface free energy density, is given by

$$\sigma(T) - \sigma(0) = E - E_0 - TS = -\tfrac{3}{4}aT^{4/3}.$$

2.5 For a long-wavelength acoustic mode $\hbar\omega \ll k_B T$ so Eq. (2.26) can be approximated to Eq. (2.28); thus $\bar\varepsilon \approx k_B T$. Roughly, for $T \ll \Theta_D$, only modes for which $\hbar\omega \leq k_B T$ will be excited, that is those with a wavenumber $k \leq k_0 = k_B T/\hbar v_S$, where v_S is the velocity of sound. We define a 'Debye wavenumber' $k_D = \omega_D/v_S = k_B\Theta_D/\hbar v_S$. Since modes are uniformly distributed in $\mathbf{k}$-space the fraction of modes excited is $(k_0/k_D)^3$ and the number is thus $3N(k_0/k_D)^3 = 3N(T/\Theta_D)^3$. If we make the approximation that modes with $k < k_0$ have energy $k_B T$ and modes with $k > k_0$ are entirely unexcited, the internal energy is $E = 3Nk_B T(T/\Theta_D)^3$ so that the heat capacity $C = dE/dT = 12Nk_B(T/\Theta_D)^3$. This has the correct dependence on temperature but is about 20 times smaller than the exact result (Eq. (2.54)). A good numerical answer cannot be expected because $C \propto k_0^3$ and k_0 is a rather arbitrary quantity.

2.6 The Debye theory estimates (Eq. (2.51)) the zero point energy as $\frac{9}{8}\hbar\omega_D$ per atom $= \frac{9}{8}k_B\Theta_D = 0.0079$ eV, about 10% of the binding energy.

2.7 The Lennard-Jones potential is $\mathscr{V}(r) = \varepsilon[(\sigma/r)^{12} - 2(\sigma/r)^6]$ where σ is the equilibrium separation. Thus evaluating the derivatives in Eq. (2.67) at $r = \sigma$ gives

$$\gamma = -\frac{\sigma}{6}\left(\frac{-1512\varepsilon/\sigma^3}{72\varepsilon/\sigma^2}\right) = 3.5.$$

2.8   (a) Assume that the phonon mean free path is about the same as the specimen diameter d; the effective thermal conductivity K_{eff} is then $\frac{1}{3}Cv_S d$. Thus inserting the given values for C and v_S gives $K_{eff} = \frac{1000}{3}dT^3$ W m^{-1} K^{-1}. The low-temperature thermal conductivities of the three specimens are fitted by $0.6T^3, 0.9T^3$ and $1.9T^3$, so we estimate the diameters as 1.8, 2.7 and 5.7 mm respectively.

(b) The thermal conductivity decreases by about a factor of 10 between 50 K and 100 K. The temperature variation in this region is $T^3 \exp(\Theta_D/bT)$ so that

$$10 = \frac{50^3 \exp(1000/50b)}{100^3 \exp(1000/100b)},$$

from which $b = 2.3$.

CHAPTER 3

3.1 The electron wavefunction satisfies

$$-\frac{\hbar^2}{2m}\frac{d^2\psi}{dx^2} = (E - V)\psi$$

outside the metal. For an electron at the Fermi surface $E - V$ is -3 eV and the solution of the Schrödinger equation therefore decays exponentially with a decay length

$$\left(\frac{\hbar^2}{2m(V-E)}\right)^{1/2} = 1.1 \text{ Å}.$$

3.2 The width of the K emission band is given by ε_F (section 3.2.4). From Eq. (3.9) and $N/V = 2/a^3$ for a body-centred cubic structure

$$\varepsilon_F = \frac{\hbar^2}{2ma^2}(6\pi^2)^{2/3} = 7.6 \times 10^{-19} \text{ J} = 4.7 \text{ eV}.$$

The width will increase *slightly* with increasing temperature as conduction electrons are excited above the Fermi energy.

3.3 The energy is $E = \int_0^{\varepsilon_F} \varepsilon g(\varepsilon) \, d\varepsilon$. Inserting $g(\varepsilon)$ (Eq. (3.7)) and integrating gives

$$E = \frac{V}{5\pi^2\hbar^3}(2m)^{3/2}\varepsilon_F^{5/2},$$

or $E = \frac{3}{5}N\varepsilon_F$ if Eq. (3.8) is used. Hence

$$p = -\frac{\partial E}{\partial V} = -\frac{3}{5}N\frac{\partial\varepsilon_F}{\partial V}.$$

Since $\varepsilon_F \propto V^{-2/3}$ (Eq. (3.9)), $\partial\varepsilon_F/\partial V = -\frac{2}{3}\varepsilon_F/V$ and thus $p = \frac{2}{5}N\varepsilon_F/V$.

From this, $p \propto V^{-5/3}$, so that $\partial p/\partial V = -\frac{5}{3}p/V$ and $B = -V\,\partial p/\partial V = \frac{5}{3}p = \frac{2}{3}N\varepsilon_F/V$. Inserting N/V and ε_F for potassium (section 3.2.1) gives $B = 0.32 \times 10^{10}\text{ N m}^{-2}$; comparison with the experimental bulk modulus $0.37 \times 10^{10}\text{ N m}^{-2}$ suggests that most of this is associated with the conduction electron pressure.

3.4 The increase in internal energy of the conduction electrons from its value at absolute zero is

$$\Delta E = \int_0^T C_V \, dT = \int_0^T \gamma' T \, dT = \frac{\gamma' T^2}{2}$$

(we use γ' for the electron specific heat coefficient to distinguish it from the Gruneisen parameter γ); the entropy of the electrons at temperature T is

$$\int_0^T \frac{C_V \, dT}{T} = \gamma' T.$$

The increase in Helmholtz free energy is therefore $\Delta E - TS = -\gamma' T^2/2$. From Eq. (2.58) this gives a temperature-dependent addition to the pressure $(\partial\gamma'/\partial V)_T T^2/2$; from Eqs. (3.9) and (3.16), $\gamma' \propto V^{2/3}$, so that $(\partial\gamma'/\partial V)_T = 2\gamma'/3V$. The additional term in the pressure, through Eq. (2.56), contributes

$$\frac{2\gamma' T}{3BV} = \frac{2}{3BV}C_{el}$$

to the thermal expansion coefficient, where C_{el} is the electronic heat capacity. Adding this contribution to that (Eq. (2.64)) from the lattice vibrations gives

$$\beta = \frac{1}{BV}(\gamma C_{\text{lattice}} + \frac{2}{3}C_{el})$$

for a free electron metal.

3.5 The change in momentum along the original direction of motion is $\delta k = \hbar k_F(1 - \cos\theta) \approx \hbar k_F \theta^2/2 \approx \hbar k_F(q/k_F)^2/2$ (for small θ).

3.6 (a) N/V for liquid ^{3}He is $\rho/m_{He} = 81/(3 \times 1.7 \times 10^{-27}) = 1.6 \times 10^{28}$ m^{-3}. Thus, from Eq. (3.9),

$$T_F = \frac{\hbar^2}{2m_{He}k_B}\left(3\pi^2\frac{N}{V}\right)^{2/3} = 4.8 \text{ K.}$$

(b) N/V for neutrons in a neutron star is $\rho/m_N = 10^{17}/1.7 \times 10^{-27} = 6 \times 10^{43}$ m^{-3}. Thus

$$T_F = \frac{\hbar^2}{2m_N k_B}\left(3\pi^2\frac{N}{V}\right)^{2/3} = 3.4 \times 10^{11} \text{ K.}$$

CHAPTER 4

4.1 Consider two arbitrary linear combinations $\phi_1 = a\,e^{ikx} + b\,e^{-ikx}$ and $\phi_2 = c\,e^{ikx} + d\,e^{-ikx}$. Orthogonality requires $\int \phi_1^* \phi_2\,dx = 0$; exploiting the fact that $\int e^{i\alpha x}\,dx = 0$ unless $\alpha = 0$, this becomes $a^*c + b^*d = 0$. Similarly $\int \phi_1^* V\phi_2\,dx = 0$ requires $a^*d + b^*c = 0$ (expand the cosine terms in the potential as complex exponentials). From these we deduce $a^*/b^* = b^*/a^* = -c/d = -d/c$ and thus either $\phi_1 \propto \sin(kx)$ and $\phi_2 \propto \cos(kx)$ or vice versa.

4.2 Consider waves in the plane of the crystal incident at angle θ to the (1 0) planes which have spacing a. The Bragg condition is satisfied if $2a\sin\theta = n\lambda = n2\pi/k$, that is if $k_x = k\sin\theta = n\pi/a$. Similarly the Bragg condition for diffraction off the (0 1) planes is satisfied if $k_y = n\pi/a$.

4.3 The substitution of ψ in the Schrödinger equation is straightforward. The term $\cos(2\pi x/a)$ can be expanded as $(e^{i2\pi x/a} + e^{-i2\pi x/a})/2$; in integrating over all space use is made of the fact that the integrals of the form $\int e^{icx}$ vanish unless $c = 0$. The resulting equations are:

(a) $\quad \left(\dfrac{\hbar^2}{2m}k^2 - \varepsilon\right)\alpha - \dfrac{V_1}{2}\beta = 0$

(b) $\quad \left[\dfrac{\hbar^2}{2m}\left(k - \dfrac{2\pi}{a}\right)^2 - \varepsilon\right]\beta - \dfrac{V_1}{2}\alpha = 0.$

These simultaneous equations for α and β have non-zero solutions only if the determinant of coefficients vanishes and this determines the energy ε.

4.4 $k_F^3 = (3\pi^2 N/V) = 3\pi^2/a^3$ for a simple cubic structure, so that $k_F = 3.09a$. The first Brillouin zone is a cube of side $2\pi/a$, so the Fermi sphere is just contained within it; the distance of closest approach is $0.05/a$.

Because the free electron Fermi sphere is so close to the zone boundary, the periodic lattice potential would almost certainly cause the Fermi surface to make contact with the zone boundary as in Fig. 4.5. However there would be more unoccupied states near the zone corner than in Fig. 4.5 and there would not be enough electrons to occupy any states in the second zone. In three dimensions such contact with the zone boundary gives a multiply connected Fermi surface: compare copper, Fig. 13.13.

4.5 The corners of a tetrahedron are in the directions $[1\ 1\ 1]$, $[\bar{1}\ \bar{1}\ 1]$, $[1\ \bar{1}\ \bar{1}]$ and $[\bar{1}\ 1\ \bar{1}]$; the sets of coefficients (a_x, a_y, a_z) must be proportional to these vectors. For normalization

$$\int |\psi|^2\, dV = a_x^2 + a_y^2 + a_z^2 = 1,$$

since the p states are orthonormal. The required sets of coefficients are therefore $(1/\sqrt{3}, 1/\sqrt{3}, 1/\sqrt{3})$, $(-1/\sqrt{3}, -1/\sqrt{3}, 1/\sqrt{3})$, $(1/\sqrt{3}, -1/\sqrt{3}, -1/\sqrt{3})$ and $(-1/\sqrt{3}, 1/\sqrt{3}, -1/\sqrt{3})$.

For the first of these we have

$$\phi_{111} = bs + \frac{c}{\sqrt{3}}(p_x + p_y + p_z).$$

For normalization

$$\int |\phi_{111}|^2\, dV = b^2 + c^2 = 1.$$

For orthogonality, for example,

$$\int \phi_{111}\phi_{-1-11}\, dV = 0.$$

That is

$$\int b^2 s^2\, dV + \int \frac{c^2}{3}(p_x + p_y + p_z)(-p_x - p_y + p_z)\, dV = b^2 - \tfrac{1}{3}c^2 = 0.$$

These are satisfied by $b = 1/2$, $c = \sqrt{3}/2$ so that $\phi_{111} = \tfrac{1}{2}(s + p_x + p_y + p_z)$. The other ϕ's can be calculated similarly.

4.6 Normalized vectors (β, γ) at $120°$ to each other are $(1, 0)$, $(-1/2, \sqrt{3}/2)$ and $(-1/2, -\sqrt{3}/2)$. Consider unnormalized states, $\chi_1 = \alpha s + p_x$ and $\chi_2 = \alpha s - (1/2)p_x + (\sqrt{3}/2)p_y$; α may be taken the same in both cases because the wavefunctions differ only in orientation. For orthogonality $\int \chi_1 \chi_2\, dV = \alpha^2 - \tfrac{1}{2} = 0$ so that $\alpha = 1/\sqrt{2}$. $\int |\chi_1|^2\, dV = \int |s/\sqrt{2} + p_x|^2\, dV = 3/2$, so that for normalization our states must be multiplied by $\sqrt{2/3}$. The required values of (α, β, γ) are $(1/\sqrt{3}, \sqrt{2/3}, 0)$, $(1/\sqrt{3}, -1/\sqrt{6}, 1/\sqrt{2})$ and $(1/\sqrt{3}, -1/\sqrt{6}, -1/\sqrt{2})$. The other orthogonality and normalization integrals may be checked in a similar way.

4.7 From Eq. (4.20), $d^2\varepsilon/dk^2 = 2Aa^2 \cos(ka)$, so that, using Eq. (4.27),

$$m_e = \frac{\hbar^2}{2Aa^2 \cos(ka)}.$$

Near $k = \pi/a$, $\cos(ka) \approx -1$ so that $m_e \approx -\hbar^2/(2Aa^2)$.

Eq. (4.20) can be written $\varepsilon = B + 2A \cos(ka - \pi)$. Thus for k near π/a,

$$\varepsilon \approx B + 2A\left(1 - \frac{(ka - \pi)^2}{2}\right) = B + 2A - Aa^2(k - \pi/a)^2 = \varepsilon_0 + \frac{\hbar^2(k - \pi/a)^2}{2m_e}.$$

CHAPTER 5

5.1 With the approximation $\hbar = 10^{-34}$ J s:
   (a) $m_h = +5 \times 10^{-32}$ kg (Eq. (5.2)),
   (b) $\varepsilon_h = +10^{-19}$ J (Eq. (5.3)),
   (c) $\mathbf{p}_h = -10^{-25}\hat{\mathbf{k}}_x$ kg m s^{-1} (Eq. (5.4)), and
   (d) $\mathbf{v}_h = -2 \times 10^6 \hat{\mathbf{k}}_x$ m s^{-1} (Eq. (5.5)).

5.2 The intrinsic carrier concentration at temperature T is (Eq. (5.23)) $AT^{3/2} \exp(-E_G/2k_B T) = 7 \times 10^{21}T^{3/2} \exp(-6400/T)$; the constant A has been chosen to obtain the correct answer at 300 K. The sample will cease to show intrinsic behaviour when the intrinsic carrier concentration becomes of the same order as the impurity concentration, i.e. $7 \times 10^{21}T^{3/2} \exp(-6400/T) \leqslant 10^{18}$, which gives $T \leqslant 360$ K.

5.3 (a) From Eq. (5.12) $E_D = -(m_e/m\varepsilon^2) \times 13.6$ eV $= 6.6 \times 10^{-4}$ eV.
   (b) From Eq. (5.13) $r = (\varepsilon m/m_e) \times 0.53$ Å $= 650$ Å.
   (c) Overlap is significant when $N_D \sim 1/(2r)^3 \sim 10^{21}$ m^{-3}. At about this concentration an impurity band of mobile states is formed; see section 5.5.1.

5.4 Consider the curve for donor concentration 10^{22} m^{-3}. The slope at low temperature (large $1/T$) corresponds to a decrease in σ by a factor of 100 for an increase in $1/T$ of about 0.04 K^{-1}; thus

$$\frac{\Delta(\ln \sigma)}{\Delta(1/T)} = -\frac{\ln 100}{0.04} = -115 \text{ K}.$$

If we assume that the temperature dependence of the conductivity is dominated by that of the electron density as given by Eq. (5.36), then $E_D/2k_B = 115$ K and $E_D = 3.2 \times 10^{-21}$ J $= 0.020$ eV.

5.5 For sodium

$$R_H = -\frac{1}{ne} = -\frac{a^3}{2e} = -2.45 \times 10^{-10} \text{ m}^3 \text{ C}^{-1}.$$

For InSb $n_i = 0.86 \times 10^{22}$ m^{-3} (from Eqs. (5.23), (5.17) and (5.21)) so $R_H = -7.2 \times 10^{-4}$ m^3 C^{-1}; electrons are the only effective carrier because their small effective mass means they have a high mobility.

The Hall voltage generated across a sample of width w and thickness t is given by $V_H/w = R_H Bi/wt$ (inserting $E_H = V_H/w$ and $j = i/wt$ in Eq. (3.36)); thus $V_H = R_H Bi/t$, independent of w. For sodium, $V_H = 2.45 \times 10^{-9}$ V; and for InSb, $V_H = 7.2 \times 10^{-3}$ V. This illustrates how the smaller number of carriers in semiconductors makes them suitable for measuring magnetic fields by the Hall effect

5.6 For a single type of carrier the Hall effect gives the sign and concentration of the carriers. Measurement of the conductivity then enables the mobility to be determined from Eq. (5.41). Cyclotron resonance gives the effective mass.

The condition for observation of cyclotron resonance is $\omega_c\tau = eB\tau/m_e \gg 1$, i.e. $B \gg m_e/e\tau$. The scattering time is $l/\bar{v} = l/(3k_B T/2m_e)^{1/2}$, where the mean free path l is related to the scattering cross section by $l = 1/N_D A$. The condition thus becomes $B/T^{1/2} \gg N_D A(1.5k_B m_e)^{1/2}/e \approx 6 \times 10^{-4}$ T K$^{-1/2}$. Thus at a temperature of 4 K a field of 0.01 T should be adequate. At room temperature the minimum field would be more like 0.1 T. At 4 K optical excitation of carriers would be necessary.

5.7 The magnetic field is at an angle $\theta_1 = 30°$ to the long axis of the z axis ellipsoids on Fig. 5.13. The direction θ_2 to the long axis of x and y axis ellipsoids is given by $\cos^2 \theta_1 + 2\cos^2 \theta_2 = 1$; thus $\cos^2 \theta_2 = 1/8$. The two electron resonances on Fig. 5.12

occur at fields 0.184 T and 0.290 T; since $m_L > m_T$ (Fig. 5.13) and the field makes a smaller angle to z than to x and y, we can identify the lower field resonance with the z axis ellipsoid. Eq. (5.58) then becomes for this resonance

$$2\pi \times 2.4 \times 10^{10} = 1.6 \times 10^{-19} \times 0.184 \left(\frac{1}{4m_L m_T} + \frac{3}{4m_T^2} \right)^{1/2}.$$

The other resonance is associated with the x and y ellipsoids and a similar equation can be written; solving the two equations simultaneously gives $m_T = 1.75 \times 10^{-31}$ kg $= 0.19m$ and $m_L = 7.8 \times 10^{-31}$ kg $= 0.86m$.

5.8 (a) For no time dependence, Eq. (5.66) becomes

$$\lambda_D^2 \frac{d^2 n'}{dx^2} = n' - p'.$$

Assume the semiconductor occupies the space $x > 0$, $n' - p'$ has a non-zero value n_0 at $x = 0$ and p' varies slowly in space on the length scale λ_D; the solution is $n' = p' + n_0 \exp(-x/\lambda_D)$, which has the stated qualitative behaviour.

(b) For no space dependence, Eq. (5.66) becomes

$$\tau_D \frac{dn'}{dt} = p' - n',$$

with solution for slowly varying p', $n'(t) = p' + n_0 \exp(-t/\tau_D)$; this again has the stated qualitative behaviour.

5.9 Differentiating Eq. (5.75) gives

$$\frac{\partial p'}{\partial t} = \left(-\frac{1}{2t} - \frac{1}{\tau_n} + \frac{x^2}{4D_h t^2} \right) p', \qquad \frac{\partial^2 p'}{\partial x^2} = \left(-\frac{1}{2D_h t} + \frac{x^2}{4D_h^2 t^2} \right) p'.$$

Substitution shows that these satisfy Eq. (5.70b) when $E = 0$.

The number of holes remaining at time t is

$$\int_{-\infty}^{\infty} p'(x, t)\, dx = P \exp(-t/\tau_n).$$

CHAPTER 6

6.1 From Eqs. (6.10) the carrier concentrations change by a factor 2 when $e\Delta\phi = 0.69k_B T$, i.e. when $\Delta\phi \approx 0.018$ V. From Eqs. (6.7) and (6.9) with $N_A = N_D$, $\phi(x) = 2\Delta\phi_0(x + w_p)^2/(w_n + w_p)^2$ on the p side of the junction; the fractional distance from the edge of the depletion layer for decrease by a factor 2 is therefore

$$\frac{x + w_p}{w_n + w_p} = \left(\frac{0.018}{2\Delta\phi_0} \right)^{1/2} = \begin{cases} 0.11 & \text{for Si } (\Delta\phi_0 = 0.7 \text{ V}), \\ 0.17 & \text{for Ge } (\Delta\phi_0 = 0.3 \text{ V}). \end{cases}$$

6.2 Using data given, $\Delta\phi_0 = 0.68$ V (Eq. (6.2)), thus, from Eq. (6.17), $C = 5.34 \times 10^{-10}$ F when $V = -1$ V and $C = 1.35 \times 10^{-10}$ F when $V = -10$ V. (Do not forget to multiply by the area of the junction.) The resonant frequency, $1/2\pi(LC)^{1/2}$, therefore changes from 689 to 1371 kHz.

6.3 The electric charge density is $e(N_D - N_A) = ekx$; inserting this in Poisson's equation (6.4) and integrating gives

$$E = -\frac{d\phi}{dx} = \frac{ekx^2}{2\varepsilon\varepsilon_0} - A, \qquad \phi(x) = -\frac{ekx^3}{6\varepsilon\varepsilon_0} + Ax + B,$$

where A and B are constants of integration. Symmetry considerations imply that the depletion layer is symmetrical about $x = 0$, with width $2w$ say. A, B and w are determined by the boundary conditions:

$$E = 0 \quad \text{at } x = \pm w, \qquad \phi(-w) = 0, \qquad \phi(+w) = \Delta\phi_0.$$

The solution is

$$\phi(x) = \frac{\Delta\phi_0}{4w^3}(w + x)^2(2w - x), \qquad \text{where } w = \left(\frac{3\varepsilon\varepsilon_0\Delta\phi_0}{2ek}\right)^{1/3}.$$

6.4 At $V = -0.15$ V and 300 K, $\exp(eV/k_B T) \ll 1$ so that $I \approx -I_0$ in Eq. (6.22); thus $I_0 \approx 5\,\mu A$. With $V = +0.15$ V, $I \approx I_0 \exp(eV/k_B T) = 1.65$ mA.

6.5 From Eq. (5.74) we can write the excess electron concentration on the p side as a function of distance x from the depletion layer edge as

$$n' = n_p(x) - n_{p0} = [n_p(0) - n_{p0}]\exp(-x/L_e),$$

where n_{p0} is the equilibrium concentration in p far from the junction. The injected current is

$$J_e = (-e)\left(-D_e \frac{\partial n'}{\partial x}\right)_{x=0} = -\frac{eD_e}{L_e}[n_p(0) - n_{p0}].$$

If Eqs. (6.10) remain valid within the depletion layer then $n_p(0) = n_{n0}\exp(-e\Delta\phi/k_B T)$, where $\Delta\phi = \Delta\phi_0 - V$ is the total potential drop across the depletion layer; we assume that the electron concentration on the n side is unchanged from its equilibrium value n_{n0}. In the absence of a bias, $n_{n0} = n_{p0}\exp(-e\Delta\phi_0/k_B T)$, so that $n_p(0) = n_{p0}\exp(eV/k_B T)$. Using this, J_e becomes

$$J_e = \frac{eD_e n_{p0}}{L_e}(e^{eV/k_B T} - 1).$$

If the electrons are injected into a base region of width w, small compared to L_e, then the exponentially increasing term in Eq. (5.73) cannot be ignored. If $w \ll L_e$ and $n' = 0$ at $x = w$, Eq. (5.73) approximates to $n' \approx [n_p(0) - n_{p0}](1 - x/w)$. The above calculation then yields the same expression for the current except that L_e is replaced by w.

6.6 From Eqs. (6.5) the maximum field $E_{max} = N_A ew_p/\varepsilon\varepsilon_0$ occurs at $x = 0$. For finite reverse bias, w_p is given by Eqs. (6.9) with $\Delta\phi_0$ replaced by $\Delta\phi_0 + |V|$; for $|V| \gg \Delta\phi_0$, therefore, $E_{max} \propto w_p \propto |V|^{1/2}$. The maximum slope of the energy bands in the depletion layer is eE_{max}; to get from the valence band to the conduction band an electron must therefore tunnel a distance of order T where $eE_{max}T \approx E_G$. Thus $T \propto 1/|V|^{1/2}$ and the tunnelling current contains a factor $\exp(-b/|V|^{1/2})$. For an explanation of the $\exp(-2\alpha T)$ factor, see French and Taylor.[4]

In time τ between collisions an electron can acquire a velocity $eE_{max}\tau/m_e$ and hence an energy $\frac{1}{2}m_e v^2 = e^2 E_{max}^2 \tau^2/2m_e$; if this exceeds E_G then avalanche breakdown can occur. Using E_{max} from above and Eq. (6.9), with $N_A = N_D$, for w_p, this condition can be written $e^3\tau^2 N_D|V|/2m_e\varepsilon\varepsilon_0 > E_G$. The critical doping level for avalanche breakdown at 100 V in a silicon diode is therefore $N_D = 8 \times 10^{18}$ m^{-3}.

CHAPTER 7

7.1 According to the Bohr model the ground state of the hydrogen atom corresponds to an electron in a circular orbit ($mv^2/r = e^2/4\pi\varepsilon_0 r^2$) with angular momentum $mvr = h$. The circulating electron is equivalent to a circular coil carrying a current $i = ev/2\pi r$ and thus produces a magnetic field at the proton $B = \mu_0 i/2r$. Inserting values for i and r from the previous equations gives $B = \pi\mu_0 m^2 e^7/8\varepsilon_0^3 h^5 = 12.6$ T.

7.2 See Fig. 7.2 for assistance. For the 4f shell, $l = 3$ and l_z can take seven possible values: 3, 2, 1, 0, -1, -2, -3. Calculating S is straightforward; note that S is symmetric about $n = 7$. For $n \leqslant 7$, $L = 3 + 2 + \cdots$ (n terms in all); this is an arithmetic series of n terms with an average value $\{3 + [3 - (n-1)]\}/2 = (7 - n)/2$. Thus $L = n(7 - n)/2$ and $L = 0$ when $n = 7$. On reaching the eighth electron the maximum L is achieved by assigning this to $l_z = 3$. For $n \geqslant 7$, therefore, $L = 3 + 2 + \cdots$; there are $n - 7$ terms of average value $(14 - n)/2$. J is obtained by use of Hund's third rule.

7.3 The components of $-\mu_B \mathbf{L}$ and $-2\mu_B \mathbf{S}$ along $\mathbf{J}$ are $-\mu_B|\mathbf{L}|\cos\theta$ and $-2\mu_B|\mathbf{S}|\cos\phi$. The effective moment is therefore

$$\boldsymbol{\mu}_{\mathrm{eff}} = \frac{\mathbf{J}}{|\mathbf{J}|}(|\mathbf{L}|\cos\theta + 2|\mathbf{S}|\cos\phi).$$

Applying the cosine rule to the triangle gives:

$$\cos\theta = (|\mathbf{L}|^2 + |\mathbf{J}|^2 - |\mathbf{S}|^2)/2|\mathbf{L}\|\mathbf{J}|, \qquad \cos\phi = (|\mathbf{S}|^2 + |\mathbf{J}|^2 - |\mathbf{L}|^2)/2|\mathbf{S}\|\mathbf{J}|.$$

Using these in the expression for $\boldsymbol{\mu}_{\mathrm{eff}}$ and replacing $|\mathbf{L}|^2$, $|\mathbf{S}|^2$ and $|\mathbf{J}|^2$ by their eigenvalues $L(L + 1)$, $S(S + 1)$ and $J(J + 1)$ gives

$$g = \frac{3}{2} - \frac{L(L+1) - S(S+1)}{2J(J+1)}.$$

7.4 A single unpaired spin can have component $+\mu_B$ or $-\mu_B$ parallel to the field; these states have energies $-\mu_B B$ and $+\mu_B B$ and thus probabilities $\exp(+\mu_B B/k_B T)$ and $\exp(-\mu_B B/k_B T)$ respectively. The magnetization of N ions per unit volume is therefore

$$M = N\frac{\mu_B \exp(+\mu_B B/k_B T) - \mu_B \exp(-\mu_B B/k_B T)}{\exp(+\mu_B B/k_B T) + \exp(-\mu_B B/k_B T)} = N\mu_B \tanh(\mu_B B/k_B T).$$

From Eq. (7.8), the internal energy E is $-\mathbf{M}.\mathbf{B}$ per unit volume, i.e. $E = -N\mu_B B \tanh(\mu_B B/k_B T)$. The magnetic work $-\mathbf{M}.\mathrm{d}\mathbf{B}$ is zero in constant B, so that the heat capacity at constant B is

$$C_B = \left(\frac{\partial E}{\partial T}\right)_B = Nk_B\left(\frac{\mu_B B}{k_B T}\right)^2 \Big/ \cosh^2\left(\frac{\mu_B B}{k_B T}\right).$$

For $k_B T \gg \mu_B B$,

$$C_B \approx Nk_B(\mu_B B/k_B T)^2.$$

For $k_B T \ll \mu_B B$,

$$C_B \approx 4Nk_B(\mu_B B/k_B T)^2 \exp(-2\mu_B B/k_B T).$$

In drawing your sketch make use of these limiting forms and also of the fact that C_B goes through a maximum value $\approx 2.2Nk_B$ at $\mu_B B/k_B T \approx 1.2$. For $CuSO_4$ the maximum occurs at a temperature of order 0.5 K in a field of 1 T.

7.5 Write the density of states (Eq. (3.7)) as $g(\varepsilon) = A\varepsilon^{1/2}$ per unit volume; in zero field B the electron concentration is given by

$$N = \int_0^{\varepsilon_F} A\varepsilon^{1/2}\, d\varepsilon$$

from which A can be conveniently written $A = 3N/2\varepsilon_F^{3/2}$. In finite B and at $T = 0$ the energy ε_F' of the highest occupied level (Fig. 7.5) is no longer equal to ε_F but must be determined from

$$N = N_\uparrow + N_\downarrow = \int_{-\mu_B B}^{\varepsilon_F'} \tfrac{1}{2}A(\varepsilon + \mu_B B)^{1/2}\, d\varepsilon + \int_{\mu_B B}^{\varepsilon_F'} \tfrac{1}{2}A(\varepsilon - \mu_B B)^{1/2}\, d\varepsilon.$$

Thus

$$N = \frac{3N}{4\varepsilon_F^{3/2}} [\tfrac{2}{3}(\varepsilon_F' + \mu_B B)^{3/2} + \tfrac{2}{3}(\varepsilon_F' - \mu_B B)^{3/2}].$$

The magnetization is

$$M = \mu_B(N_\uparrow - N_\downarrow) = \frac{3N\mu_B}{4\varepsilon_F^{3/2}} [\tfrac{2}{3}(\varepsilon_F' + \mu_B B)^{3/2} - \tfrac{2}{3}(\varepsilon_F' - \mu_B B)^{3/2}].$$

From these equations $\varepsilon_F' \pm \mu_B B = \varepsilon_F(1 \pm M/M_s)^{2/3}$ where $M_s = N\mu_B$ is the saturation magnetization corresponding to N perfectly aligned spins. Hence

$$2\mu_B B/\varepsilon_F = (1 + M/M_s)^{2/3} - (1 - M/M_s)^{2/3}.$$

The field required to achieve saturation ($M = M_s$) is therefore $B_s = \varepsilon_F/2^{1/3}\mu_B$. From Eq. (3.11), ε_F for potassium is 2.12 eV so that $B_s \approx 3 \times 10^4$ T, much higher than fields that can be applied in a laboratory. For the ^{3}He/^{4}He mixture the ^{3}He concentration is $0.001 \times 130/(4 \times 1.67 \times 10^{-27}) = 1.95 \times 10^{25}$ m^{-3}; the ^{3}He mass is $3 \times 1.67 \times 10^{-27}$ kg, so that, from Eq. (3.9), $\varepsilon_F = 4.8\ \mu$eV. B_s is therefore 2.9 T and is achievable in a laboratory; note however that the mixture must be cooled to below its degeneracy temperature $T_F \approx 0.06$ K.

7.6 $|\mathbf{A}|$ is constant along C and equal to $\Phi/2\pi\rho$, where ρ is the radius of C. We can conveniently write $\mathbf{A} = \hat{\boldsymbol{\theta}}\Phi/2\pi\rho$, where $\hat{\boldsymbol{\theta}}$ is a unit vector parallel to C. Using the expression for div in cylindrical polar coordinates (ρ, θ, z) gives

$$\text{div } \mathbf{A} = \left(\frac{1}{\rho}\frac{\partial}{\partial\rho}(\rho A_r) + \frac{1}{\rho}\frac{\partial}{\partial\theta}A_\theta + \frac{\partial}{\partial z}A_z\right) = 0.$$

Note that, in terms of the vector $\mathbf{r} = (x, y, z)$ in Fig. 7.7, Eq. (7.35) can be written $\mathbf{A} = -\tfrac{1}{2}\mathbf{r} \times \mathbf{B} = \tfrac{1}{2}B(-y, x, 0)$; thus curl $\mathbf{A} = (0, 0, B)$ and div $\mathbf{A} = 0$ for constant B.

7.7 The solution satisfies $d^2\mathbf{A}/dx^2 = \mathbf{A}/\lambda^2$ as required; the solution $\propto \exp(+x/\lambda)$ can be discounted on the grounds that it diverges as $x \to \infty$. The magnetic field is curl $\mathbf{A} = \hat{z}\,\partial A_y/\partial x = -A_0\hat{z}\exp(-x/\lambda)/\lambda$. The current density, obtained using the Maxwell equation curl $\mathbf{B} = \mu_0\mathbf{j}$, is $-A_0\hat{y}\exp(-x/\lambda)/\mu_0\lambda^2$.

7.8 We apply Eq. (7.39) to the six electrons per molecule with extended wavefunctions; the number of such electrons per unit volume is

$$N = 6 \times 6.02 \times 10^{23} \times \frac{0.88}{78} \times 10^6 = 4.1 \times 10^{28}\ \text{m}^{-3}$$

and for randomly oriented planar molecules $\langle r^2 \rangle = \frac{2}{3}\langle \rho^2 \rangle = \frac{2}{3}(1.4 \text{ Å})^2$. Hence Eq. (7.39) gives

$$\chi = -\frac{4.1 \times 10^{28} \times (1.6 \times 10^{-19})^2 \times 4\pi \times 10^{-7}}{4 \times 9.1 \times 10^{-31}} \times \frac{4}{9}(1.4 \times 10^{-10})^2$$

$$= -3.2 \times 10^{-6}.$$

The relatively large contribution of the electrons in extended orbitals to the total susceptibility is due to the large $\langle \rho^2 \rangle$ for these electrons.

CHAPTER 8

8.1 s_1^2 and s_2^2 have the eigenvalue $\frac{1}{2}(\frac{1}{2} + 1) = \frac{3}{4}$; S^2 has the eigenvalue 0 for a singlet state and $1(1 + 1) = 2$ for a triplet state. Thus for a singlet state $s_1 . s_2 = (S^2 - s_1^2 - s_2^2)/2 = (0 - \frac{3}{4} - \frac{3}{4})/2 = -\frac{3}{4}$; similarly for a triplet state $s_1 . s_2 = (2 - \frac{3}{4} - \frac{3}{4})/2 = \frac{1}{4}$. The energy difference between singlet and triplet states is therefore $E_T - E_S = -2\mathcal{J}[\frac{1}{4} - (-\frac{3}{4})] = -2\mathcal{J}$.

8.2 The major assumption of the Weiss model is that a spin is subject to a molecular field $\lambda\mu_0 M$ that depends only on the *average* magnetization of the sample.

Substituting Eqs. (8.5) in Eq. (8.1) and ignoring the term quadratic in the deviation of S from $\langle S \rangle$ gives

$$H \approx -\sum_i \sum_j \mathcal{J}_{ij}[\langle S \rangle . \langle S \rangle + \langle S \rangle . (S_j - \langle S \rangle) + \langle S \rangle . (S_i - \langle S \rangle)]$$

$$= +N\langle S \rangle . \langle S \rangle \sum_j \mathcal{J}_{ij} - 2\sum_i \left(\sum_j \mathcal{J}_{ij}\right) S_i . \langle S \rangle = \frac{1}{2}\lambda\mu_0 M^2 - \sum_i \lambda\mu_0 \mu_i . M,$$

where we have used $\mu_i = -g\mu_B S_i$, $M = -Ng\mu_B \langle S \rangle$ and defined λ by Eq. (8.4). The internal energy E is the average value of H. Thus

$$E = \langle H \rangle = \frac{1}{2}\lambda\mu_0 M^2 - \sum_i \lambda\mu_0\langle \mu_i \rangle . M = -\frac{1}{2}\lambda\mu_0 M^2.$$

The magnetic contribution to the heat capacity is

$$C_m = \frac{dE}{dT} = -\frac{1}{2}\lambda\mu_0 \frac{d(M^2)}{dT}.$$

Using the limiting forms of Eqs. (8.13) and (8.14) for the magnetization gives the following mean field predictions for C_m:

$$C_m = 3Nk_B \frac{T}{T_C}\left(\frac{3T}{2T_C} - 1\right) \qquad \text{for } T \to T_C,$$

$$C_m = 4Nk_B \left(\frac{T_C}{T}\right)^2 \exp\left(-\frac{2T_C}{T}\right) \qquad \text{for } T \to 0,$$

where we have used Eq. (8.10). The heat capacity increases discontinuously by $\frac{3}{2}Nk_B$ on cooling through T_C. We deduce that the entropy $\int C_m \, dT/T$ is continuous and the latent heat zero; mean field theory therefore predicts that the ferromagnetic transition is second order in zero applied field. In practice the above results are modified by fluctuation effects near T_C and spin waves near $T = 0$ (Eq. (8.37)).

8.3 Using Eqs. (8.2) and (8.10) enables Eq. (8.7) to be written

$$m = \tanh\left(\frac{\mu_B B}{k_B T} + \frac{m}{t}\right) = \frac{h + \tanh{(m/t)}}{1 + h\tanh{(m/t)}} \qquad \text{where } h = \tanh\left(\frac{\mu_B B}{k_B T}\right).$$

(We have used $\tanh(A + B) = (\tanh A + \tanh B)/(1 + \tanh A \tanh B)$.) Put $t = 1$, solve for h and use the small-argument expansion $\tanh x \approx x - \frac{1}{3}x^3$ to obtain $\mu_B B/k_B T_C = \frac{1}{3}m^3$ to lowest order in m. Hence $M \approx N\mu_B(3\mu_B B/k_B T)^{1/3}$. The mean field exponent $\frac{1}{3}$ is modified by fluctuations.

8.4 $\lambda_a/\lambda_p = z_a/z_p$ follows from Eq. (8.16). Inserting expressions for $\mathbf{B}_{eff}^A$ and $\mathbf{B}_{eff}^B$ ($= \mathbf{B}_{loc} - \lambda_a \mathbf{M}_A - \lambda_p \mathbf{M}_B$) into Eqs. (8.17) and using $\tanh x \approx x$ gives, instead of Eqs. (8.18),

$$M_A(2T/C + \lambda_p) + \lambda_a M_B = H \qquad \text{and} \qquad M_B(2T/C + \lambda_p) + \lambda_a M_A = H. \quad \text{(i)}$$

Solving for the high-temperature susceptibility gives $\chi = M/H = (M_A + M_B)/H = C/(T + \theta)$, where $\theta = (\lambda_a + \lambda_p)C$. The Néel temperature T_N is the highest temperature at which a solution of Eqs. (i) with M_A, $M_B \neq 0$ exists in zero H; thus equating the determinant of coefficients to zero gives $T_N = C(\lambda_a - \lambda_p)$. Hence $\theta/T_N = (\lambda_a + \lambda_p)/(\lambda_a - \lambda_p) = (z_a + z_p)/(z_a - z_p) = 3$ for $z_a = 8$ and $z_p = 4$.

8.5 *For* **H** *parallel to* $\mathbf{M}_A$

From Eqs. (8.17), the change in M_A induced by a small field is

$$\delta M_A \approx \left(\frac{\partial M_A}{\partial B_{eff}^A}\right)_{H=0} \delta B_{eff}^A = \frac{N\mu_B^2}{2k_B T}\operatorname{sech}^2\left(\frac{\mu_0 \lambda M_0}{k_B T}\right)\delta B_{eff}^A$$

$$= \frac{C}{2\mu_0 T}\operatorname{sech}^2\left(\frac{\mu_0 \lambda M_0}{k_B T}\right)\delta B_{eff}^A,$$

where M_0 is the value of M_A in zero field and C is the Curie constant. Similarly for the B sublattice

$$\delta M^B \approx \frac{C}{2\mu_0 T}\operatorname{sech}^2\left(\frac{\mu_0 \lambda M_0}{k_B T}\right)\delta B_{eff}^B.$$

From Eqs. (8.15), $\delta B_{eff}^A = \mu_0(H - \lambda\delta M_B)$ and $\delta B_{eff}^B = \mu_0(H - \lambda\delta M_A)$; inserting these into the above equations and solving for δM_A and δM_B gives

$$\delta M_A = \delta M_B = \frac{\alpha H}{\lambda(1 + \alpha)} \qquad \text{where } \alpha = \frac{C\lambda}{2T}\operatorname{sech}^2\left(\frac{\mu_0 \lambda M_0}{k_B T}\right)$$

$$= \frac{T_N}{T}\operatorname{sech}^2\left(\frac{\mu_0 \lambda M_0}{k_B T}\right).$$

Hence

$$\chi = \frac{M}{H} = \frac{\delta M_A + \delta M_B}{H} = \frac{2\alpha}{\lambda(1 + \alpha)} = \frac{C}{T_N}\frac{\alpha}{(1 + \alpha)}.$$

As $T \to 0$, $\alpha \to 0$ and hence $\chi \to 0$. As $T \to T_N$, $\alpha \to 1$ and $\chi \to C/2T_N$.

For **H** *perpendicular to* $\mathbf{M}_A$

To first order in **H**, $|\mathbf{M}_A|$ and $|\mathbf{M}_B|$ do not change; instead they rotate so as to attain a component parallel to the field. By symmetry and the definition of χ, $\delta\mathbf{M}_A = \delta\mathbf{M}_B =$

$\chi H/2$. Hence, from Eqs. (8.15), $\delta \mathbf{B}_{\text{eff}}^{A} = \mu_0(\mathbf{H} - \lambda\delta\mathbf{M}_B) = \mu_0(\mathbf{H} - \lambda\chi H/2)$. Since $\mathbf{M}_A$ and $\mathbf{B}_{\text{eff}}^{A}$ must remain parallel

$$\frac{|\delta\mathbf{B}_{\text{eff}}^{A}|}{|\mathbf{B}_{\text{eff}}^{A}|} = \frac{\mu_0(1 - \lambda\chi/2)H}{\mu_0\lambda M_0} = \frac{|\delta\mathbf{M}_A|}{|\mathbf{M}_A|} = \frac{\chi H/2}{M_0},$$

from which $(1 - \lambda\chi/2) = \lambda\chi/2$, i.e. $\chi = 1/\lambda = C/2T_N$ and is independent of T.

8.6 From Eq. (8.31), $\hbar\omega = 4\mathscr{J}S[1 - \cos(ka)]$. From Fig. 12.10, $\hbar\omega = 0.0012$ eV when $ka/2\pi = 0$ and $\hbar\omega = 0.045$ eV when $ka/2\pi = 0.2$. If the dispersion relation correctly describes the difference between these values, then

$$0.045 - 0.0012 = 4\mathscr{J}S[1 - \cos(0.4\pi)] \qquad \text{or} \qquad \mathscr{J}S = 0.016 \text{ eV}.$$

The finite value of $\hbar\omega$ at $k = 0$ suggests that there is a preferred direction for the spontaneous magnetization within the crystal so that a finite energy is needed to rotate a spatially uniform (i.e. $k = 0$) magnetization.

8.7 The straight line shows that the heat capacity $C = AT^3 + BT^{3/2}$. The first term is the lattice heat capacity (section 2.6) and the second term is the spin wave contribution. The slope of the graph gives information on the sound velocity (Eq. (2.48)) and the intercept gives $\mathscr{J}S$ (Eq. (8.37)).

8.8 Eq. (8.24) is valid for both $\uparrow$ and $\downarrow$ spins. The linearized equations for $\sigma_n^{\uparrow}$ and $\sigma_{n+1}^{\downarrow}$ are:

$$\hbar\frac{d\sigma_n^{\uparrow}}{dt} = 2\mathscr{J}S\hat{\mathbf{z}} \times (\sigma_{n-1}^{\downarrow} + \sigma_{n+1}^{\downarrow} - 2\sigma_n^{\uparrow})$$

$$\hbar\frac{d\sigma_{n+1}^{\downarrow}}{dt} = -2\mathscr{J}S\hat{\mathbf{z}} \times (\sigma_n^{\uparrow} + \sigma_{n+2}^{\uparrow} - 2\sigma_{n+1}^{\downarrow}).$$

Taking components, writing equations for $d\sigma_n^{\uparrow-}/dt$ and $d\sigma_{n+1}^{\downarrow-}/dt$, and inserting solutions as indicated gives

$$(-i\omega)i\hbar u = 2\mathscr{J}Su(\alpha e^{-ika} + \alpha e^{ika} - 2)$$
$$(-i\omega)i\hbar\alpha u = -2\mathscr{J}Su(e^{-ika} + e^{ika} - 2\alpha),$$

which can be written (cf. Eq. (2.19))

$$\alpha = \frac{(1 + \hbar\omega/4\mathscr{J}S)}{\cos(ka)} = \frac{\cos(ka)}{(1 - \hbar\omega/4\mathscr{J}S)}.$$

The dispersion relation for antiferromagnetic spin waves is thus

$$1 - (\hbar\omega/4\mathscr{J}S)^2 = \cos^2(ka) \qquad \text{or} \qquad \hbar\omega = 4\mathscr{J}S|\sin(ka)|.$$

For small ka, $\hbar\omega = 4\mathscr{J}Ska$ so that the dispersion relation is linear, just as for long-wavelength phonons. Hence the spin wave contribution to the heat capacity should vary as T^3 at low temperatures.

8.9 The film thickness corresponds to an odd number of half-wavelengths $d = (2n + 1)\lambda/2$, so that $k = 2\pi/\lambda = \pi(2n + 1)/d$ and, from Eqs. (8.45) and (8.34),

$$\gamma\delta B_e = \frac{2\mathscr{J}Sa^2}{\hbar}\frac{\pi^2}{d^2}\delta(2n + 1)^2 = \frac{\hbar}{2m^*}\frac{\pi^2}{d^2}\delta(2n + 1)^2.$$

From Fig. 8.12, $(2n + 1) = 3$ to $(2n + 1) = 21$ (corresponding to $\delta(2n + 1)^2 = 432$) occurs for $\delta B_e \approx 0.27$ T. Thus, inserting $\gamma = e/m$, we obtain $m^* \approx 14m$.

CHAPTER 9

9.1 Maxwell's equations governing the propagation of electromagnetic waves through a dielectric medium are: curl $\mathbf{H} = \dot{\mathbf{D}}$ and curl $\mathbf{E} = -\mu_0 \dot{\mathbf{H}}$ (assume $\mu = 1$). Inserting solutions $\mathbf{D} = \varepsilon(\omega)\varepsilon_0 i E_0 \exp[i(kz - \omega t)]$ and $\mathbf{H} = jH_0 \exp[i(kz - \omega t)]$ gives $\omega/k = 1/[\varepsilon(\omega)\varepsilon_0\mu_0]^{1/2} = c/[\varepsilon(\omega)]^{1/2}$, corresponding to $n(\omega) = [\varepsilon(\omega)]^{1/2}$.

Let $k = k' + ik''$; the amplitude of the wave then decays as $\exp(-k''z)$ and the energy as $\exp(-2k''z)$. The distance in which the energy decays by $1/e$ is thus $(2k'')^{-1} = c/2\omega n''$, where n'' is the imaginary part of $n(\omega)$. To evaluate n'' use $[n(\omega)]^2 = n'^2 - n''^2 + 2in'n'' = \varepsilon + i\varepsilon'$; equating real parts and imaginary parts on either side and solving for n'' gives $n'' = \{[(\varepsilon'^2 + \varepsilon''^2)^{1/2} - \varepsilon']/2\}^{1/2}$ and hence the answer given for the decay length.

9.2 From Eq. (9.19), the refractive index $n = \varepsilon^{1/2} \approx 1 - NZe^2/2\varepsilon_0 m\omega^2$ (assuming $\varepsilon - 1$ is small); write this as $n = 1 - A/\omega^2$. The phase velocity v_p is $c/(1 - A/\omega^2)$. Using $k = \omega n/c = (\omega - A/\omega)/c$ gives $dk/d\omega = (1 + A/\omega^2)/c = v_g^{-1}$. Hence $v_g v_p \approx c^2$ to terms of order A/ω^2.

9.3 The Bohr theory result for the ground state energy of an electron bound to a nucleus of charge Ze is $E_0 = Z^2 me^4/32\pi^2\varepsilon_0^2\hbar^2$; the radius of the orbit is $a = 4\pi\varepsilon_0\hbar^2/Zme^2$. E_0 can therefore be written $E_0 = (Ze^2\hbar^2/16\pi\varepsilon_0 ma^3)^{1/2}$. Comparison with Eq. (9.18) shows that $\hbar\omega_0 = E_0$ if we identify $r = 4^{1/3}a$.

9.4 The peaks in $-\varepsilon''$ occur when $\omega\tau = 1$ (see Fig. 9.5); identifying -6.5, -22.9 and $-32.9°C$ as the positions of the peaks at 5000, 1000 and 300 Hz, we deduce that the relaxation times are 31.8, 159.2 and 530.5 μs at these temperatures. These are fitted well by a temperature dependence of the form $\tau = \tau_0 \exp(T_0/T)$ with $\tau_0 = 2.408 \times 10^{-16}$ s and $T_0 = 6827$ K. Such a temperature dependence arises if rotation occurs by thermal activation over an energy barrier; the probability that a molecule has sufficient energy to rotate is then given by a Boltzmann factor.

9.5 The reflection occurs between the frequencies, ω_0 and ω_L, defined in section 9.1.4. We use the value of Young's modulus Y to estimate the spring constant K. The spacing of ions in the sodium chloride structure (Fig. 1.13) is $a/2$ so a slice perpendicular to $[1\,0\,0]$ cuts $(2/a)^2$ springs per unit area. The force per unit area required to extend each spring by δx in the $[1\,0\,0]$ direction is therefore $(2/a)^2 K\delta x = Y \times$ (extension)/(original length) $= Y\delta x/(a/2)$; hence $K = Ya/2$ (note that we are ignoring the lateral contraction that occurs when a crystal is stretched in one direction). The wavelength λ_0 of light corresponding to the frequency ω_0 is

$$\lambda_0 = 2\pi c/\omega_0 = 2\pi c/(2K/M^*)^{1/2} = 2\pi c(M^*/Ya)^{1/2} = 53 \ \mu m.$$

From Eq. (9.34),

$$\frac{\varepsilon(\infty)}{\varepsilon(0)} = 1 - \frac{Ne^2}{2K\varepsilon_0\varepsilon(0)} = 1 - \frac{(4/a^3)e^2}{Ya\varepsilon_0\varepsilon(0)} = 0.67;$$

from Eq. (9.40), the wavelength λ_L of light corresponding to the frequency ω_L is

$$\lambda_L = 2\pi c/\omega_L = 2\pi c/\{\omega_0[\varepsilon(0)/\varepsilon(\infty)]^{1/2}\} = \lambda_0[\varepsilon(\infty)/\varepsilon(0)]^{1/2} = 43 \ \mu m.$$

This corresponds reasonably well to the experimental data of Fig. 9.9(b).

9.6 If α is positive, β negative and γ positive then F has two minima: one at $P = 0$, the other where $P_m^2 = [|\beta| + (|\beta|^2 - 3\gamma\alpha)^{1/2}]/3\gamma$. The value of F at this other minimum can be written $F = F_0 + P_m^2(2\alpha/3 - |\beta|P_m^2/3)$; this minimum is thus lower than that at $P = 0$ if $P_m^2 > 2\alpha/|\beta|$, a condition that can be rewritten $4\alpha\gamma < |\beta|^2$. $4\alpha\gamma$ and $|\beta|^2$ are therefore equal at T_C and P_m^2 is then given by $P_m^2 = (\alpha/\gamma)^{1/2} = |\beta|/2\gamma = 2\alpha/|\beta|$.

CHAPTER 10

10.1   The decay of the current is governed by $I = I_0 \exp(-tR/L)$, where R is the resistance and L the inductance. Since $I/I_0 > 0.98$ after a time of $7 \times 3600 = 25\,200$ s we deduce that $L/R > -25\,200/\ln 0.98 = 1.25 \times 10^6$ s.

The current flows on the inner walls of the tube. To estimate L we calculate the field due to two parallel current sheets of dimensions 0.003 m and 0.02 m parallel and perpendicular to the current; using Ampere's law $\oint \mathbf{B} \cdot d\mathbf{l} = 0.02B = \mu_0 I$ gives $B = \mu_0 I/0.02$. The flux linked with the tube is $\Phi = B \times 0.003 \times d$ where d is the effective distance between the sheets, which we estimate as the actual gap $(5 \times 10^{-7}$ m) plus two penetration depths $(10^{-7}$ m); hence $L = \Phi/I = \mu_0 \times 0.003 \times 6 \times 10^{-7}/0.02$. To relate R to the resistivity we use $R = \rho l/A = \rho \times 2 \times 0.003/(0.02 \times \lambda)$; $0.02 \times \lambda$ is the effective area since the current flows only within λ of the surface.

Thus $L/R = \mu_0 \times 6 \times 10^{-7} \times 10^{-7}/2\rho > 1.25 \times 10^6$ s,   so   that $\rho < 4 \times 10^{-26}$ Ω m. The value for copper at room temperature is 1.6×10^{-8} Ω m.

10.2   The field B at the surface of a wire of radius a carrying a current I is $\mu_0 I/2\pi a$. The critical field of tin at 2 K is (Eq. (10.1)) $B_c(2) \approx B_c(0)[1 - (T/T_c)^2] = 21.7$ mT. The critical current that gives this field is $I_c = 2\pi a B_c/\mu_0 = 54$ A. Since $I_c \propto a$, a wire of diameter $100 \times 1/54$ mm $= 1.85$ mm would have $I_c = 100$ A.

10.3   Inserting Eq. (10.1) into Eqs. (10.6), (10.7) and (10.8) gives

$$G_N(0, T) - G_S(0, T) = \frac{B_c^2(0)}{2\mu_0}\left[1 - \left(\frac{T}{T_c}\right)^2\right]^2,$$

$$S_S - S_N = -\frac{2B_c^2(0)}{\mu_0 T_c}\frac{T}{T_c}\left[1 - \left(\frac{T}{T_c}\right)^2\right],$$

$$C_S - C_N = -\frac{2B_c^2(0)}{\mu_0 T_c}\frac{T}{T_c}\left[1 - 3\left(\frac{T}{T_c}\right)^2\right].$$

At $T = T_c$,

$$C_S - C_N = \frac{4B_c^2(0)}{\mu_0 T_c}.$$

10.4   We assume that the flux penetration in the mixed state is almost perfect so that $\mathbf{M} \approx 0$ and, from Eq. (10.4), $G_S(B_e, T) \approx G_S(0, T)$. The Cooper pairing of opposite-spin electrons reduces the Pauli spin susceptibility (section 7.2.4) of the supercon-ductor to zero at low temperatures. For the normal (N) state $\mathbf{M} = \chi_p \mathbf{H} = \chi_p \mathbf{B}/\mu_0$ so that Eq. (10.4) gives $G_N(B_e, T) = G_N(0, T) - \chi_p B^2/2\mu_0$. The upper critical field B_{c2} is then given by

$$G_N(0, T) - G_S(0, T) = \chi_p B_{c2}^2/2\mu_0.$$

The free energy difference $G_N - G_S$ corresponds approximately to a fraction $k_B T_c/\varepsilon_F$ of the N electrons per unit volume having their energy reduced by $k_B T_c$ so that $G_N(0, T) - G_S(0, T) \approx N(k_B T_c)^2/\varepsilon_F$. Using χ_p from Eq. (7.23) then allows us to write the Clogston limiting field as $B_{c2} \approx k_B T_c/\mu_B$.

10.5   For a thin film, both terms in Eq. (10.12) must be retained and the constants a and b chosen to satisfy the boundary conditions $B = B_e$ at $z = \pm d/2$. $\mathbf{B} = B_e \cosh(z/\lambda)/\cosh(d/2\lambda)$ is the appropriate solution.

There is a screening current density $\mathbf{j} = \text{curl } \mathbf{B}/\mu_0$ in the plane and this is equivalent to a magnetization $\mathbf{M}$, where $\text{curl } \mathbf{M} = \mathbf{j}$; we deduce $\mu_0 \mathbf{M} = \mathbf{B} - \mathbf{B}_e$ ($\mathbf{B}_e$ is the integration constant chosen to ensure that $\mathbf{M} = 0$ when $\mathbf{B} = \mathbf{B}_e$). With a spatially varying magnetization, Eq. (10.4) becomes

$$\Delta G = G_S(B_e, T) - G_S(0, T) = -\frac{1}{d} \int_{-d/2}^{d/2} dz \int_0^{B_e} \mathbf{M} \cdot d\mathbf{B}_e,$$

where the $1/d$ factor ensures that the free energy difference is per unit volume. Inserting $\mathbf{M} = (\mathbf{B} - \mathbf{B}_e)/\mu_0$ with $\mathbf{B}$ as obtained above and integrating gives

$$\Delta G = \frac{B_e^2}{2\mu_0} \left[1 - \frac{2\lambda}{d} \tanh\left(\frac{d}{2\lambda}\right) \right].$$

The free energies of the S and N states are equal when ΔG is given by Eq. (10.6). The critical field for the film is therefore given by

$$B_e = B_c / \left[1 - \frac{2\lambda}{d} \tanh\left(\frac{d}{2\lambda}\right) \right]^{1/2}$$

where B_c is the bulk critical field. For $d \ll \lambda$ this becomes $B_e \approx B_c(d/\lambda)^2/\sqrt{12}$ so that the penetration of the field into the film increases the critical field. Our calculation is not very good quantitatively because it ignores the effect of the field on the density of superconducting electrons.

10.6   Eq. (10.10) becomes

$$\mathbf{j} = -\frac{n_S e^2 \lambda}{m\xi} \mathbf{A}$$

so that, proceeding as in section 10.3, Eq. (10.14) for the $T = 0$ penetration depth is modified to

$$\lambda = \left(\frac{m\xi}{\mu_0 n e^2 \lambda}\right)^{1/2} = \lambda_L(0)\left(\frac{\xi}{\lambda}\right)^{1/2}.$$

Hence $\lambda^3 = \lambda_L^2(0)\xi$.

10.7   (a) Photons of wavelength 0.9 mm are energetic enough to break Cooper pairs; photons of wavelength 1.1 mm are not. Using Eq. (10.16), we deduce that the energy gap satisfies $0.56 \text{ meV} < \Delta < 0.69 \text{ meV}$.

(b) The superconducting (paired) electrons are highly ordered and cannot transport entropy. The conduction electron contribution to the thermal conductivity is therefore absent in superconductors for $T \ll T_c$.

(c) The critical field is given by Eq. (10.6) where the free energy difference corresponds approximately to a fraction $k_B T_c/\varepsilon_F$ of the N electrons per unit volume having their energy reduced by $k_B T_c$ so that $G_N(0, T) - G_S(0, T) \approx N(k_B T_c)^2/\varepsilon_F$. Since N/ε_F is approximately the same for metals we deduce that B_c is approximately proportional to T_c.

(d) This is an indication that lattice vibrations are involved in the superconducting transition. Lattice vibration frequencies are lower in heavier isotopes.

10.8   Substituting the given order parameter into Eq. (10.27) gives $\mathbf{j} = -en_S\hbar\mathbf{q}/2m - n_S e^2 \mathbf{A}/m$. Maxwell's equation $\text{curl } \mathbf{H} = \mathbf{j}$ can be written $-\nabla^2 \mathbf{A} = \mu_0 \mathbf{j}$ if $\text{div } \mathbf{A} = 0$, and inserting the expression for $\mathbf{j}$ into this enables $\mathbf{A}$ to be calculated. If we assume that the second term in $\mathbf{j}$ is small then $\mathbf{A} \approx \mu_0 en_S\hbar z^2 \mathbf{q}/4m$ (this satisfies $\text{div } \mathbf{A} = 0$ and $\mathbf{A} = 0$ at $z = 0$). Inserting this in the expression for $\mathbf{j}$ gives the ratio of the first and second terms as $1:\mu_0 n_S e^2 z^2/2m$, which is $1:z^2/\lambda_L^2(T)$ ($\lambda_L(T)$ is given by Eq. (10.14) with $n \to n_S$).

10.9   The requirement div $A = 0$ follows from applying charge conservation, div $j = 0$, to Eq. (10.10). If the change $A \rightarrow A + \nabla\chi$ is to leave j unchanged then $\nabla\theta$ in Eq. (10.26) must change in accordance with $\nabla\theta \rightarrow \nabla\theta - 2e\nabla\chi/\hbar$, that is $\theta \rightarrow \theta - 2e\chi/\hbar$. The change in order parameter is thus $\psi(r) \rightarrow \psi(r)\,e^{-2ie\chi/\hbar}$.

10.10   (a) 1108 J m^{-3} (insert $B_c = 53$ mT in Eq. (10.6)).
  (b) From Fig. 10.6, $\Delta C = C_S - C_N = 2.1$ mJ mol^{-1} K^{-1} = 210 J m^{-3} K^{-1}. From Eq. (10.8), $dB_c/dT = -(\mu_0 \Delta C/T_c)^{1/2} = -0.015$ T K^{-1} at $T = T_c$.
  (c) The 'average' slope of Fig. 10.16(b) corresponds to 15.2 μT per flux quantum. The area is therefore $\Phi_0/15.2 \times 10^{-6} = 1.36 \times 10^{-10}$ m^2.
  (d) The onset of single-particle tunnelling at $V = 2.7$ mV corresponds to $eV = 2\Delta$. Hence $\Delta = 1.35$ meV $= 2.16 \times 10^{-22}$ J.
  (e) From Eq. (10.40), the steps occur at voltage intervals $\Delta V = h\nu/2e = \nu\Phi_0$. Thus $\Phi_0 = \Delta V/\nu = 150 \times 10^{-6}/72 \times 10^9 = 2.08 \times 10^{-15}$ T m^2.
  (f) For two junctions with different critical currents the total current is $I = I_a \sin\delta_a + I_b \sin\delta_b$; $\delta_a - \delta_b$ is equal to $2\pi\Phi/\Phi_0$ (Eq. (10.42)) but $\delta_a + \delta_b$ adjusts itself to match the input current. It is straightforward to show that the maximum supercurrent is $I_a + I_b$ when $\Phi = n\Phi_0$ and the minimum supercurrent is $|I_a - I_b|$ when $\Phi = (n + \tfrac{1}{2})\Phi_0$. The distance between maxima on Fig. 10.23(b) is 4.04 μT and since this corresponds to Φ_0 we deduce the area of the loop is 5.1×10^{-10} m^2.

CHAPTER 11

11.1 Solving the equations, $mu = mv + MV$ and $\tfrac{1}{2}mu^2 = \tfrac{1}{2}mv^2 + \tfrac{1}{2}MV^2$, for conservation of momentum and energy we find: fraction of momentum carried away by M is

$$\frac{MV}{mu} = \frac{2M}{M + m} \approx 2 \qquad \text{for } M \gg m,$$

fraction of energy carried away by M is

$$\frac{MV^2}{mu^2} = \frac{4Mm}{(M + m)^2} \approx \frac{4m}{M} \ll 1 \qquad \text{for } M \gg m.$$

11.2 To prove that

$$\frac{2\pi(b^* \times c^*)}{a^* . (b^* \times c^*)} = a,$$

insert a^*, b^* and c^* from Eqs. (11.9) and use the vector identities: $A \times (B \times C) = (A.C)B - (A.B)C$, $A.(B \times C) = B.(C \times A)$ and $A \times A = 0$.

11.3 The four possible values of $|S|^2$ are:

(a) $|S|^2 = 0$ if $\begin{cases} (h + k + l) \text{ is even and } (h - k) \text{ an odd multiple of 3,} \\ (h + k + l) \text{ is odd and } (h - k) \text{ an even multiple of 3;} \end{cases}$

(b) $|S|^2 = f^2$ if $\begin{cases} (h + k + l) \text{ is even, } (h - k) \text{ is even and not a multiple of 3,} \\ (h + k + l) \text{ is odd, } (h - k) \text{ is odd and not a multiple of 3;} \end{cases}$

(c) $|S|^2 = 3f^2$ if $\begin{cases} (h + k + l) \text{ is even, } (h - k) \text{ is odd and not a multiple of 3,} \\ (h + k + l) \text{ is odd, } (h - k) \text{ is even and not a multiple of 3;} \end{cases}$

(d) $|S|^2 = 4f^2$ if $\begin{cases} (h + k + l) \text{ is even and } (h - k) \text{ an even multiple of 3,} \\ (h + k + l) \text{ is odd and } (h - k) \text{ an odd multiple of 3.} \end{cases}$

11.4 The structure factor, obtained by summing over atoms at $\mathbf{r}_1 = 0$ and $\mathbf{r}_2 = \frac{1}{2}(\mathbf{a} + \mathbf{b} + \mathbf{c})$ is $S = f\{1 + \exp[-i\pi(h + k + l)]\}$. Thus $S = 0$ if $(h + k + l)$ is odd and $S = 2f$ if $(h + k + l)$ is even. This eliminates alternate points in the simple cubic reciprocal lattice and turns it into the fcc reciprocal lattice of the bcc structure.

11.5 Let $\mathbf{a}_p^*, \mathbf{b}_p^*, \mathbf{c}_p^*$ and $\mathbf{a}_c^*, \mathbf{b}_c^*, \mathbf{c}_c^*$ be the reciprocal lattice vectors of the primitive and conventional lattice vectors as given by Eqs. (11.16) and (11.14) respectively. The (hkl) reflection referred to the primitive lattice vectors has scattering vector $\Delta\mathbf{k} = h\mathbf{a}_p^* + k\mathbf{b}_p^* + l\mathbf{c}_p^*$. Using Eqs. (11.16) and (11.14), this can be written $\Delta\mathbf{k} = (-h + k + l)\mathbf{a}_c^* + (h - k + l)\mathbf{b}_c^* + (h + k - l)\mathbf{c}_c^*$. Thus:

Primitive cell	$(1\,0\,0)$	$(\bar{1}\,0\,0)$	$(1\,1\,1)$	$(\bar{1}\,1\,1)$	$(1\,1\,0)$	$(\bar{1}\,1\,0)$
Conventional cell	$(\bar{1}\,1\,1)$	$(1\,\bar{1}\,\bar{1})$	$(1\,1\,1)$	$(3\,\bar{1}\,\bar{1})$	$(0\,0\,2)$	$(2\,\bar{2}\,0)$
$a(\sin\theta)/\lambda$	$\sqrt{3}/2$	$\sqrt{3}/2$	$\sqrt{3}/2$	$\sqrt{11}/2$	1	$\sqrt{2}$

To calculate the values of $a(\sin\theta)/\lambda$, we have used $2d\sin\theta = \lambda$ with $d = a/(h^2 + k^2 + l^2)^{1/2}$ (problem 1.4), where the Miller indices are for the conventional unit cell. The conventional labels have the advantage that the Miller indices are similar for reflections related by symmetry.

11.6 Taking a Cs^+ ion of form factor f_+ at $(0, 0, 0)$ and a Cl^- ion of form factor f_- at $(\frac{1}{2}, \frac{1}{2}, \frac{1}{2})$ gives $S = f_+ + f_- \exp[-i\pi(h + k + l)]$. Hence $S = f_+ + f_-$ if $(h + k + l)$ is even and $S = f_+ - f_-$ if $(h + k + l)$ is odd.

11.7 The spacing of the (hkl) planes is $d = a/(h^2 + k^2 + l^2)^{1/2}$ (problem 1.4). The Bragg law can therefore be written $a = (h^2 + k^2 + l^2)^{1/2}\lambda/2\sin\theta$. The average value of a obtained using this formula is $a = 4.05$ Å.

11.8 The diagram below shows the mapping of the third (left) and fourth (right) Brillouin zones into the first:

The diagram below shows the mapping into the first zone of the nearly free electron energy contours from the third zone. Maxima (M), minima (m) and saddle points (S) are identified:

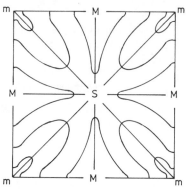

11.9 The shortest reciprocal lattice vector in the bcc reciprocal lattice (unit cell side $4\pi/a$) of an fcc real space lattice is from the cube corner to the body-centred position and has length $\sqrt{3}(2\pi/a)$; $k_m = \sqrt{3}\pi/a$ is half this. Since there are four electrons per cubic unit cell $k_F^3 = 3\pi^2 N/V = 12\pi^2/a^3$. Thus, for fcc,

$$k_F/k_m = (12\pi^2)^{1/3}/\sqrt{3}\pi = 0.901.$$

The shortest reciprocal lattice vector in the fcc reciprocal lattice (unit cell side $4\pi/a$) of a bcc real space lattice is from the cube corner to the face-centred position and has length $\sqrt{2}(2\pi/a)$ so that $k_m = \sqrt{2}\pi/a$. With two electrons per cubic unit cell $k_F^3 = 3\pi^2 N/V = 6\pi^2/a^3$. Thus, for bcc,

$$k_F/k_m = (6\pi^2)^{1/3}/\sqrt{2}\pi = 0.879.$$

The lower value of the ratio for bcc accounts (at least in part) for the fact that sodium has an almost spherical Fermi surface, but that of copper touches the hexagonal faces of the Brillouin zone (see Fig. 13.13).

CHAPTER 12

12.1 (a) f is proportional to $\int n \exp(-i\mathbf{K} \cdot \mathbf{r}) \, dV$, where $n = 3Z/4\pi R^3$ is the electron density, the integral is over the volume occupied by the atom, $\mathbf{K}$ is the scattering vector and the phase factor arises in the same way as in Eq. (11.2). To evaluate the integral use spherical polar coordinates centred on the atom with $\mathbf{K}$ along z; thus $\mathbf{K} \cdot \mathbf{r} = Kr \cos \theta$, $dV = 2\pi r^2 \sin \theta \, dr \, d\theta$ and

$$f \propto n \int_0^R 2\pi r^2 \, dr \int_0^\pi \sin \theta \exp(-iKr \cos \theta) \, d\theta$$

$$= -\frac{4\pi n}{K} \int_0^R r \sin(Kr) \, dr$$

$$= \frac{3Z}{R^3 K^3} [\sin(KR) - KR \cos(KR)] \qquad \text{(integrating by parts).}$$

The dependence on scattering angle follows from Fig. 11.4, which shows that $K = 2k \sin \theta_B = 4\pi(\sin \theta_B)/\lambda$; we use the notation θ_B to distinguish the Bragg angle θ_B from the spherical polar angle θ used in evaluating the integral. The form factor f is thus a function of $(\sin \theta_B)/\lambda$ of the form of Fig. 11.7:

(b) b is proportional to $\int n' \exp(-i\mathbf{K} \cdot \mathbf{r}) \, dA$, where $n' = \mu/4\pi R^2$ and the integral is over the surface of the sphere. Introducing spherical polar coordinates as above,

$$b \propto n' 2\pi R^2 \int_0^\pi \sin \theta \exp(-iKR \cos \theta) \, d\theta = \frac{4\pi n' R}{K} \sin(KR) = \frac{\mu \sin(KR)}{RK}.$$

Inserting $K = 2k \sin \theta_B$ shows that this falls off in a similar way to the form factor for the scattering of x-rays.

12.2 Momentum conservation, $\mathbf{k}_1 = \mathbf{k}_2 + \mathbf{q}$, gives

$$q^2 = k_1^2 + k_2^2 - 2k_1 k_2 \cos \theta \approx (k_1 - k_2)^2 + k_1 k_2 \theta^2 \qquad \text{for small } \theta.$$

Energy conservation gives $\hbar q c = \hbar^2 (k_1^2 - k_2^2)/2m$, where c is the velocity of sound and m the neutron mass. Elimination of q gives

$$(k_1 - k_2)^2 \left[\left(\frac{\hbar(k_1 + k_2)}{2mc}\right)^2 - 1\right] = k_1 k_2 \theta^2.$$

For 0.02 eV neutrons $k_1 = 3.1 \times 10^{10} \text{ m}^{-1}$ so that $\hbar k_1/mc \approx 6.8$, which implies $k_1 \approx k_2$ and $q \approx k_1 \theta$. Hence $\Delta E \approx -\hbar k_1 \theta c = -1.71 \times 10^{-22} \text{ J} = -1.07 \times 10^{-3}$ eV.

Initial time of flight $t = 5.2$ ms. Since t is inversely proportional to velocity, the fractional change $\Delta t/t = -\frac{1}{2}(\Delta E/E) = +1.07/(2 \times 20)$ and $\Delta t = +0.14$ ms.

For sapphire $\hbar k_1 < mc$ and a small-angle solution to the above equations with $k_1 \approx k_2$ cannot be found, and there is no scattering near the origin of reciprocal space. But if we choose a larger scattering angle so that $\mathbf{k}_1 - \mathbf{k}_2 \approx \mathbf{G}$ a phonon of small wavevector and hence small energy can be created. That is why inelastic neutron scattering experiments are usually done near reciprocal lattice points other than the origin.

12.3 The observed reflections and their relative intensities obey the rules derived in section 11.2.4 for the diamond structure. The Bragg angle for the peaks is $45°$ so that the Bragg law is $\sqrt{2}d = n\lambda$, which can be written $a = (h^2 + k^2 + l^2)^{1/2} \lambda \sqrt{2}$ (see solution to problem 11.7). The average value of a obtained from this formula is 5.385 Å.

12.4 The cubic unit cell of the fcc reciprocal lattice of the bcc structure has side $4\pi/a$; the first Brillouin zone is the rhombic dodecahedron (Fig. 1.9) with boundaries at distances $2\pi/a$ ($=|\mathbf{G}_{100}|/2$), $\sqrt{3}\pi/a$ ($=|\mathbf{G}_{111}|/4$) and $\sqrt{2}\pi/a$ ($=|\mathbf{G}_{110}|/4$) from the origin in the [1 0 0], [1 1 1] and [1 1 0] directions respectively. The magnitudes of the reciprocal lattice vectors can be calculated from information given in the previous sentence and thus the sound velocities $2\pi v/q$ can be obtained from the initial ($q \to 0$) slopes of the dispersion curves. Hence:

[1 0 0] $v_T = 1750 \text{ m s}^{-1}, v_L = 2400 \text{ m s}^{-1}$;
[1 1 1] $v_T = 1190 \text{ m s}^{-1}, v_L = 3080 \text{ m s}^{-1}$;
[1 1 0] $v_{T1} = 620 \text{ m s}^{-1}, v_{T2} = 1620 \text{ m s}^{-1}, v_L = 2620 \text{ m s}^{-1}$.

(i) The atomic displacements are unchanged if any reciprocal lattice vector is added to $\mathbf{q}$. If we add the reciprocal lattice vector $(2\pi/a)(\mathbf{j} - \mathbf{i})$ (this gives one of the face-centred reciprocal lattice points) to $\mathbf{q} = (2\pi/a)\mathbf{i}$ (this is the point on the first Brillouin zone boundary in the [1 0 0] direction) we obtain $\mathbf{q}' = (2\pi/a)\mathbf{j}$, which must give the same atomic displacements as $\mathbf{q}$. Since $\mathbf{q}$ and $\mathbf{q}'$ are orthogonal such a change converts a longitudinal mode into a transverse mode and vice versa. Longitudinal and transverse modes cannot therefore be distinguished at the point $\mathbf{q} = (2\pi/a)\mathbf{i}$.

(ii) $\mathbf{q} = \frac{1}{2}\mathbf{G}_{111} = (2\pi/a)(\mathbf{i} + \mathbf{j} + \mathbf{k})$ and $\mathbf{q} = \frac{1}{2}\mathbf{G}_{100} = (2\pi/a)\mathbf{i}$ are equivalent points in reciprocal space separated by the reciprocal lattice vector $(2\pi/a)(\mathbf{j} + \mathbf{k})$.

(iii) Only the component of $\nabla\omega$ perpendicular to the Brillouin zone boundary must vanish; the direction [1 1 1] does not intersect the Brillouin zone boundary at right angles.

12.5 The helical modulation of the magnetic structure has a similar effect on neutron scattering to that of a spin wave. Thus additional (satellite) peaks occur close to the Bragg peaks at angles determined by Eq. (12.11), $\mathbf{k}' = \mathbf{k} + \mathbf{q} + \mathbf{G}$. The wavevector $\mathbf{q}$ of the spiral is in the z direction and the pitch of the spiral is $2\pi/|\mathbf{q}|$; the details of the magnetic structure can therefore be deduced from the positions of the satellites. Since the spiral structure is static, the scattering is elastic.

12.6 The Bragg angle θ is half the scattering angle and $\sin^2 \theta$ is proportional to $(h^2 + k^2 + l^2)$ for a cubic structure (problem 11.7). Since the reflection at $2\theta = 11.9°$ is identified as the (1 1 1) reflection of the magnetic unit cell, we deduce $(h^2 + k^2 + l^2) \approx 19$ and 27 for the other two peaks; these are the (3 1 1) and (5 1 1) reflections from the magnetic cell. The chemical unit cell has half the linear dimension of the magnetic cell; the Miller indices for the chemical cell are thus $(\frac{1}{2}, \frac{1}{2}, \frac{1}{2})$, $(\frac{3}{2}, \frac{1}{2}, \frac{1}{2})$ and $(\frac{5}{2}, \frac{1}{2}, \frac{1}{2})$.

12.7 Suppose a monolayer consists of one O_2 molecule per $(3 \text{ Å})^2$, i.e. about 10^{19} molecules/m^2. Molecules collide with the surface at a rate of order $\frac{1}{4}n\bar{c}$ per second where n is the concentration in the gas and $\bar{c}$ the mean speed. Using $p = nk_B T$ for the pressure and $\bar{c} \approx (k_B T/M)^{1/2}$ we obtain times of order: 6×10^{-9} s, 6×10^{-3} s and 6×10^3 s at pressures of 1, 10^{-6} and 10^{-12} bar.

CHAPTER 13

13.1
$$g(\varepsilon_F) = \frac{V}{4\pi^3} \int \frac{1}{\hbar v_F} \, dS_F = \frac{Vm}{4\pi^3 \hbar^2 k_F} 4\pi k_F^2 = \frac{V}{2\pi^2} \frac{2m}{\hbar^2 k_F^2} k_F^3$$

$$= \frac{3N}{2\varepsilon_F} \qquad \text{(using Eq. (3.10)).}$$

Take $\mathbf{E}$ along z and use spherical polars; $\mathbf{j}$ is along z so we need the z component of dS_F, which is $|dS_F| \cos \theta = k_F^2 \sin \theta \cos \theta \, d\theta \, d\phi$. Also $v_F . \hat{\mathbf{E}}$ is $v_F \cos \theta$. Thus

$$\sigma = \frac{e^2\tau}{4\pi^3 \hbar} v_F \int_0^\pi d\theta \int_0^{2\pi} d\phi \, k_F^2 \cos^2 \theta \sin \theta = \frac{e^2\tau}{4\pi^3 m} k_F^3 \frac{4\pi}{3}$$

$$= \frac{Ne^2\tau}{Vm} \qquad \text{(using Eq. (3.10)).}$$

13.2 (a) The area of a constant energy circle in a plane perpendicular to B is $A_k = \pi k_\perp^2 = \pi(k^2 - k_z^2)$; the energy is $\varepsilon = \hbar^2 k^2/2m_e$. Thus

$$m_C = \frac{\hbar^2}{2\pi} \left(\frac{\partial A_k}{\partial \varepsilon} \right)_{k_z} = \frac{\hbar^2}{2\pi} \left(\frac{\partial A_k}{\partial k^2} \right) \left(\frac{\partial k^2}{\partial \varepsilon} \right) = \frac{\hbar^2}{2\pi} \frac{2m_e \pi}{\hbar^2} = m_e.$$

Note that this is independent of which cross section perpendicular to the field is considered.

(b) $(1/m_e)_{xx} = (1/m_e)_{yy} = 1/m_T$, $(1/m_e)_{zz} = 1/m_L$. All other components (e.g. $(1/m_e)_{xy}$) vanish. Magnetic field is included in Eq. (13.12) by replacement $E_j \rightarrow E_j + (\mathbf{v} \times \mathbf{B})_j$; writing the resulting equation in component form with $\mathbf{B} = B(\sin \theta, 0, \cos \theta)$, and substituting a solution of the form $v_x = v_1 e^{i\omega t}$, $v_y = v_2 e^{i\omega t}$, $v_z = v_3 e^{i\omega t}$, leads to Eq. (5.58).

13.3 From Eqs. (13.23) and (13.25), $\delta(1/B) = e/\omega m_C$. From the data given $\delta(1/B) = 0.345 \text{ T}^{-1}$, whence $m_C = 1.08 \times 10^{-30}$ kg.

13.4 Consider the cyclotron orbit corresponding to the $n = 0$ Landau level. From Eq. (13.28), its radius in $\mathbf{k}$-space is $(eB/\hbar)^{1/2}$. From section 13.3.1 the radius r_C in real space is $\hbar/eB$ times this; thus $r_C = (\hbar/eB)^{1/2}$. For free electrons $\hbar\omega_C \approx \varepsilon_F$ implies $\hbar eB/m \approx \varepsilon_F$ and hence $r_C \approx (\hbar^2/m\varepsilon_F)^{1/2} \approx k_F^{-1}$. A wavepacket with a well defined energy of order ε_F must have a size $\gg k_F^{-1}$.

13.5 Use Eq. (C2), $\hat{K} = (-i\hbar\nabla + e\mathbf{A})^2/2m$, to obtain the Schrödinger equation in the form given. Since

$$\left(\frac{\partial}{\partial y} - \frac{ieBx}{\hbar}\right)^2 \psi = \left(\frac{\partial}{\partial y} - \frac{ieBx}{\hbar}\right)\left(i\beta - \frac{ieBx}{\hbar}\right)\psi = -\left(\beta - \frac{eBx}{\hbar}\right)^2 \psi$$

and $\partial^2\psi/\partial z^2 = -k_z^2\psi$, substituting the proposed solution and cancelling $\exp[i(\beta y + k_z z)]$ gives the required equation for $u(x)$. The substitution $x' = x - \hbar\beta/eB$ shows that the particle is moving in a potential $\frac{1}{2}m(eB/m)^2 x'^2 = \frac{1}{2}m\omega_C^2 x'^2$, where $\omega_C = eB/m$ is the cyclotron frequency. Comparison with the potential energy of a simple harmonic oscillator identifies ω_C as the frequency; the energy levels E' are therefore $(n + \frac{1}{2})\hbar\omega_C$ and the total energy E is given by Eq. (13.26).

13.6 Estimating 108 oscillations between fields 10.70 and 10.93 T and using Eq. (13.30), $\delta(1/B) = -\delta B/B^2 = 0.23/(108 \times 10.8^2) = 2\pi e/\hbar A_F$. Thus $A_F = 5.2 \times 10^{20}$ m^{-2}. For free electrons, $k_F = (3\pi^2 N/V)^{1/3}$ so that $A_F = \pi k_F^2 = 5.8 \times 10^{20}$ m^{-2}. The Landau level spacing $\hbar\omega_C$ must be large compared to $k_B T$; thus $T \ll \hbar\omega_C/k_B = eB\hbar/m_C k_B \approx 13$ K. Electrons must complete many cyclotron orbits between collisions, $\omega_C\tau \gg 1$; thus $1/\tau = n_i/10^{14} \ll \omega_C$, whence $n_i \ll 1.8 \times 10^{26}$ m^{-3}.

13.7 Averaging the peak separations gives $\hbar\omega_P = 10.7$ eV so that $\omega_P = 1.62 \times 10^{16}$ s^{-1}. The conduction electron density is $2 \times 4.3 \times 10^{28}$ m^{-3} so that, from Eq. (13.41), $\omega_P = 1.65 \times 10^{16}$ s^{-1}.

13.8 Including a collision term $-mn\mathbf{v}/\tau$ on the left-hand side of Eq. (13.34) means that Eq. (13.40) is modified to $d^2n/dt^2 + (1/\tau)\,dn/dt + \omega_P^2(n - n_0) = 0$. This is the equation of motion of a damped simple harmonic oscillator; the plasma oscillations have a frequency $(\omega_P^2 - 1/4\tau^2)^{1/2}$ and their amplitude decays as $\exp(-t/\tau)$. Critical damping corresponds to zero frequency; thus, for oscillatory behaviour, $\omega_P^2 > 1/4\tau^2$ or, using Eq. (13.41), $n_0 > m\varepsilon_0/4e^2\tau^2$. For $\tau = 10^{-12}$ s, the critical concentration is 8×10^{19} m^{-3}. The limit $\omega_P\tau \ll 1$ is that discussed for semiconductors in section 5.6.2; at such low concentrations the carriers are no longer degenerate at room temperature.

13.9
$$\frac{1}{r^2}\frac{d}{dr}\left(r^2\frac{dn}{dr}\right) = \frac{1}{r^2}\frac{d}{dr}\left[\frac{r^2}{4\pi\lambda^2}\left(\frac{1}{r^2} + \frac{1}{\lambda r}\right)e^{-r/\lambda}\right]$$

$$= \frac{1}{4\pi\lambda^2 r^2}\left(-\frac{1}{\lambda} + \frac{1}{\lambda} - \frac{r}{\lambda^2}\right)e^{-r/\lambda}$$

$$= (n - n_0)/\lambda^2.$$

The point charge $-e$ contributes a charge density $-e\delta(r)$, where $\delta(r)$ is the Dirac delta function (zero everywhere except $r = 0$ and infinite at $r = 0$ so that $\int(-e)\delta(r)4\pi r^2\,dr = -e$). The point charge contributes a term $-e\delta(r)/\varepsilon_0$ to the right-hand side of Eq. (13.35); hence it contributes $\delta(r)/\lambda^2$ to the right-hand side of Eq. (13.42) and integrating this over a small sphere of radius r gives $\int \delta(r)4\pi r^2\,dr/\lambda^2 = 1/\lambda^2$. Integrating the left-hand side of Eq. (13.42) gives

$$\underset{r\to 0}{\text{Lt}}\int \nabla^2 n\,dV = \underset{r\to 0}{\text{Lt}}\int \nabla n\cdot d\mathbf{S} = \underset{r\to 0}{\text{Lt}}\left(\frac{dn}{dr}4\pi r^2\right)$$

$$= \underset{r\to 0}{\text{Lt}}\left(\frac{1}{4\pi\lambda^2}\left(\frac{1}{r^2} + \frac{1}{\lambda r}\right)e^{-r/\lambda}4\pi r^2\right) = \frac{1}{\lambda^2}$$

using the value of dn/dr calculated above.

13.10 The response of the electrons to a point charge $+e$ is the negative of that, Eq. (13.44), to a point charge $-e$. Thus $n - n_0 = \exp(-r/\lambda)/4\pi\lambda^2 r$. The potential associated with this charge distribution is given by Poisson's equation $\nabla^2\phi = -\rho/\varepsilon_0 = -[-e(n - n_0)]/\varepsilon_0 = e\exp(-r/\lambda)/4\pi\varepsilon_0\lambda^2 r$. We have seen in solution 13.9 that $\nabla^2[\exp(-r/\lambda)/r] = \exp(-r/\lambda)/r\lambda^2$ so that the potential that satisfies Poisson's equation is $\phi = e\exp(-r/\lambda)/4\pi\varepsilon_0 r$. Our discussion in the previous problem shows that ϕ also obeys Poisson's equation at $r = 0$, where $\rho(r) = +e\delta(r)$.

13.11 Since the carriers in a semiconductor are non-degenerate, we use the classical gas result $p = nk_B T$ rather than Eq. (13.36) in deriving the equation for n; we also include the dielectric constant ε in Eq. (13.35). The screening length λ of Eq. (13.43) is modified to $\lambda = (k_B T\varepsilon\varepsilon_0/n_0 e^2)^{1/2}$, which is just the Debye length of Eq. (5.68). The 'Bohr radius' r_1 for the impurity bound state is given by Eq. (5.13). The condition for metallic behaviour is thus $\lambda < r_1$, which gives $n_0 > 1.3 \times 10^{22}\text{ m}^{-3}$ at $T = 2.5$ K, in reasonable agreement with Fig. 13.18.

CHAPTER 14

14.1 Using Eq. (14.6) for an infinite well, the photon energy for a transition between the nth bound states in the conduction and valence bands is

$$E_G + \frac{n^2 h^2}{8md^2}\left(\frac{m}{m_e} + \frac{m}{m_h}\right) = \begin{cases} 1.519 + 0.0331n^2\text{ eV} & \text{for } d = 140\text{ Å}. \\ 1.519 + 0.0147n^2\text{ eV} & \text{for } d = 210\text{ Å}. \end{cases}$$

The calculation of the bound states of a finite well is discussed in chapter 4 of French and Taylor.[4]

14.2 The binding energy is $E = \mu e^4/8h^2\varepsilon^2\varepsilon_0^2$, where $\mu = m_e m_h/(m_e + m_h)$ is the reduced mass; thus $E = 0.0046$ eV.

14.3 The slope of the dashed line on Fig. 14.7 is 1.623 kΩ T^{-1}. From Eq. (14.13) the slope is $1/n_A e$; we deduce $n_A = 3.85 \times 10^{15}\text{ m}^{-2}$. From Eq. (14.13) the scattering time is $m_e/n_A e^2\rho_L = 7.9 \times 10^{-12}$ s. Using the constant density of states $g(\varepsilon) = m_e/\pi\hbar^2$ of Eq. (14.9), $\varepsilon_F = n_A/g(\varepsilon) = 0.0132$ eV and hence $v_F = (2\varepsilon_F/m_e)^{1/2} = 2.57 \times 10^5\text{ m s}^{-1}$. The cyclotron frequency ω_C is eB/m_e, so that $v_C\tau = \omega_C\tau/2\pi = 1$ when $B = 0.3$ T. Departures from the classical Hall effect begin to appear on Fig. 14.7 at a field of this order. The separation $\hbar\omega_C$ between Landau levels must exceed $k_B T$ for observation of the steps; this requires $B > 0.01$ T at $T = 8$ mK and $B > 10$ T at $T = 8$ K.

Index

The use of **bold** type for a page number indicates where information on the meaning of an indexed term can be found.